Handbook of Heat Transfer

Edited by **Nathan Rice**

Clanrye International
New Jersey

Published by Clanrye International,
55 Van Reypen Street,
Jersey City, NJ 07306, USA
www.clanryeinternational.com

Handbook of Heat Transfer
Edited by Nathan Rice

International Standard Book Number: 978-1-63240-273-8 (Hardback)

Printed in China.

Contents

Preface

Heat transfer is fundamentally described as the energy in transit due to temperature difference. Heat transfer plays a central role in various industrial technologies. This book deals with the joint action of heat transmission and concomitant procedures. It discusses topics such as heat effects due to laser, ion and plasma-solid interaction; engineering procedures of heat conduction equations and curing reaction kinetics in manufacturing procedure, their consolidation with ohmic and di-electric losses; and heat conduction in metallic porous media and power cables. Evaluation of the safety of mine hoist under the influence of heat produced by mechanical friction, transfer of heat in boilers and internal combustion engine chambers, the management for ultrahigh strength steel manufacturing, and air cooling of electronic devices are also discussed within the book.

All of the data presented henceforth, was collaborated in the wake of recent advancements in the field. The aim of this book is to present the diversified developments from across the globe in a comprehensible manner. The opinions expressed in each chapter belong solely to the contributing authors. Their interpretations of the topics are the integral part of this book, which I have carefully compiled for a better understanding of the readers.

At the end, I would like to thank all those who dedicated their time and efforts for the successful completion of this book. I also wish to convey my gratitude towards my friends and family who supported me at every step.

Editor

Part 1

Laser- Plasma- and Ion-Solid Interaction

1

Pulsed Laser Heating and Melting

David Sands
University of Hull
UK

1. Introduction

Modification of surfaces by laser heating has become a very important aspect of modern materials science. The author's own interests in laser processing have been involved in the main with laser processing of semiconductors, especially II-VI materials such as CdTe, but also amorphous silicon. Applications of laser processing are diverse and include, in addition to the selective recrystallisation of amorphous semiconductors, the welding of metals and other functional materials, such as plastic, case hardening in tool steels, and phase changes in optical data storage media. In addition there are numerous ideas under investigation in research laboratories around the world that have not yet become commercial applications and perhaps never will, but make use of the advantages and flexibility afforded by laser irradiation for both fundamental research into materials as well as small scale fabrication and technological innovation.

The range of laser types is truly staggering, but from the perspective of heat conduction it is possible to regard the laser simply as black box that provides a source of heat over a period of time that can range from femtoseconds to many tens of seconds, the latter effectively corresponding to continuous heating. The laser is therefore an incredibly versatile tool for effecting changes to the surfaces of materials, with the depth of material affected ranging from a few nanometres to several hundreds of microns, and possibly even millimetres. Strictly, a full discussion of surface modification by laser processing should include laser ablation and marking, but, interesting though the physics and technology undoubtedly are, attention will instead be restricted to the range of temperatures up to and beyond melting but excluding material removal by ablation as the loss of material from the surface represents for the purposes of this chapter a significant loss of energy which is then no longer available for heat conduction into the bulk.

Despite this apparent diversity in both the lasers and the possible processes, models of heat conduction due to laser irradiation share many common features. For pulse durations longer than a nanosecond or so thermal transport is essentially based on Fourier's law whilst for shorter pulses the models need to account for the separate contributions from both electrons and phonons. Geometry can also greatly simplify the modelling. Many processes typically involve moving a work-piece, or target, against a stationary beam (figure 1a), which can be either pulsed repetitively or continuous. The intensity of the beam may vary laterally or not (figure 1b), so a model of a moving work-piece might have to account for either a temporally varying intensity as the material encounters first the low-intensity leading edge, then the high-intensity middle, and finally the low-intensity trailing edge, or the cumulative effect of a number of pulses of equal intensity. In much of materials research the aim is to determine the effect of laser radiation on some material property and this kind of moving geometry can represent an unnecessary complication. The material is therefore heated statically for a limited time and the intensity is often, but not always, assumed to be constant over this time. These simplifications are often essential in order to make the problem mathematically tractable or in order to reduce the computation time for numerical models. A number of such models will be described, starting with analytical models of thermal transport.

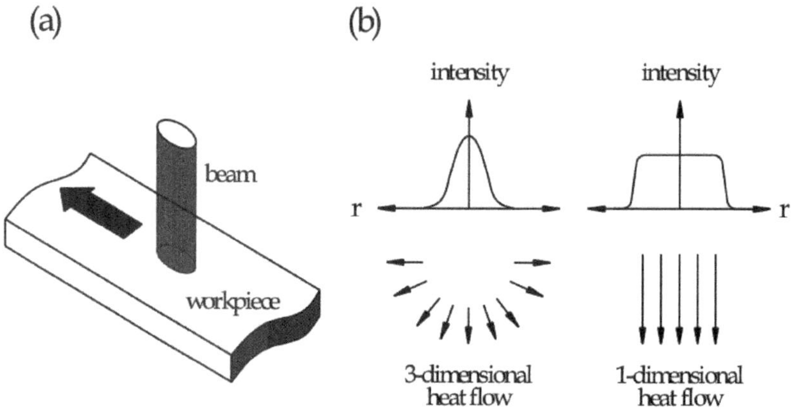

Fig. 1. Typically a workpiece, or target, is scanned relative to a stationary laser beam (a), which might have a non-uniform intensity profile across the beam radius, r, leading to 3-dimensional heat flow or a "top hat" profile leading to 1-dimensional heat flow (b).

2. Laser heating basics

As indicated above, analytical models of laser processing represent a simplification of the real process, but they can nonetheless provide valuable insight into the response of the material. Attention is restricted in this section to pulses longer than a nanosecond, for which the energy of the lattice can be assumed to follow the laser pulse. As described by von Allmen (von Allmen & Blatter, 1995) electrons which gain energy by absorbing a photon will relax back to the ground state by giving up their energy to the lattice within a few picoseconds, so at the time scales of interest here, which for convenience we shall call the long-pulse condition, the details of this process can be ignored and it can safely be assumed that any optical energy absorbed from the laser beam manifests itself as heat. This heat propagates through the material according to Fourier's law of heat conduction:

$$\frac{dQ}{dt} = -k\nabla T \tag{1}$$

The coefficient k is known as the thermal conductivity and ∇ is the gradient operator.
Fourier's law is empirical and essentially describes thermal diffusion, analogous to Fick's first law of diffusion. Heat is a strange concept and was thought at one stage by early thermodynamicists to represent the total energy contained in a body. This is not the case and in a thermodynamic sense heat represents an interaction; an exchange of energy not in

the form of work that changes the total internal energy of the body. There is no sense in modern thermodynamics of the notion of the heat contained in a body, but in the present context the energy deposited within a material by laser irradiation manifests itself as heating, or a localised change in temperature above the ambient conditions, and it seems on the face of it to be a perfectly reasonable idea to think of this energy as a quantity of heat. Thermodynamics reserves the word enthalpy, denoted by the symbol *H*, for such a quantity and henceforth this term will be used to describe the quantity of energy deposited within the body. A small change in enthalpy, ΔH, in a mass of material, *m*, causes a change in temperature, ΔT, according to.

$$\Delta H = mc_p \Delta T \tag{2}$$

The quantity c_p is the specific heat at constant pressure. In terms of unit volume, the mass is replaced by the density ρ and

$$\Delta H_V = \rho c_p \Delta T \tag{3}$$

Equations (2) and (3) together represent the basis of models of long-pulse laser heating, but usually with some further mathematical development. Heat flows from hot to cold against the temperature gradient, as represented by the negative sign in eqn (1), and heat entering a small element of volume ΔV must either flow out the other side or change the enthalpy of the volume element. Mathematically, this can be represented by the divergence operator

$$\nabla \bullet \tilde{Q} = -\frac{dH_V}{dt} \tag{4}$$

where $\tilde{Q} = {}^{dQ}\!/\!_{dt}$ is the rate of flow of heat. The negative sign is required because the divergence operator represents in effect the difference between the rate of heat flow out of a finite element and the rate of heat flow into it. A positive divergence therefore means a nett loss of heat within the element, which will cool as a result. A negative divergence, ie. more heat flowing into the element than out of it, is required for heating.

If, in addition, there is an extra source of energy, *S(z)*, in the form of absorbed optical radiation propagating in the z-direction normal to a surface in the x-y plane, then this must contribute to the change in enthalpy and

$$S(z) - \nabla \bullet \tilde{Q} = \frac{dH_V}{dt} \tag{5}$$

Expanding the divergence term on the left,

$$-\nabla \bullet \tilde{Q} = \nabla \bullet (k\nabla T) = k\nabla^2 T + \nabla k \bullet \nabla T \tag{6}$$

In Cartesian coordinates, and taking into account equations (3), (4) and (5)

$$\frac{dT}{dt} = \frac{k}{\rho c_p}\left(\frac{\partial^2 T}{\partial x^2} + \frac{\partial^2 T}{\partial y^2} + \frac{\partial^2 T}{\partial z^2}\right) + \frac{1}{\rho c_p}\left(\frac{\partial k}{\partial x}\frac{\partial T}{\partial x} + \frac{\partial k}{\partial y}\frac{\partial T}{\partial y} + \frac{\partial k}{\partial z}\frac{\partial T}{\partial z}\right) + \frac{S(z)}{\rho c_p} \tag{7}$$

The source term in (7) can be derived from the laws of optics. If the intensity of the laser beam is I_0, in Wm^{-2}, then an intensity, I_T, is transmitted into the surface, where

$$I_T = I_0(1-R) \tag{8}$$

Here R is the reflectivity, which can be calculated by well known methods for bulk materials or thin film systems using known data on the refractive index. Even though the energy density incident on the sample might be enormous compared with that used in normal optical experiments, for example a pulse of 1 J cm^{-2} of a nanosecond duration corresponds to a power density of 10^9 Wcm^{-2}, significant non-linear effects do not occur in normal materials and the refractive index can be assumed to be unaffected by the laser pulse.

The optical intensity decays exponentially inside the material according to

$$I(z) = I_T \exp(-\alpha z) \tag{9}$$

where α is the optical absorption coefficient. Therefore

$$S(z) = \alpha I(z) = \alpha I_0(1-R)\exp(-\alpha z) \tag{10}$$

Analytical and numerical models of pulsed laser heating usually involve solving equation (7) subject to a source term of the form of (10). There have been far too many papers over the years to cite here, and too many different models of laser heating and melting under different conditions of laser pulse, beam profile, target geometry, ambient conditions, etc. to describe in detail. As has been described above, analytical models usually involve some simplifying assumptions that make the problem tractable, so their applicability is likewise limited, but they nonetheless can provide a valuable insight into the effect of different laser parameters as well as provide a point of reference for numerical calculations. Numerical calculations are in some sense much simpler than analytical models as they involve none of the mathematical development, but their implementation on a computer is central to their accuracy. If a numerical calculation fails to agree with a particular analytical model when run under the same conditions then more than likely it is the numerical calculation that is in error.

3. Analytical solutions

3.1 Semi-infinite solid with surface absorption

Surface absorption represents a limit of very small optical penetration, as occurs for example in excimer laser processing of semiconductors. The absorption depth of UV nm radiation in silicon is less than 10 nm. Although it varies slightly with the wavelength of the most common excimer lasers it can be assumed to be negligible compared with the thermal penetration depth. Table 1 compares the optical and thermal penetration in silicon and gallium arsenide, two semiconductors which have been the subject of much laser processing research over the years, calculated using room temperature thermal and optical properties at various wavelengths commonly used in laser processing.

It is evident from the data in table 1 that the assumption of surface absorption is justified for excimer laser processing in both semiconductors, even though the thermal penetration depth in GaAs is just over half that of silicon. However, for irradiation with a Q-switched Nd:YAG laser, the optical penetration depth in silicon is comparable to the thermal penetration and a different model is required. GaAs has a slightly larger band gap than silicon and will not absorb at all this wavelength at room temperature.

Laser	Wavelength (nm)	Typical pulse length τ (ns)	Thermal penetration depth, $(D\tau)^{1/2}$ (nm)		Optical penetration depth, α^{-1} (nm)	
			silicon	Gallium arsenide	silicon	Gallium arsenide
XeCl excimer	308	30	1660	973	6.8	12.8
KrF excimer	248	30	1660	973	5.5	4.8
ArF excimer	192	30	1660	973	5.6	10.8
Q-switched Nd:YAG	1060	6	743	435	1000	N/A

Table 1. The thermal and optical penetration into silicon and gallium arsenide calculated for commonly used pulsed lasers.

Assuming, then, surface absorption and temperature-independent thermo-physical properties such as conductivity, density and heat capacity, it is possible to solve the heat diffusion equations subject to boundary conditions which define the geometry of the sample. For a semi-infinite solid heated by a laser with a beam much larger in area than the depth affected, corresponding to 1-D thermal diffusion as depicted in figure 1b, equation (7) becomes

$$\frac{dT}{dt} = D\frac{\partial^2 T}{\partial z^2} \tag{11}$$

Here k is the thermal conductivity and $D = \frac{k}{\rho c_p}$ the thermal diffusivity. Surface absorption implies

$$S(0) = \alpha I(0) = \alpha I_0(1 - R) \tag{12}$$

$$S(z) = 0, z > 0 \tag{13}$$

Solution of the 1-D heat diffusion equation (11) yields the temperature, T, at a depth z and time t shorter than the laser pulse length, τ, (Bechtel, 1975)

$$T(z, t < \tau) = \frac{2\alpha I_0(1-R)}{k}(Dt)^{\frac{1}{2}} ierfc\left[\frac{z}{2(Dt)^{\frac{1}{2}}}\right] \tag{14}$$

The integrated complementary error function is given by

$$ierfc(z) = \int_z^{\infty} erfc(\xi)d\xi \tag{15}$$

with

$$erfc(z) = 1 - erf(z) = 1 - \frac{2}{\sqrt{\pi}}\int_0^x e^{-t^2}dt \tag{16}$$

The surface (z=0) temperature is given by,

$$T(0,t<\tau)=\frac{2\alpha I_0(1-R)}{k}(Dt)^{\frac{1}{2}}\left(\frac{1}{\pi}\right)^{\frac{1}{2}} \tag{17}$$

For times greater than the pulse duration, τ, the temperature profile is given by a linear combination of two similar terms, one delayed with respect to the other. The difference between these terms is equivalent to a pulse of duration τ (figure 2).

$$T(z,t>\tau)=\frac{2\alpha I_0(1-R)}{k}\left\{(Dt)^{\frac{1}{2}}ierfc\left[\frac{z}{2(Dt)^{\frac{1}{2}}}\right]-[D(t-\tau)]^{\frac{1}{2}}ierfc\left[\frac{z}{2[D(t-\tau)]^{\frac{1}{2}}}\right]\right\} \tag{18}$$

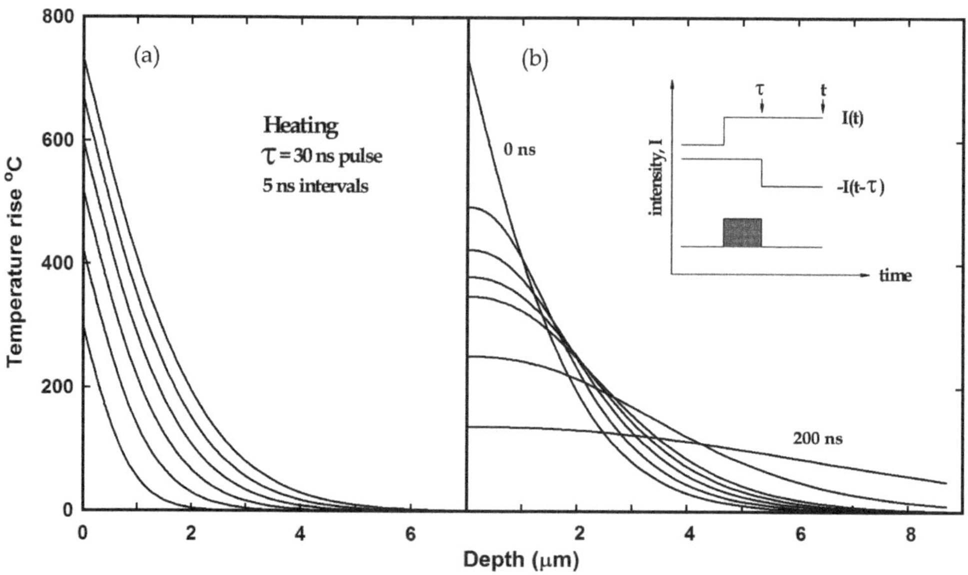

Fig. 2. Solution of equations (14) and (18) for a 30 ns pulse of energy density 400 mJ cm^{-2} incident on crystalline silicon with a reflectivity of 0.56. The heating curves (a) are calculated at 5 ns intervals up to the pulse duration and the cooling curves are calculated for 5, 10, 15, 20, 50 and 200 ns after the end of the laser pulse according to the scheme shown in the inset.

3.2 Semi-infinite solid with optical penetration

Complicated though these expressions appear at first sight, they are in fact simplified considerably by the assumption of surface absorption over optical penetration. For example, for a spatially uniform source incident on a semi-infinite slab, the closed solution to the heat transport equations with optical penetration, such as that given in Table 1 for Si heated by pulsed Nd:YAG, becomes (von Allmen & Blatter, 1995)

$$T(z,t)=\frac{\alpha I_0(1-R)}{k}\left\{\begin{array}{l}2(Dt)^{\frac{1}{2}}ierfc\left[\dfrac{z}{2(Dt)^{\frac{1}{2}}}\right]-\dfrac{1}{\alpha}e^{-\alpha z}\\ +\dfrac{1}{2\alpha}e^{\alpha^2(Dt)}\times\left(e^{-\alpha z}erfc[\alpha(Dt)^{\frac{1}{2}}-\dfrac{z}{(Dt)^{\frac{1}{2}}}]+e^{\alpha z}erfc[\alpha(Dt)^{\frac{1}{2}}+\dfrac{z}{(Dt)^{\frac{1}{2}}}]\right)\end{array}\right\} \quad (19)$$

3.3 Two layer heating with surface absorption

The semi-infinite solid is a special case that is rarely found within the realm of high technology, where thin films of one kind or another are deposited on substrates. In truth such systems can be composed of many layers, but each additional layer adds complexity to the modelling. Nonetheless, treating the system as a thin film on a substrate, while perhaps not always strictly accurate, is better than treating it as a homogeneous body. El-Adawi et al (El-Adawi et al, 1995) have developed a two-layer of model of laser heating which makes many of the same assumptions as described above; surface absorption and temperature independent thermophysical properties, but solves the heat diffusion equation in each material and matches the solutions at the boundary. We want to find the temperature at a time t and position $z=z_f$ within a thin film of thickness Z, and the temperature at a position $z_s = z - Z$ within the substrate. If the thermal diffusivity of the film and substrate are α_f and α_s respectively then the parabolic diffusion equation in either material can be written as

$$\left.\begin{array}{l}\dfrac{\partial T_f(z_f,t)}{\partial t}=D_f\dfrac{\partial^2 T_f(z_f,t)}{\partial {z_f}^2},0\le z_f\le Z\\ \dfrac{\partial T_s(z_s,t)}{\partial t}=D_s\dfrac{\partial^2 T_s(z_s,t)}{\partial {z_s}^2},0\le z_s\le\infty\end{array}\right\} \quad (20)$$

These are solved by taking the Laplace transforms to yield a couple of similar differential equations which in general have exponential solutions. These can be transformed back once the coefficients have been found to give the temperatures within the film and substrate.

If $0\le n\le\infty$ is an integer, then the following terms can be defined:

$$\left.\begin{array}{l}a_n=2Z(1+n)-z_f\\ b_n=2nZ+z_f\\ g_n=(1+2n)Z+z_s\sqrt{\dfrac{D_f}{D_s}}\end{array}\right\} \quad (21a)$$

$$L_f^2=4D_f t \quad (21b)$$

The temperatures within the film and substrate are then given by

$$
\left.
\begin{aligned}
T_f(z_f,t) &= \sum_{n=0}^{\infty} \frac{I_0 A_f}{k_f} B^{n+1} \left[\frac{L_f}{\sqrt{\pi}} \exp\left(-\frac{a_n^2}{L_f^2} \right) - a_n.erfc\left(\frac{a_n}{L_f} \right) \right] \\
&+ \sum_{n=0}^{\infty} \frac{I_0 A_f}{k_f} B^{n} \left[\frac{L_f}{\sqrt{\pi}} \exp\left(-\frac{b_n^2}{L_f^2} \right) - b_n.erfc\left(\frac{b_n}{L_f} \right) \right] \\
T_s(z_s,t) &= \sum_{n=0}^{\infty} \frac{2 I_0 A_f}{k_f} \frac{B^n}{(1+\varepsilon)} \left[\frac{L_f}{\sqrt{\pi}} \exp\left(-\frac{g_n^2}{L_f} \right) - g_n.erfc\left(\frac{g_n}{L_f} \right) \right]
\end{aligned}
\right\} \tag{22}
$$

Here I_0 is the laser flux, or power density, A_f is the surface absorptance of the thin film material, k_f is the thermal conductivity of the film and

$$B = \frac{1-\varepsilon}{1+\varepsilon} < 1 \tag{23}$$

It follows, therefore, that higher powers of B rapidly become negligible as the index increases and in many cases the summation above can be curtailed for $n>10$. The parameter ε is defined as

$$\varepsilon = \frac{k_s}{k_f} \sqrt{\frac{D_f}{D_s}} \tag{24}$$

Despite their apparent simplicity, at least in terms of the assumptions if not the final form of the temperature distribution, these analytical models can be very useful in laser processing. In particular, El-Adawi's two-layer model reduces to the analytical solution for a semi-infinite solid with surface absorption (equation 14) if both the film and the substrate are given the same thermal properties. This means that one model will provide estimates of the temperature profile under a variety of circumstances. The author has conducted laser processing experiments on a range of semiconductor materials, such as Si, CdTe and other II-VI materials, GaAs and SiC, and remarkably in all cases the onset of surface melting is observed to occur at an laser irradiance for which the surface temperature calculated by this model lies at, or very close to, the melting temperature of the material. Moreover, by the simple expedient of subtracting a second expression, as in equation (18) and illustrated in the inset of figure 2b, the temperature profile during the laser pulse and after, during cooling, can also be calculated. El-Adawi's two-layer model has thus been used to analyse time-dependent reflectivity in laser irradiated thin films of ZnS on Si (Hoyland et al, 1999), calculate diffusion during the laser pulse in GaAs (Sonkusare et al, 2005) and CdMnTe (Sands et al, 2000), and examine the laser annealing of ion implantation induced defects in CdTe (Sands & Howari, 2005).

4. Analytical models of melting

Typically, analytical models tend to treat simple structures like a semi-infinite solid or a slab. Equation (22) shows how complicated solutions can be for even a simple system comprising only two layers, and if a third were to be added in the form of a time-dependent molten layer, the mathematics involved would become very complicated. One of the earliest models of melting considered the case of a slab either thermally insulated at the rear or thermally connected to some heat sink with a predefined thermal transport coefficient. Melting times either less than the transit time (El-Adawi, 1986) or greater than the transit time (El-Adawi & Shalaby, 1986) were considered separately. The transit time in this instance refers to the time required for temperature at the rear interface to increase above ambient, ie. when heat reaches the rear interface, located a distance l from the front surface, and has a clear mathematical definition.

The detail of El-Adawi's treatment will not be reproduced here as the mathematics, while not especially challenging in its complexity, is somewhat involved and the results are of limited applicability. Partly this is due to the nature of the assumptions, but it is also a limitation of analytical models. As with the simple heating models described above, El-Adawi assumed that heat flow is one-dimensional, that the optical radiation is entirely absorbed at the surface, and that the thermal properties remain temperature independent. The problem then reduces to solving the heat balance equation at the melt front,

$$I_0 A(1-R) + k\frac{dT}{dz} = \rho_s L\frac{dZ}{dt} \tag{25}$$

Here Z represents the location of the melt front and any value of $Z \le z \le l$ corresponds to solid material. The term on the right hand side represents the rate at which latent heat is absorbed as the melt front moves and the quantity L is the latent heat of fusion. Notice that optical absorption is assumed to occur at the liquid-solid interface, which is unphysical if the melt front has penetrated more than a few nanometres into the material. The reason for this is that El-Adawi fixed the temperature at the front surface after the onset of melting at the temperature of the phase change, T_m. Strictly, there would be no heat flow from the absorbing surface to the phase change boundary as both would be at the same temperature, so in effect El-Adawi made a physically unrealistic assumption that molten material is effectively evaporated away leaving only the liquid-solid interface as the surface which absorbs incoming radiation.

El-Adawi derived quadratic equations in both Z and dZ/dt respectively, the coefficients of which are themselves functions of the thermophysical and laser parameters. Computer solution of these quadratics yields all necessary information about the position of the melt front and El-Adawi was able to draw the following conclusions. For times greater than the critical time for melting but less than the transit time the rate of melting increases initially but then attains a constant value. For times greater than the critical time for melting but longer than the transit time, both Z and dZ/dt increase almost exponentially, but at rates depending on the value of h, the thermal coupling of the rear surface to the environment. This can be interpreted in terms of thermal pile-up at the rear surface; as the temperature at the rear of the slab increases this reduces the temperature gradient within the remaining solid, thereby reducing the flow of heat away from the melt front so that the rate at which material melts increases with time.

The method adopted by El-Adawi typifies mathematical approaches to melting in as much as simplifying assumptions and boundary conditions are required to render the problem tractable. In truth one could probably fill an entire chapter on analytical approaches to melting, but there is little to be gained from such an exercise. Each analytical model is limited not only by the assumptions used at the outset but also by the sort of information that can be calculated. In the case of El-Adawi's model above, the temperature profile within the molten region is entirely unknown and cannot be known as it doesn't feature in the formulation of the model. The models therefore apply to specific circumstances of laser processing, but have the advantage that they provide approximate solutions that may be computed relatively easily compared with numerical solutions. For example, El-Adawi's model of melting for times less than the transit time is equivalent to treating the material as a semi-infinite slab as the heat has not penetrated to the rear surface. Other authors have treated the semi-infinite slab explicitly. Xie and Kar (Xie & Kar, 1997) solve the parabolic heat diffusion equation within the liquid and solid regions separately and use similar heat balance equations. That is, the liquid and solid form a coupled system defined by a set of equations like (20) with Z again locating the melt front rather than an interface between two different materials. The heat balance equation at the interface between the liquid and solid becomes

$$k_l \frac{\partial T_l(z,t)}{\partial z} = k_s \frac{\partial T_s(z,t)}{\partial z} - \rho_s L \frac{dZ(t)}{dt} \tag{26}$$

At the surface the heat balance is defined by

$$I_0 A(1-R) + k_l \frac{\partial T(0,t)}{\partial z} = 0 \tag{27}$$

The solution proceeds by assuming a temperature within the liquid layer of the form

$$T_l(z,t) = T_m - \frac{AI}{k_l}[z - Z(t)] + \psi(t)[z^2 - Z(t)^2] \tag{28}$$

The heat balance equation at z=0 then determines $\psi(t)$. Similarly the temperature in the solid is assumed to be given by

$$T_s(z,t) = T_m - (T_m - T_o)[1 - \exp(-b(t)[z - Z(t)])] \tag{29}$$

The boundary conditions at $z=Z(t)$ then determine $b(t)$. Some further mathematical manipulation is necessary before arriving at a closed form which is capable of being computed. Comparison with experimental data on the melt depth as a function of time shows that this model is a reasonable, if imperfect, approximation that works quite well for some metals but less so for others.

Other models attempt to improve on the simplifying assumption by incorporating, for example, a temperature dependent absorption coefficient as well as the temporal variation of the pulse energy (Abd El-Ghany, 2001; El-Nicklawy et al, 2000) . These are some of the simplest models; 1-D heat flow after a single pulse incident on a homogeneous solid target with surface absorption. In processes such as laser welding the workpiece might be scanned across a fixed laser beam (Shahzade et al, 2010), which in turn might well be Gaussian in profile (figure 1) and focussed to a small spot. In addition, the much longer exposure of the surface to laser irradiation leads to much deeper melting and the possibility of convection currents within the molten material (Shuja et al, 2011). Such processes can be treated analytically (Dowden, 2009), but the models are too complicated to do anything more than mention here. Moreover, the models described here are heating models in as much as they deal with the system under the influence of laser irradiation. When the irradiation source is removed and the system begins to cool, the problem then is to decide under what conditions the material begins to solidify. This is by no means trivial, as melting and solidification appear to be asymmetric processes; whilst liquids can quite readily be cooled below the normal freezing point the converse is not true and materials tend to melt once the melting point is attained.

Models of melting are, in principle at least, much simpler than models of solidification, but the dynamics of solidification are just as important, if not more so, than the dynamics of melting because it is upon solidification that the characteristic microstructure of laser processed materials appears. One of the attractions of short pulse laser annealing is the effect on the microstructure, for example converting amorphous silicon to large-grained polycrystalline silicon. However, understanding how such microstructure develops is impossible without some appreciation of the mechanisms by which solid nuclei are formed from the liquid state and develop to become the recrystallised material. Classical nucleation theory (Wu, 1997) posits the existence of one or more stable nuclei from which the solid grows. The radius of a stable nucleus decreases as the temperature falls below the equilibrium melt temperature, so this theory favours undercooling in the liquid. In like manner, though the theory is different, the kinetic theory of solidification (Chalmers and Jackson, 1956; Cahoon, 2003) also requires undercooling. The kinetic theory is an atomistic model of solidification at an interface and holds that solidification and melting are described by different activation energies. At the equilibrium melt temperature, T_m, the rates of solidification and melting are equal and the liquid and solid phases co-exist, but at temperatures exceeding T_m the rate of melting exceeds that of solidification and the material melts. At temperatures below T_m the rate of solidification exceeds that of melting and the material solidifies. However, the nett rate of solidification is given by the difference between the two rates and increases as the temperature decreases. The model lends itself to laser processing not only because the transient nature of heating and cooling leads to very high interface velocities, which in turn implies undercooling at the interface, but also because the common theory of heat conduction, that is, Fourier's law, across the liquid-solid interface implies it.

A common feature of the analytical models described above is the assumption that the interface is a plane boundary between solid and liquid that stores no heat. The idea of the interface as a plane arises from Fourier's law (equation 1) in conjunction with coexistence, the idea that liquid and solid phases co-exist together at the melt temperature. It follows that if a region exists between the liquid and solid at a uniform temperature then no heat can be conducted across it. Therefore such a region cannot exist and the boundary between the liquid and solid must be abrupt. An abrupt boundary implies an atomistic crystallization model; the solid can only grow as atoms within the liquid make the transition at the interface to the solid, which is of course the basis of the kinetic model. However, there has been growing recognition in recent years that this assumption might be wanting, especially in the field of laser processing where sometimes the melt-depth is only a few nanometres in extent. This opens the way to consideration of other recrystallisation mechanisms.

One possibility is transient nucleation (Shneidman, 1995; Shneidman and Weinberg, 1996), which takes into account the rate of cooling on the rate of nucleation. Most of Shneidman's work is concerned with nucleation itself rather than the details of heat flow during crystallisation, but Shneidman has developed an analytical model applicable to the solidification of a thin film of silicon following pulsed laser radiation (Shneidman, 1996). As with most analytical models, however, it is limited by the assumptions underlying it, and if details of the evolution of the microstructure in laser melted materials are required, this is much better done numerically. We shall return to the topic of the liquid-solid interface and the mechanism of re-crystallization after describing numerical models of heat conduction.

5. Numerical methods in heat transfer

Equations (1), (3) and (11), which form the basis of the analytical models described above, can also be solved numerically using a forward time step, finite difference method. That is, the solid target under consideration is divided into small elements of width δz, with element 1 being located at the irradiated surface. The energy deposited into this surface from the laser in a small interval of time, δt, is, in the case of surface absorption,

$$\delta E = \alpha I_0(1-R)\delta t \tag{30}$$

and

$$\delta E = S(z)\delta t = \alpha I_0(1-R)\exp(-\alpha z).\delta t \tag{31}$$

in the case of optical penetration. If the adjacent element is at a mean temperature T_2, assumed to be constant across the element, the heat flowing out of the first element within this time interval is

$$\delta Q_{12} = -k\frac{(T_2 - T_1)}{\delta z}.\delta t \tag{32}$$

The enthalpy change in element 1 is therefore

$$\delta H = \delta E - \delta Q_{12} = (\rho \delta z)c_p \delta T_1 \tag{33}$$

In this manner the temperature rise in element 1, δT_1, can be calculated. The heat flowing out of element 1 flows into element 2. Together with any optical power absorbed directly within the element as well as the heat flowing out of element 2 and into 3, this allows the temperature rise in element 2 to be calculated. This process continues until an element at the ambient temperature is reached, and conduction stops. In practice it might be necessary to specify some minimum value of temperature below which it is assumed that heat conduction does not occur because it is a feature of Fourier's law that the temperature distribution is exponential and in principle very small temperatures could be calculated. However the matter is decided in practice, once heat conduction ceases the time is stepped on by an amount δt and the cycle of calculations is repeated again. In this way the temperature at the end of the pulse can be calculated or, if the incoming energy is set to zero, the calculation can be extended beyond the duration of the laser pulse and the system cooled.

This is the essence of the method and the origin of the name "forward time step, finite difference", but in practice calculations are often done differently because the method is slow; the space and time intervals are not independent and the total number of calculations is usually very large, especially if a high degree of spatial accuracy is required. However, this is the author's preferred method of performing numerical calculations for reasons which will become apparent. The calculation is usually stable if

$$\delta z^2 > 2D.\delta t \tag{34}$$

but the stability can be checked empirically simply by reducing δt at a fixed value of δz until the outcome of the calculation is no longer affected by the choice of parameters.

In order to overcome the inherent slowness of this technique, which involves explicit calculations of heat fluxes, alternative schemes based on the parabolic heat diffusion equation are commonly reported in the literature. It is relatively straightforward to show that between three sequential elements, say j-1, j and j+1, with temperature gradients

$$\frac{dT_{j-1,j}}{dz} \approx \frac{(T_j - T_{j-1})}{\delta z}. \tag{35a}$$

$$\frac{dT_{j,j+1}}{dz} \approx \frac{(T_{j+1} - T_j)}{\delta z}. \tag{35b}$$

the second differential is given by

$$\frac{d^2T}{dz^2} \approx \frac{(T_{j+1} - 2T_j + T_{j-1})}{\delta z^2}. \tag{36}$$

Hence the parabolic heat diffusion equation becomes

$$\frac{dT_j}{dt} \approx \frac{\delta T_j}{\delta t} = D\frac{(T_{j+1} - 2T_j + T_{j-1})}{\delta z^2}. + \frac{1}{\rho c_p}\frac{(k_{j+1} - k_{j-1})}{2\delta z}.\frac{(T_{j+1} - T_{j-1})}{2\delta z} \tag{37}$$

with appropriate source terms of the form of equation (31) for any optical radiation absorbed within the element. Thus if the temperature of any three adjacent elements is known at any given time the temperature of the middle element can be calculated at some time δt in the future without calculating the heat fluxes explicitly. This particular scheme is known as the forward-time, central-space (FTCS) method, but there are in fact several different schemes and a great deal of mathematical and computational research has been conducted to find the fastest and most efficient methods of numerical integration of the parabolic heat diffusion equation (Silva et al, 2008; Smith, 1965).

The difficulty with this equation, and the reason why the author prefers the more explicit, but slower method, lies in the second term, which takes into account variations in thermal conductivity with depth. Such changes can arise as a result of using temperature-dependent thermo-physical properties or across a boundary between two different materials, including a phase-change. However, Fourier's law itself is not well defined for heat flow across a junction, as the following illustrates. Mathematically, Fourier's law is an abstraction that describes heat flow across a temperature gradient at a point in space. A point thus defined has no spatial extension and strictly the problem of an interface, which can be assumed to be a 2-dimensional surface, does not arise in the calculus of heat flow. Besides, in simple problems the parabolic equation can be solved on both sides of the boundary, as was described earlier in El-Adawi's two-layer model, but in discrete models of heat flow, the location of an interface relative to the centre of an element assumes some importance. Within the central-space scheme the interface coincides with the boundary between two elements, say j and j+1 with thermal conductivities k_j and k_{j+1} and temperatures T_j and T_{j+1}.

The thermal gradient can be defined according to equation (35), but the expression for the rate of flow of heat requires a thermal conductivity which changes between the elements. Which conductivity do we use; k_j , k_{j+1} or some combination of the two?
This difficulty can be resolved by recognising that the temperatures of the elements represent averages over the whole element and therefore represent points that lie on a smooth curve. The interface between each element therefore lies at a well defined temperature and the heat flow can be written in terms of this temperature, T_i , as

$$\frac{dQ_{j+1,j}}{dt} = \frac{dQ_{j,i}}{dt} = -k_j \frac{T_i - T_j}{\delta z/2} \quad (38a)$$

$$\frac{dQ_{j+1,j}}{dt} = \frac{dQ_{i,j+1}}{dt} = -k_{j+1} \frac{T_{j+1} - T_i}{\delta z/2} \quad (38b)$$

Solving for T_i in terms of T_j and T_{j+1}, it can be shown that

$$T_i = \frac{k_j T_j + k_{j+1} T_{j+1}}{k_j + k_{j+1}} \quad (39)$$

Substituting back into either of equations (38a) or (38b) yields

$$\frac{dQ_{j,j+1}}{dt} = -2.\frac{k_j k_{j+1}}{k_j + k_{j+1}} \left[\frac{T_{j+1} - T_j}{\delta z} \right] \quad (40)$$

The correct thermal conductivity in the discrete central-space method is therefore a composite of the separate conductivities of the adjacent cells. This is in fact entirely general, and applies even if the interface between the two cells does not coincide with the interface between two different materials. For example, if the two conductivities, k_j and k_{j+1} are identical the effective conductivity reduces simply to the conductivity $k_j = k_{j+1}$. If, however, the two cells, j and j+1, comprise different materials such that the thermal conductivity of one vastly exceeds the other the effective conductivity reduces to twice the small conductivity and the heat flow is limited by the most thermally resistive material. For small changes in k such that $k_j = k_{j-1} + \delta k$ and $k_{j+1} = k_j + \delta k = k_{j-1} + 2\delta k$, the difference in heat flow between the three elements can be written in terms of k_{j-1} and δk. After some manipulation it can be shown that

$$\nabla \bullet \left[\frac{dQ}{dt} \right] \approx \frac{k_{j-1}}{\delta z^2} \left[T_{j+1} - 2T_j - T_{j-1} \right] + \frac{\delta k}{\delta z} . \frac{\delta T}{\delta z} \quad (41)$$

with

$$T_{j+1} - T_j \approx T_j - T_{j-1} = \delta T \quad (42)$$

This is equivalent to equation (6) in one dimension. If, however, the change in thermal conductivity arises from a change in material such that $k_{j+1} = k_j + \delta k$ and $k_j = k_{j-1}$, and δk need not be small in relation to k_j , then it can be shown that

$$\nabla \bullet \left[\frac{dQ}{dt} \right] \approx \frac{k_{j-1}}{\delta z^2} \left[T_{j+1} - 2T_j - T_{j-1} \right] + \frac{k_{j-1}}{\left(2k_{j-1} + \delta k \right)} \frac{\delta k}{\delta z} \cdot \frac{(T_{j+1} - T_j)}{\delta z} \tag{43}$$

We can consider two limiting cases. First, if $k_{j-1} >> k_{j+1}$, such that $\delta k \approx -k_{j-1}$ then

$$\frac{k_{j-1}}{2k_{j-1} + \delta k} \approx \frac{k_{j-1}}{k_{j-1}} = 1 \tag{44}$$

In this case equation (43) approximates to equation (37). Secondly, if $k_{j-1} << k_{j+1}$, such that $\delta k \approx k_{j+1}$ then

$$\frac{k_{j-1}}{2k_{j-1} + \delta k} \approx \frac{k_{j-1}}{k_{j+1}} << 1 \tag{45}$$

In this case the contribution from the second term in (43) is very small, but more importantly, equation (43) is shown not to be equivalent to (37). Likewise, if we choose some intermediate value, say $k_{j-1} = 2\,k_{j+1}$ or conversely $2k_{j-1} = k_{j+1}$ this term becomes respectively 2/3 or 1/3. The precise value of this ratio will depend on the relative magnitudes of k_{j-1} and k_{j+1}, but we see that in general equation (43) is not numerically equivalent to (37). The difference might only be small, but the cumulative effect of even small changes integrated over the duration of the laser pulse can turn out to be significant. For this reason the author's own preference for numerical solution of the heat diffusion equation involves explicit calculation of the heat fluxes into and out of an element according to equation (40) and explicit calculation of the temperature change within the element according to equation (3). As described, the method is slow, but the results are sure.

5.1 Melting within numerical models

The advantage of numerical modelling over analytical solutions of the heat diffusion equation is the flexibility in terms of the number of layers within the sample, the use of temperature dependent thermo-physical and optical properties as well as the temporal profile of the laser pulse. This advantage should, in principle, extend to treatments of melting, but self-consistent numerical models of melting and recrystallisation present considerable difficulty. Chalmers and Jackson's kinetic theory of solidification described previously implies that a fast rate of solidification, as found, for example, in nano-second laser processing, should be accompanied by significant undercooling of the liquid-solid interface. However, tying the rate of cooling to the rate of solidification within a numerical model presents considerable difficulties. Moreover, it might not be necessary.

In early work on laser melting of silicon it was postulated that an interface velocity of approximately 15 ms^{-1} is required to amorphise silicon. Amorphous silicon is known to have a melting point some 200°C below the melting point of crystalline silicon so it was assumed that in order to form amorphous silicon from the melt the interface must cool by at least this amount, which requires in turn such high interfacial velocities. By implication, however, the converse would appear to be necessary; that high rates of melting should be accompanied by overheating, yet the evidence for the latter is scant. Indeed, extensive modelling work in the 1980s on silicon (Wood & Jellison, 1984) , and GaAs (Lowndes, 1984) showed that very high interface velocities arise from the rate of heating supplied by the laser rather than any change in the temperature of the interface. These authors held the liquid-solid interface at the equilibrium melt temperature and calculated curves of the kind shown in figure 3a. Differentiation of the melt front position with respect to time (figure 3b) shows that the velocity during melting can exceed 20 ms^{-1} and during solidification can reach as high as 6 ms^{-1}, settling at 3 ms^{-1}. The fact of such large interface velocities does not, of itself, invalidate the notion of undercooling but it does mean that undercooling need not be a pre-requisite for, or indeed a consequence of, a high melt front velocity.

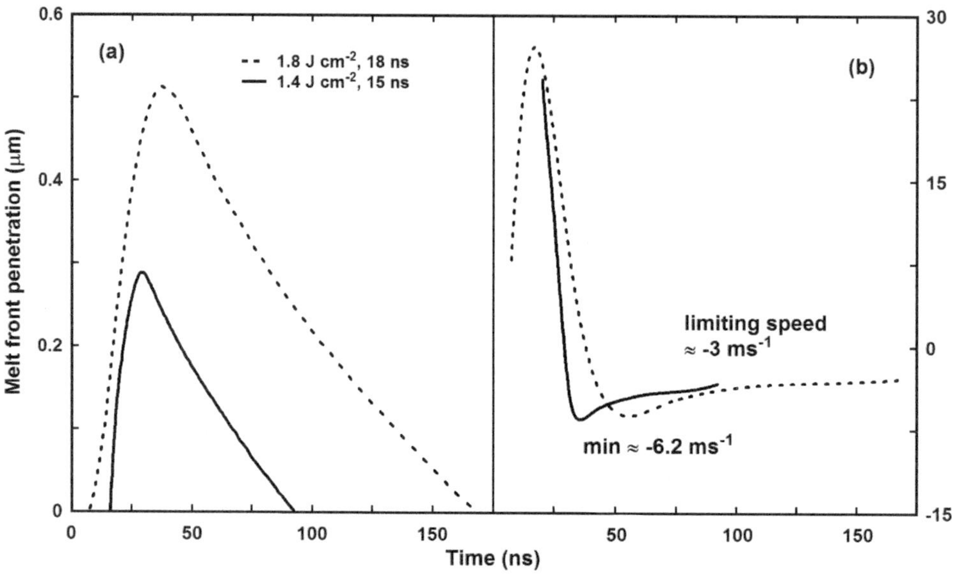

Fig. 3. Typical curves of the melt front penetration (a) taken from figures 4 and 6 of Wood and Jellison (1984) and the corresponding interface velocity (b).

If undercooling is not necessary for large interface velocities then the requirement that the interface be sharp, which is required by both the kinetic model of solidification and Fourier's law, might also be unnecessary. Various attempts have been made over the years to define an interface layer but the problem of ascribing a temperature to it is not trivial. The essential difficulty is that we have no knowledge of the thermal properties of materials in this condition, nor indeed a fully satisfactory theory of melting and solidification. One idea that has gained a lot of ground in recent years is the "phase field", a quantity, denoted by ϕ, constructed within the theory of non-equilibrium thermodynamics that has the properties of a field but takes a value of *either 0 or 1* for solid and liquid phases respectively and $0< \phi <1$ for the interphase region (Qin & Bhadeshia, 2010; Sekerka, 2004). In essence, gradients within the thermodynamic quantities drive the process of crystallisation.

The phase field method was originally proposed for equilibrium solidification and has been very successful in predicting the large scale structure, such as the growth of dendrites, often seen in such systems. It has also been applied to rapid solidification (Kim & Kim, 2001), including excimer laser processing of silicon (La Magna, 2004; Shih et al, 2006; Steinbach &

Apel, 2007). Despite its success in replicating many experimentally observed features in solidification (see for example, Pusztai, 2008, and references therein) and the phase field itself is not necessarily associated with any physical property of the interface (Qin & Bhadeshia, 2010). Moreover, even though it can be adapted to apply to the numerical solution of the 1-D heat diffusion equation, it is essentially a method for looking at 2-D structures such as dendrites and is not well suited to planar interfaces. For example, in the work of Shih et al mentioned above, it was necessary to introduce a spherical droplet within the solid in order to initiate melting.

The author's own approach to this problem is to question the validity of Fourier's law in the domain of melting (Sands, 2007). It is necessary to state that either Fourier's law is invalid or a liquid and solid cannot co-exist at exactly the same temperature because the two concepts are mutually exclusive. Coexistence at the equilibrium melt temperature is, of course, a macroscopic idea that might, or might not apply at the microscopic level. It is difficult to imagine an experiment with sufficient resolution to measure the temperature either side of an interface, but if even a small difference exists it is sufficient for heat to flow according to Fourier's law. If no difference exists heat cannot flow across the interface. Of course, we know in practice that heat must flow in order to supply, or conduct away, the latent heat. This tension between the microscopic and macroscopic domains also applies to the model of the interface. The idea of a plane sharp interface that stores no heat arises in essence from mathematical models in which the heat diffusion equation is solved on either wide of the interface and the solutions matched. By definition, heat flowing out of one side flows into the other and the interface does nothing more than mark the point at which the phase changes. However, the idea of a fuzzy interface, as represented for example in phase field models, implies that interfaces do not behave like this at the microscopic level. More fundamental, however, is the question of whether a formulation of heat flow in terms of temperature or enthalpy per unit volume, H_v, is the more fundamental.

The parabolic heat diffusion equation arises from equations (1) and (4), with equation (3) being used to convert the rate of change of volumetric enthalpy to a rate of change of temperature. However, equation (3) can also be used to convert equation (1) to an expression for heat flow in terms of H_v, which now resembles Fick's first law of diffusion. Application of continuity, as expressed by equation (4), now leads to a parabolic equation in H_v rather than T. Both forms of heat diffusion are mathematically valid, but they do not lead to the same outcome except in the case of a homogeneous material heated below the melting point. Whichever is the primary variable in the parabolic equation becomes continuous; temperature in one case, volumetric enthalpy in the other. Experience would seem to suggest that temperature is the more fundamental variable as thermal equilibrium between two different materials is expressed in terms of the equality of temperature rather than volumetric enthalpy, and indeed this is a weakness of the enthalpy formulation, but we have already seen in the derivation of equation (40) how the mathematical form of Fourier's law breaks down in numerical computation of heat flow across a junction. On the other hand, expressing the heat flow in terms of Fick's law of diffusion would seem to bring the idea of thermal diffusion in line with a host of other diffusion phenomena, thereby seeming to make this a more fundamental formulation. Moreover, it leads naturally to a diffuse model of the interface.

The width of the diffuse interface generated by the enthalpy model is much greater than the width of the liquid-solid interface observed in phase field models, but this is not in itself a difficulty. Nor does it imply that the normally accepted idea of coexistence is invalid, but it does require a different model of melting and solidification. The details are discussed by Sands (2007), but the ideas can be summarised as follows. Molecular dynamics simulations suggest that beyond a certain degree of disorder a defective solid changes abruptly into a liquid, so a continuous enthalpy can be interpreted as a mixture of liquid and solid regions co-existing together in a ratio that gives the average enthalpy of the mixture. However, co-existence is probably dynamic as the material fluctuates between solid and liquid forms.

The latter is more than just a fanciful idea. The classical theory of solidification is based on the idea of forming a stable nucleus of solid material at some temperature below the equilibrium melt temperature, but in fact such a model is not supported by the experimental evidence. It can be shown (Sands, 2007) that typically only 11% of the latent heat released during the formation of a solid nucleus is expended in the form of work done in forming the nucleus, meaning that the remainder must go into heating up the solid nucleus and the surrounding material. Indeed, it is a common observation in electrostatic levitation experiments (Sung et al, 2003) that once an undercooled melt begins to solidify the temperature rises to the equilibrium melt temperature. The phenomenon is known as recalescence. There is no requirement for solid nuclei to be stable under these conditions. Indeed, if a solid nucleus does in fact heat up its local environment then by implication the formation of a liquid nucleus within a solid matrix must cool the surrounding material. There is thus the possibility of large fluctuations in the temperature field, which could in turn trigger fluctuations between phases.

The advantage of the enthalpy formulation is that it is intuitive. The idea that an interface must achieve a certain velocity before amorphous solids can be formed is, in this author's view, no longer tenable. First, we have shown that computation using a fixed temperature lead to significant non-zero interface velocities, thereby breaking the link with undercooling. Secondly, the phenomenon of recalescence would tend to raise the temperature of the interface. Thirdly, it is well known that glassy metals are produced by rapid cooling which essentially freezes in the disorder associated with the liquid state. Although amorphous silicon is not a glass it is characterised by a similar disorder and it is not immediately clear why a similar mechanism cannot be responsible for its formation. The enthalpy formulation allows for this because thermal transport is determined essentially by the laser parameters; if the material does not crystallise before too much heat is lost amorphous silicon forms, but if crystallisation does occur and the rate of release of latent heat is faster than the rate at which heat is transported away then recalescence occurs. Whether crystallisation occurs or not is determined essentially by probability. Based on the author's work (Sands, 2007) nucleation in silicon would seem to require something in the region of 8-10 ns to initiate and amorphous silicon is formed because the system heats and cools within that timescale.

The weakness of the enthalpy formulation is undoubtedly its lack of self consistency, as it is necessary to switch to a temperature formulation at the interface between different materials. Indeed, it might be possible to treat the above ideas within a temperature formulation, because a model of the phase transition which accounts for recalescence could, in principle, be treated using Fourier's law. Unfortunately, such a general model has not yet been formulated and it is not clear to the author even that such ideas have been widely accepted. The difficulty lies in finding a simple formulation for the change in temperature associated with a transition from liquid to solid and *vice versa*, but if the microscopic variations in temperature could be formulated then large variations would exist across and within the interfacial layer and the resulting flow of heat would be quite complex. Finding a formulation of melting that is physically sensible and can be treated self consistently within an appropriate model of heat conduction is one of the most interesting problems in pulsed laser heating and melting.

6. Short pulse heating

The final topic of this chapter concerns heating using laser pulses of much shorter duration than the nanosecond pulses typical of excimer lasers and Q-switched YAG lasers. Lasers capable of delivering sub-picosecond pulses are now routinely used in materials processing, especially for ablation and marking, and a pre-requisite for modelling such processes is an accurate model of thermal transport, which is quite different from the mechanisms considered above. All the models of heating discussed above are predicated on the assumption that the temperature of the lattice follows the laser pulse, but with sub-picosecond pulses that is no longer true; the electron and lattice temperatures have to be considered separately.

Models of short pulse heating have been around since the late 1980s, but papers continue to be published. One of the problems is the lack of data for different materials. Data for Ni and Au has been known for some time (Yilbas & Shuja, 2000) and experiments on these materials continue to be performed (Chen, 2005), but data on silicon is more recent (Lee, 2005). The mechanisms of energy transport in semiconductors are also different from those in metals, where a two-temperature model is often invoked to explain ultra-fast heating (Yilbas & Shuja, 1999). Photons are absorbed by electrons, which have a non-equilibrium energy distribution as a result. Electron-electron collisions establish thermal equilibrium within femtoseconds and electron-phonon (lattice vibrations) interactions transfer the energy from the electrons to the lattice in timescales of the order of picoseconds. A lot of theoretical papers are concerned with the nature of the electron-electron collisions (electron kinetic theory) and there is a suggestion (Rethfield et al, 2002b) that in fact the two-temperature model is inappropriate under conditions of weak excitation, as it takes longer than a picosecond for the electrons to establish an equilibrium distribution. On the other hand, for very strong excitation the electrons behave as if they follow a well defined Fermi-Dirac distribution and a temperature can be defined. The temperatures of the lattice and the electron distribution are then linked (Rethfield et al, 2002) by;

$$c_p \frac{\partial T_p}{dt} = -c_e \frac{\partial T_e}{dt} = \alpha (T_e - T_p) \tag{46}$$

The subscripts p and e refer to phonons (lattice) and electrons respectively, c_p and c_e are the respective heat capacities and α is the energy exchange rate between phonons and electrons. For semiconductors, however, three temperatures are needed to account for thermal transport (Lee, 2005); those of the electrons, the optical mode phonons and the acoustic mode phonons. Two-temperature models of semiconductor heating can be found (Medvedev & Rethfeld, 2010), but this is only valid if thermal transport is neglected. That is, if the semiconductor is thin enough to be heated uniformly without thermal conduction. Otherwise, a three-stage process ought to be considered; electron-electron collisions, electron-optical phonon interactions, and phonon-phonon interactions. Electrons lose energy to optical mode phonons, but these have a low group velocity and do not contribute much to thermal conduction. The optical mode phonons exchange energy with the lower frequency acoustic mode phonons and these contribute to thermal conduction. Over long timescales these different models reduce to Fourier transport.

As with longer pulse interactions, some of the most interesting physics is to be found in phase transitions. Siwick (2003) has performed time-resolved electron diffraction experiments to demonstrate that very thin aluminium films undergo thermal melting within 3.5 ps. Siwick discusses some of the earlier work that led to the view that melting occurred within 500 fs, which is faster than the transfer of energy to the lattice. This led to the view that melting is non-thermal; the cohesive forces between the atoms are essentially reduced by the formation of the electron hole plasma. However, Siwick shows through an analysis of time-dependent diffraction data that even though the solid is highly disordered after 1.5 ps it is nonetheless still a solid. The disorder is due to very large amplitude oscillations caused by superheating of the lattice to temperatures well above the equilibrium melting point and the material does not become fully liquid until after 3.5 ps when, in Siwick's words, "The solid shakes itself apart".

This mechanism of melting is quite different from the mechanisms operating at longer timescales, where evidence for a picture of vacancy mediated melting is beginning to emerge. The selective evaporation of volatile components of compounds semiconductors has long been recognised (Sands & Howari, 2005) but the role of vacancies in the formation of amorphous silicon is beginning to be recognized (Lulli et al, 2006). The corresponding generation of such defects during excimer laser processing has also been observed, both after melting and solidification (La Magna et al, 2007) and immediately prior to the melting of crystalline silicon (Maekawa & Kawasuso, 2009). These vacancies are generated primarily at the surface and melting is initiated there. At the timescales involved in ultra fast melting, however, there is no evidence as yet that vacancies are involved. Indeed, the evidence of Siwick et al points to superheating of the lattice and the homogeneous nucleation of liquid droplets within the material (Rethfield at el, 2002a). The melting time is determined in this model by the electron-phonon relaxation time.

7. Conclusion

Pulsed laser heating of mainly metals and semiconductors has been discussed within the framework of Fourier's law of heat conduction. A number of analytical solutions of the 1-dimensional heat conduction equations have been considered. In the pre-melting regime these include the simple semi-infinite solid with surface absorption as well as a two-layer model, and analytical models of melting have also been examined. However, analytical models are limited and numerical methods of solving the heat diffusion equation have been discussed. In particular, it has been shown that Fourier's law is not well defined for abrupt changes in material properties and that the effective thermal conductivity across the interface is given by a combination of the two different conductivities. The usual parabolic form of the heat diffusion equation can give rise to errors in such circumstances, although it can be used when the spatial variations in the thermal conductivity are small, almost linear. Analytical models, such as El-Adawi's two-layer model, do not suffer this difficulty as the parabolic heat diffusion equation is usually solved on either side of the junction.

Of particular interest is the formulation of melting within numerical models. Classical thermodynamic models of melting and solidification have been discussed and shown to be found wanting, especially in relation to a diffuse interface. In this regard, phase field models have been discussed, but these are not well suited to 1-D heat conduction and other approaches would appear to be needed. However, a completely satisfactory and consistent numerical model of rapid heat conduction, melting and solidification has yet to be formulated, though progress has been made in this direction. For ultra-fast heating, it has been shown that heat conduction requires consideration of separate electron and phonon temperatures as well as an interaction between the two. The experimental evidence on ultra-fast melting suggests that the process is thermal, being limited by the electron-phonon interaction, but models of heat conduction which incorporate melting have not been discussed. It does not seem likely at the present time that such models of melting will have much impact on models of melting on nanosecond or longer timescales as the two processes appear to be very different. On the other hand, mechanisms of solidification, which have not been discussed for ultra-fast heating, might have more in common, particularly with regard to the evolution of microstructure.

In the author's view, the problems of melting, and the associated issues of thermal conduction, constitute some of the most promising and fruitful areas of research in this field. A great deal of work remains to be done, not only in experimental studies to gather data and to characterize these processes, but also in the development of theoretical and computational models.

8. References

Abd El-Ghany, S. E. -S. (2001) . The temperature profile in the molten layer of a semi-infinite target induced by irradiation using a pulsed laser, *Optics and Laser Technology*, Vol. 33 , No. 8, (November, 2001), pp. (539-551), ISSN 0030-3992

von Allmen, Martin & Blatter, Andreas. (1995). *Laser Beam Interactions with Materials: physical principles and applications*, Springer, ISBN 3-540-59401-9, Berlin.

Bechtel, J. H. (1975). Heating of solid targets with laser pulses, *Journal of Applied Physics*, Vol. 46, No.4 , (April, 1975), pp. (1585-1593), ISSN: 0021-8979

Cahoon, J. R (2003); On the atomistic theory of solidification, *Metallurgical and Materials Transactions A* Vol. 34, No. 11, (November, 2003) pp. (2683-2688), ISSN: 1073-5623

Chen, J.K., Tzou, D. Y., Beraun, J. E. (2006). A semiclassical two-temperature model for ultrafast laser heating, *International Journal Of Heat And Mass Transfer*, Vol. 49, No. 1-2 (January, 2006), pp. (307-316) , ISSN: 0017-9310

Dowden, J. (Ed.). (2009). *The Theory of Laser Materials Processing*, Springer Series in Materials Science 119, Springer , ISBN 978-1-4020-9339-5, Dordrecht, Netherlands

El-Adawi, M. K. (1986). Laser Melting of Solids - An exact solution for time intervals less or equal to the transit time, *Journal of Applied Physics*, Vol 60, No. 7, (October,1986), pp 2256-2259, ISSN: 0021-8979

El-Adawi, M. K. & Shalaby, S. A. (1986). Laser Melting of Solids – An exact solution for time intervals greater than the transit time, *Journal of Applied Physics, Vol.* 60, No. 7, (October,1986), pp. (2260-2265), ISSN: 0021-8979

El-Adawi, M. K., Abdel-Naby, M. A. and Shalaby, S. A. (1995). Laser heating of a two-layer system with constant surface absorption: an exact solution, *International Journal of Heat and Mass Transfer*, Vol. 38, No. 5, (March, 1995), pp. (947-952), ISSN 0017-9310

El-Nicklawy, M.M., Hassan, A. F., -S Abd El-Ghany, S.E. (2000). On melting a semi-infinite target using a a pulsed laser, *Optics and Laser Technology*, Vol. 32, No. 3, (April, 2000), pp. (157-164), ISSN 0030-3992

Howari, H., Sands, D., Nicholls, J. E., Hogg, J. H. C., Hagston, W. E., Stirner, T. (2000). Excimer laser induced diffusion in magnetic semiconductor quantum wells. *Journal of Applied Physics,* Vol. 88 , No. 3, (August, 2000) pp. (1373-1379), ISSN: 0021-8979

Jackson, K. A. and Chalmers, B. (1956). Kinetics of Solidification, *Canadian Journal of Physics,* Vol. 34, No. 5, (May, 1956) pp. (473-490), ISSN 0008-4204

Kim, S. G., and Kim, W. T., (2001). Phase-field modeling of rapid solidification, *Materials Science and Engineering A,* Vol. 304, Sp. Iss. SI (May, 2001) pp. (281–286), ISSN: 0921-5093

La Magna, A., Alippi, P., Privitera, V., Fortunato, G., Camalleri, M., Svensson, B. (2004). A phase-field approach to the simulation of the excimer laser annealing process in Si, *Journal Of Applied Physics* Vol.95, No. 9, (May, 2004), pp. (4806-4814), ISSN: 0021-8979

La Magna A., Privitera V., Mannino G. (2007). Defect generation and evolution in laser processing of Si, in *15th IEEE International Conference On Advanced Thermal Processing Of Semiconductors - RTP 2007,* pp.(245-250), IEEE, New York, ISBN: 978-1-4244-1227-3

Lee, S. H. (2005). Nonequilibrium heat transfer characteristics during ultrafast pulse laser heating of a silicon microstructure, *Journal Of Mechanical Science And Technology* Vol. 19 , No. 6, (June, 2005), pp.(1378-1389) ISSN: 1738-494X

Lowndes, D. H. , (1984). Pulsed Beam Processing of Gallium Arsenide, in *Pulsed Laser Processing of Semiconductors, Semiconductors and Semimetals,* Volume 23, Wood, R.F., White C.W., Young, R. T. , pp. (471-553), Academic Press, Orlando

Lulli G., Albertazzi E., Balboni S., Colombo L. (2006). Defect-induced homogeneous amorphization of silicon: the role of defect structure and population, *Journal of Physics: Condensed Matter,* Vol 18, No 6 (February, 2006) pp. (2077-2088) ISSN: 0953-8984

Maekawa M. & Kawasuso A. (2009). Vacancy Generation in Si During Solid-Liquid Transition Observed by Positron Annihilation Spectroscopy, *Japanese Journal Of Applied Physics* Vol. 48, No. 3, (March 2009), Art. No. 030203, ISSN: 0021-4922

Medvedev N., Rethfeld B. (2010). A comprehensive model for the ultrashort visible light irradiation of semiconductors, *Journal Of Applied Physics* Vol. 108 No. 10 (November, 2010), Art. Num. 103112 , ISSN: 0021-8979

Pusztai, T., Tegze, G., Tóth, G. I., Környei, L., Bansel, G., Fan, Z. Y., Gránásy, L. (2008). Phase-field approach to polycrystalline solidification including heterogeneous and homogeneous nucleation, *Journal of Physics: Condensed Matter* Vol. 20, No. 40 (October, 2008) 404205, ISSN: 0953-8984

Qin, R. S. and Bhadeshia, H. K. (2010). Phase field method, *Materials Science and Technology,* Vol. 26 No. 7 (July, 2010) pp. (803-811), ISSN 0267-0836

Rethfeld B., Sokolowski-Tinten K., von der Linde D. (2002a). Ultrafast thermal melting of laser-excited solids by homogeneous nucleation, *Physical Review B,* Vol. 65 No. 9 , (March 2002), Art. No. 092103, ISSN: 1098-0121

Rethfeld B., Kaiser A., Vicanek M., Simon G. (2002b). Ultrafast dynamics of nonequilibrium electrons in metals under femtosecond laser irradiation, *Physical Review B,* Vol. 65, No. 21 (June, 2002) Art. No. 214303 , ISSN: 1098-0121

Sands, D., Key, P. H., Hoyland, J. D., (1999). In-situ measurements of excimer laser irradiated zinc sulphide films on silicon, *Applied Surface Science*, Vol. 138, Nos. 1-2, (January, 1999) pp. (240-243), ISSN 0169-4332

Sands, D., Howari, H. (2005). The kinetics of point defects in low-power pulsed laser annealing of ion-implanted CdTe/CdMnTe double quantum well structures, *J Journal of Applied Physics* Vol. 98, No. 8, (October, 2005), Art. No. 083506, ISSN: 0021-8979

Sands, D. (2007). New theory of undercooling during rapid solidification: application to pulsed laser heated silicon, *Applied Physics A-Materials Science & Processing*, Vol. 88, No. 1, (July, 2007), pp. (179-189), ISSN: 0947-8396

Sekerka, Robert F. (2004). Morphology, from sharp interface to phase field models, *Journal of Crystal Growth*, Vol. 264, No.4 (March, 2004) pp. (530-540), ISSN 0022-0248

Shahzade, Z., Shuja, Y., Bekir S. and Momin, O. (2010). Laser Heating of Moving Solid: Influence of Workpiece Speed on Melt Size, *AIChE (American Institute of Chemical Engineers) Journal* Vol. 56, No. 11 (November, 2010) pp (2997-3004)

Shih, C. J., Fang, C. H. , Lu, C. C. , Wang, M. H., Lee, M. H. and Lana, C. W. (2006). Phase field modeling of excimer laser crystallization of thin silicon films on amorphous substrates, *Journal Of Applied Physics*,Vol. 100, No. 5, (September 2006) art. 053504, ISSN: 0021-8979

Shneidman, Vitaly A. (1995). Theory of time-dependent nucleation and growth during a rapid quench, *Journal of Chemical Physics*, Vol. 103, No. 22, (December, 1995), pp. (9772-9781), ISSN 0021-9606

Shneidman, Vitaly A. (1996). Interplay of latent heat and time-dependent nucleation effects following pulsed-laser melting of a thin silicon film, *Journal of Applied Physics*, Vol. 80, No. 2, (July, 1996), pp. (803-811), ISSN: 0021-8979

Shneidman, Vitaly A. & Weinberg, Michael C. (1996). Crystallization of rapidly heated amorphous solids, *Journal of Non-Crystalline Solids*, Vol. 194, Nos. 1-2, (January, 1996) pp. (145-154), ISSN: 0022-3093

Shuja, S. Z., Yilbas, B. S. and Momin, O. (2011). Laser repetitive pulse heating and melt pool formation at the surface, *Journal of Mechanical Science and Technology*, Vol. 25, No. 2 (February, 2011) pp. (479-487), ISSN: 1738-494X

Siwick, B.J., Dwyer, J. R., Jordan, R. E., Miller, R. J. D. (2003). An atomic-level view of melting using femtosecond electron diffraction, *Science*, Vol. 302, No 5649, (November, 2003) pp. (1382-1385), ISSN: 0036-8075

Silva, J. B. C., Romão, E. C. , de Moura, L. F. M. (2008). A comparison of time discretization methods in the solution of a parabolic equation, *7th Brazilian Conference on Dynamics, Control and Applications*, May07-09, 2008, Retrieved from < http://www4.fct.unesp.br/ dmec/dincon2008/artigos/02/01%20-%2002-Campos_Silva.pdf>

Smith, G. D. (1965). *Numerical Solution of Partial Differential Equations*, Oxford University Press, London

Sonkusare, A., Sands, D., Rybchenko, S. I., Itskevich, I. (2005). Photoluminescence from Ion Implanted and Low-Power-Laser Annealed GaAs/AlGaAs Quantum Wells, *AIP Conference Proceedings* Vol. 772, (June, 2005), pp. (965-966), PHYSICS OF SEMICONDUCTORS - 27th International Conference on the Physics of Semiconductors (ICPS 27)

Steinbach, I., and Apel, M. (2007). Phase-field simulation of rapid crystallization of silicon on substrate, *Materials Science and Engineering A*, Vol. 449 (March, 2007) pp. (95–98), ISSN: 0921-5093

Sung, Y.H., Takeya, H., Hirata, K., Togano, K. (2003). Specific heat capacity and hemispherical total emissivity of liquid Si measured in electrostatic levitation, *Applied Physics Letters* Vol. 83, No. 6, (August 2003), pp. (1122-1124), **ISSN:** 0003-6951

Wood R. F. and Jellison, G. E., Jr., (1984). Melting Model of Pulsed Laser processing, in *Pulsed Laser Processing of Semiconductors, Semiconductors and Semimetals*, Volume 23, Wood, R.F., White C.W., Young, R. T. , pp. (165-250), Academic Press, Orlando.

Wu, David T. (1997), Nucleation Theory, *Solid State Physics*, volume 50, H. Ehrenreich and F. Spaepen, eds. pp. 37-187, Academic Press, San Diego

Xie, J. & Kar, A. (1997). Mathematical modelling of melting during laser materials processing, *Journal of Applied Physics*, Vol. 81, No. 7 pp. (3015-3022), ISSN: 0021-8979

Yilbas, B. S., Shuja, S. Z. (1999). Laser short-pulse heating of surfaces, *Journal of Physics D: Applied Physics*, Vol. 32 No. 16, (August, 1999), pp. (1947-1954), ISSN: 0022-3727

Yilbas, B. S., Shuja, S. Z. (2000). Electron kinetic theory approach for sub-nanosecond laser pulse heating, *Proceedings of the Institution of Mechanical Engineers, Part C: Journal of Mechanical Engineering Science*, Vol. 214 No. 10, (2000) pp. (1273-1284), ISSN: 0954-4062

2

Energy Transfer in Ion– and Laser–Solid Interactions

Alejandro Crespo-Sosa
Instituto de Física, Universidad Nacional Autonoma de México
México

1. Introduction

While the fundamentals of ion beam interaction with solids had been studied as early as the 1930s, its utility in the modification of materials was not fully recognized until the 60's and 70's. About the same time, the fabrication of high–power lasers permitted their application in the processing of materials, especially the use of short-pulsed lasers. Both techniques are nowadays widely used in a great variety of applications. The electromagnetic radiation (or photons, from a quantum mechanical point of view) from lasers interact with the electrons of the materials, transferring energy to them within femtoseconds. Energetic ions also transfer part of their energy to the electrons of the solid, but they can also interact directly with the nuclei in elastic collisions. The primary energy transferred involved in these precesses is not thermal and some assumptions must be made before treating the problem as a thermal one. Furthermore, these processes take place in very short periods of time and are localized in the nanometer range. This means that the system can hardly satisfy the condition of thermodynamic equilibrium. Despite the complexity of these processes, many of the effects on the materials can be understood by using simple classical concepts contained in the heat equation.

During the last decades, different aspects of the ion–solid interaction have been incorporated in the calculation of the temperature evolution in the so–called thermal spike. This implementation has been possible in part, by the development of fast computers, but also by the availability of ultra short laser pulses that have given a great amount of information about the dynamics of electronic processes Elsayed-Ali et al. (1987); Schoenlein et al. (1987); Sun et al. (1994)). From a thermal point of view, these processes are very similar either for ions or for lasers pulses. The results obtained in one case can be applied most of the times to the other. For many of these experimental phenomena, the estimation of the temperature is only the first step and supplementary diffusion or stress equations must be solved, consistent with the spatial temperature evolution in order to describe them.

From another point of view, nano–structures are nowadays of great interest in technology. Nano–structured materials have opened the possibility to fabricate smaller, more efficient and faster devices. Thus, the fabrication and characterization of new nano–structured materials has become very important and the use of ion beams and short laser pulses have proved to be quite appropriate tools for that purpose (Klaumunzer (2006); Meldrum et al. (2001); Takeda & Kishimoto (2003)). Thus, their modeling and understanding is very important.

It is shown, in this chapter, firstly, how the the "thermal spike" model has recently incorporated detailed aspects of the ion–solid interaction, as well as from the dynamics of the electronic system up to a high grade of sophistication. Then some experimental effects of ion beams on nano–structured materials are presented and discussed from a point of view of the thermal evolution of the system. Finally some examples of the effects of short laser pulses on nano–structured materials are also discussed.

2. The thermal spike

The concept of thermal spike in ion–solid interaction, is the result of assuming that the ion deposits an amount of energy F_D, increasing the local temperature and that thereafter it obeys the classical laws of heat diffusion. The temperature is therefore, a function of time and location and can be calculated with the aid of the heat equation:

$$\frac{\partial T}{\partial t} = \frac{1}{\rho c_p} \nabla \left[\kappa \nabla T \right] + \frac{1}{\rho c_p} s(t, \vec{r})) \tag{1}$$

where T is the temperature as function of time t and position $\vec{r}$, and $s(t, \vec{r})$ is, in general, a source or a sink of heat, that can also be a function of time t and position $\vec{r}$. In the simplest model, the source $s(\vec{r}, t)$ is taken as a Dirac delta function in time and space. If it is assumed that the energy is deposited at a point, spherical thermal spike comes to one's mind, while if it is deposited along a straight line, the spike is said to be cylindrical. Vineyard (Vineyard (1976)) solved this equation and further calculated the total number of atomic jumps produced by the ion within the spike using the temperature evolution within it and an jumping rate proportional to $\exp(-\frac{E}{k_B T})$. Because of its simplicity, this model is still widely used to estimate the "temperature" of the thermal spike, whether it is an elastic spike due to nuclear stopping power or the so-called inelastalic spike due to electronic interaction.

In the two-temperature model, the energy transfer from the electrons to the lattice is considered with a second equation coupled with the first through an interaction term $g(t, \vec{r})$:

$$\frac{\partial T_e}{\partial t} = \frac{1}{\rho c_{p_e}} \nabla \left[\kappa_e \nabla T_e \right] + \frac{1}{\rho c_{p_e}} s_e(t, \vec{r})) - \frac{1}{\rho c_{p_e}} g(t, \vec{r}) \tag{2}$$

$$\frac{\partial T_l}{\partial t} = \frac{1}{\rho c_{p_l}} \nabla \left[\kappa_l \nabla T_l \right] + \frac{1}{\rho c_{p_l}} s_l(t, \vec{r})) + \frac{1}{\rho c_{p_l}} g(t, \vec{r}) \tag{3}$$

here, the subscript e stands for electron, while l for lattice, and $g(t, \vec{r})$ is the electron–phonon coupling term that allows the heat transfer from the electronic subsystem to the lattice via electron–phonon scattering (Lin & Zhigilei (2007); Toulemonde (2000); Wang et al. (1994)). Free electrons contribute at most to electronic conductivity, so that it is larger for metals than for semiconductors or dielectrics.

At higher ion energies, that is when the electronic interaction prevails, the geometry of the spike is that of the global spike along the whole ion's path, however the energy deposition cannot be considered instantaneous nor one-dimensional (Waligórski et al. (1986))Katz & Varma (1991). The energy of the ejected electrons is high and therefore their range (tens of nanometers) is needed to be taken into account. For an ion with velocity v, the radial energy distribution density is given by:

$$D(r) = \frac{Ne^4 Z^{*2}}{am_e c^2 \beta^2} \left[\frac{\left(1 - \frac{r+R}{T+R}\right)^{1/a}}{(w+I)^2} \right] \tag{4}$$

where R is the range of an electron with energy I and T is the maximum range, corresponding to the maximum possible energy transfer. Finally, the temporal component can be included (Toulemonde (2000); Toulemonde et al. (2003; 1992); Wang et al. (1994)) and then, the source of heat $s_e(r,t)$ in Eq. 2 is:

$$s_e(r,t) = s_0 D(r) \exp\left(-\frac{(t-t_0)^2}{2t_0{}^2}\right) \tag{5}$$

here, t_0 is the mean flight time of the electrons and the width of the gaussian function has also been set to t_0 ($\approx 10^{-15}$ s).
The main effect of the energy deposition and subsequent temperature rise is the formation of tracks in dielectrics and some metal alloys (Toulemonde et al. (2004)). As the ion moves along the material, the heat provokes melting of the matrix with a corresponding expansion and structure change. Even though the material cools down again, the quenching rate is too fast for a full reconstruction and an amorphous volume is left, if the original structure was crystalline, or else, with an important amount of defects.
The description of the formation of tracks in insulators has been successfully described by means of Eq. 2 and considering the energy input given by Eq. 5. With this model, it is possible to explain quantitatively the dimensions of the latent tracks left in insulators, as well as sputtering observed in this regime (Toulemonde (2000); Toulemonde et al. (2003)). It has been compared, in a rather complete calculation (Awazu et al. (2008)), that because gold's electronic heat conduction is very high, no melting occurs and no track is left when irradiated with 110 MeV Br ions, contrary to the case of SiO_2, where tracks are formed. This, is in agreement with experimental observations.
The implementation of the two-temperature model in metals is straightforward, as far as the electronic subsystem is composed mostly by free electrons, for which kinetic theory can give good estimates of the thermal properties. The model, as mentioned above, has also been adapted and widely used to semi–conductor and insulator materials (Chettah et al. (2009)), Recently, the model has been treated with further detail for semi-conductors (Daraszewicz & Duffy (2010)) by incorporating the fact, that the total number of conduction electrons equals the number of holes:

$$\frac{\partial N}{\partial t} + \nabla J = G_e - R_e \tag{6}$$

Here, N is the concentration of electron–hole pairs, J is the carrier current density and G_e an R_e are the source and sink of conduction electrons. The carrier current density is related with the electronic temperature by:

$$J = -D\left(\nabla N + \frac{2N}{2k_B T}\nabla E_g + \frac{N}{2T_e}\nabla T_e\right) \tag{7}$$

where D is the ambipolar diffusivity and E_g is the value of the band gap. While the validity of the additional hypothesis is beyond any doubt, its solution becomes very complicated and additional simplifications must also be added. The magnitude of the resulting correction is still to be investigated.

Another approach has been proposed by Duffy and co-workers (Duffy et al. (2008); Duffy & Rutherford (2007)). They have coupled a molecular dynamics simulation for the lattice subsystem, while the electronic one is described by Eq. 2. In this equation, $s_e(t,\vec{r})$ is then the term corresponding to the electronic stopping power and $g(t,\vec{r})$ is the usual electron–phonon coupling term between the two subsystems. This approach permits a direct quantification or the radiation damage and has allowed to show, with an Fe target, that the material can be above the melting temperature without actually melting.

3. Ion beam effects on nano–structured materials

It is known, that ion irradiation causes dielectrics and some metal alloys to expand in a direction transversal to the ion's path as a consequence of the track formation (Ryazanov et al. (1995); Toulemonde et al. (2004); Trinkaus (1998); van Dillen et al. (2005)). This effect is particularly noticeable in dielectric nano–particles and is very important as it offers an effective method to tailor the shape of dielectric nano–particles by controlling the ion's energy, fluence and irradiation angle. The modeling of this effect has been performed at the nanoscale by solving the equation of mechanical stress that results from the increase of temperature inside the ion's track (Klaumunzer (2006); Schmidt et al. (2009); van Dillen et al. (2005)). As these equation are complex by themselves, a uniform, effective track temperature is considered to solve them, nevertheless it is possible to quantitatively reproduce the experimentally measured deformation rates.

On the contrary, when metallic nano-particles embedded in SiO_2 are irradiated with high energy ions, they are deformed in the direction of the ion's path (D'Orléans et al. (2003); Oliver et al. (2006); Penninkhof et al. (2003); Ridgway et al. (2011)).

The deformation of metallo–dielectric core–shell colloids under MeV ion irradiation was extensively studied by Penninkhof and co-workers (Penninkhof et al. (2003; 2006)). They found, that while the dielectric shell deformed perpendicularly to the ion beam, the metallic core was elongated along the ion's path. They proposed a kind of passive deformation mechanism, in which the metallic core was a consequence of the well-known deformation of the dielectric material. However, when they studied the deformation of Ag, and Au nano–particles embedded in thin soda–lime films, they observed no deformation of the Ag nano–particles, suggesting that thermodynamic parameters of the metal were also involved in the deformation mechanism.

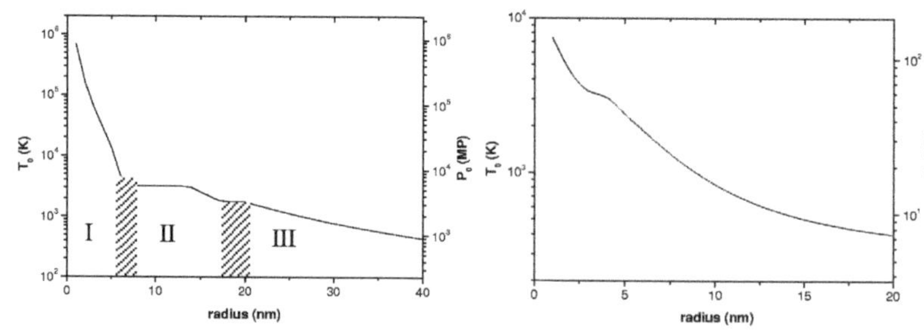

Fig. 1. (a) Temperature reached by the nano–particle as function of its radius, after the ion has passed through it. (b) Temperature of the substrate at a distance r from the ion's path after 10^{-12} s according to D'Orléans et al. (2003).

D'Orleans et. al. (D'Orléans et al. (2003; 2004); D'Orléans, et al. (2004)) have studied thoroughly the elongation of Co nano–particles by 200 MeV Iodine ions. They proposed a simple model, in which the ion goes through the nano–particle. As the stopping power is larger in the metal than in the matrix, and as metal conductivity is also larger, they considered that the energy deposited at the nano–particle with radius r is transformed entirely within it into heat to raise its temperature. So, they found that very small nano–particles reach evaporation temperature, while very large do not melt. The intermediate sized nano–particles that reach up to liquid temperature are subjected to thermal stress. Fig. 1 shows their calculation in which they determine, that after 10^{-12} s the temperature at the center of the track is higher than the nano–particles temperature, but that the thermal stress (right axis), $\Delta P = \frac{\alpha}{\chi}\Delta T$, is lower due to differences in the expansion coefficient α and the compressibility χ with the metal. The time 10^{-12} s is taken for the comparison, as it is the typical time for reaching equilibrium between the electronic and lattice systems. Simple though this model might seem, it has the virtue of putting emphasis firstly on the active role that nano–particles play (it is not a consequence of the transversal expansion of the matrix), and secondly on the importance of thermal stress and thermal parameters diferences of the materials involved.
Awazu has performed calculations for 110 MeV Br, 100 MeV Cu and 90 MeV Cl ions impinging on Au rods embedded in SiO_2 based on the two-temperature model (Eq. 2), and has shown some details of the thermal conduction process (Awazu et al. (2008)). Even though the electronic conductivity of the metal is very high, the border of the nano–particle limits the conduction, allowing the temperature to raise above the melting point, a requirement that has been established to be necessary for the deformation to take place.
Silver and gold nano–particles, embedded in silica, irradiated with 8–10 MeV Si ions exhibit a similar behavior (Oliver et al. (2006); Rodríguez-Iglesias et al. (2010); Silva-Pereyra et al. (2010)), This case, is similar to the previous one, but in a smaller scale. The energy deposited by the ion is lower, and so are the nano–particles that can be elongated, as well as the track radius, as shown in Fig. 2. The difference in thermal stress remains and therefore the same explanation is applicable. It had been observed before that silver ions migrate and evaporate out of the matrix when the samples are annealed at temperatures close to 1000 °C (Cheang-Wong et al. (2000)), therefore, it is suggested that not only does the thermal stress plays an important role, but also the increase of the metal solubility in the matrix, allowing ions to move preferentially through the track and aggregating again at the cooling stage of the thermal spike.
Because of the high electrical conductivity of silver and gold, the surface plasmon resonance is very well defined, so that the shape of the nano–particles can be characterized optically by it and by its splitting when they are ellipsoid instead of spheres. Fig. 3 (a) shows this effect as function of the geometry. When the wave vector is parallel to the major axis and the electric field is therefore perpendicular, only the minor axis mode can be excited and only one resonance observed. Otherwise, two resonances are observed, with relative intensities depending on the angle of orientation. In Fig. 3 (b) the selection of the excitation mode is obtained by changing the light polarization to excite either the minor or the major axis. If we know the geometry used for the irradiation, we can determine that there are two minor axes and one major axis.
Electron microscopy has given additional evidence of the deformation and detailed information on the microstructure of the nano–particles, as shown in Fig. 4. Furthermore, it has also allowed the in-situ observation of the effect that the electron beam has on the

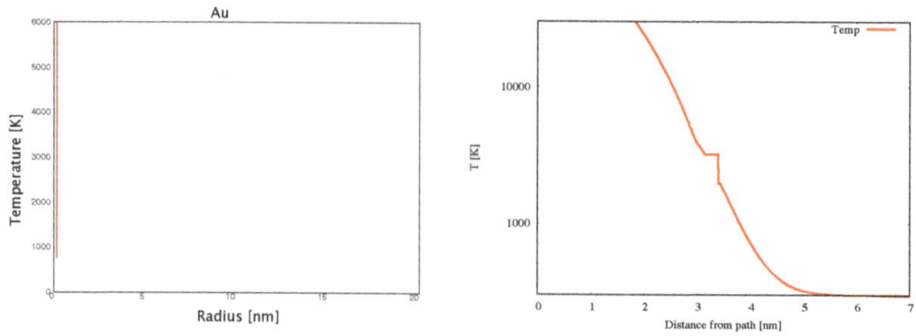

(a) Nano–particles temperature (b) Spike temperature

Fig. 2. (a) Temperature reached by gold nano–particles as function of its radius, after the ion has passed through it. (b) Temperature of the SiO_2 substrate at a distance r from the ion's path after 10^{-12} s.

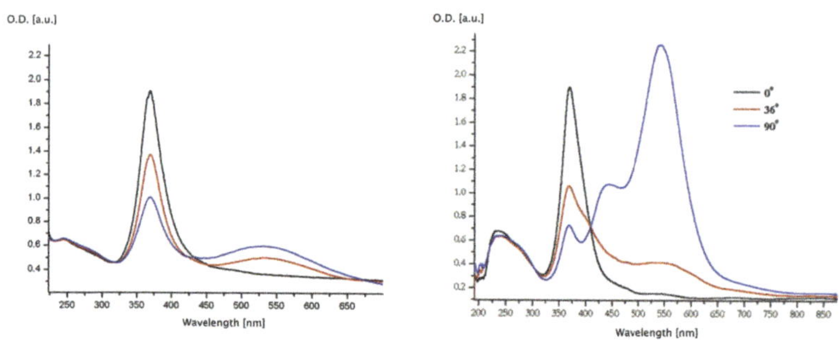

(a) Surface plasmon resonances, as function to the orientation angle. (b) Surface plasmon resonances, as function to the polarization angle.

Fig. 3. Optical analysis of metallic nano–particles embedded inside SiO_2. When they are prolate ellipsoids, two resonant modes appear instead of one

elongated nano–particle: they recover their spherical shape without melting (Silva-Pereyra (2011); Silva-Pereyra et al. (2010)).

Ridgway and co-workers (Giulian et al. (2008); Kluth et al. (2009; 2007); Ridgway et al. (2011)) have shown that this effect is also present when Pt, Cu and Au nano–particles are irradiated with Sn and Au ions at different energies above 100 MeV. They have shown many new features of the phenomena, from which two attract attention. Firstly, that when nano–particles elongate, the minor axis reaches a limiting value, less than, but in correspondence with the track radius formed by the ion. And secondly, that nuclear interaction can also activate the elongation or can cause structure transitions in the nano–particles. They have even provided significant evidence of an amorphous Cu phase (Johannessen et al. (2008)).

Sapphire is a harder material and it has an expansion coefficient higher than SiO_2. For this reason, thermal stress is higher and becomes comparable to that of the metallic nano–particles. We have begun studying the induced anisotropic deformation caused by 6–10 MeV Si ions. Samples were prepared using the same methodology as above (Mota-Santiago et al. (2011; 2012)). Our preliminary results show that nano–particles do expand along the ion's path.

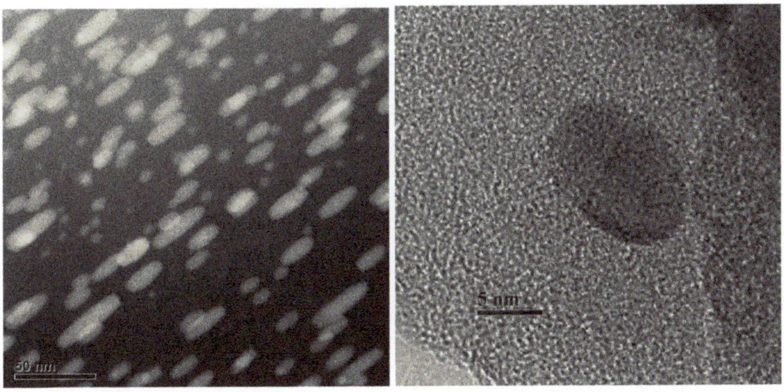

(a) Panoramic veiw (b) Single nano–particle

Fig. 4. Electron microscopy images showing the elongation of gold nano–particles by Si ions. (Rangel-Rojo et al. (2010))

Nevertheless, as the refractive index is higher (1.76), the separation of the two plasmon resonancess is not complete (Fig. 5) and the preparation of samples for microscopy is still in course of obtaining the appropriate images.

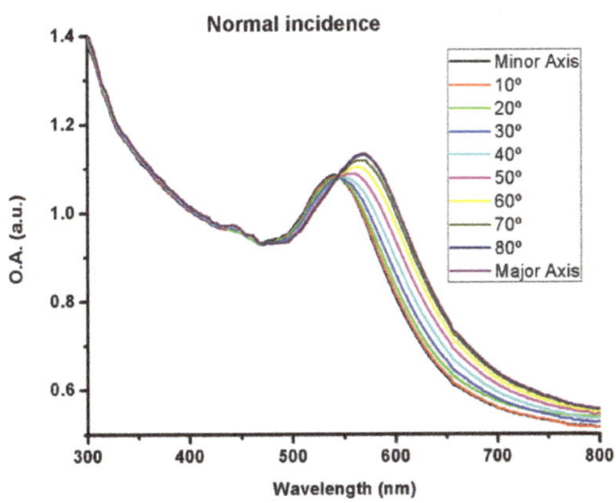

Fig. 5. Extinction spectra of Gold nano–particles in sapphire taken at different polarization angles.

Apart from being a very interesting problem from the fundamental point of view, it is worth mentioning that the control of the shape of the nano–particles by means of ion irradiation is of interest for their potential technological applications. Ag and Au nano–particles could be used in photo–electronic devices, while Co nano–particles could have magnetic applications. This reason is an additional motivation to further study the mechanisms of the deformation. Currently, many groups are working experimentally as well as theoretically to better describe the effect.

4. Laser effects on nano–structured materials

The fact that ballistic effects are minimal in laser–solid interactions, simplifies greatly the theoretical description of laser–solid interaction and its effects, while the probability to control the duration of the pulse allows the experimental determination of the characteristics of the process. If the irradiance of the laser pulse is given by $I = I_0 \exp((t - t_0)^2/2\sigma^2)$, the energy absorbed at $\vec{r}$ is simply

$$s(\vec{r}, t) = \alpha(\vec{r}) I(\vec{r}) \exp((t - t_0)^2/2\sigma^2) \tag{8}$$

where α is the absorption coefficient of the material. A deeper description of these phenomena is to be found elsewhere in this book (Sands (2011)). An important parameter to be considered when describing laser effects, is the heat diffusion length l, that tells us how long heat has travelled after time t. It is defined as:

$$l = \sqrt{\frac{\kappa t}{\rho c_p}} \tag{9}$$

For example, when this length for the duration of the laser pulse is longer than the radius of the nano–particle, the calculation can be done by considering the system as homogeneous, as shown previously in (Crespo-Sosa et al. (2007)), where the thermal effects of excimer laser pulses on metallic nano–particles embedded in a transparent matrix was studied.
Silver and gold nano–particles where fabricated by firstly implanting 2×10^{16} ions/cm^2, 2 Mev ions in high quality fused silica substrates. Thereafter, the samples were annealed at 600 (Ag) and 1000 (Au) °C to obtain the nano–particles with known characteristics. During annealing, also most of the radiation defects are bleached so that the substrate is transparent at the laser's wavelength. In this case, we used a XeCl excimer laser with 55 ns FWHM width. Absorption occurs entirely at the nano–particles, by intraband transitions of the metal. However, as the heat conductivity is much larger for the metal, as the filling fraction of the metal is low (less than 2.5 %), and as the heat diffusion length (Eq. 9) for 55 ns is much larger than the mean separation distance between nano–particles, the temperature can be considered transversally homogenous and only a one dimensional heat transport problem considered. So, the source in Eq. 1 is due to the nano–particles. And as the nano–particles are not uniformly distributed as function of depth, the absorption coefficient α is considered to be a function of the depth and proportional to the amount of ions implanted. On the other hand, the conductivity and the heat capacity are governed by the matrix. The numerical solution of Eq. 1 shows that a minimum laser fluence (2 J/cm^2 in the Ag samples) is needed to take the system above the melting point of the metal. The maximum temperature in the sample occurs where the maximum density of nano–particles is found. The temperature profile for a 2.8 J/cm^2 pulse is shown in Fig. 6 (a). The maximum laser irradiance took place at $t_0 = 70$ ns. It can be observed that the temperature of the system is well above the melting point of silver. And this agrees well with the experimental results shown in Fig. 6 (b), where one can see that with laser fluences above 2 J/cm^2, the surface plasmon resonance broadens, indicating that the nano–particles become smaller. Increment of the thermal stress above the tensile strength occurs at the same time, and therefore, parts of the surface fall down leaving a square well behind, instead of the usual crater in normal surface ablation. The depth of the step matches the ion range.The same effect was observed with gold nano–particles, and could be explained in the same way.

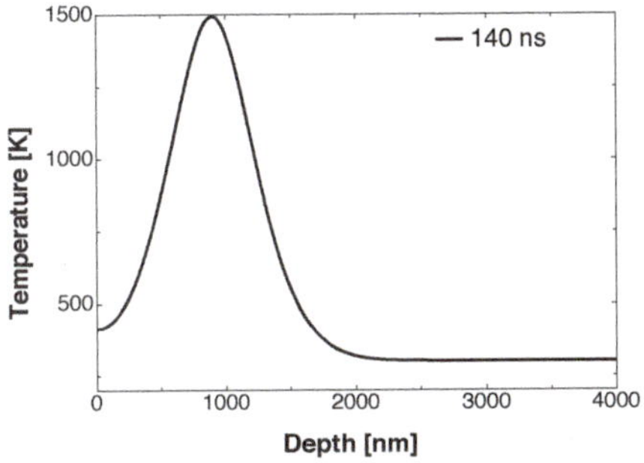

(a) Temperature profile.

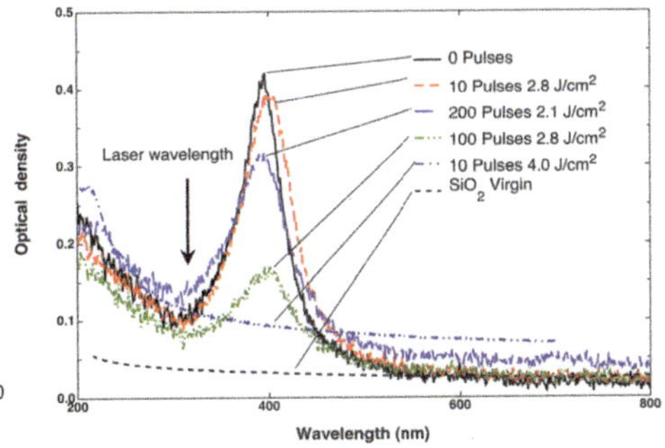

(b) Extinction spectra.

Fig. 6. Effects of excimer laser on silver nano–particles embedded in SiO_2: (a) Temperature profile as function of depth, 70 ns after the maximum irradiance of a 2.8 J/cm^2 pulse. (b) Extinction spectra of samples treated with increasing laser fluences.

By means of a 6 ns FWHM pulsed Nd:YAG laser at 1064 nm and at 532 nm (Crespo-Sosa & Schaaf (n.d.)), samples containing Ag and Au nano–particles, prepared with the same method described above, were also irradiated. At this wavelength, energy is absorbed mainly by the matrix and little or no reduction is observed in the nano–particles size as they do not melt. On the contrary, in Fig. 7, one can see, that the first 10 pulses remove the surface carbon deposited (few nanometers below the surface) during Ag and Au implantation, and therefore the "background" drops. After 100 pulses, the resonance has turned narrower, indicating a slight growth of the nano–particles, but this growth does not continue after 1000 or 10000 pulses. In this case, the calculation of the temperature evolution indicates no significant increment. This means that this slight growth is not produced by a thermal process, and that another mechanism must be present.

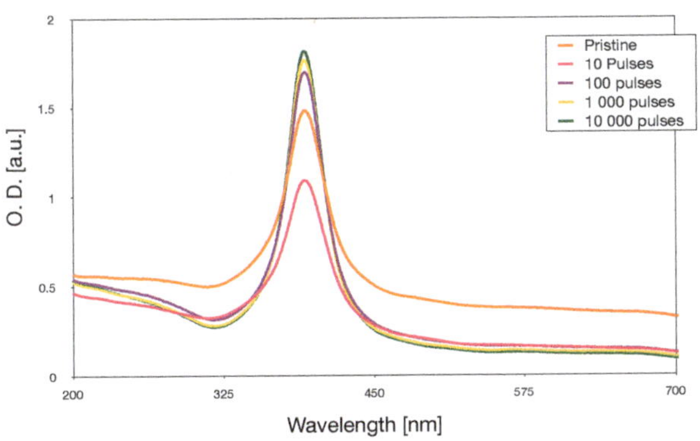

Fig. 7. Effects of infrared laser on Ag nano–particless embedded in SiO_2: Extinction spectra of samples treated with increasing number of pulses.

When irradiating these samples with a wavelength of 532 nm, we observed opposite effects between silver and gold nano–particles. This is because the resonance of gold nano–particles falls very close to the irradiation wavelength, while the resonance for silver is around 400 nm. In other words, the system with Ag nano–particles absorbs the energy uniformly by the matrix, whereas Au nano–particles absorb the energy in the other case. By tunning the wavelength, one can select whether to provoke effects directly on the nano–particles or onto the matrix.

Nano–particles decomposition and accompanying surface ablation is usually related to the energy absorbed, the location and the duration of the pulse. The shorter the pulse is, the higher the temperature that the nano–particles can reach and therefore the lower the ablation threshold. This has been experimentally verified with nanosecond pulses, but with picosecond pulses, non thermal effects may appear. For example, when Ag nano–particles are irradiated with 26 ps pulses at 355 nm , a surprisingly high ablation threshold is found (Torres-Torres et al. (2010)). The cause for this, is not fully understood. The measured non-linear absorption coefficient is, from the thermal point of view, negligible to account for such an effect. On the other hand, it has been reported that two–photon absorption, (an equally improbable event) can be important in the determination of the melting threshold of silicon by ps laser pulses at 1064 nm (van Driel (1987)).

From a merely thermal point of view, the use of shorter laser pulses can be treated "locally" as the heat diffusion length becomes shorter. Xia and co–workers have, for example, modeled the temperature evolution of a nano–particle embedded in a transparent matrix by means of Eq. 2. And from this calculation , they showed that the corresponding thermal stress and phase transformations are important in the description of surface ablation and of nano–particles fragmentation (Xia et al. (2006)). Picosecond and femtosecond pulses can provoke damage in materials that can also be treated thermally. It has been mentioned above, that typically, hot electrons transfer their energy to the lattice in times shorter than few picoseconds. When pulses shorter than this time are used, the dynamics of the electrons must be taken into account. Today's main interest in such pulses is precisely the possibility of studying the dynamic evolution of the system. In this case, Eq. 2 is used to test if the fundamental parameters of the electron-electron and electron-phonon interactions are properly reproduced by the proposed model (Bertussi et al. (2005); Bruzzone & Malvaldi (2009); Dachraoui & Husinsky (2006); Muto et al. (2008); Zhang & Chen (2008)). It is in a certain way the inverse problem where the thermal properties are to be determined. Another fine example, where the calculation of the electronic temperature by means of Eq. 2 plays an important role, is the determination of the contribution of the hot electrons to the third–order non–linear susceptibility of gold nano–particles (Guillet et al. (2009)).

5. Discussion

As seen above, the methodology for studying the temperature increase in the material due to laser– or to ion–irradiation has been well established using the heat equation. However, let us make a few remarks on it:

Even though calculations are not too sensitive to changes in the values of the thermal properties, the uncertainty of them should always be a concern. The processes involved occur and also cause high pressure regions, where a state equation of the system can hardly be known. Additionally, the possibility of a change in these values in nano–structures must also be considered (Buffat & Borel (1976)). Also, the possibility of non–Fourier's heat conduction has not been discussed enough (Cao & Guo (2007); Rashidi-Huyeh et al. (2008)). Indeed, it is not always clear how important a variation in such parameters is or how important the consideration of a particular effect is.

Another problem to be considered, is the cumulative nature of the effects. Most of the calculations are based on single events, an ion or a pulse, and then scaled, while events might be cumulative. Neither are charge effects considered in these kinds of calculation and they might, in some cases, have an important influence on the effects observed. Also, most of the calculations have been simplified to solve the one dimensional heat equation (Awazu et al. (2008)).

The process in which the ion deposits its energy to the nuclei of the target is highly stochastic. The ion does not follow a straight line and the energy deposition density (F_d) is not uniform. The process described by the heat equation, must be then considered as an "average" event, as in an statistical point of view. Furthermore, the description through the heat equation assumes thermal equilibrium and energy transfer, but during the first stages of the process, the energy is limited to only few atoms, that move with high kinetic energy, that might be better described by a ballistic approach. Indeed, there are effects (in ion beam mixing, for instance), that are directly related to the primary knock-on collisions, that cannot be described by the thermal equation.

The interaction of the ion with the electrons can be thought as more uniform because the electron density is much higher, but additional parameters arise, like the coupling function g in Eq. 2 and the thermal properties of the electronic cloud. In this case, the consideration of the "ballistic" range of the ejected electrons by the ion is important to input correctly the spatial deposition of energy.

Though in principle simpler, the interaction of high power lasers with matter also present interesting challenges to consider, first, the effects that raise due to high intensity pulses, in which the absorption and conductive processes might be altered within the same pulse, and the effects due to the ultrashort pulses that might be even faster than the system thermalization.

6. Conclusions

In this chapter, it has been reviewed how the simple, yet powerful concepts of classical heat conduction theory have been extended to phenomena like ion beam and laser effects on materials. These phenomena are characterized by the wide range of temperatures involved, extreme short times and high annealing and cooling rates, as well as by the nanometric spaces in which they occur. In consequence, there is a high uncertainty in the values of the thermal properties that must be used for the calculations. Nevertheless, the calculations done up-today have proved to be very useful to describe the effects of them. They also agree with other methods like Monte Carlo and molecular dynamics simulations. In the future these parameters must be better determined (theoretically and experimentally) and further applied to more complex systems, like nano–structured materials as well as to femto and atosecond processes. The knowledge of the fundamentals of radiation interaction behind these processes will benefit a lot from thess new experimental, theoretical and computational tools.

7. Aknowledgments

The author would like to thank all the colleagues, technicians and students that have participated in the experiments described above. And to the following funding organizations: CONACyT, DGAPA-UNAM, ICyTDF and DAAD.

8. References

Awazu, K., Wang, X., Fujimaki, M., Tominaga, J., Aiba, H., Ohki, Y. & Komatsubara, T. (2008). Elongation of gold nanoparticles in silica glass by irradiation with swift heavy ions, *Physical Review B* 78(5): 1–8.
URL: *http://link.aps.org/doi/10.1103/PhysRevB.78.054102*

Bertussi, B., Natoli, J., Commandre, M., Rullier, J., Bonneau, F., Combis, P. & Bouchut, P. (2005). Photothermal investigation of the laser-induced modification of a single gold nano-particle in a silica film, *Optics Communications* 254(4-6): 299–309.
URL: *http://linkinghub.elsevier.com/retrieve/pii/S0030401805005377*

Bruzzone, S. & Malvaldi, M. (2009). Local Field Effects on Laser-Induced Heating of Metal Nanoparticles, *The Journal of Physical Chemistry C* 113(36): 15805–15810.
URL: *http://pubs.acs.org/doi/abs/10.1021/jp9003517*

Buffat, P. & Borel, J. (1976). Size effect on melting temperature of gold particles, *Physical Review A* 13(6): 2287.
URL: *http://pra.aps.org/abstract/PRA/v13/i6/p2287_1*

Cao, B.-Y. & Guo, Z.-Y. (2007). Equation of motion of a phonon gas and non-Fourier heat conduction, *Journal of Applied Physics* 102(5): 053503.
URL: *http://link.aip.org/link/JAPIAU/v102/i5/p053503/s1&Agg=doi*

Cheang-Wong, J. C., Oliver, A., Crespo-Sosa, A., Hernández, J. M., Muñoz, E. & Espejel-Morales, R. (2000). Dependence of the optical properties on the ion implanted depth profiles in fused quartz after a sequential implantation with Si and Au ions, *Nuclear Instruments and Methods in Physics Research Section B: Beam Interactions with Materials and Atoms* 161-163: 1058–1063.
URL: *http://linkinghub.elsevier.com/retrieve/pii/S0168583X99009192*

Chettah, a., Kucal, H., Wang, Z., Kac, M., Meftah, a. & Toulemonde, M. (2009). Behavior of crystalline silicon under huge electronic excitations: A transient thermal spike description, *Nuclear Instruments and Methods in Physics Research Section B: Beam Interactions with Materials and Atoms* 267(16): 2719–2724.
URL: *http://linkinghub.elsevier.com/retrieve/pii/S0168583X09006569*

Crespo-Sosa, A. & Schaaf, P. (n.d.). Unpublished.

Crespo-Sosa, A., Schaaf, P., Reyes-Esqueda, J. A., Seman-Harutinian, J. A. & Oliver, A. (2007). Excimer laser absorption by metallic nano-particles embedded in silica, *Journal of Physics D: Applied Physics* 40(7): 1890–1895.
URL: *http://stacks.iop.org/0022-3727/40/i=7/a=008?key=crossref.f57509912f821b768966f484bf900042*

Dachraoui, H. & Husinsky, W. (2006). Fast electronic and thermal processes in femtosecond laser ablation of Au, *Applied Physics Letters* 89(10): 104102.
URL: *http://link.aip.org/link/APPLAB/v89/i10/p104102/s1&Agg=doi*

Daraszewicz, S. & Duffy, D. (2010). Extending the inelastic thermal spike model for semiconductors and insulators, *Nuclear Instruments and Methods in Physics Research Section B: Beam Interactions with Materials and Atoms* 269(14): 1646–1649.
URL: *http://linkinghub.elsevier.com/retrieve/pii/S0168583X10008566*

D'Orléans, C., Stoquert, J., Estournès, C., Cerruti, C., Grob, J., Guille, J., Haas, F., Muller, D. & Richard-Plouet, M. (2003). Anisotropy of Co nanoparticles induced by swift heavy ions, *Physical Review B* 67(22): 10–13.
URL: *http://link.aps.org/doi/10.1103/PhysRevB.67.220101*

D'Orléans, C., Stoquert, J., Estournes, C., Grob, J., Muller, D., Cerruti, C. & Haas, F. (2004). Deformation yield of Co nanoparticles in SiO_2 irradiated with 200 MeV ^{127}I ions, *Nuclear Instruments and Methods in Physics Research Section B: Beam Interactions with Materials and Atoms* 225(1-2): 154–159.
URL: *http://linkinghub.elsevier.com/retrieve/pii/S0168583X04007852*

D'Orléans, C., Stoquert, J., Estournes, C., Grob, J., Muller, D., Guille, J., Richardplouet, M., Cerruti, C. & Haas, F. (2004). Elongated Co nanoparticles induced by swift heavy ion irradiations, *Nuclear Instruments and Methods in Physics Research Section B: Beam Interactions with Materials and Atoms* 216(1-2): 372–378.
URL: *http://linkinghub.elsevier.com/retrieve/pii/S0168583X03021736*

Duffy, D. M., Itoh, N., Rutherford, a. M. & Stoneham, a. M. (2008). Making tracks in metals, *Journal of Physics: Condensed Matter* 20(8): 082201.
URL: *http://stacks.iop.org/0953-8984/20/i=8/a=082201?key=crossref.47f54125ceffe4f9f25b4fd4082dde60*

Duffy, D. M. & Rutherford, a. M. (2007). Including the effects of electronic stopping and electron-ion interactions in radiation damage simulations, *Journal of Physics: Condensed Matter* 19(1): 016207.
URL: *http://stacks.iop.org/0953-8984/19/i=1/a=016207?key=crossref.cc3ab92c89a9b411156b1a3956294e00*

Elsayed-Ali, H., Norris, T., Pessot, M. & Mourou, G. (1987). Time-Resolved Observation of Electron-Phonon Relaxation in Copper, *Physical Review Letters* 58(12): 1212–1215.
URL: *http://link.aps.org/doi/10.1103/PhysRevLett.58.1212*

Giulian, R., Kluth, P., Araujo, L., Sprouster, D., Byrne, A., Cookson, D. & Ridgway, M. (2008). Shape transformation of Pt nanoparticles induced by swift heavy-ion irradiation, *Physical Review B* 78(12): 1–8.
URL: *http://link.aps.org/doi/10.1103/PhysRevB.78.125413*

Johannessen, B., Kluth, P., Giulian, R., Araujo, L., Leewelin, D.J., Foran, G.J., Cookson, D. & Ridgway, M. (2008). Modification of embedded Cu Nano–particles: Ion irradiation at room temperature., *Nuclear Instruments and Methods in Physics Research Section B: Beam Interactions with Materials and Atoms* 257(1-2): 37–41.
URL:

Guillet, Y., Rashidi–Huyeh, M. & Palpant. B. (2009). Influence of laser Pulse characteristics on the hot electron contribution to the third–order nonlinear optical response of gold nanoparticles., *Physical Review B* 79: 045410.
URL:

Katz, R. & Varma, M. N. (1991). Radial distribution of dose., *Basic life sciences* 58: 163–79; discussion 179–80.
URL: *http://www.ncbi.nlm.nih.gov/pubmed/1811472*

Klaumunzer, S. (2006). Modification of nanostructures by high-energy ion beams, *Nuclear Instruments and Methods in Physics Research Section B: Beam Interactions with Materials and Atoms* 244(1): 1–7.
URL: *http://linkinghub.elsevier.com/retrieve/pii/S0168583X05019002*

Kluth, P., Giulian, R., Sprouster, D. J., Schnohr, C. S., Byrne, a. P., Cookson, D. J. & Ridgway, M. C. (2009). Energy dependent saturation width of swift heavy ion shaped embedded Au nanoparticles, *Applied Physics Letters* 94(11): 113107.
URL: *http://link.aip.org/link/APPLAB/v94/i11/p113107/s1&Agg=doi*

Kluth, P., Johannessen, B., Giulian, R., Schnohr, C. S., Foran, G. J., Cookson, D. J., Byrne, A. P. & Ridgway, M. C. (2007). Ion irradiation effects on metallic nanocrystals, *Radiation Effects and Defects in Solids* 162(7): 501–513.
URL: *http://www.informaworld.com/openurl?genre=article&doi=10.1080/10420150701472221&magic=crossref| | D404A21C5BB053405B1A640AFFD44AE3*

Lin, Z. & Zhigilei, L. (2007). Temperature dependences of the electron-phonon coupling, electron heat capacity and thermal conductivity in Ni under femtosecond laser irradiation, *Applied Surface Science* 253(15): 6295–6300.
URL: *http://linkinghub.elsevier.com/retrieve/pii/S0169433207000815*

Meldrum, A., Boatner, L. & White, C. (2001). Nanocomposites formed by ion implantation: Recent developments and future opportunities, *Nuclear Instruments and Methods in Physics Research Section B: Beam Interactions with Materials and Atoms* 178(1-4): 7–16.
URL: *http://linkinghub.elsevier.com/retrieve/pii/S0168583X00005012*

Mota-Santiago, P. E., Crespo-Sosa, A., Jiméenez-Hernáandez, J. L., Silva-Pereyra, H.-G., Reyes-Esqueda, J. A. & Oliver, A. (2011). Noble-Metal Nano-Crystal Aggregation in Sapphire by Ion Irradiation And Subsequent Thermal Annealing, *Journal of Physics D: Applied Physics* submitted.

Mota-Santiago, P. E., Crespo-Sosa, A., Jiméenez-Hernáandez, J. L., Silva-Pereyra, H.-G., Reyes-Esqueda, J. A. & Oliver, A. (2012). Ion beam induced deformation of gold nano-particles embedded in Sapphire.

Muto, H., Miyajima, K. & Mafune, F. (2008). Mechanism of Laser-Induced Size Reduction of Gold Nanoparticles As Studied by Single and Double Laser Pulse Excitation, *Journal of Physical Chemistry C* 112(15): 5810–5815.
URL: *http://pubs.acs.org/cgi-bin/doilookup/?10.1021/jp711353m*

Oliver, A., Reyes-Esqueda, J. A., Cheang-Wong, J. C., Román-Velázquez, C., Crespo-Sosa, A., Rodríguez-Fernández, L., Seman-Harutinian, J. A. & Noguez, C. (2006). Controlled anisotropic deformation of Ag nanoparticles by Si ion irradiation, *Physical Review B* 74(24): 1–6.
URL: *http://link.aps.org/doi/10.1103/PhysRevB.74.245425*

Penninkhof, J. J., Polman, A., Sweatlock, L. a., Maier, S. a., Atwater, H. a., Vredenberg, a. M. & Kooi, B. J. (2003). Mega-electron-volt ion beam induced anisotropic plasmon resonance of silver nanocrystals in glass, *Applied Physics Letters* 83(20): 4137.
URL: *http://link.aip.org/link/APPLAB/v83/i20/p4137/s1&Agg=doi*

Penninkhof, J. J., van Dillen, T., Roorda, S., Graf, C., Vanblaaderen, A., Vredenberg, a. M. & Polman, A. (2006). Anisotropic deformation of metallo-dielectric core-shell colloids under MeV ion irradiation, *Nuclear Instruments and Methods in Physics Research Section B: Beam Interactions with Materials and Atoms* 242(1-2): 523–529.
URL: *http://linkinghub.elsevier.com/retrieve/pii/S0168583X05015922*

Rangel-Rojo, R., Reyes-Esqueda, J. A., Torres-Torres, C., Oliver, A., Rodríguez-Fernández, L., Crespo-Sosa, A., Cheang-Wong, J. C., McCarthy, J., Bookey, H. & Kar, A. (2010). Linear and nonlinear optical properties of aligned elongated silver nanoparticles embedded in silica, *in* D. Pozo Perez (ed.), *Silver Nanoparticles*, InTech, pp. 35 – 62.
URL: *http://www.intechopen.com/articles/show/title/linear-and-nonlinear-optical-properties-of-aligned-elongated-silver-nanoparticles-embedded-in-silica*

Rashidi-Huyeh, M., Volz, S. & Palpant, B. (2008). Non-Fourier heat transport in metal-dielectric core-shell nanoparticles under ultrafast laser pulse excitation,

Physical Review B 78(12): 1–8.
URL: *http://link.aps.org/doi/10.1103/PhysRevB.78.125408*

Ridgway, M., Giulian, R., Sprouster, D., Kluth, P., Araujo, L., Llewellyn, D., Byrne, a., Kremer, F., Fichtner, P., Rizza, G., Amekura, H. & Toulemonde, M. (2011). Role of Thermodynamics in the Shape Transformation of Embedded Metal Nanoparticles Induced by Swift Heavy-Ion Irradiation, *Physical Review Letters* 106(9): 1–4.
URL: *http://link.aps.org/doi/10.1103/PhysRevLett.106.095505*

Rodríguez-Iglesias, V., Peña Rodríguez, O., Silva-Pereyra, H.-G., Rodríguez-Fernández, L., Cheang-Wong, J. C., Crespo-Sosa, A., Reyes-Esqueda, J. A. & Oliver, A. (2010). Tuning the aspect ratio of silver nanospheroids embedded in silica., *Optics letters* 35(5): 703–5.
URL: *http://www.ncbi.nlm.nih.gov/pubmed/20195325*

Ryazanov, A., Volkov, A. & Klaumünzer, S. (1995). Model of track formation, *Physical Review B* 51(18): 12107–12115.
URL: *http://link.aps.org/doi/10.1103/PhysRevB.51.12107*

Sands, D. (2011). Pulsed laser heating and melting, *in* Vyacheslav S. Vikhrenko (ed.), *Heat Conduction / Book 2*, InTech, pp. .
URL: *http://www.intechopen.com/*

Schmidt, B., Heinig, K.-H., Mücklich, A. & Akhmadaliev, C. (2009). Swift-heavy-ion-induced shaping of spherical Ge nanoparticles into disks and rods, *Nuclear Instruments and Methods in Physics Research Section B: Beam Interactions with Materials and Atoms* 267(8-9): 1345–1348.
URL: *http://linkinghub.elsevier.com/retrieve/pii/S0168583X09000834*

Schoenlein, R., Lin, W., Fujimoto, J. & Eesley, G. (1987). Femtosecond Studies of Nonequilibrium Electronic Processes in Metals, *Physical Review Letters* 58(16): 1680–1683.
URL: *http://link.aps.org/doi/10.1103/PhysRevLett.58.1680*

Silva-Pereyra, H.-G. (2011). *Estudio de los mecanismos de deformacióon de nano-partíiculas de oro embebidas en síilice, producidas por implantacióon de iones.*, PhD thesis, Universidad Nacional Autóonoma de Méexico.

Silva-Pereyra, H.-G., Arenas-Alatorre, J., Rodríguez-Fernández, L., Crespo-Sosa, A., Cheang-Wong, J. C., Reyes-Esqueda, J. A. & Oliver, A. (2010). High stability of the crystalline configuration of Au nanoparticles embedded in silica under ion and electron irradiation, *Journal of Nanoparticle Research* 12(5): 1787–1795.
URL: *http://www.springerlink.com/index/10.1007/s11051-009-9735-6*

Sun, C., Vallée, F., Acioli, L., Ippen, E. & Fujimoto, J. (1994). Femtosecond-tunable measurement of electron thermalization in gold., *Physical review. B, Condensed matter* 50(20): 15337–15348.
URL: *http://www.ncbi.nlm.nih.gov/pubmed/9975886*

Takeda, Y. & Kishimoto, N. (2003). Nonlinear optical properties of metal nanoparticle composites for optical applications, *Nuclear Instruments and Methods in Physics Research Section B: Beam Interactions with Materials and Atoms* 206: 620–623.
URL: *http://linkinghub.elsevier.com/retrieve/pii/S0168583X03007973*

Torres-Torres, C., Peréa-López, N., Reyes-Esqueda, J. A., Rodríguez-Fernández, L., Crespo-Sosa, A., Cheang-Wong, J. C. & Oliver, A. (2010). Ablation and optical third-order nonlinearities in Ag nanoparticles., *International journal of nanomedicine* 5: 925–32.

URL: *http://www.pubmedcentral.nih.gov/articlerender.fcgi?artid=3010154&tool=pmcentrez&rendertype=abstract*

Toulemonde, M. (2000). Transient thermal processes in heavy ion irradiation of crystalline inorganic insulators, *Nuclear Instruments and Methods in Physics Research Section B: Beam Interactions with Materials and Atoms* 166-167: 903–912.
URL: *http://linkinghub.elsevier.com/retrieve/pii/S0168583X99007995*

Toulemonde, M., Assmann, W., Trautmann, C., Gruner, F., Mieskes, H., Kucal, H. & Wang, Z. (2003). Electronic sputtering of metals and insulators by swift heavy ions, *Nuclear Instruments and Methods in Physics Research Section B: Beam Interactions with Materials and Atoms* 212: 346–357.
URL: *http://linkinghub.elsevier.com/retrieve/pii/S0168583X0301721X*

Toulemonde, M., Dufour, C. & Paumier, E. (1992). Transient thermal process after a high-energy heavy-ion irradiation of amorphous metals and semiconductors, *Physical Review B* 46(22): 14362–14369.
URL: *http://link.aps.org/doi/10.1103/PhysRevB.46.14362*

Toulemonde, M., Trautmann, C., Balanzat, E., Hjort, K. & Weidinger, a. (2004). Track formation and fabrication of nanostructures with MeV-ion beams, *Nuclear Instruments and Methods in Physics Research Section B: Beam Interactions with Materials and Atoms* 216: 1–8.
URL: *http://linkinghub.elsevier.com/retrieve/pii/S0168583X03021025*

Trinkaus, H. (1998). dynamics of viscoelastic flow in ion tracks: origin of platic deformation of amorphous materials, *Nuclear Instruments and Methods in Physics Research B* 146: 204–216.

van Dillen, T., Polman, A., Onck, P. & van der Giessen, E. (2005). Anisotropic plastic deformation by viscous flow in ion tracks, *Physical Review B* 71(2): 1–12.
URL: *http://link.aps.org/doi/10.1103/PhysRevB.71.024103*

van Driel, H. (1987). Kinetics of high-density plasmas generated in Si by 1.06- and 0.53-μm picosecond laser pulses, *Physical Review B* 35(15): 8166–8176.
URL: *http://prb.aps.org/abstract/PRB/v35/i15/p8166_1 http://link.aps.org/doi/10.1103/PhysRevB.35.8166*

Vineyard, G. H. (1976). Thermal spikes and activated processes, *Radiation Effects and Defects in Solids* 29(4): 245–248.
URL: *http://www.informaworld.com/openurl?genre=article&doi=10.1080/00337577608233050&magic=crossref||D404A21C5BB053405B1A640AFFD44AE3*

Waligórski, M. P. R., Hamm, R. N. & Katz, R. (1986). The Radial Distribution of Dose around the Path of a Heavy Ion in Liquid Water, *Nuclear Tracks and Radiation Measurements* 11(6): 309–319.

Wang, Z., Dufour, C., Paumier, E. & Toulemonde, M. (1994). The Se sensitivity of metals under swift-heavy-ion irradiation: a transient thermal process, *Journal of Physics: Condensed Matter* 6: 6733.
URL: *http://iopscience.iop.org/0953-8984/6/34/006*

Xia, Z., Shao, J., Fan, Z. & Wu, S. (2006). Thermodynamic damage mechanism of transparent films caused by a low-power laser., *Applied optics* 45(32): 8253–61.
URL: *http://www.ncbi.nlm.nih.gov/pubmed/17068568*

Zhang, Y. & Chen, J. K. (2008). Ultrafast melting and resolidification of gold particle irradiated by pico- to femtosecond lasers, *Journal of Applied Physics* 104(5): 054910.
URL: *http://link.aip.org/link/JAPIAU/v104/i5/p054910/s1&Agg=doi*

3

Mathematical Models of Heat Flow in Edge-Emitting Semiconductor Lasers

Michał Szymanski
Institute of Electron Technology
Poland

1. Introduction

Edge-emitting lasers started the era of semiconductor lasers and have existed up to nowadays, appearing as devices fabricated out of various materials, formed sometimes in very tricky ways to enhance light generation. However, in all cases radiative processes are accompanied by undesired heat-generating processes, like non-radiative recombination, Auger recombination, Joule effect or surface recombination. Even for highly efficient laser sources, great amount of energy supplied by pumping current is converted into heat.

High temperature leads to deterioration of the main laser parameters, like threshold current, output power, spectral characteristics or lifetime. In some cases, it may result in irreversible destruction of the device via catastrophic optical damage (COD) of the mirrors. Therefore, deep insight into thermal effects is required while designing the improved devices.

From the thermal point of view, the laser chip (of dimensions of 1-2 mm or less) is a rectangular stack of layers of different thickness and thermal properties. This stack is fixed to a slightly larger heat spreader, which, in turn, is fixed to the huge heat-sink (of dimensions of several cm), transferring heat to air by convection or cooled by liquid or Peltier cooler. Schematic view of the assembly is shown in Fig. 1. Complexity and large size differences between the elements often induce such simplifications like reduction of the dimensionality of equations, thermal scheme geometry modifications or using non-uniform mesh in numerical calculations.

Mathematical models of heat flow in edge-emitting lasers are based on the heat conduction equation. In most cases, solving this equation provides a satisfactory picture of thermal behaviour of the device. More precise approaches use in addition the carrier diffusion equation. The most sophisticated thermal models take into consideration variable photon density found by solving photon rate equations.

The heat generated inside the chip is mainly removed by conduction and, in a minor degree, by convection. Radiation can be neglected. Typical boundary conditions for heat conduction equation are the following: isothermal condition at the bottom of the device, thermally insulated side walls, convectively cooled upper surface. It must be said that obtaining reliable temperature profiles is often impossible due to individual features of particular devices, which are difficult to evaluate within the quantitative analysis. Mounting imperfections

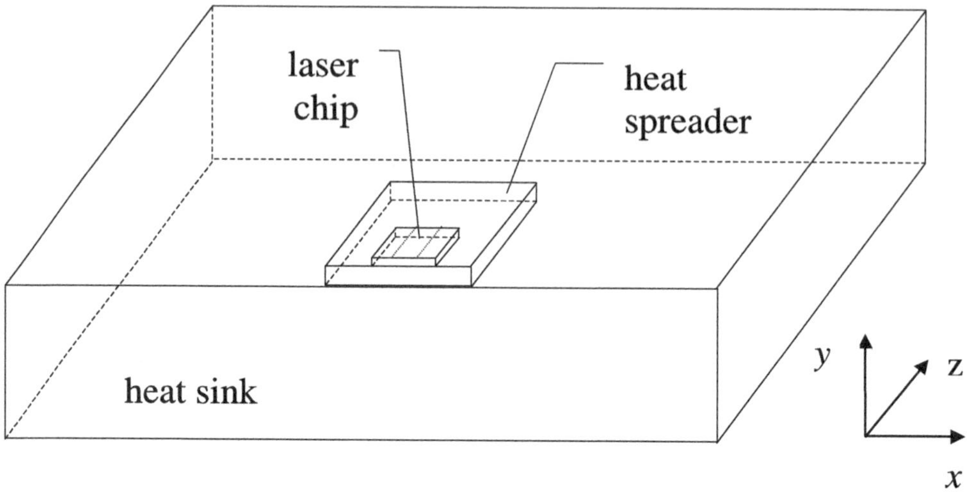

Fig. 1. Schematic view of the laser chip or laser array mounted on the heat spreader and heat sink (not in scale).

like voids in the solder or overhang (the chip does not adhere to the heat-spreader entirely) may significantly obstruct the heat transfer. Surface recombination, the main mirror heating mechanism in bipolar devices, strongly depends on facet passivation.

Since quantum cascade lasers (QCL's) exploit superlattices (SL's) as active layers, they have brought new challenges in the field of thermal modelling. Numerous experiments show that the thermal conductivity of a superlattice is significantly reduced. The phenomenon can be explained in terms of phonon transport across a stratified medium. As a consequence, mathematical models of heat flow in quantum cascade lasers resemble those created for standard edge-emitting lasers, but the stratified active region is replaced by an equivalent layer described by anisotropic thermal conductivity. In earlier works, the cross-plane and in-plane values of this parameter were obtained by arbitrary reduction of bulk values or treated as fitting parameters. Recently, some theoretical methods of assessing the thermal conductivity of superlattices have been developed.

The present chapter is organised as follows. In sections 2, 3 and 4, one can find the description of static thermal models from the simplest to the most complicated ones. Section 5 provides a discussion of the non-standard boundary condition assumed at the upper surface. Dynamical issues of thermal modelling are addressed in section 6, while section 7 is devoted to quantum cascade lasers. In greater part, the chapter is a review based on the author's research supported by many other works. However, Fig. 7, 8, 12 and 13 present the unpublished results dealing with facet temperature reduction techniques and dynamical thermal behaviour of laser arrays. Note that section 8 is not only a short revision of the text, but contains some additional information or considerations, which may be useful for thermal modelling of edge-emitting lasers. The most important mathematical symbols are presented in Table 1. Symbols of minor importance are described in the text just below the equations, in which they appear.

Symbol	Description
A_{nr}	non-radiative recombination coefficient
B	bi-molecular recombination coefficient
b	chip width (see Fig. 2)
C_A	Auger recombination coefficient
c_h	specific heat
D	diffusion coefficient
d_n	total thickness of the n-th medium
g	heat source function
I	driving current
L	resonator length
n_{eff}	effective refractive index
n_i	number of interfaces
N, N_{tr}	carrier concentration, transparency carrier concentration
R_f, R_b	power reflectivity of the front and back mirror
$r_{\mathrm{Bd}}(1 \rightarrow 2)$	TBR for the heat flow from medium 1 to 2
S	total photon density
S_f, S_b	photon density of the forward and backward travelling wave
S_{av}	averaged photon density
t	time
T	temperature
T_{up}	temperature of the upper surface
V	voltage
v_{sur}	surface recombination velocity
w	contact width (see Fig. 2)
y_t	top of the structure (see Fig. 2)
x, y, z	spatial coordinates (see Fig. 1)
α	convection coefficient
α_{gain}	linear gain coefficient
α_{int}	internal loss within the active region
β	spontaneous emission coupling coefficient
Γ	confinement factor
λ	thermal conductivity
$\lambda_{\perp}, \lambda_{\parallel}$	thermal conductivity of QCL's active layer in the direction perpendicular and parallel to epitaxial layers, respectively
ν	frequency
ρ_n	density, subscript n (if added) denotes the medium number
τ, τ_{av}	carrier lifetime, averaged carrier lifetime
c, e, h, k_B	physical constants: light velocity, elementary charge, Planck and Boltzmann constants, respectively.

Table 1. List of symbols.

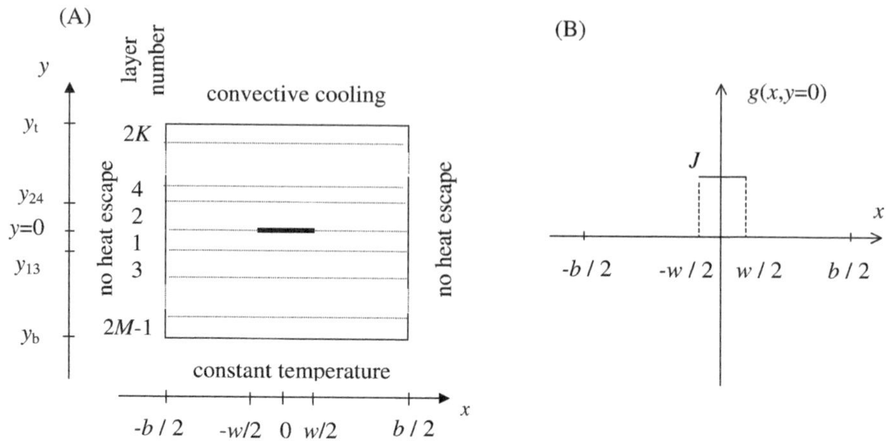

Fig. 2. Schematic view of a laser chip cross-section (A). Function describing the heat source (B).

2. Models based on the heat conduction equation only

Basic thermal behaviour of an edge-emitting laser can be described by the stationary heat conduction equation:

$$\nabla(\lambda(y)\nabla T(x,y)) = -g(x,y) \tag{1}$$

accepting the following assumptions (see Fig. 2):

— the laser is a rectangular stack of layers of different thickness and thermal conductivities; [1]

— there is no heat escape from the top and side walls, while the temperature of the bottom of the structure is constant;

— the active layer is the only heat source in the structure and it is represented by infinitely thin stripe placed between the waveguide layers.

The heat power density is determined according to the crude approximation:

$$g(x,y) = \frac{VI - P_{out}}{Lw}, \tag{2}$$

which physically means that the difference between the total power supplied to the device and the output power is uniformly distributed over the surface of the selected region.[2] The problem was solved analytically by Joyce & Dixon (1975). Further works using this model introduced convective cooling at the top of the laser, considered extension and diversity of heat sources or changed the thermal scheme in order to take into account the non-ideal heat sink (Bärwolff et al. (1995); Puchert et al. (1997); Szymański et al. (2007; 2004)). Such approach allows to calculate temperature inside the resonator, while the temperature in the vicinity of

[1] Note that the thermal scheme can be easily generalised to laser array by periodic duplication of stack along the x axis.

[2] In a three-dimensional case the surface is replaced by the volume.

mirrors is reliable only in the near-threshold regime. The work by Szymański et al. (2007) can be regarded as a recent version of this model and will be briefly described below.
Assuming no heat escape from the side walls:

$$\frac{\partial}{\partial x}T(\pm\frac{b}{2},y) = 0 \tag{3}$$

and using the separation of variables approach (Bärwolff et al. (1995); Joyce & Dixon (1975)), one obtains the solution for T in two-fold form. In the layers above the active layer (n - even) temperature is described by

$$T_n(x,y) = A_{2K}^{(0)}(w_{A,n}^{(0)} + w_{B,n}^{(0)}y) + \sum_{k=1}^{\infty} A_{2K}^{(k)}[w_{A,n}^{(k)}exp(\mu_k y) + w_{B,n}^{(k)}exp(-\mu_k y)]cos(\mu_k x), \tag{4}$$

while under the active layer (n - odd) it takes the form:

$$T_n(x,y) = A_{2M-1}^{(0)}(w_{A,n}^{(0)} + w_{B,n}^{(0)}y) + \sum_{k=1}^{\infty} A_{2M-1}^{(k)}[w_{A,n}^{(k)}exp(\mu_k y) + w_{B,n}^{(k)}exp(-\mu_k y)]cos(\mu_k x). \tag{5}$$

In (4) and (5) $\mu_k = 2k\pi/b$ is the separation constant and thus it appears in both directions (x and y). Integer number k numerates the heat modes. Coefficients $w_{A,n}^{(k)}$ and $w_{B,n}^{(k)}$ and relation between $A_{2K}^{(k)}$ and $A_{2M-1}^{(k)}$ can be found in Szymański (2007). They are determined by the bottom boundary condition, continuity conditions for the temperature and heat flux at the layer interfaces and the top boundary condition.

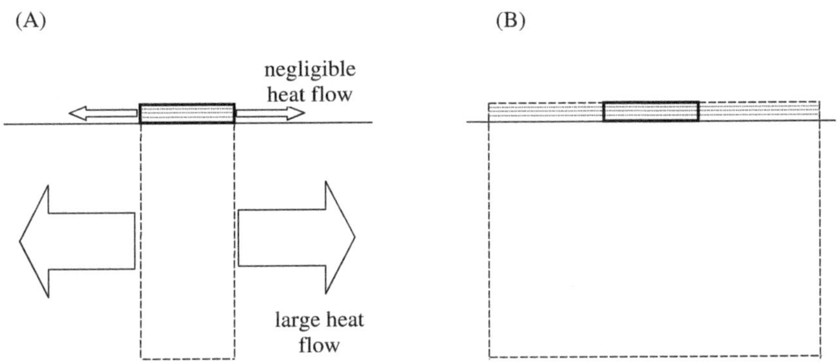

Fig. 3. Thermal scheme modification. Assuming larger b allows to keep the rectangular cross-section of the whole assembly and hence equations (4) and (5) can be used.

The results obtained according to the model described above are presented in Table 2. The calculated values are slightly underestimated due to bonding imperfections, which elude

Device number	Heterostructure A	Heterostructure B	Heterostructure C
1	12.03/7.38	11.23/7.31	8.9/8.24
2	13.35/7.38	12.17/7.31	7.0/4.76

Table 2. Measured/calculated thermal resistances in K/W (Szymański et al. (2007)).

qualitative assessment. A similar problem was described in Manning (1981), where even greater discrepancies between theory and experiment were obtained. For the properly mounted device C1 excellent convergence is found.
Improving the accuracy of calculations was possible due to taking into account the finite thermal conductivity of the heat sink material by thermal scheme modifications (see Fig. 3). Assuming constant temperature at the chip-heat spreader interface leads to significant errors, especially for p-side-down mounting (see Fig. 4).
The analytical approach presented above has been described in detail since it has been developed by the author of this chapter. However, it should not be treated as a favoured one. In recent years, numerical methods seem to prevail. Pioneering works using Finite Element Method (FEM) in the context of thermal investigations of edge-emitting lasers have been described by Sarzała & Nakwaski (1990; 1994). Broader discussion of analytical vs. numerical methods is presented in 8.3.

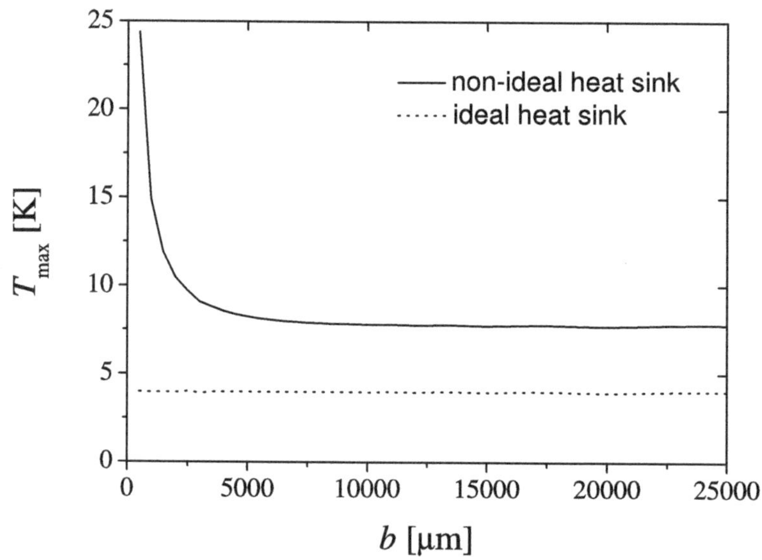

Fig. 4. Maximum temperature inside the laser for *p*-side down mounting. It is clear that the assumption of ideal heat sink leads to a 50% error in calculations (Szymański et al. (2007)).

Thermal effects in the vicinity of the laser mirror are important because of possible COD during high-power operation. Unfortunately, theoretical investigations of these processes, using the heat conduction only, is rather difficult. There are two main mirror heating mechanisms (see Rinner et al. (2003)): surface recombination and optical absorption. Without including additional equations, like those described in sections 3 and 4, assessing the heat source functions may be problematic. An interesting theoretical approach dealing with mirror heating and based on the heat conduction only, can be found in Nakwaski (1985; 1990). However, both works consider the time-dependent picture, so they will be mentioned in section 6.

3. Models including the diffusion equation

Generation of heat in a semiconductor laser occurs due to: (A) non-radiative recombination, (B) Auger recombination, (C) Joule effect, (D) spontaneous radiative transfer, (E) optical absorption and (F) surface recombination. The effects (A)—(C) and (E,F) are discussed in standard textbooks (see Diehl (2000) or Piprek (2003)). Additional interesting information about mirror heating mechanisms (E,F) can be found in Rinner et al. (2003). The effect (D) will be briefly described below.

Apart from stimulated radiation, the laser active layer is a source of spontaneous radiation. The photons emitted in this way propagate isotropically in all directions. They penetrate the wide-gap layers and are absorbed in narrow-gap layers (cap or substrate) creating the additional heat sources (see Nakwaski (1979)). Temperature calculations by Nakwaski (1983a) showed that the considered effect is comparable to Joule heating in the near-threshold regime. On the other hand, it is known that below the threshold spontaneous emission grows with pumping current and saturates above the threshold. Thus, the radiative transfer may be recognised as a minor effect and will be neglected in calculations presented in this chapter.

Note that processes (A)—(C) and (F) involve carriers, so $g(x,y,z)$ should be a carrier dependent function. To avoid crude estimations, like equation (2), a method of getting to know the carrier distribution in regions essential for thermal analysis is required.

3.1 Carrier distribution in the laser active layer

An edge-emitting laser is a p-i-n diode operating under forward bias and in the plane of junction the electric field is negligible. Therefore, the movement of the carriers is governed by diffusion. Bimolecular recombination and Auger process engage two and three carriers, respectively. Such quantities like pumping or photon density are spatially inhomogeneous. Far from the pumped region, the carrier concentration falls down to zero level. At the mirrors, surface recombination occurs. Taking all these facts into account, one concludes that carrier concentration in the active layer can be described by a nonlinear diffusion equation with variable coefficients and mixed boundary conditions. Solving such an equation is really difficult, but the problem can often be simplified to 1-dimensional cases. For example, if problems of beam quality (divergence or filamentation) are discussed, considering the lateral direction only is a good enough approach. In the case of a thermal problem, since surface recombination is believed to be a very efficient facet heating mechanism responsible for COD, considering the axial direction is required and the most useful form of the diffusion equation can be written as

$$D\frac{d^2N}{dz^2} - \frac{c}{n_{eff}}\Gamma G(N)S(z) - \frac{N}{\tau} + \frac{I}{eV} = 0, \tag{6}$$

where linear gain $G(N) = \alpha_{gain}(N - N_{tr})$ and non-linear carrier lifetime $\tau(N) = (A_{nr} + BN + C_A N^2)^{-1}$ have been assumed. The surface recombination at the laser facets is expressed through the boundary conditions:

$$D\frac{dN(0)}{dz} = v_{sur}N(0), \qquad D\frac{dN(L)}{dz} = -v_{sur}N(L). \tag{7}$$

The problem of axial carrier concentration in the active layer of an edge-emitting laser was investigated by Szymański (2010). Three cases were considered:

(i) the nonlinear diffusion equation with variable coefficients (equation (6) and boundary conditions (7)) ;

(ii) the linear diffusion equation with constant coefficients derived form equation (6) by assuming the averaged carrier lifetime τ_{av} and averaged photon density S_{av};

(iii) the algebraic equation derived form equation (6) by neglecting the diffusion ($D = 0$).

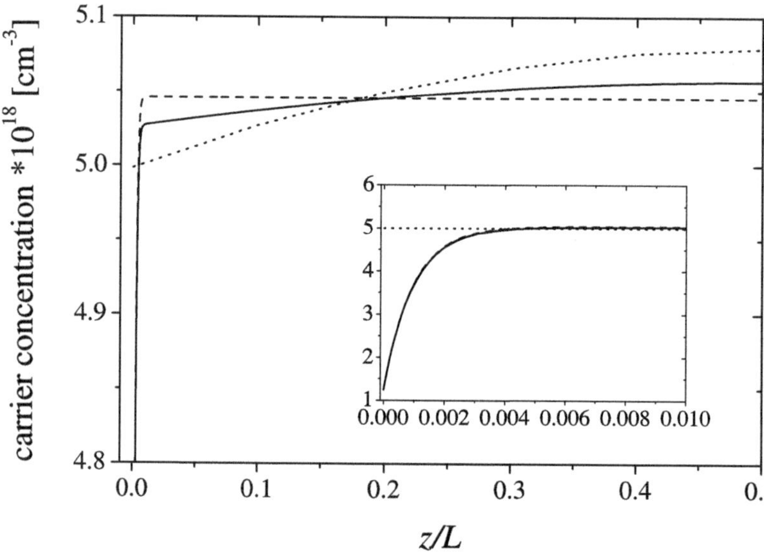

Fig. 5. Axial (mirror to mirror) carrier concentration in the active layer calculated according to algebraic equation (dotted line), linear diffusion equation with constant coefficients (dashed line) and nonlinear diffusion equation with variable coefficients (solid line) (Szymański (2010)).

The results are shown in Fig. 5. It is clear that the approach (iii) yields a crude estimation of the carrier concentration in the active layer. However, for thermal modelling, where phenomena in the vicinity of facets are crucial due to possible COD processes, the diffusion equation must be solved. In many works (see for example Chen & Tien (1993), Schatz & Bethea (1994), Mukherjee & McInerney (2007)), the approach (ii) is used. It seems to be a good approximation for a typical edge-emitting laser, which is an almost axially homogeneous device in the sense that the depression of the photon density does not vary too much or temperature differences along the resonator are not so significant to dramatically change the non-linear recombination terms B and C_A. The approach (i) is useful in all the cases where the above-mentioned axial homogeneity is perturbed. In particular, the approach is suitable for edge-emitting lasers with modified regions close to facets. These modifications are meant to achieve mirror temperature reduction through placing current blocking layers (Rinner et al. (2003)), producing non-injected facets (so called NIFs) (Pierscińska et al. (2007)) or generating larger band gaps (Watanabe et al. (1995)).

3.2 Carrier-dependent heat source function

The knowledge of axial carrier concentration opens up the possibility to write the heat source function more precisely compared to equation (2), namely

$$g(x,y,z) = g_a(x,y,z) + g_J(x,y,z), \tag{8}$$

where the first term describes the heat generation in the active layer and the second - Joule heating. According to Romo et al. (2003):

$$g_a(x,y,z) = [(A_{nr} + C_A N^2(z))N(z) + \frac{c}{n_{eff}}\alpha_{int}S_{av} + \frac{\mu_{sur}N(z=0)}{d_{sur}}\Pi_{sur}(z)]h\nu\Pi_a(x,y,z). \tag{9}$$

The terms in the right hand side of equation (9) are related to non-radiative recombination, Auger processes, absorption of laser radiation and surface recombination at the facets, respectively. Assessing the value of S_{av} was widely discussed by Szymański (2010). The Π's are positioning functions:

$$\Pi_{sur}(z) = \begin{cases} 1, & \text{for } 0 < z < d_{sur}; \\ 0, & \text{for } z > d_{sur}, \end{cases} \tag{10}$$

expresses the assumption that the defects in the vicinity of the facets are uniformly distributed within a distance $d_{sur} = 0.5\mu m$ from the facet surface (Nakwaski (1990); Romo et al. (2003)), while $\Pi_a(x,y,z) = 1$ for x,y,z within the active layer and $\Pi_a(x,y,z) = 0$ elsewhere.
The effect of Joule heating is strictly related to the electrical resistance of a particular layer. High values of this parameter are found in waveguide layers, substrate and p-doped cladding due to the lack of doping, large thickness and low mobility of holes, respectively Szymański et al. (2004). Thus, it is reasonable to calculate the total Joule heat and assume that it is uniformly generated in layers mentioned above of total volume V_{hr}:

$$g_J(x,y,z) = \frac{I^2 R_s}{V_{hr}}, \tag{11}$$

where R_s is the device series resistance.

3.3 Selected results

Axial (mirror to mirror) distribution of relative temperature[3] in the active layer of the edge-emitting laser is shown in Fig. 6. It has been calculated numerically solving the

[3] The temperature exceeding the ambient temperature.

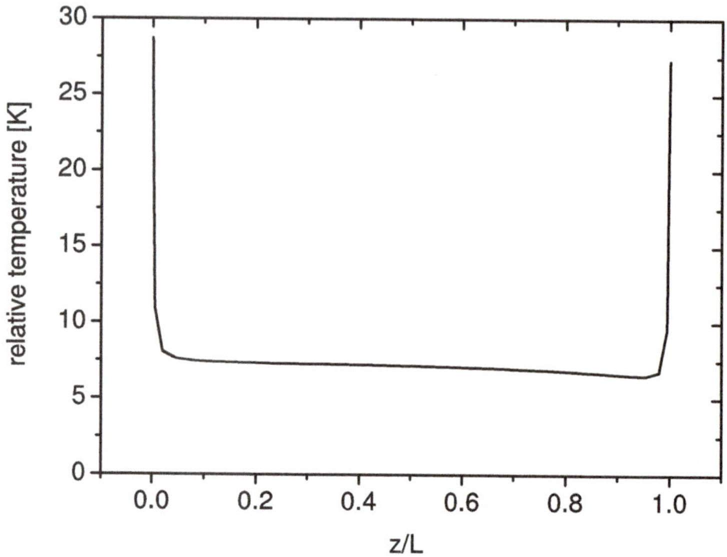

Fig. 6. Axial (mirror to mirror) distribution of relative temperature in the active layer.

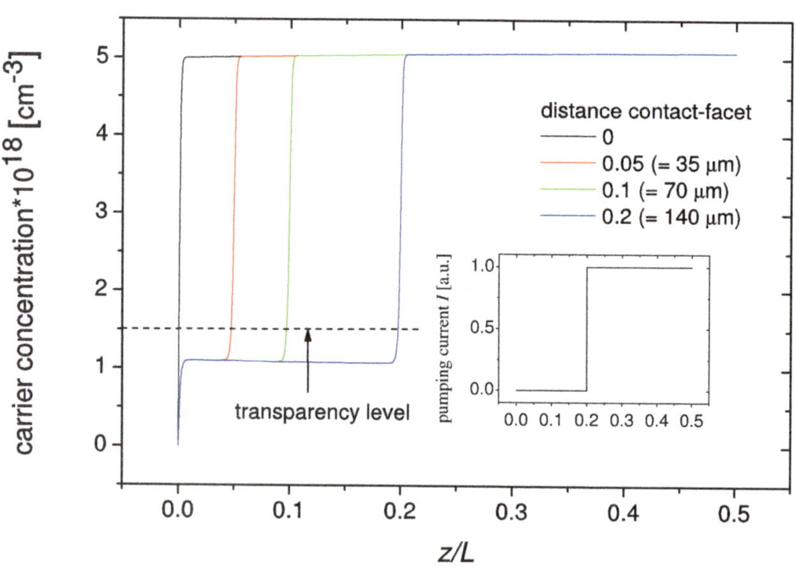

Fig. 7. Axial distribution of carriers in the active layer for the laser with non-injected facets. The inset shows the step-like pumping profile.

three-dimensional heat conduction equation.[4] Heat source has been inserted according to (8)-(11), where $N(z)$ has been calculated analytically from the linear diffusion equation with constant coefficients (approach (ii) from section 3.1). Fig. 6 is in qualitative agreement with plots presented by Chen & Tien (1993); Mukherjee & McInerney (2007); Romo et al. (2003), where similar or more advanced models were used. Note that the temperature along the resonator axis is almost constant, while it rises rapidly in the vicinity of the facets. The small asymmetry is caused by the location of the laser chip: the front facet is over the edge of the heat sink, so the heat removal is obstructed.

Facet temperature reduction techniques are often based on the idea of suppressing the surface recombination by preventing the current flow in the vicinity of facets. It can be realised by placing current blocking layers (Rinner et al. (2003)) or producing non-injected facets (so called NIFs) (Pierscińska et al. (2007)). To investigate such devices the author has solved the equation (6) numerically[5] inserting step-like function $I(z)$. Fig. 7 shows that, in the non-injected region, the carrier concentration rapidly decreases to values lower than transparency level, which is an undesired effect and may disturb laser operation. A solution to this problem, although technologically difficult, can be producing a device with segmented contact. Even weak pumping near the mirror drastically reduces the length of the non-transparent region, which is illustrated in Fig. 8.

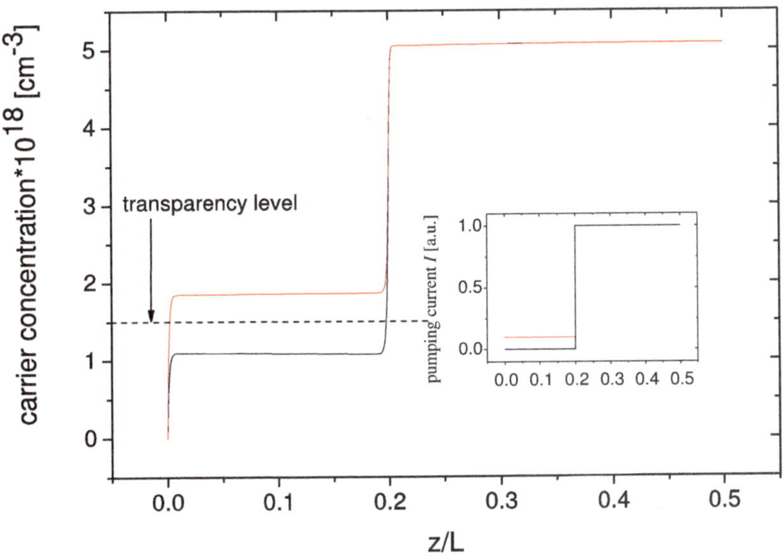

Fig. 8. Axial distribution of carriers in the active layer for the laser with non- and weakly-pumped near-facet region. The inset shows the pumping profile for both cases.

[4] Calculations have been done by Zenon Gniazdowski using the commercial software CFDRC (http://www.cfdrc.com/).

[5] The commercial software FlexPDE (http://www.pdesolutions.com/) has been used.

4. Models including the diffusion equation and photon rate equations

The most advanced thermal model is described by Romo et al. (2003). It takes into account electro-opto-thermal interactions and is based on 3-dimensional heat conduction equation

$$\nabla(\lambda(T)\nabla T) = -g(x,y,z,T), \tag{12}$$

1-dimensional diffusion equation (6), and photon rate equations

$$\frac{c}{n_{eff}}\frac{dS_f}{dz} = \frac{c}{n_{eff}}[\Gamma G(N,T) - \alpha_{int}]S_f + \beta B(T)N^2, \tag{13}$$

$$-\frac{c}{n_{eff}}\frac{dS_b}{dz} = \frac{c}{n_{eff}}[\Gamma G(N,T) - \alpha_{int}]S_b + \beta B(T)N^2. \tag{14}$$

The mirrors impose the following boundary conditions:

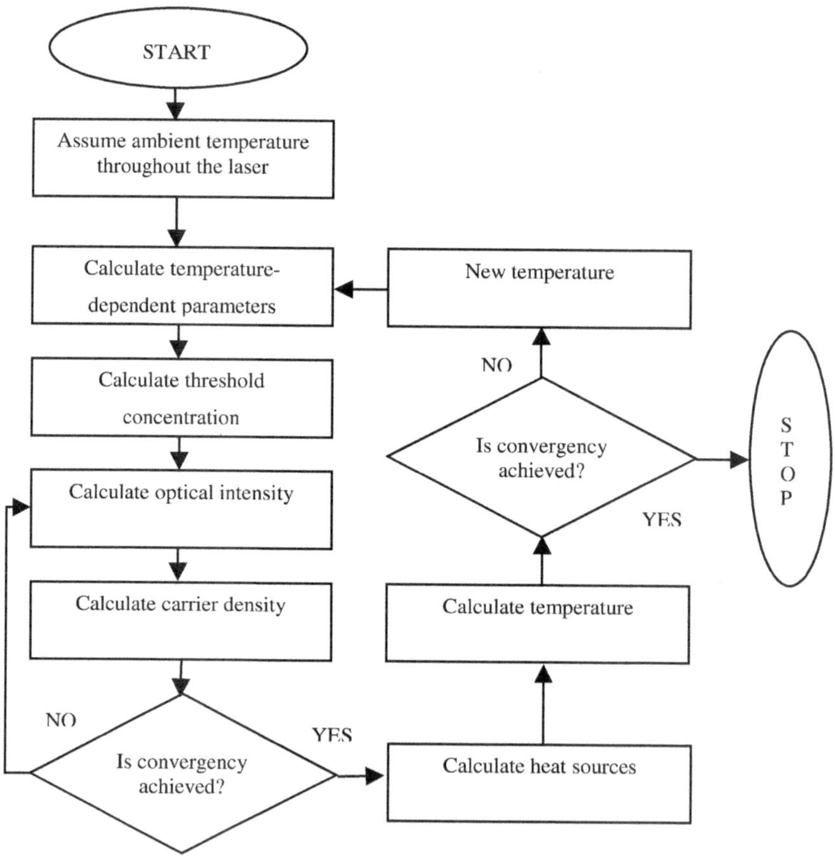

Fig. 9. Self-consistent algorithm (Romo et al. (2003)).

$$S_f(z=0) = R_f S_b(z=0), \; S_b(z=L) = R_b S_f(z=L). \quad (15)$$

Note the quadratic terms in equations (13) and (14), which describe the spontaneous radiation. To avoid problems with estimating the spatial distribution and extent of heat sources related to radiative transfer, Romo et al. (2003) have 'squeezed' the effect to the active layer. Such assumption resulted in inserting the term $(1-2\beta)B(T)N^2 h\nu$ into equation (9).

The set of four differential equations mentioned above was solved numerically in the self-consistent loop, as schematically presented in Fig. 9. Several interesting conclusions formulated by Romo et al. (2003) are worth presenting here:

— the calculations confirmed that the temperature along the resonator axis is almost constant, while it rises rapidly in the vicinity of the facets (cf. Fig. 6);

— taking into account the non-linear temperature dependence of thermal conductivity significantly improves the accuracy of predicted temperature;

— using the 1-dimensional (axial direction) diffusion or photon rate equations is a good enough approach;

— heat conduction equation should be solved in 3 dimensions, reducing it to 2 dimensions is acceptable, while using the 1-dimensional form leads to significant overestimations of temperature in the vicinity of facets.

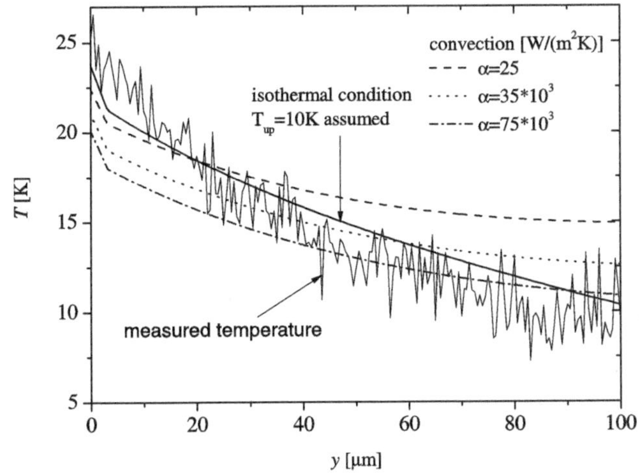

Fig. 10. Transverse temperature profiles across the substrate at $x = 0$ (Szymański (2007)).

5. Discussion of the upper boundary condition

Typical thermal models for edge-emitting lasers assume convectively cooled or thermally insulated (which is the case of zero convection coefficient) upper surface. In Szymański (2007), using the isothermal condition

$$T(x, y_t) = T_{up}. \quad (16)$$

instead of convection is proposed. The model is based on the solution of equation (1) obtained

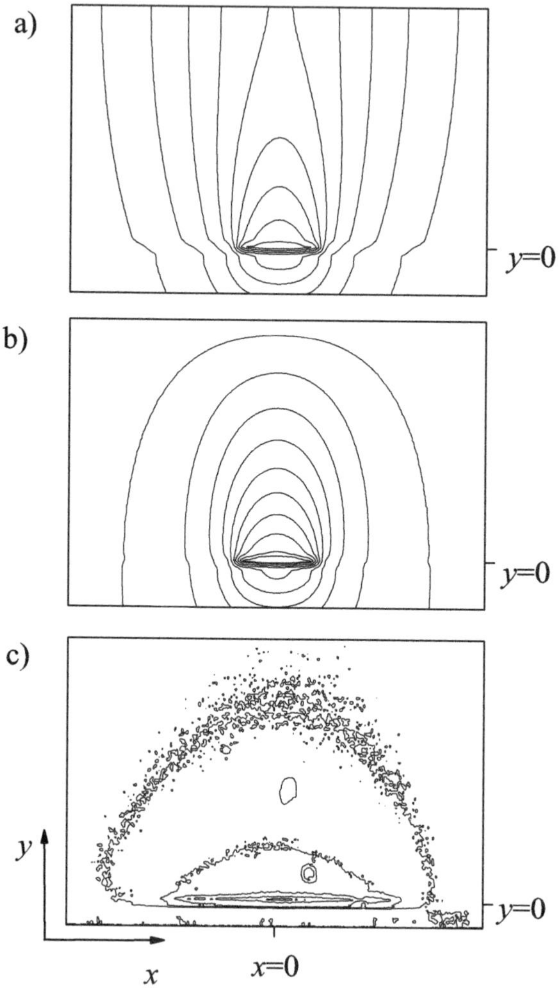

Fig. 11. Contour plot of temperature calculated under the assumption of convective cooling at the top surface (a), isothermal condition at the top surface (b) and measured by thermoreflectance method (c) (Szymański (2007)).

by separation-of-variables approach. Due to (16), the expression (4) describing temperature in the layers above the active layer must be modified in the following way:

$$T_n(x,y) = (\overline{w}_{A,n}^{(0)} A_{2K}^{(0)} + \overline{\overline{w}}_{A,n}^{(0)}) + (\overline{w}_{B,n}^{(0)} A_{2K}^{(0)} + \overline{\overline{w}}_{B,n}^{(0)})y + \sum_{k=1}^{\infty} A_{2K}^{(k)} [w_{A,n}^{(k)} exp(\mu_k y) + w_{B,n}^{(k)} exp(-\mu_k y)] cos(\mu_k x). \quad (17)$$

Full analytical expressions can be found in Szymański (2007).
The investigations have been inspired by temperature maps obtained by thermoreflectance method (Bugajski et al. (2006); Wawer et al. (2005)) for p-down mounted devices. These maps suggest the presence of the region of constant temperature in the vicinity of the n-contact. Besides, the isothermal lines are rather elliptic, surrounding the hot active layer, than directed upward as calculated for convectively cooled surface. The results are presented in Fig. 10 and 11. It is clear that assuming the isothermal condition and convection at the top surface one gets nearly the same device thermal resistances, but with the first assumption closer convergence with thermoreflectance measurements is found.

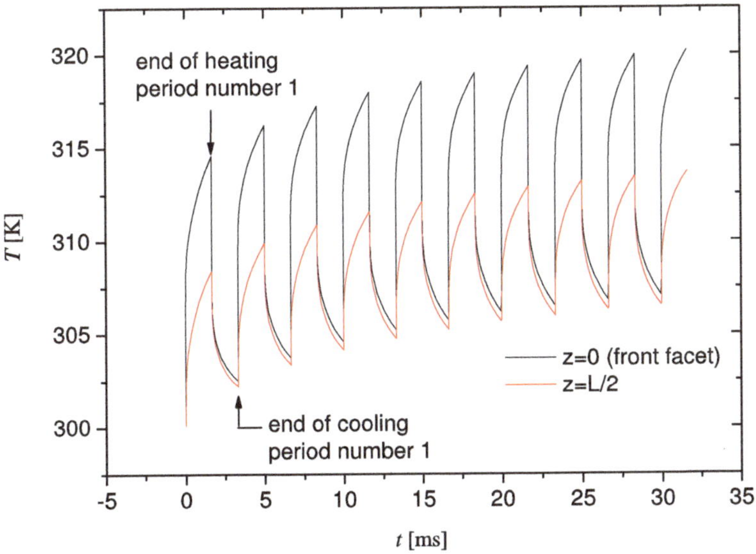

Fig. 12. Time evolution of central emitter active layer temperature.

6. Dynamical picture of thermal behaviour

Time-dependent models of edge-emitting lasers are considered rather seldom for two main reasons. First, edge-emitting lasers are predominantly designed for continuous-wave operation, so there is often no real need to investigate transient phenomena. Second, the complicated geometry of these devices, different kinds of boundary conditions and uncertain values of material parameters make that even static cases are difficult to solve.

The authors who consider dynamical models usually concentrate on initial heating (temperature rise during the first current pulse) of the laser inside the resonator (Nakwaski (1983b)) or at the mirrors (Nakwaski (1985; 1990)). The papers mentioned above developed analytical solutions of time-dependent heat conduction equation using sophisticated mathematical methods, like for example Green function formalism or Kirchhoff transformation. Numerical approach to this class of problems appeared much later. As an example see Puchert et al. (2000), where laser array was investigated. It is noteworthy that the heat source function was obtained by a rate equation model. Remarkable agreement with experimental temperature values showed the importance of the concept of distributed heat sources. The author has theoretically investigated the dynamical thermal behaviour of

Layer	thickness $[\mu m]$	λ[W/(mK)]	c_h[J/(kgK)]	ρ[kg/m^3]	heat source
substrate	100	44	327	5318	yes - equation (11)
$Al_{0.6}Ga_{0.4}As$ (n-cladding)	1.5	11.4	402	4384	no
$Al_{0.4}Ga_{0.6}As$ (waveguide)	0.35	11.1	378	4696	yes - equation (11)
active layer	0.007	44	327	5318	yes - equation (9)
$Al_{0.4}Ga_{0.6}As$ (waveguide)	0.59	11.1	378	4696	yes - equation (11)
$Al_{0.6}Ga_{0.4}As$ (p-cladding)	1.5	11.4	402	4384	yes - equation (11)
GaAs (cap)	0.2	44	327	5318	no
p-contact	1	318	128	19300	no
In (solder)	1	82	230	7310	no

Table 3. Transverse structure of the investigated laser array.

p-down mounted 25-emitter laser array. Temperature profiles during first 10 pulses have been calculated (Fig. 12 and 13). The transverse structure of the device, material parameters and the distribution of heat sources are presented in Table 3. The time-dependent heat conduction equation

$$\rho(x,y,z)c_h(x,y,z)\frac{\partial T}{\partial t} = \nabla(\lambda(x,y,z)\nabla T) + g(x,y,z,t), \tag{18}$$

has been solved numerically.[6] The following boundary and initial conditions have been assumed:

— constant temperature $T = 300K$ at the heat spreader-heat sink interface;[7]

— convective cooling of the top surface ($\alpha = 35 * 10^3 WK^{-1}m^{-2}$);

— all side walls (including mirrors) thermally insulated;

— $T(x,y,z,t=0) = 300K$.

Note that the considered laser array has been driven by rectangular pulses of period 3.33 ms and 50% duty cycle, while the carrier lifetime is of the order of several nanoseconds. Thus, the electron response for the applied voltage can be regarded as immediate[8] and equation (8) can be transformed to the time-dependent function in the following way:

$$g(x,y,z,t) = [g_a(x,y,z) + g_J(x,y,z)]\Theta(t), \tag{19}$$

[6] Calculations have been done by Zenon Gniazdowski using the commercial software CFDRC (http://www.cfdrc.com/).

[7] To simplify the problem, the ideal heat sink $\lambda = \infty$ has been assumed.

[8] This statement is common for all standard edge-emitting devices.

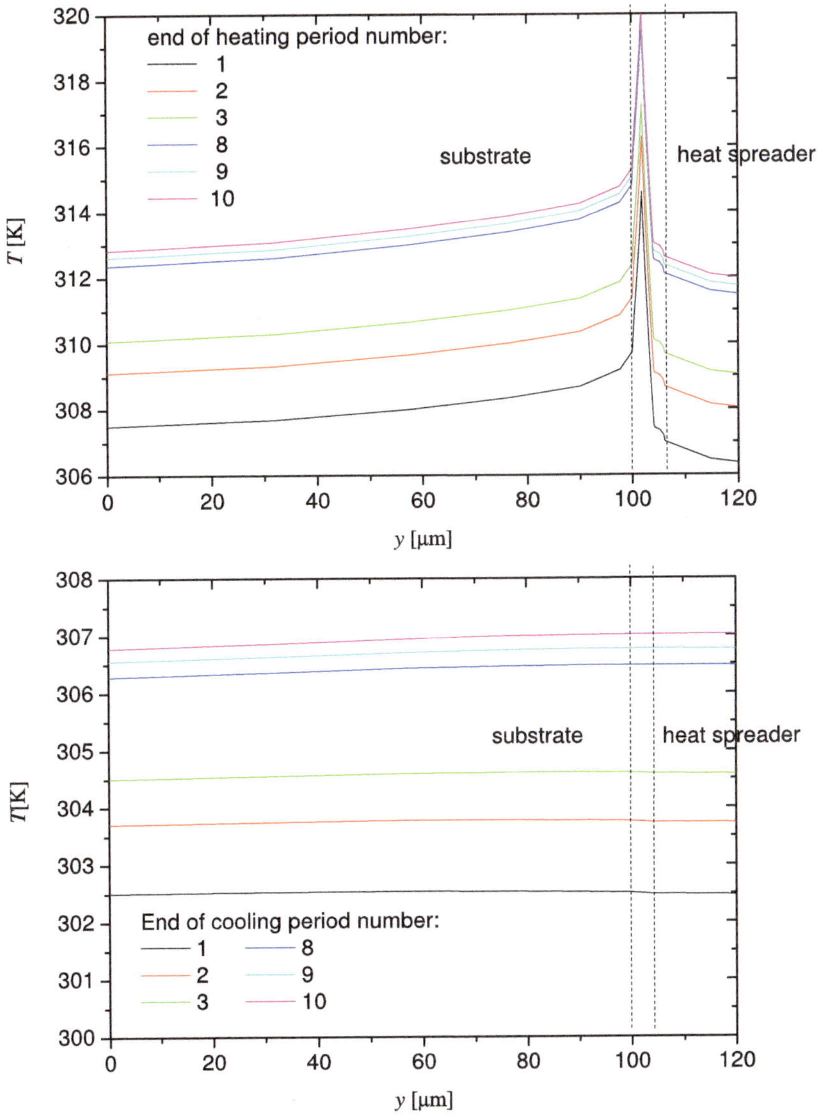

Fig. 13. Transverse temperature profiles at the front facet of the central emitter. Dashed vertical lines indicate the edges of heat spreader and substrate.

where $\Theta(t) = 1$ or 0 exactly reproduces the driving current changes.

7. Heat flow in a quantum cascade laser

Quantum-cascade lasers are semiconductor devices exploiting superlattices as active layers. In numerous experiments, it has been shown that the thermal conductivity λ of a superlattice

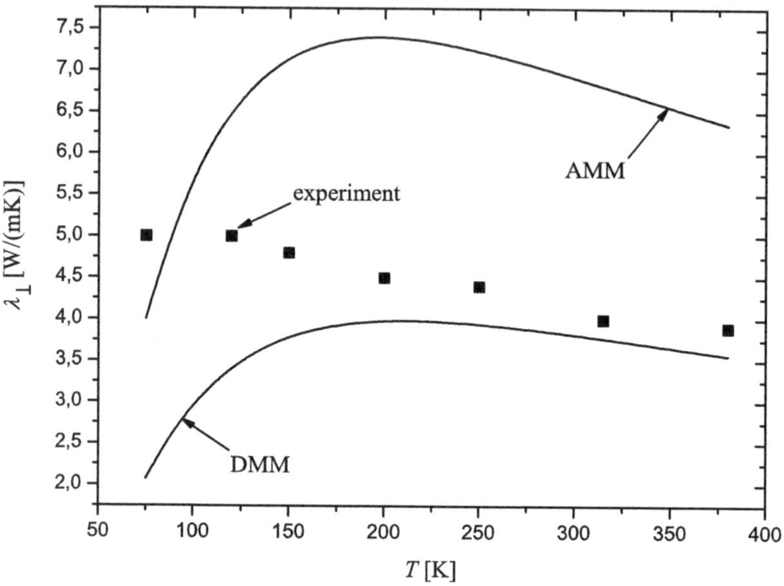

Fig. 14. Calculated cross-plane thermal conductivity for the active region of THz QCL (Szymański (2011)). Square symbols show the values measured by Vitiello et al. (2008).

is significantly reduced (Capinski et al. (1999); Cahill et al. (2003); Huxtable et al. (2002)). Particularly, the cross-plane value $\lambda_\perp$ may be even order-of-magnitude smaller than than the value for constituent bulk materials. The phenomenon is a serious problem for QCLs, since they are electrically pumped by driving voltages over 10 V and current densities over 10 kA/cm^2. Such a high injection power densities lead to intensive heat generation inside the devices. To make things worse, the main heat sources are located in the active layer, where the density of interfaces is the highest and—in consequence—the heat removal is obstructed. Thermal management in this case seems to be the key problem in design of the improved devices.

Theoretical description of heat flow across SL's is a really hard task. The crucial point is finding the relation between phonon mean free path Λ and SL period D Yang & Chen (2003). In case $\Lambda > D$, both wave- and particle-like phonon behaviour is observed. The thermal conductivity is calculated through the modified phonon dispersion relation obtained from the equation of motion of atoms in the crystal lattice (see for example Tamura et al. (1999)). In case $\Lambda < D$, phonons behave like particles. The thermal conductivity is usually calculated using the Boltzmann transport equation with boundary conditions involving diffuse scattering.

Unfortunately, using the described methods in the thermal model of QCL's is questionable. They are very complicated on the one hand and often do not provide satisfactionary results on the other. The comprehensive comparison of theoretical predictions with experiments for

nanoscale heat transport can be found in Table II in Cahill et al. (2003). This topic was also widely discussed by Gesikowska & Nakwaski (2008). In addition, the investigations in this field usually deal with bilayer SL's, while one period of QCL active layer consists of dozen or so layers of order-of-magnitude thickness differences.
Consequently, present-day mathematical models of heat flow in QCLs resemble those created for standard edge emitting lasers: they are based on heat conduction equation, isothermal condition at the bottom of the structure and convective cooling of the top and side walls are assumed. QCL's as unipolar devices are not affected by surface recombination. Their mirrors may be hotter than the inner part of resonator only due to bonding imperfections (see 8.4). Colour maps showing temperature in the QCL cross-section and illustrating fractions of heat flowing through particular surfaces can be found in Lee et al. (2009) and Lops et al. (2006). In those approaches, the SL's were replaced by equivalent layers described by anisotropic values of thermal conductivity $\lambda_{\perp}$ and $\lambda_{\|}$ arbitrarily reduced (Lee et al. (2009)) or treated as fitting parameters (Lops et al. (2006)).

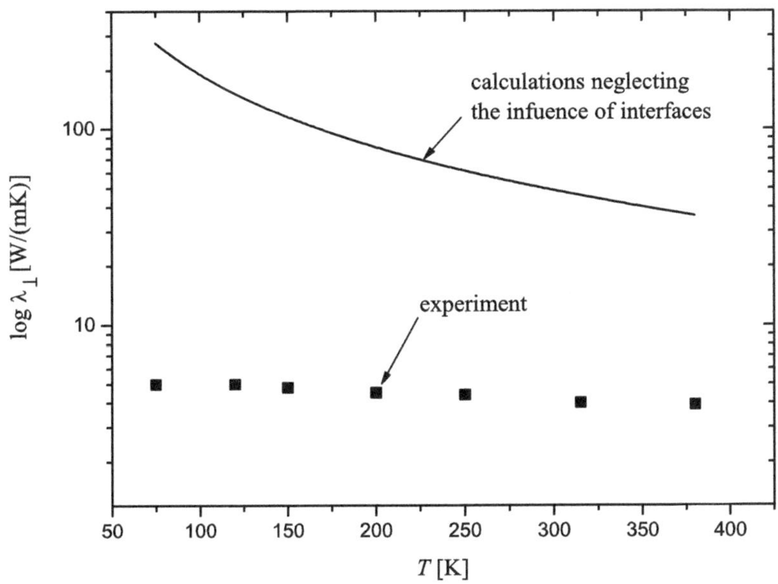

Fig. 15. Illustration of significant discrepancy between values of $\lambda_{\perp}$ measured by Vitiello et al. (2008) and calculated according to equation (20), which neglects the influence of interfaces (Szymański (2011)).

Proposing a relatively simple method of assessing the thermal conductivity of QCL active region has been a subject of several works. A very interesting idea was mentioned by Zhu et al. (2006) and developed by Szymański (2011). The method will be briefly described below.

The thermal conductivity of a multilayered structure can be approximated according to the rule of mixtures Samvedi & Tomar (2009); Zhou et al. (2007):

$$\lambda^{-1} = \sum_n f_n \lambda_n^{-1}, \tag{20}$$

where f_n and λ_n are the volume fraction and bulk thermal conductivity of the n-th material. However, in case of high density of interfaces, the approach (20) is inaccurate because of the following reason. The interface between materials of different thermal and mechanical properties obstructs the heat flow, introducing so called 'Kapitza resistance' or thermal boundary resistance (TBR) Swartz & Pohl (1989). The phenomenon can be described by two phonon scattering models, namely the acoustic mismatch model (AMM) and the diffuse mismatch model (DMM). Input data are limited to such basic material parameters like Debye temperature, density or acoustic wave speed. Thus, the thermal conductivity of the QCL active region can be calculated as a sum of weighted average of constituent bulk materials reduced by averaged TBR multiplied by the number of interfaces:

$$\lambda_{\perp}^{-1} = \frac{d_1}{d_1+d_2} r_1 + \frac{d_2}{d_1+d_2} r_2 + \frac{n_i}{d_1+d_2} r_{\mathrm{Bd}}^{(\mathrm{av})}, \tag{21}$$

where TBR has been averaged with respect to the direction of the heat flow

$$r_{\mathrm{Bd}}^{(\mathrm{av})} = \frac{r_{\mathrm{Bd}}(1 \to 2) + r_{\mathrm{Bd}}(2 \to 1)}{2}. \tag{22}$$

The detailed prescription on how to calculate $r_{\mathrm{Bd}}^{(\mathrm{av})}$ can be found in Szymański (2011).
The model based on equations (21) and (22) was positively tested on bilayer $Si_{0.84}Ge_{0.16}/Si_{0.74}Ge_{0.26}$ SL's investigated experimentally by Huxtable et al. (2002). Then, $GaAs/Al_{0.15}Ga_{0.85}As$ THz QCL was considered. Results of calculations exhibit good convergence with measurements presented by Vitiello et al. (2008) as shown in Fig. 14. On the contrary, values of $\lambda_{\perp}$ calculated according to equation (20), neglecting the influence of interfaces, show significant discrepancy with the measured ones (Fig. 15).

8. Summary

Main conclusions or hints dealing with thermal models of edge-emitting lasers will be aggregated in the form of the following paragraphs.

8.1 Differential equations

A classification of thermal models is presented in Table 4. Basic thermal behaviour of an edge-emitting laser can be described according to Approach 1. It is assumed that the heat power is generated uniformly in selected regions: mainly in active layer and, in minor degree, in highly resistive layers. Considering the laser cross-section parallel to mirrors' surfaces and reducing the dimensionality of the heat conduction equation to 2 is fully justified. For calculating the temperature in the entire device (including the vicinity of mirrors) Approach 2 should be used. The main heat sources may be determined as functions of carrier concentration calculated from the diffusion equation. It is recommended to use three-dimensional heat conduction equation. The diffusion equation can be solved in the

Approach	Equations(s)	Calculated T inside the resonator	Calculated T in the vicinity of mirrors	Application	Example references
1	HC	yes	near-threshold regime	basic thermal behaviour of a laser	Joyce & Dixon (1975), Puchert et al. (1997), Szymański et al. (2007)
2	HC+D	yes	low-power operation	thermal behaviour of a laser including the vicinity of mirrors	Chen & Tien (1993), Mukherjee & McInerney (2007)
3	HC+D+PR	yes	high-power operation	facet temperature reduction	Romo et al. (2003)

Table 4. A classification of thermal models. Abbreviations: HC-heat conduction, D-diffusion, PR -photon rate.

plane of junction (2 dimensions) or reduced to the axial direction (1 dimension). Approach 3 is the most advanced one. It is based on 4 differential equations, which should be solved in self-consisted loop (see Fig. 9). Approach 3 is suitable for standard devices as well as for lasers with modified close-to-facet regions.

8.2 Boundary conditions

The following list presents typical boundary conditions (see for example Joyce & Dixon (1975), Puchert et al. (1997), Szymański et al. (2007)):

— isothermal condition at the bottom of the device,

— thermally insulated side walls,

— convectively cooled or thermally insulated (which is the case of zero convection coefficient) upper surface.

In Szymański (2007), it was shown that assuming isothermal condition at the upper surface is also correct and reveals better convergence with experiment.

Specifying the bottom of the device may be troublesome. Considering the heat flow in the chip only, i.e. assuming the ideal heat sink, leads to significant errors (Szymański et al. (2007)). On the other hand taking into account the whole assembly (chip, heat spreader and heat sink) is difficult. In the case of analytical approach, it significantly complicates the geometry of the thermal scheme. In order to avoid that tricky modifications of thermal scheme (like in Szymański et al. (2007)) have to be introduced. In case of numerical approach, using non-uniform mesh is absolutely necessary (see for example Puchert et al. (2000)).

In Ziegler et al. (2006), an actively cooled device was investigated. In that case a very strong convection ($\alpha = 40 * 10^4 W/(mK)$) at the bottom surface was assumed in calculations.

8.3 Calculation methods

Numerous works dealing with thermal modelling of edge-emitting lasers use analytical approaches. Some of them exploit highly sophisticated mathematical methods. For example,

Kirchhoff transformation (see Nakwaski (1980)) underlied further pioneering theoretical studies on the COD process by Nakwaski (1985) and Nakwaski (1990), where solutions of the three-dimensional time-dependent heat conduction equation were found using the Green function formalism. Conformal mapping has been used by Laikhtman et al. (2004) and Laikhtman et al. (2005) for thermal optimisation of high power diode laser bars. Relatively simple separation-of-variables approach was used by Joyce & Dixon (1975) and developed in many further works (see for example Bärwolff et al. (1995) or works by the author of this chapter).

Analytical models often play a very helpful role in fundamental understanding of the device operation. Some people appreciate their beauty. However, one should keep in mind that edge-emitting devices are frequently more complicated. This statement deals with the internal chip structure as well as packaging details. Analytical solutions, which can be found in widely-known textbooks (see for example Carslaw & Jaeger (1959)), are usually developed for regular figures like rectangular or cylindrical rods made of homogeneous materials. Small deviation from the considered geometry often leads to substantial changes in the solution. In addition, as far as solving single heat conduction equation in some cases may be relatively easy, including other equations enormously complicates the problem. Recent development of simulation software based on Finite Element Method creates the temptation to relay on numerical methods. In this chapter, the commercial software has been used for computing dynamical temperature profiles (Fig. 12 and 13)[9] and carrier concentration profiles (Fig. 7 and 8).[10] Commercial software was also used in many works, see for example Mukherjee & McInerney (2007); Puchert et al. (2000); Romo et al. (2003). In Ziegler et al. (2006; 2008), a self-made software based on FEM provided results highly convergent with sophisticated thermal measurements of high-power diode lasers. Thus, nowadays numerical methods seem to be more appropriate for thermal analysis of modern edge-emitting devices. However, one may expect that analytical models will not dissolve and remain as helpful tools for crude estimations, verifications of numerical results or fundamental understanding of particular phenomena.

8.4 Limitations

While using any kind of model, one should be prepared for unavoidable inaccuracies of the temperature calculations caused by factors characteristic for individual devices, which elude qualitative assessment. The paragraphs below briefly describe each factor.

Real solder layers may contain a number of voids, such as inclusions of air, clean-up agents or fluxes. Fig. 12 in Bärwolff et al. (1995) shows that small voids in the solder only slightly obstruct the heat removal from the laser chip to the heat sink unless their concentration is very high. In turn, the influence of one large void is much bigger: the device thermal resistance grows nearly linearly with respect to void size.

The laser chip may not adhere to the heat sink entirely due to two reasons: the metallization may not extend exactly to the laser facets or the chip can be inaccurately bonded (it can extend over the heat sink edge). In Lynch (1980), it was shown that such an overhang may contribute to order of magnitude increase of the device thermal resistance.

---

[9] CFDRC software (http://www.cfdrc.com/) used used by Zenon Gniazdowski.

[10] FlexPDE software (http://www.pdesolutions.com/) used by Michal Szymański.

In Pipe & Ram (2003) it was shown that convective cooling of the top and side walls plays a significant role. Unfortunately, determining of convective coefficient is difficult. The values found in the literature differ by 3 order-of-magnitudes (see Szymański (2007)).
Surface recombination, one of the two main mirror heating mechanisms, strongly depends on facet passivation. The significant influence of this phenomenon on mirror temperature was shown in Diehl (2000). It is noteworthy that the authors considered values v_{sur} of one order-of-magnitude discrepancy.[11]
Modern devices often consist of multi-compound semiconductors of unknown thermal properties. In such cases, one has to rely on approximate expressions determining particular parameter upon parameters of constituent materials (see for example Nakwaski (1988)).

8.5 Quantum cascade lasers

Present-day mathematical models of heat flow in QCL resemble those created for standard edge emitting lasers: they are based on heat conduction equation, isothermal condition at the bottom of the structure and convective cooling of the top and side walls are assumed. The SL's, which are the QCLs' active regions, are replaced by equivalent layers described by anisotropic values of thermal conductivity $\lambda_{\perp}$ and $\lambda_{\parallel}$ arbitrarily reduced (Lee et al. (2009)), treated as fitting parameters (Lops et al. (2006)) or their parameters are assessed by models considering microscale heat transport (Szymański (2011)).

9. References

Bärwolff A., Puchert R., Enders P., Menzel U. and Ackermann D. (1995) Analysis of thermal behaviour of high power semiconductor laser arrays by means of the finite element method (FEM), *J. Thermal Analysis*, Vol. 45, No. 3, (September 1995) 417-436.

Bugajski M., Piwonski T., Wawer D., Ochalski T., Deichsel E., Unger P., and Corbett B. (2006) Thermoreflectance study of facet heating in semiconductor lasers, *Materials Science in Semiconductor Processing* Vol. 9, No. 1-3, (February-June 2006) 188-197.

Capinski W S, Maris H J, Ruf T, Cardona M, Ploog K and Katzer D S (1999) Thermal-conductivity measurements of GaAs/AlAs superlattices using a picosecond optical pump-and-probe technique, *Phys. Rev. B*, Vol. 59, No. 12, (March 1999) 8105-8113.

Carslaw H. S. and Jaeger J. C. (1959) *Conduction of heat in solids*, Oxford University Press, ISBN, Oxford.

Cahill D. G., Ford W. K., Goodson K. E., Mahan G. D., Majumdar A., Maris H. J., Merlin R. and Phillpot S. R. (2003) Nanoscale thermal transport, *J. Appl. Phys.*, Vol. 93, No. 2, (January 2003) 793-818.

Chen G. and Tien C. L. (1993) Facet heating of quantum well lasers, *J. Appl. Phys.*, Vol. 74, No. 4, (August 1993) 2167-2174.

Diehl R. (2000) *High-Power Diode Lasers. Fundamentals, Technology, Applications*, Springer, ISBN, Berlin.

[11] Surface recombination does not deal with QCL's as they are unipolar devices. In turn, inaccuracies related to assessing $\lambda_{\perp}$ and $\lambda_{\parallel}$ may occur.

Gesikowska E. and Nakwaski W. (2008) An impact of multi-layered structures of modern optoelectronic devices on their thermal properties, *Opt. Quantum Electron.*, Vol. 40, No. 2-4, (August 2008) 205-216.

Huxtable S. T., Abramson A. R., Chang-Lin T., and Majumdar A. (2002) Thermal conductivity of Si/SiGe and SiGe/SiGe superlattices *Appl. Phys. Lett.* Vol. 80, No. 10, (March 2002) 1737-1739.

Joyce W. B. & Dixon R. (1975). Thermal resistance of heterostructure lasers, *J. Appl. Phys.*, Vol. 46, No. 2, (February 1975) 855-862.

Laikhtman B., Gourevitch A., Donetsky D., Westerfeld D. and Belenky G. (2004) Current spread and overheating of high power laser bars, *J. Appl. Phys.*, Vol. 95, No. 8, (April 2004) 3880-3889.

Laikhtman B., Gourevitch A., Westerfeld D., Donetsky D. and Belenky G., (2005) Thermal resistance and optimal fill factor of a high power diode laser bar, *Semicond. Sci. Technol.*, Vol. 20, No. 10, (October 2005) 1087-1095.

Lee H. K., Chung K. S., Yu J. S. and Razeghi M. (2009) Thermal analysis of buried heterostructure quantum cascade lasers for long-wave-length infrared emission using 2D anisotropic, heat-dissipation model, *Phys. Status Solidi A*, Vol. 206, No. 2, (February 2009) 356-362.

Lops A., Spagnolo V. and Scamarcio G. (2006) Thermal modelling of GaInAs/AlInAs quantum cascade lasers, *J. Appl. Phys.*, Vol. 100, No. 4, (August 2006) 043109-1-043109-5.

Lynch Jr. R. T. (1980) Effect of inhomogeneous bonding on output of injection lasers, *Appl. Phys. Lett.*, Vol. 36, No. 7, (April 1980) 505-506.

Manning J. S. (1981) Thermal impedance of diode lasers: Comparison of experimental methods and a theoretical model, *J. Appl. Phys.*, Vol. 52, No. 5, (May 1981) 3179-3184.

Mukherjee J. and McInerney J. G. (2007) Electro-thermal Analysis of CW High-Power Broad-Area Laser Diodes: A Comparison Between 2-D and 3-D Modelling, *IEEE J. Sel. Topics in Quantum Electron.* , Vol. 13, No. 5, (September/October 2007) 1180-1187.

Nakwaski W. (1979) Spontaneous radiation transfer in heterojunction laser diodes, *Sov. J. Quantum Electron.*, Vol. 9, No. 12, (December 1979) 1544-1546.

Nakwaski W. (1980) An application of Kirchhoff transformation to solving the nonlinear thermal conduction equation for a laser diode, *Optica Applicata*, Vol. 10, No. 3, (?? 1980) 281-283.

Nakwaski W. (1983) Static thermal properties of broad-contact double heterostructure GaAs-(AlGa)As laser diodes, *Opt. Quantum Electron.*, Vol. 15, No. 6, (November 1983) 513-527.

Nakwaski W. (1983) Dynamical thermal properties of broad-contact double heterostructure GaAs-(AlGa)As laser diodes, *Opt. Quantum Electron.*, Vol. 15, No. 4, (July 1983) 313-324.

Nakwaski W. (1985) Thermal analysis of the catastrophic mirror damage in laser diodes, *J. Appl. Phys.*, Vol. 57, No. 7, (April 1985) 2424-2430.

Nakwaski W. (1988) Thermal conductivity of binary, ternary and quaternary III-V compounds, *J. Appl. Phys.*, Vol. 64, No. 1, (July 1988) 159-166.

Nakwaski W. (1990) Thermal model of the catastrophic degradation of high-power stripe-geometry GaAs-(AlGa)As double-heterostructure diode-lasers, *J. Appl. Phys.*, Vol. 67, No. 4, (February 1990) 1659-1668.

Pierscińska D., Pierscińsk K., Kozlowska A., Malag A., Jasik A. and Poprawe R. (2007) Facet heating mechanisms in high power semiconductor lasers investigated by spatially resolved thermo-reflectance, *MIXDES*, ISBN, Ciechocinek, Poland, June 2007

Pipe K. P. and Ram R. J. (2003) Comprehensive Heat Exchange Model for a Semiconductor Laser Diode, *IEEE Photonic Technology Letters*, Vol. 15, No. 4, (April 2003) 504-506.

Piprek J. (2003) *Semiconductor optoelectronic devices. Introduction to physics and simulation*, Academic Press, ISBN 0125571909, Amsterdam.

Puchert R., Menzel U., Bärwolff A., Voß M. and Lier Ch. (1997) Influence of heat source distributions in GaAs/GaAlAs quantum-well high-power laser arrays on temperature profile and thermal resistance, *J. Thermal Analysis*, Vol. 48, No. 6, (June 1997) 1273-1282.

Puchert R., Bärwolff A., Voß M., Menzel U., Tomm J. W. and Luft J. (2000) Transient thermal behavior of high power diode laser arrays, *IEEE Components, Packaging, and Manufacturing Technology Part A* Vol. 23, No. 1, (January 2000) 95-100.

Rinner F., Rogg J., Kelemen M. T., Mikulla M., Weimann G., Tomm J. W., Thamm E. and Poprawe R. (2003) Facet temperature reduction by a current blocking layer at the front facets of high-power InGaAs/AlGaAs lasers, *J. Appl. Phys.*, Vol. 93, No. 3, (February 2003) 1848-1850

Romo G., Smy T., Walkey D. and Reid B. (2003) Modelling facet heating in ridge lasers, *Microelectronics Reliability*, Vol. 43, No. 1, (January 2003) 99-110.

Samvedi V. and Tomar V. (2009) The role of interface thermal boundary resistance in the overall thermal conductivity of Si-Ge multilayered structures, *Nanotechnology*, Vol. 20, No. 36, (September 2009) 365701.

Sarzała R. P. and Nakwaski W. (1990) An appreciation of usability of the finite element method for the thermal analysis of stripe-geometry diode lasers, *J. Thermal Analysis*, Vol. 36, No. 3, (May 1990) 1171-1189.

Sarzała R. P. and Nakwaski W. (1994) Finite-element thermal model for buried-heterostructure diode lasers, *Opt. Quantum Electron.* , Vol. 26, No. 2, (February 1994) 87-95.

Schatz R. and Bethea C. G. (1994) Steady state model for facet heating to thermal runaway in semiconductor lasers, *J. Appl. Phys.* , Vol. 76, No. 4, (August 1994) 2509-2521.

Swartz E. T. and Pohl R. O. (1989) Thermal boundary resistance *Rev. Mod. Phys.*, Vol. 61, No. 3, (July 1989) 605-668.

Szymański M., Kozlowska A., Malag A., and Szerling A. (2007) Two-dimensional model of heat flow in broad-area laser diode mounted to the non-ideal heat sink, *J. Phys. D: Appl. Phys.*, Vol. 40, No. 3, (February 2007) 924-929.

Szymański M. (2010) A new method for solving nonlinear carrier diffusion equation in axial direction of broad-area lasers, *Int. J. Num. Model.*, Vol. 23, No. 6, (November/December 2010) 492-502.

Szymański M. (2011) Calculation of the cross-plane thermal conductivity of a quantum cascade laser active region *J. Phys. D: Appl. Phys.*, Vol. 44, No. 8, (March 2011) 085101-1-085101-5.

Szymański M., Zbroszczyk M. and Mroziewicz B. (2004) The influence of different heat sources on temperature distributions in broad-area lasers *Proc. SPIE* Vol. 5582, (September 2004) 127-133.

Szymański M. (2007) Two-dimensional model of heat flow in broad-area laser diode: Discussion of the upper boundary condition *Microel. J.* Vol. 38, No. 6-7, (June-July 2007) 771-776.

Tamura S, Tanaka Y and Maris H J (1999) Phonon group velocity and thermal conduction in superlattices *Phys. Rev. B* , Vol. 60, No. 4, (July 1999) 2627-2630.

Vitiello M. S., Scamarcio G. and Spagnolo V. 2008 Temperature dependence of thermal conductivity and boundary resistance in THz quantum cascade lasers *IEEE J. Sel. Top. in Quantum Electron.*, Vol. 14, No. 2, (March/April 2008) 431-435 .

Watanabe M., Tani K., Takahashi K., Sasaki K., Nakatsu H., Hosoda M., Matsui S., Yamamoto O. and Yamamoto S. (1995) Fundamental-Transverse-Mode High-Power AlGaInP Laser Diode with Windows Grown on Facets, *IEEE J. Sel. Topics in Quantum Electron.*, Vol. 1, No. 2, (June 1995) 728-733.

Wawer D., Ochalski T.J., Piwoński T., Wójcik-Jedlińska A., Bugajski M., and Page H. (2005) Spatially resolved thermoreflectance study of facet temperature in quantum cascade lasers, *Phys. Stat. Solidi (a)* Vol. 202, No. 7, (May 2005) 1227-1232.

Yang B and Chen G (2003) Partially coherent phonon heat conduction in superlattices, *Phys. Rev. B*, Vol. 67, No. 19, (May 2003) 195311-1-195311-4.

Zhu Ch., Zhang Y., Li A. and Tian Z. 2006 Analysis of key parameters affecting the thermal behaviour and performance of quantum cascade lasers, *J. Appl. Phys.*, Vol. 100, No. 5, (September 2006) 053105-1-053105-6.

Zhou Y., Anglin B. and Strachan A. 2007 Phonon thermal conductivity in nanolaminated composite metals via molecular dynamics, *J. Chem. Phys.*, Vol. 127, No. 18, (November 2007) 184702-1-184702-11.

Ziegler M., Weik F., Tomm J.W., Elsaesser T., Nakwaski W., Sarzała R.P., Lorenzen D., Meusel J. and Kozlowska A. (2006) Transient thermal properties of high-power diode laser bars *Appl. Phys. Lett.* Vol. 89, No. 26, (December 2006) 263506-1-263506-3.

Ziegler M., Tomm J.W., Elsaesser T., Erbert G., Bugge F., Nakwaski W. and Sarzała R.P. (2008) Visualisation of heat flows in high-power diode lasers by lock-in thermography *Appl. Phys. Lett.* Vol. 92, No. 10, (March 2008) 103513-1-103513-3.

4

Temperature Rise of Silicon Due to Absorption of Permeable Pulse Laser

Etsuji Ohmura
Osaka University
Japan

1. Introduction

Blade dicing is used conventionally for dicing of a semiconductor wafer. Stealth dicing (SD) was developed as an innovative dicing method by Hamamatsu Photonics K.K. (Fukuyo et al., 2005; Fukumitsu et al., 2006; Kumagai et al., 2007). The SD method includes two processes. One is a "laser process" to form a belt-shaped modified-layer (SD layer) into the interior of a silicon wafer for separating it into chips. The other is a "separation process" to divide the wafer into small chips. A schematic illustration of the laser process is shown in Fig. 1.

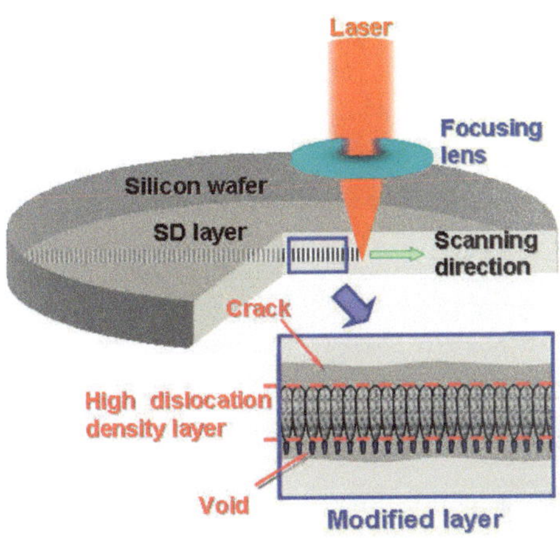

Fig. 1. Schematic illustration of "laser process" in Stealth Dicing (SD)

When a permeable nanosecond laser is focused into the interior of a silicon wafer and scanned in the horizontal direction, a high dislocation density layer and internal cracks are formed in the wafer. Fig. 2 shows the pictures of a wafer after the laser process and small chips divided through the separation process. The internal cracks progress to the surfaces by applying tensile stress due to tape expansion without cutting loss. An example of the photographs of divided face of the SD processed silicon wafer is shown in Fig. 3.

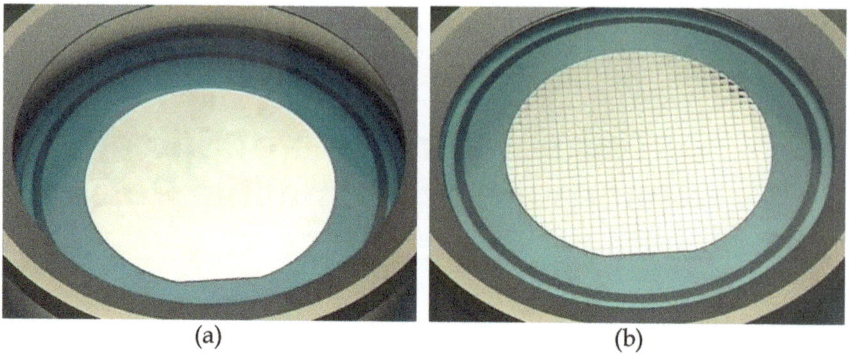

Fig. 2. A wafer after the laser process (a) and small chips divided through the separation process (b) (Photo: Hamamatsu Photonics K.K.)

Fig. 3. Internal modified layer observed after division by tape expansion

As the SD is a noncontact processing method, high speed processing is possible. Fig. 4 shows a comparison of edge quality between blade dicing and SD. In the SD, there is no chipping and no cutting loss, so there is no pollution caused by the debris. The advantage of using the SD method is clear. Fig. 5 shows an example of SD application to actual MEMS device. This device has a membrane structure whose thickness is 2 μm, but it is not damaged. A complete dry process of dicing technology has been realized and problems due to wet processing have been solved.

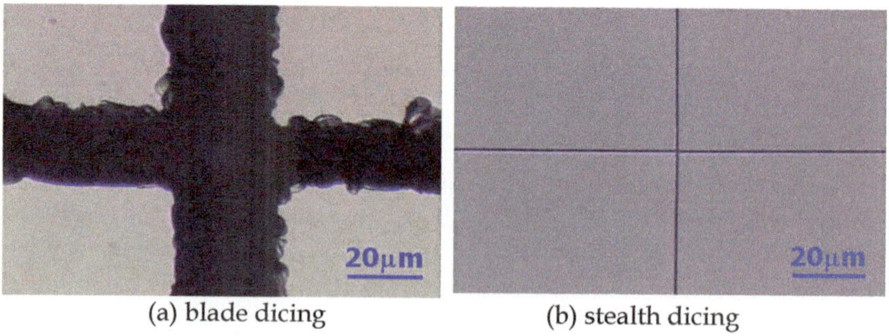

Fig. 4. Comparison of edge quality between blade dicing and SD (Photo: Hamamatsu Photonics K.K.)

In this chapter, heat conduction analysis by considering the temperature dependence of the absorption coefficient is performed for the SD method, and the validity of the analytical result is confirmed by experiment.

Fig. 5. SD application to actual MEMS device (Photo: Hamamatsu Photonics K.K.)

2. Analysis method

A 1,064 nm laser is considered here, and the internal temperature rise of Si by single pulse irradiation is analyzed (Ohmura et al., 2006). Considering that a laser beam is axisymmetric, we introduce the cylindrical coordinate system $O-rz$ whose z-axis corresponds to the optical axis of laser beam and r-axis is taken on the surface of Si. The heat conduction equation which should be solved is

$$\rho C_{\mathrm{p}} \frac{\partial T}{\partial t} = \frac{1}{r}\frac{\partial}{\partial r}\left(rK\frac{\partial T}{\partial r} \right) + \frac{\partial}{\partial z}\left(K\frac{\partial T}{\partial z} \right) + w \tag{1}$$

where T is temperature, ρ is density, C_{p} is isopiestic specific heat, K is thermal conductivity, and w is internal heat generation per unit time and unit volume. The finite difference method based on the alternating direction implicit (ADI) method was used for numerical calculation of Eq. (1). The temperature dependence of isopiestic specific heat (Japan Society for Mechanical Engineers ed., 1986) and thermal conductivity (Touloukian et al. ed., 1970) is considered.

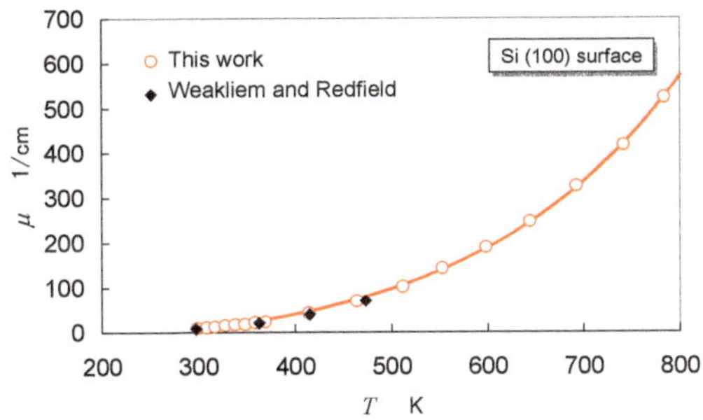

Fig. 6. Temperature dependence of absorption coefficient of silicon single crystal for 1,064 nm

Figure 6 (Fukuyo et al., 2007; Weakliem & Redfield, 1979) shows temperature dependence of the absorption coefficient of single crystal silicon for a wavelength 1,064 nm. The absorption coefficient $\mu(T_{i,j})$ in a lattice (i, j) whose temperature is $T_{i,j}$ is expressed by $\mu_{i,j}$.

When the Lambert law is applied between a small depth Δz from depth $z = z_{j-1}$ to $z = z_j$, the laser intensity $I'_{i,j}$ at the depth $z = z_j$ is expressed by

$$I'_{i,j} = I_{i,j} e^{-\mu_{i,j}\Delta z}, \quad i = 1, 2, \cdots, i_{\max}, \; j = 1, 2, \cdots, j_{\max} \tag{2}$$

where $I_{i,j}$ is the laser intensity at the depth $z = z_{j-1}$. The measurement values of Fig. 6 are approximated by

$$\mu = 12.991\exp(0.0048244T) - 52.588\exp(-0.0002262T) \;\left[\mathrm{cm}^{-1}\right] \tag{3}$$

The absorption coefficient of molten silicon is 7.61×10^5 cm^{-1} (Jellison, 1987). Therefore, this value is used for the upper limit of applying Eq. (3).

The $1/e^2$ radius at the depth z of a laser beam which is focused with a lens is expressed by $r_e(z)$. In propagation of light waves from the depth $z = z_{j-1}$ to $z = z_j$, focusing or divergence of a beam can be evaluated by a parameter

$$\gamma_j = \frac{r_e(z_j)}{r_e(z_{j-1})}, \quad j = 1, 2, \cdots, j_{\max} \tag{4}$$

The beam is focused when γ_j is less than 1, and is diverged when γ_j is larger than 1. Now, the laser intensity $I_{i,j}$ at the depth $z = z_{j-1}$ of a finite difference grid (i, j) can be expressed by the energy conservation as follows:

1. For $\gamma_{j-1} < 1$

$$I_{i,j} = \frac{\left(\gamma_{j-1}^2 r_i^2 - r_{i-1}^2\right) I'_{i,j-1} + \left(1 - \gamma_{j-1}^2\right) r_i^2 I'_{i+1,j-1}}{\gamma_{j-1}^2 \left(r_i^2 - r_{i-1}^2\right)}, \quad i = 1, 2, \cdots, i_{\max} \tag{5}$$

$$I_{0,j} = I'_{0,j-1} + \left(\frac{1}{\gamma_{j-1}^2} - 1\right) I'_{1,j-1} \tag{6}$$

2. For $\gamma_{j-1} > 1$

$$I_{i,j} = \frac{\left(\gamma_{j-1}^2 - 1\right) r_{i-1}^2 I'_{i-1,j-1} + \left(r_i^2 - \gamma_{j-1}^2 r_{i-1}^2\right) I'_{i,j-1}}{\gamma_{j-1}^2 \left(r_i^2 - r_{i-1}^2\right)}, \quad i = 1, 2, \cdots, i_{\max} \tag{7}$$

$$I_{0,j} = \frac{I'_{0,j-1}}{\gamma_{j-1}^2} \tag{8}$$

Considering Eq. (2), the internal heat generation per unit time and unit volume in the grid (i, j) is given by

$$w_{i,j} = \frac{\left(1 - e^{-\mu_{i,j}\Delta z}\right) I_{i,j}}{\Delta z} \tag{9}$$

In addition, the calculation of the total power at the depth $z = z_{j-1}$ by Eqs. (5) to (8) yields

$$\pi r_0^2 I_{0,j+1} + \sum_{i=1}^{\infty} \pi\left(r_i^2 - r_{i-1}^2\right) I_{i,j+1} = \pi r_0^2 I'_{0,j} + \sum_{i=1}^{\infty} \pi\left(r_i^2 - r_{i-1}^2\right) I'_{i,j} \tag{10}$$

and it can be confirmed that energy is conserved in the both cases of $\gamma_{j-1} < 1$ and $\gamma_{j-1} > 1$.

3. Analysis results and discussions

3.1 The formation mechanism of the inside modified layer

Concrete analyses are conducted under the irradiation conditions that the pulse energy, $E_{\mathrm{p}0}$, is 6.5 μJ, the pulse width (FWHM), τ_{p}, is 150 ns and the minimum spot radius, r_0, is 485 nm. The pulse shape is Gaussian. The pulse center is assumed to occur at $t = 0$. The intensity distribution (spatial distribution) of the beam is assumed to be Gaussian. It is supposed that the thickness of single crystal silicon is 100 μm and the depth of focal plane z_0 is 60 μm. The initial temperature is 293 K.

The analysis region of silicon is a disk such that the radius is 100 μm and the thickness is 100 μm. In the numerical calculation, the inside radius of 20 μm is divided into 400 units at a width 50 nm evenly, and its outside region is divided into 342 units using a logarithmic grid. The thickness is divided into 10,000 units at 10 nm increments evenly in the depth direction. The time step is 20 ps. The boundary condition is assumed to be a thermal radiation boundary.

For comparison with the following analysis results, the temperature dependence of the absorption coefficient is ignored at first, and a value of $\mu = 8.1$ cm^{-1} at room temperature is used. In this case, the time variation of the intensity distribution inside the silicon is given by

$$I(r, z, t) = \sqrt{\frac{\ln 2}{\pi}} \frac{4E_{\mathrm{p}}}{\pi r_e^2(z)\tau_{\mathrm{p}}} \exp\left[-4\ln 2\frac{t^2}{t_{\mathrm{p}}^2} - \frac{2r^2}{r_e^2(z)} - \mu z\right] \tag{11}$$

where E_{p} is an effective pulse energy penetrating silicon and $r_e(z)$ is the spot radius of the Gaussian beam at depth z.

The time variation of temperature at various depths along the central axis is shown in Fig. 7. The maximum temperature distribution is shown in Fig. 8. It is understood from Fig. 7 that the temperature becomes the maximum at time 20 ns at depth of 60 μm which corresponds to the focal position. In Fig. 8, due to reflecting laser absorption, the temperature of the side that is shallower than the focal point of the laser beam is slightly higher. However, the maximum temperature distribution becomes approximately symmetric with respect to the focal plane. At any rate the maximum temperature rise is about 360 K, which is much smaller than the melting point of 1,690 K under atmospheric pressure (Parker, 2004). It is concluded that polycrystallization after melting and solidification does not occur at all, if the absorption coefficient is independent of the temperature and is the value at the room temperature.

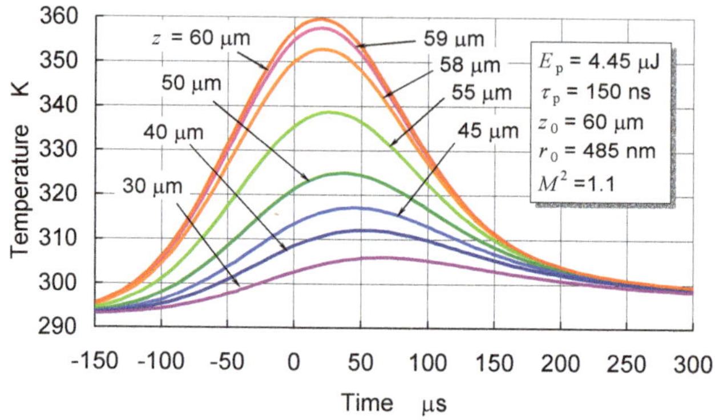

Fig. 7. Time variation of temperature at various depths along the central axis when temperature dependence of absorption coefficient is ignored

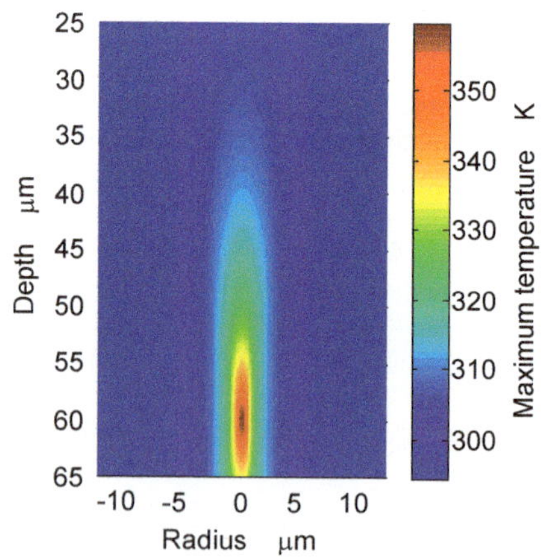

Fig. 8. Maximum temperature distribution when temperature dependence of absorption coefficient is ignored

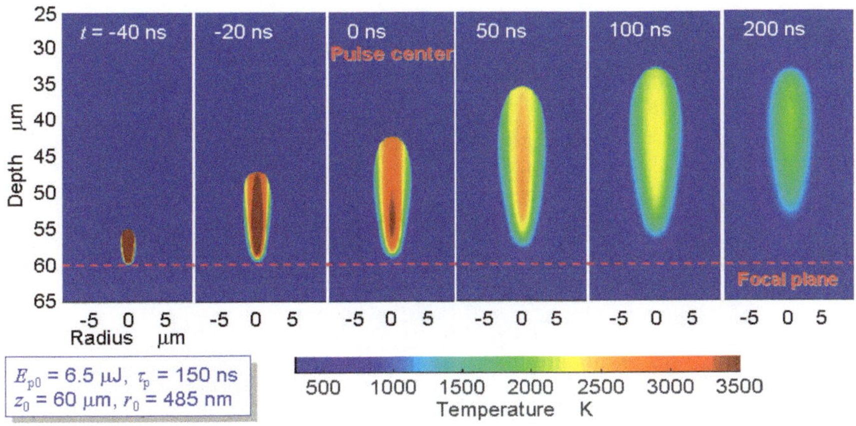

Fig. 9. Time variation of temperature distribution obtained by heat conduction analysis considering the temperature dependence of the absorption coefficient

When the temperature dependence of absorption coefficient (Eq. (3)) is taken into account, the time variation of temperature distribution is shown in Fig. 9. Figure 10 shows the time variation of the temperature distribution along the central axis in Fig. 9. It can be understood from these figures that laser absorption begins suddenly at a depth of $z = 59$ μm at about $t = -45$ ns and the temperature rises to about 20,000 K instantaneously. The region where the temperature rises beyond 10,000 K will be instantaneously vaporized and a void is formed. High temperature region of about 2,000 K propagates in the direction of the laser irradiation from the vicinity of the focal point as a thermal shock wave. The region where the thermal shock wave propagates becomes a high dislocation density layer due to the shear stress caused by the very large compressive stress.

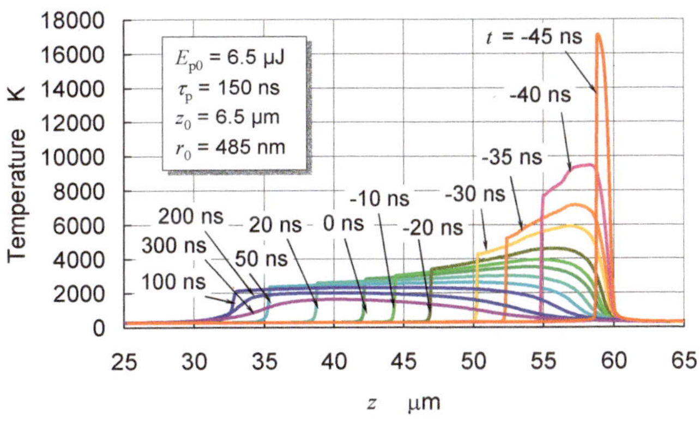

Fig. 10. Time variation of temperature distribution along the central axis

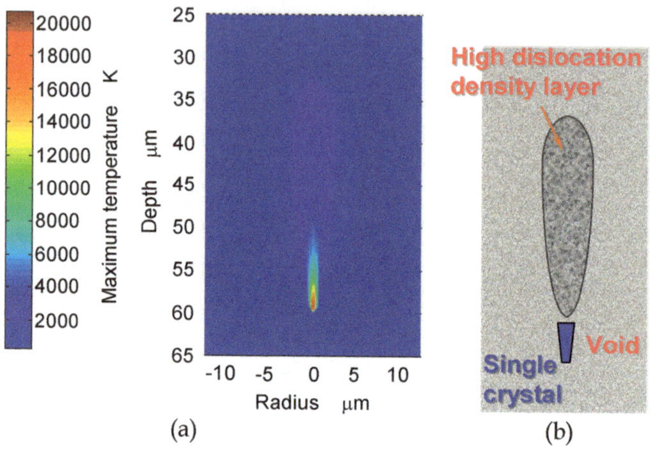

Fig. 11. The maximum temperature distribution (a) and a schematic of SD layer formation (b)

Figure 11 shows the maximum temperature distribution and a schematic of SD layer formation. SD layer looks like an exclamation mark "!". As a result, a train of the high dislocation density layer and void is generated as a belt in the laser scanning direction as shown schematically in Fig. 1. When the thermal shock wave caused by the next laser pulse propagates through part of the high dislocation density layer produced by previous laser pulse, a crack whose initiation is a dislocation progresses. Figure 12 shows a schematic of crack generation by the thermal shock wave. Analyses of internal crack propagation in SD were conducted later using stress intensity factor (Ohmura et al., 2009, 2011).

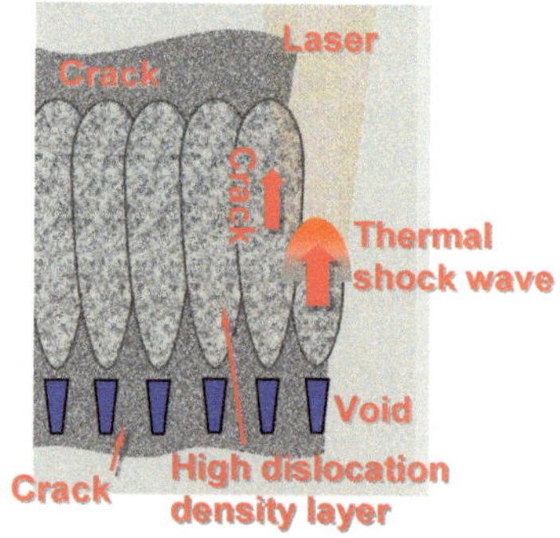

Fig. 12. Schematic of crack generation

Figure 13 shows an inside modified-layer observed by a confocal scanning infrared laser microscope OLYMPUS OLS3000-IR before division (Ohmura et al., 2009). It is confirmed that a train of the high dislocation density layer and void is generated as a belt as estimated in the previous studies. It also can be understood that the internal cracks have been already generated before division.

Fig. 13. Confocal scanning IR laser microscopy image before division

3.2 Stealth Dicing of ultra thin silicon wafer

Here heat conduction analysis is performed for the SD method when applied to a silicon wafer of 50 μm thick, and the difference in the processing result depending on the depth of focus is investigated (Ohmura et al., 2007, 2008). Furthermore, the validity of the analytical result is confirmed by experiment. In the analysis, the pulse energy, E_{p0}, is 4 μJ, the pulse width, τ_p, is 150 ns, and the pulse shape is Gaussian. The intensity distribution of the beam is assumed to be Gaussian. It is supposed that the depth of focal plane z_0 is 30 μm, 15 μm and 0 μm. The initial temperature is 293 K.

The analysis region of silicon is a disk such that the radius is 111 μm and the thickness is 50 μm. In the numerical calculation, the inside radius of 11 μm is divided into 440 units at a width 25 nm evenly, and its outside region is divided into 622 units using a logarithmic grid. The thickness is divided into 10,000 units at 5 nm increments evenly in the depth direction. The time step is 10 ps. The boundary condition is assumed to be a thermal radiation boundary.

3.2.1 In the case of focal plane depth 30 μm

The time variation the temperature distribution along the central axis is shown in Fig.14. Figure 14(b) shows the temperature change on a two-dimensional plane of depth and time by contour lines.

It can be understood from Fig. 14(a) that laser absorption begins suddenly at a depth of $z = 29$ μm at about $t = -8$ ns and the temperature rises to about 12,000 K instantaneously. The region where the temperature rises beyond 8,000 K will be instantaneously vaporized and a void is formed. The high temperature area beyond 2,000 K then expands rapidly in the surface direction until $t = 100$ ns as shown in Fig. 14(b). The contour at the leading edge of this high temperature area is clear in this figure. Also the temperature gradient is steep as shown in Fig. 14(a). Therefore, this high-temperature area is named a thermal shock wave as well. It is calculated that the thermal shock wave travels at a mean speed of about 300 m/s.

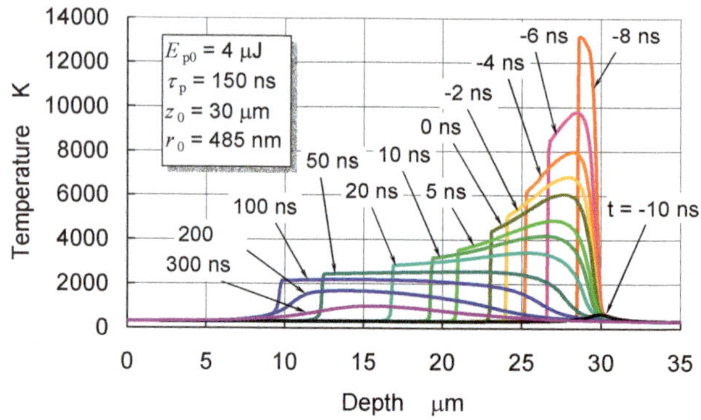

(a)

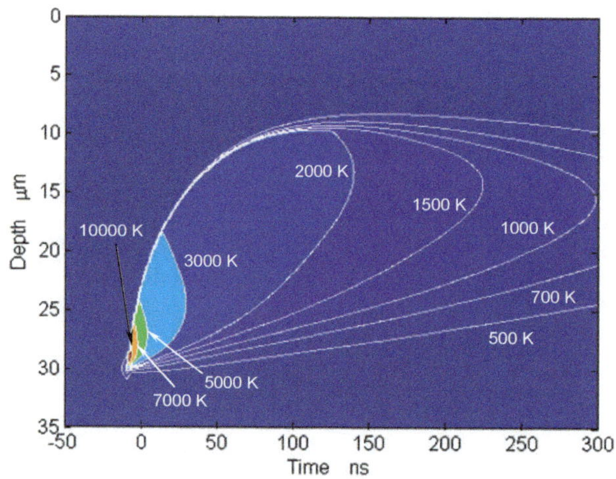

(b)

Fig. 14. Time variation of temperature distribution along the central axis ($z_0 = 30\,\mu\text{m}$)

Propagation of the thermal shock wave is shown in Fig. 15 by a time variation of the two-dimensional temperature distribution. The contour of the high-temperature area is comparatively clear until $t = 50$ ns, because the traveling speed of the thermal shock wave is much higher than the velocity of thermal diffusion. The contour of the high temperature area becomes gradually vague at $t = 100$ ns when the thermal shock wave propagation is finished. Because the temperature history is similar to the case of thickness 100 μm, the inside modified layer such as Fig. 3 is expected to be generated.

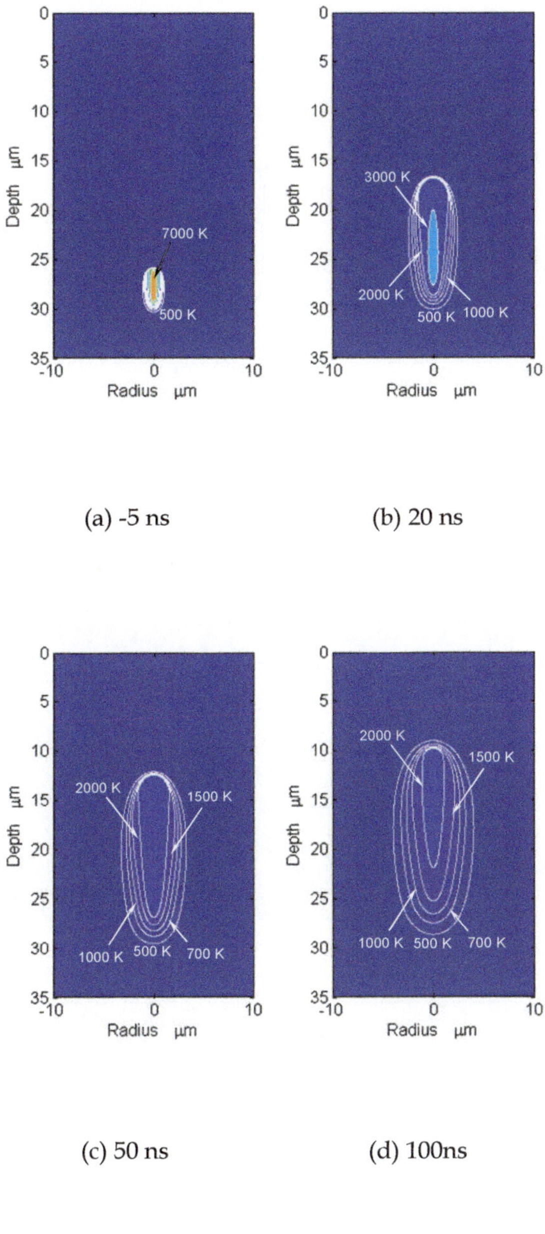

Fig. 15. Time variation of temperature distribution ($z_0 = 30\,\mu m$)

3.2.2 In the case of focal plane depth 15 μm

The time variation of the temperature distribution along the central axis in case of focal plane depth 15 μm is shown in Fig. 16.

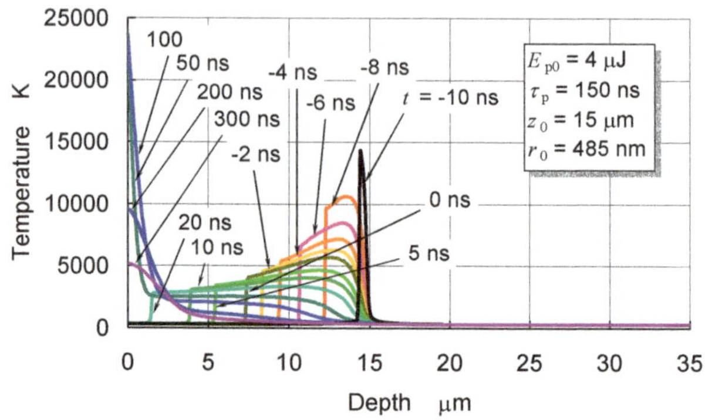

(a)

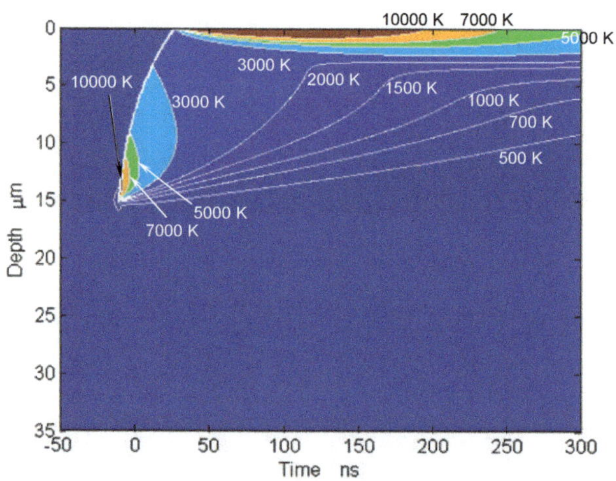

(b)

Fig. 16. Time variation of temperature distribution along the central axis ($z_0 = 15$ μm)

It can be understood from Fig. 16(a) that laser absorption begins suddenly at a depth of $z = 14$ μm at about $t = -10$ ns and the temperature rises to about 12,000 K instantaneously. As well as the case of focal plane depth 30 μm, the region where the temperature rises beyond 8,000 K will be instantaneously vaporized and a void is formed. Then the thermal shock wave propagates in the surface direction until about 25 ns.

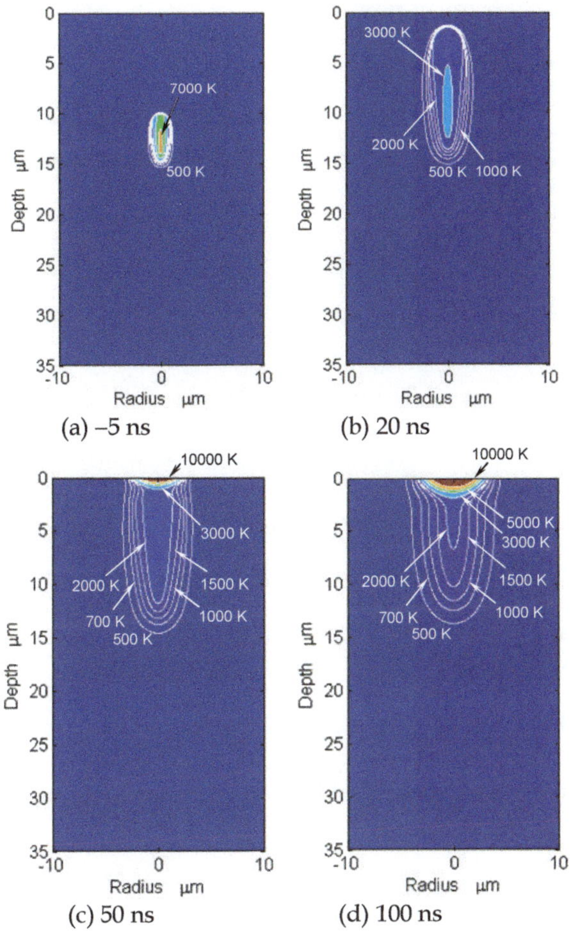

Fig. 17. Time variation of temperature distribution ($z_0 = 15\,\mu m$)

It is understood from Fig. 16(b) that laser absorption suddenly begins at the surface, once the thermal shock wave reaches the surface. Though the laser power already passes the peak, and gradually decreases, the surface temperature rises beyond 20000 K, which is higher than the maximum temperature which is reached at the inside. Although the thermal diffusion velocity is fairly slower than the thermal shock wave velocity, the internal heat is diffused to the surrounding. However, because the heat in the neighborhood of the surface is diffused only in the inside of the lower half, the surface temperature becomes very high and is maintained comparatively for a long time. Ablation occurs of course in such a high-temperature state. As a result, it is expected that not only is an inside modified layer generated, but also the surface is removed by ablation. Figure 17 shows that the surface temperature rises suddenly after the thermal shock wave propagates in the inside of the silicon, and reaches the surface, by the time variation of two dimensional temperature distribution.

3.2.3 In the case of focal plane depth 0 μm

When the laser is focused at the surface, as shown in Fig. 18, laser absorption begins suddenly at the surface at $t = -35$ ns, and the maximum surface temperature in the calculation reaches 6×10^5 K. It is estimated that violent ablation occurs when such an ultra-high temperature is reached. Because of the pollution of the device area by the scattering of the debris and thermal effect, the ablation at the surface is quite unfavorable.

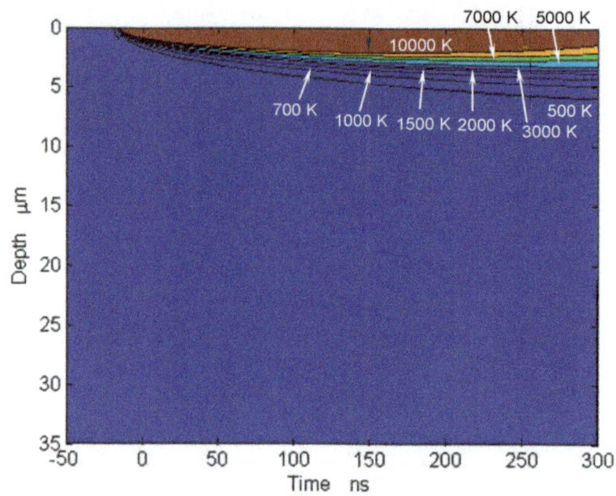

Fig. 18. Time variation of temperature distribution along the central axis ($z_0 = 0$ μm)

3.2.4 Comparison of the maximum temperature distributions and the experimental results

The maximum temperature distributions at the focal plane depths of 30 μm, 15 μm and 0 μm are shown in Fig. 19 in order to compare the previous analysis results at a glance.

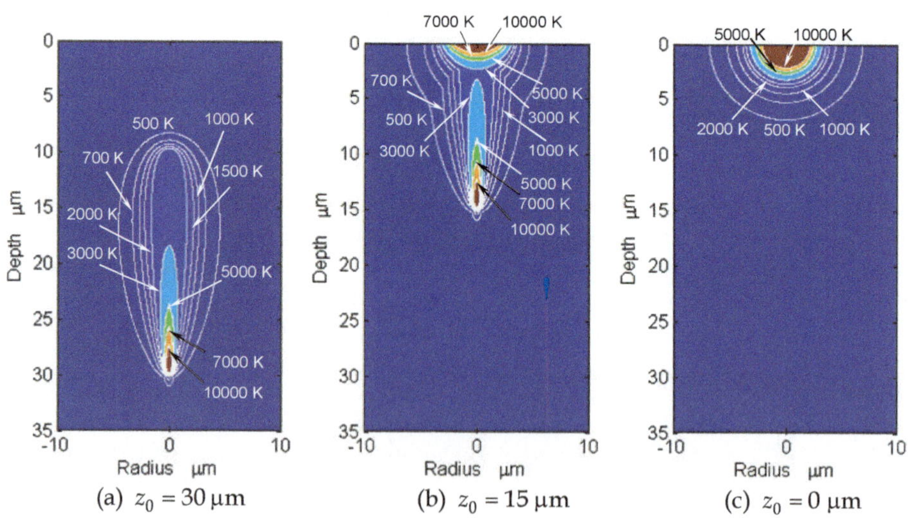

Fig. 19. Comparison of the maximum temperature distribution

Because high-temperature area stays in the inside of the wafer when z_0 is 30 μm, it was estimated that the inside modified layer as shown in Fig. 3 will be generated. In the case of $z_0 = 15$ μm, it was estimated that the surface is ablated although the modified layer is generated inside. In the case of $z_0 = 0$ μm, it was estimated that the surface was ablated intensely. It is concluded from the above analysis results that the laser irradiation condition for SD processing should be selected at a suitable focal plane depth so that the thermal shock wave does not reach the surface.

In order to verify the validity of the estimated results, laser processing experiments were conducted under the same irradiation condition as the analysis condition. The repetition rate in the experiments was 80 kHz. The results are shown in Fig. 20. Optical microscope photographs of the top views of the laser-irradiated surfaces and the divided faces are shown in the middle row and the bottom row, respectively. Figures 20 (a), (b) and (c) are results in the case of $z_0 = 30$ μm, $z_0 = 15$ μm, $z_0 = 0$ μm, respectively.

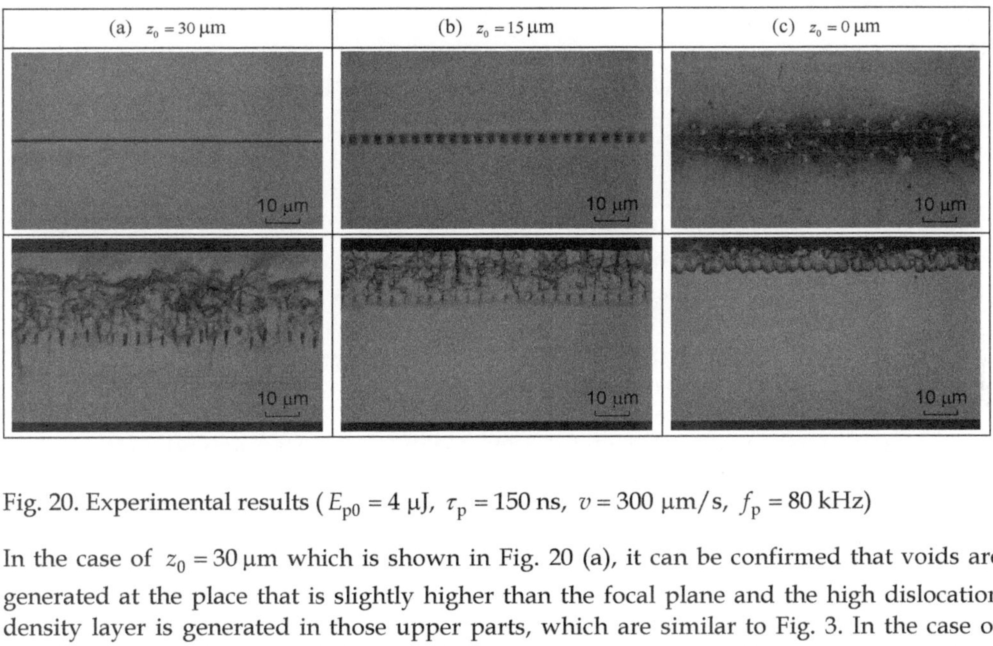

Fig. 20. Experimental results ($E_{p0} = 4$ μJ, $\tau_p = 150$ ns, $v = 300$ μm/s, $f_p = 80$ kHz)

In the case of $z_0 = 30$ μm which is shown in Fig. 20 (a), it can be confirmed that voids are generated at the place that is slightly higher than the focal plane and the high dislocation density layer is generated in those upper parts, which are similar to Fig. 3. In the case of $z_0 = 15$ μm which is shown in Fig. 20 (b), it is recognized that voids are generated at the place that is slightly higher than the focal plane and the high dislocation density layer is generated in those upper parts. However, it is observed that the surface is ablated and holes are opened from the photograph of the laser irradiated surface. In the case of $z_0 = 0$ μm which is shown in Fig. 20 (c), it is seen that strong ablation occurs and debris is scattered to the surroundings. Voids and the high dislocation density layer are not recognized in the divided face. Only the cross section of the hole caused by ablation is seen. These experimental results agree fairly well with the estimation based on the previous analysis result. Therefore, the validity of the analytical model, the analysis method, and the analysis results of this study are proven. The processing results can be estimated to some extent by using the analysis model and the analysis method in the present study. It is useful in optimization of the laser irradiation condition.

4. Conclusion

In the stealth dicing (SD) method, the laser beam that is permeable for silicon is absorbed locally in the vicinity of the focal point, and an interior modified layer (SD layer), which consists of voids and high dislocation density layer, is formed. In this chapter, it was clarified by our first analysis that the above formation was caused by the temperature dependence of the absorption coefficient and the propagation of a thermal shock wave. Then, the SD processing results of an ultra thin wafer of 50 μm in thickness were estimated based on this analytical model and analysis method. Particularly we paid attention to the difference in the results depending on the focal plane depth. Furthermore, in order to compare with the analysis results, laser processing experiments were conducted with the same irradiation condition as the analysis conditions.

In the case of focal plane depth $z_0 = 30\,\mu\text{m}$, the analysis result of temperature history was similar to the case when the wafer thickness is 100 μm and the focal plane depth is 60 μm. Therefore, it was predicted that a similar inside modified layer will be generated. In the case of $z_0 = 15\,\mu\text{m}$, it was estimated that not only the inside modified layer is generated, but also the surface is ablated. Because the thermal shock wave reached the surface, remarkable laser absorption occurred at the surface. In the case of $z_0 = 0\,\mu\text{m}$, it was estimated that the surface is ablated intensely. These estimation results agreed well with experimental results. Therefore, the validity of the analytical model, the analysis method and the analysis results of this study was proven.

As conclusion of this chapter, the following points became clear:

1. When the analytical model and the analysis method of the present study are used, the processing mechanism can be understood well, and the processing results can be estimated to some extent. It is useful in optimization of the laser irradiation condition.
2. There is a suitable focal plane depth in the SD processing, and it is necessary to select the laser irradiation condition so that the thermal shock wave does not reach the surface.

5. References

Fukumitsu, K., Kumagai, M., Ohmura, E., Morita, H., Atsumi, K., Uchiyama, N. (2006). The Mechanism of Semi-Conductor Wafer Dicing by Stealth Dicing Technology, *Online Proceedins of 4th International Congress on Laser Advanced Materials Processing (LAMP2006)*, Kyoto, Japan, May 16-19, 2006

Fukuyo, F., Fukumitsu, K., Uchiyama, N. (2005). The Stealth Dicing Technologies and Their Application, *Proceedings of 6th Internaitonal Symposium onLaser Precision Micro-Fabrication (LPM2005)*, Williamsburg, USA, April 4-7, 2005

Fukuyo, F., Ohmura, E., Fukumitsu, K., Morita, H. (2007). Measurement of Temperature Dependence of Absorption Coefficient of Single Crystal Silicon, *Journal of Japan Laser Processing Society*, Vol.14, No.1 (January 2007), pp. 24-29, ISSN 1881-6797 (in Japanese)

Japan Society for Mechanical Engineers (Eds.) (1986). *JSME Data Book: Heat Transfer, 4th ed.*, Japan Society for Mechanical Engineers, ISBN 978-4-88898-041-8, Tokyo, Japan (in Japanese)

Jellison, Jr. G.E. (1987). Measurements of the Optical Properties of Liquid Silicon and Germanium Using Nanosecond Time-Eesolved Ellipsometry, *Applied Physics Letters*, Vol.51, No.5 (August 1987), pp. 352-354, ISSN 0003-6951

Kumagai, M., Uchiyama, N., Ohmura, E., Sugiura, R., Atsumi, K., Fukumitsu, K. (2007). Advanced Dicing Technology for Semiconductor Wafer —Stealth Dicing—, *IEEE Transactions on Semiconductor Manufacturing*, Vol.20, No.3 (August 2007) pp. 259-265, ISSN 0894-6507

Ohmura, E., Fukumitsu, K., Uchiyama, N., Atsumi, K., Kumagai, M., Morita, H. (2006). Analysis of Modified Layer Formation into Silicon Wafer by Permeable Nanosecond Laser, *Proceedings of the 25th International Congress on Application of Laser and Electro-Optics (ICALEO2006)*, pp. 24-31, ISBN #0-912035-85-4, Scottsdale, USA, October 30-November 2, 2006

Ohmura, E., Fukuyo, F., Fukumitsu, K., Morita, H. (2006). Internal Modified-Layer Formation Mechanism into Silicon with Nanosecond Laser, *Journal of Achievements in Materials and Manufacturing Engineering*, Vol.17, No.1/2 (July 2006), pp. 381-384, ISSN 1734-8412

Ohmura, E., Kawahito, Y., Fukumitsu, K., Okuma, J., Morita, H. (2011). Analysis of Internal Crack Propagation in Silicon Due to Permeable Laser Irradiation —Study on Processing Mechanism of Stealth Dicing, *Journal of Materials Science and Engineering, A*, Vol.1, No.1, (June 2011), pp. 46-52, ISSN 2161-6213

Ohmura, E., Kumagai, M., Nakano, M., Kuno, K., Fukumitsu, K., Morita, H. (2007). Analysis of Processing Mechanism in Stealth Dicing of Ultra Thin Silicon Wafer, *Proceedings of the International Conference on Leading Edge Manufacturing in 21st Century (LEM21)*, pp. 861-866, Fukuoka Japan, November 7-9, 2007

Ohmura, E., Kumagai, M., Nakano, M., Kuno, K., Fukumitsu, K., Morita, H. (2008). Analysis of Processing Mechanism in Stealth Dicing of Ultra Thin Silicon Wafer, *Journal of Advanced Mechanical Design, Systems, and Manufacturing*, Vol.2, No.4 (March 2008) pp. 540-549, ISSN 1881-3054

Ohmura, E., Ogawa, K., Kumagai, M., Nakano, M., Fukumitsu, K., Morita, H. (2009). Analysis of Crack Propagation in Stealth Dicing Using Stress Intensity Factor, *On-line Proceedings of the 5th International Congress on Laser Advanced Materials Processing (LAMP2009)*, Kobe, Japan, June 29-July 2, 2009

Parker, S.P. et al. (Eds.) (2004). *Dictionary of Physics, 2nd ed.*, McGraw-Hill, ISBN 0-07-052429-7, New York, USA

Touloukian, Y.S., Powell, R.W., Ho, C.Y., Klemens, P.G. (Eds.) (1970). *Thermal Conductivity: Metallic Elements and Alloys*, IFI/Plenum, ISBN 306-67021-6, New York, USA

Weakliem, H.A. & Redfield, D. (1979). Temperature Dependence of the Optical Properties of Silicon, *Journal of Applied Physics*, Vol.50, No.3 (March 1979), pp. 1491-1493, ISSN 0021-8979

5

Temperature Measurement of a Surface Exposed to a Plasma Flux Generated Outside the Electrode Gap

Nikolay Kazanskiy and Vsevolod Kolpakov
Image Processing Systems Institute, Russian Academy of Sciences,
S.P. Korolev Samara State Aerospace University (National Research University)
Russia

1. Introduction

Plasma processing in vacuum is widely applied in optical patterning, formation of micro- and nanostructures, deposition of films, etc. on the material surface (Orlikovskiy, 1999a; Soifer, 2002). Surface–plasma interaction raises the temperature of the material, causing the parameters of device features to deviate from desired values. To improve the accuracy of micro- and nanostructure fabrication, it is necessary to control the temperature at the site where a plasma flux is incident on the surface. However, such a control is difficult, since the electric field of the plasma affects measurements. Pyrometric (optical) control methods are inapplicable in the high-temperature range and also suffer from nonmonochromatic self-radiation of gas-discharge plasma excited species.

At the same time, in the plasma-chemical etching setups that have been used until recently, the plasma is generated by a gas discharge in the electrode gap (see, for example (Orlikovskiy, 1999b; Raizer, 1987)). Low-temperature plasma is produced in a gas discharge, such as glow discharge, high-frequency, microwave, and magnetron discharge (Kireyev & Danilin, 1983). The major disadvantages of the above-listed discharges are: etch velocity is decreased with increasing relative surface area (Doh Hyun-Ho et al., 1997; Kovalevsky et al., 2002); the gas discharge parameters and properties show dependence on the substrate's material and surface geometry (Woodworth et al., 1997; Hebner et al., 1999); contamination of the surface under processing with low-active or inactive plasma particles leads to changed etching parameters (Miyata Koji et al., 1996; Komine Kenji et al., 1996; McLane et al., 1997); the charged particle parameters are affected by the gas-discharge unit operation modes; process equipment tends to be too complex and bulky, and reactor designs are poorly compatible with each other in terms of process conditions; these factors hinder integration (Orlikovskiy, 1999b); plasma processes are power-consuming and use expensive gases; hence high cost of finished product.

This creates considerable problems when generating topologies of the integrated circuits and diffractive microreliefs, and optimizing the etch regimes for masking layer windows.

The above problems could be solved by using a plasma stream satisfying the following conditions: (i) The electrodes should be outside the plasma region. (ii) The charged and reactive plasma species should not strike the chamber sidewalls. (iii) The plasma stream should be uniform in transverse directions. It is also desired to reduce the complexity, dimensions, mass, cost, and power consumption of plasma sources. Furthermore, these should be compatible with any type of vacuum machine in industrial use. Published results suggest that the requirements may be met by high-voltage gas-discharge plasma sources (Kolpakov & V.A. Kolpakov, 1999; V.A. Kolpakov, 2002; Komov et al., 1984; Vagner et al., 1974).

In (Kazanskiy et al., 2004), a reactor (of plasma-chemical etching) was used for the first time; in this reactor, a low-temperature plasma is generated by a high-voltage gas discharge outside the electrode gap (Vagner et al., 1974). Generators of this type of plasma are effectively used in welding (Vagner et al., 1974), soldering of elements in semiconducting devices (Komov et al., 1984), purification of the surface of materials (Kolpakov et al., 1996), and enhancement of adhesion in thin metal films (V.A. Kolpakov, 2006).

This study is devoted to elaborate upon a technique for measuring the temperature of a surface based on the studies into mechanisms of interaction a surface and a plasma flux generated outside the electrode gap.

2. Experimental conditions

Experiments were performed in a reactor shown schematically in Fig. 1a. The high-voltage gas discharge is an anomalous modification of a glow discharge, which emerges when the electrodes are brought closer up to the Aston dark space; the anode must have a through hole in this case. Such a design leads to a considerable bending of electric field lines in this region (Fig. 1b) (Vagner et al., 1974). The electric field distribution exhibits an increase in the length of the rectilinear segment of the field line in the direction of the symmetry axis of the aperture in the anode. Near the edge of the aperture, the length of the rectilinear segment is smaller than the electron mean free path, and a high-voltage discharge is not initiated.

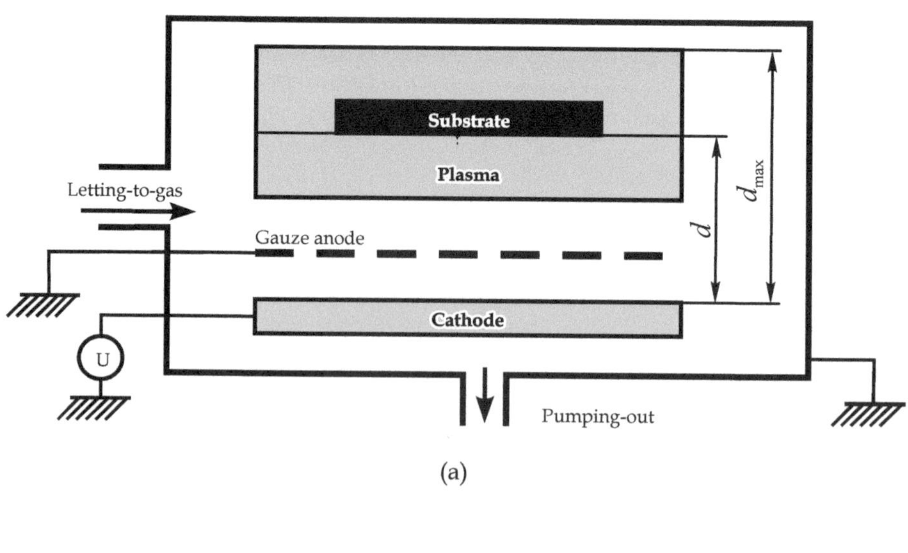

(a)

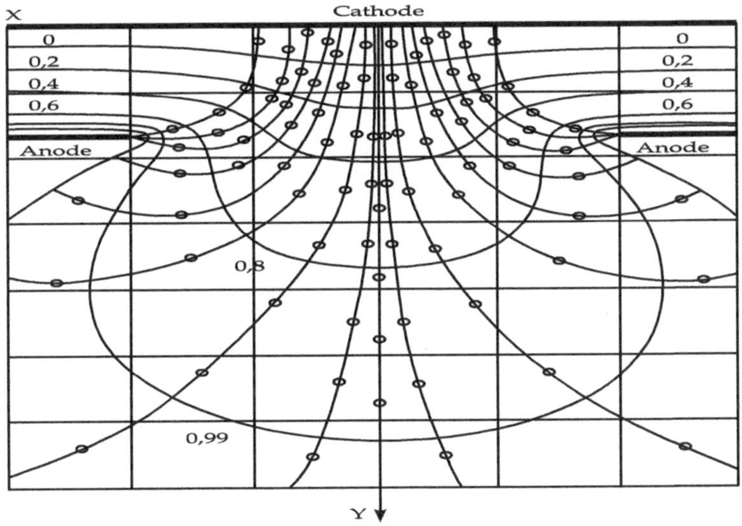

(b)

Fig. 1. (a) Schematic of the reactor and (b) field distribution in the near -electrode region of a gas-discharge tube; the mesh size is 0.0018 × 0.0018 m

The electrons emitted from the cathode under the action of the field gradient and moving along the rectilinear segments of field lines acquire an energy sufficient for ionizing the residue gas outside the electrode gap. The majority of positive ions is formed on the rectilinear segments of field lines in the axial zone in the anode aperture and reaches the cathode surface at the points of electron emission. This is confirmed by the geometrical parameters of the spots formed by positive ions on the cathode surface (see Fig. 2). The shape of the spots corresponds to the gauze mesh geometry, while their size is half the mesh size, which allows us to treat this size as the size of the axial region participating in self-sustaining of the charge.

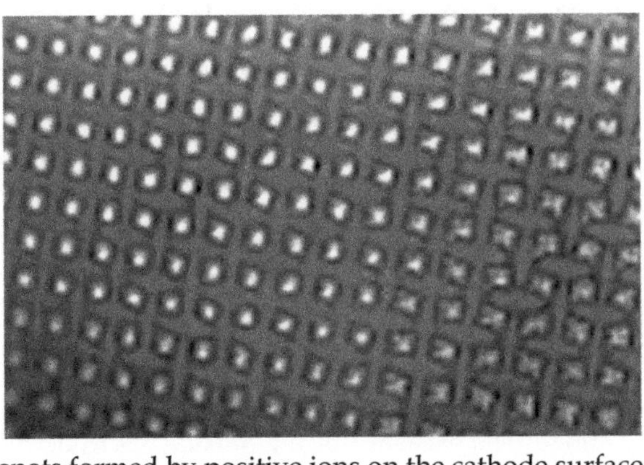

Fig. 2. The shape of spots formed by positive ions on the cathode surface; the spot size is 0.0009 × 0.0009 m

The plasma parameters were measured using collector (Molokovsky & Sushkov, 1991) and rotating probe (Rykalin et al., 1978) methods. To exclude sputtering, the probe was fabricated from a tungsten wire of diameter 0.1 mm, thus practically eliminating any impact on the plasma parameters.

To increase the electron emission, an aluminum cathode was used (Rykalin et al., 1978). To improve the energy distribution uniformity of plasma particles a stainless-steel-wire grid anode of a 1.8 x 1.8 mm cell and 0.5 mm diameter was used, which resulted in a significantly weaker chemical interaction with plasma particles and an increased resistance to thermal heating. This statement can be supported by the analysis of a gas-discharge device described in Ref. (Vagner et al., 1974), with each cell of the anode grid representing a hole and the entire flux of the charged particles being composed of identical micro-fluxes. The microflux parameters are determined by the cell size and the cathode surface properties, which are identical in the case under study and, so are the parameters of the individual microflux. As a result, the charged particle distribution over the flux cross-section will also be uniform, with the nonuniformity resulting only from the edge effect of the anode design, whose area is minimal. For the parameters under study, the uniformity of the charge particle distribution over the flux cross-section was not worse than 98% (Kolpakov & V.A. Kolpakov, 1999). The discharge current and the accelerating voltage were 0-140 mA and 0-6 kV. The process gases are CF_4, CF_4-O_2 mixture, O_2 and air. The sample substrates were made up of silicon dioxide of size 20x20 mm^2, with/without a photoresist mask in the form of a photolithograpically applied periodic grating, polymer layers of the DNQ based on diazoquinone and FP-383 metacresol novolac deposited on silicon dioxide plates with a diameter of up to 0.2 m (Moreau, 1988a). Before the formation of the polymer layer, the surface of the substrates was chemically cleaned and finished to 10^{-8} kg/m^2 (10^{-9} g/cm^2) in a plasma flow with a discharge current of I = 10 mA, accelerating voltage U = 2 kV, and a cleaning duration of 10 s (Kolpakov et al., 1996). The profile and depth of etched trenches were determined with the Nanoink Nscriptor Dip Pen Nanolithography System, Carl Zeiss Supra 25 Field emission Scanning Electron Microscopes and a "Smena" scanning-probe microscope operated in the atomic-force mode. Cathode deposit was analyzed with a x-ray diffractometer. Surface temperature was measured by a precision chromel–copel thermocouple.

3. Experimental results and discussion of the high-voltage gas discharge characteristics

The high-voltage gas discharge is an abnormal variety of the glow discharge and, therefore, while featuring all benefits of the latter, is devoid of its disadvantages, such as the correlation between the gas discharge parameters and the substrate's location and surface properties.

When the cathode and anode are being brought together to within Aston space, the glow discharge is interrupted because of fulfillment of the inequality $nG<1$, where n and G are the number of electrons and ions, respectively. However, if a through hole is arranged in the anode, in its region there is no more ban on the fulfillment of the inequality $nG\geq1$ (Vagner et al., 1974). Physically, this means that this inequality is valid when one or more electrons take part in generating one or several pairs of positive ions, thus providing conditions for a gas discharge outside the anode. The existence of the outside-electrode discharge suggests the conclusion that the discharge particles are in free motion (Vagner et al., 1974). This sharply reduces the impact of the discharge unit operation modes on the parameters of the particles, practically eliminating the loading effect and cathode protection from sputtering. Free motion of the particles and sharp boundaries of the discharge suggest that outside the anode the particles move straight and perpendicularly to its surface. Actually, Fig. 3 shows that the distribution of the charged particles across the plasma flow is uniform, with its motion toward the sample surface being perpendicular.

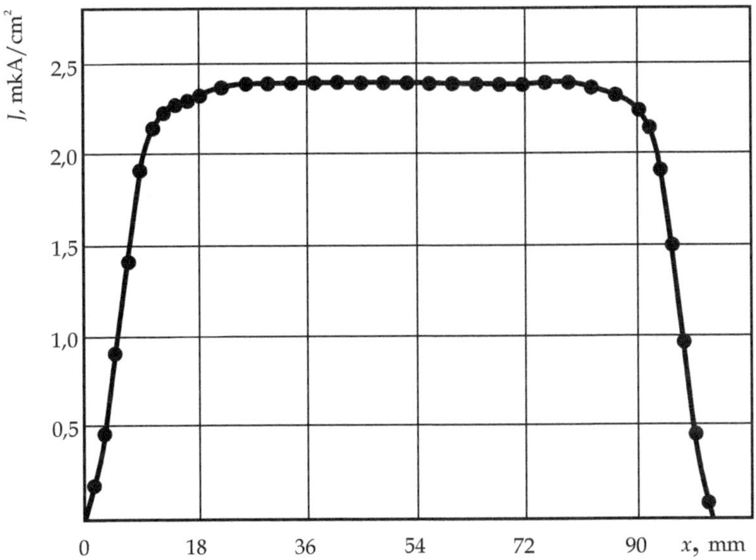

Fig. 3. Distribution of the charged particles across the plasma flux

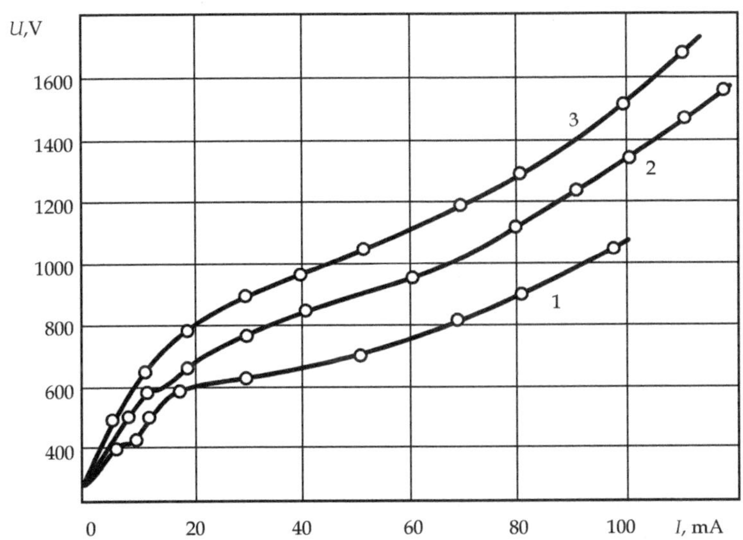

Fig. 4. The *V-I* curve of the high-voltage gas discharge at various pressures in the chamber: 1-1.5·10^{-1}torr; 2-1.2·10^{-1} torr; 3-9·10^{-2} torr.

Analysis of the *V-I* curve of the discharge (Fig. 4) shows that its formation is due to the ionization process of atoms of the working gas (a -process) and the cathode material (γ-process) (Chernetsky, 1969). It is noteworthy that in the range of voltages $300 \le U \le 1000$ V the working gas atoms ionization is predominant, whereas at $U \ge 1000$ V the intense cathode sputtering takes place, thus leading to the ion-electron emission responsible for the remaining section of the *V-I* curve.

However, in the region of relatively low pressure ($p \le 1.5 \cdot 10^{-1}$ torr), in the range $20 \le I \le 50$ mA, there is a pronounced *I-V* curve section where the *I*-dependence is weak. This suggests that for the above voltage range and high pressures, the electrons still manage to gain sufficient energies for the working gas atom ionization, thus actively contributing to the current increase even at a small voltage increase.

The assumption made is in good agreement with the plot shown in Fig. 5: the voltage saturation in the pressure range $1.8 \cdot 10^{-1}$ torr $\ge p \ge 9 \cdot 10^{-2}$ torr in the case of a clean (new) cathode proves that the working gas ionization capabilities have been exhausted, with sputtering and ionization of the cathode atoms (ion-electron emission) being responsible for the curve rise at $p < 9 \cdot 10^{-2}$ torr.

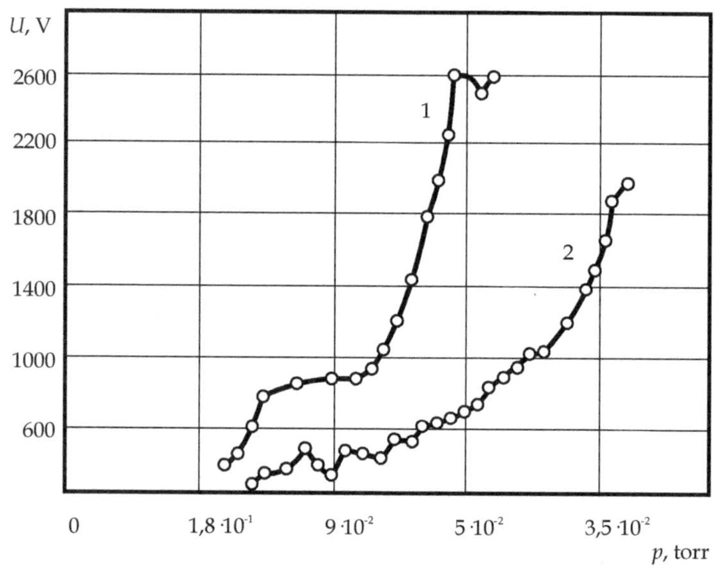

Fig. 5. The cathode voltage vs the chamber pressure: 1 - clean (new) cathode, 2 - contaminated cathode (after a long period of work)

To prove the above statements we will estimate the parameters of mechanisms that provide the gas discharge existence. It has been known that the ionization of the working gas atoms can result from the electron (a-process) and positive ion (β-process) action. The secondary electron emission can be caused by the ion bombardment (γ-process) and radiation-induced surface ionization (δ-process) (Chernetsky, 1969). Let us elucidate which of the above-listed processes are predominant in the emergence and maintenance of the high-voltage gas discharge.

The volume ionization coefficient that characterized the α-process is given by (Raizer, 1987)

$$a_i = \frac{1}{l_i} = \frac{E}{\varphi_i}, \tag{1}$$

where l_i is the ion range, cm, φ_i is the ionization potential, V, and E is the strength of the nonuniform electric field, V/cm, derived from the relation (Kolpakov & Rastegayev, 1979)

$$E(y) = \frac{4cU/\pi}{(1 + h/c\pi)4c^2 + y^2}, \tag{2}$$

where U is the cathode voltage, V, c is a constant derived from a set of equations (Kolpakov & Rastegayev, 1979), which equals c=0.08 cm for a 1.8 x 1.8 mm anode hole, and h is the cathode-to-anode distance, cm. To derive the strength of the electric field acting upon a charged particle at the first length of its free path λ, cm, we must replace y in (2) with the value of λ derived from

$$\lambda = \frac{4\sqrt{2}}{n_0 \sigma}, \tag{3}$$

where n_0 is the concentration of molecules of the hladon-14 gas, which equals $n_0 = 0.29 \cdot 10^{16}$ cm^{-3} for the pressure of $9 \cdot 10^{-2}$ torr and σ is the effective cross-section of the chladon-14 molecule. According to the calculation based on Eq. (3), we find λ = 1.3 cm. Substituting the known discharge ignition voltage of U=300 V, as well as the h=0.5 cm and c=0.08 cm, into Eq. (2) we obtain E=15.4 V/cm. Substituting the derived value of the electric field strength into Eq. (1) yields a_i =1 cm^{-1}, which corresponds to the condition for the outside-anode gas discharge ($nG \geq 1$). Also, the comparison of the values of λ and l_i at the above voltage has shown that $\lambda > l_i$, suggesting the ionization possibility of the remaining gas molecules (Chernetsky, 1969).

The efficiency of the positive-ion-induced ionization of the working gas molecules is small and, therefore, the β-process can be disregarded when studying the gas discharge (Raizer, 1987). Because the high-voltage discharge is independent, with no extra irradiation sources found in the discharge vacuum camera, the δ-process can also be disregarded. Hence, the positive ions are the major source of cathode-emitted secondary electrons. The contribution of the positive ions to the production of the secondary electrons is characterized by the secondary emission coefficient, which equals $\gamma = 7.16 \cdot 10^{-5}$ for U=300 V (Izmailov, 1939). Given the cathode voltage of 1000 V, the above-discussed calculation techniques give the following values of the coefficients (Izmailov, 1939): $a_i \approx 4{,}8$, $\gamma = 0{,}66$. From comparison of the two values, we can see that there is only a three-fold increase in the volume ionization of the working gas molecules, whereas the ionization due to ion-electron emission has increased by a factor of 10^4. Thus, for the cathode volume in the range $300 \leq U \leq 1000$ V the working gas ionization is mainly due to the volume ionization by electron impact. For $U \geq 1000$ V, the major ionization mechanism is ion-electron emission, which complies well with the plots shown in Figs. 2 and 3.

The violation of the exponential dependence in Fig. 3 in the range $p = 5.5 \cdot 10^{-2}$ - $4.8 \cdot 10^{-2}$ torr is due to emergence of unstable microarch discharges between the cathode and anode, seen with naked eye. The conditions for emergence of this type of parasite discharge in the above range of values and pressures become similar to those for the high-voltage discharge and, therefore, the two emerge practically simultaneously. With further increase of voltage, one of the discharges starts to prevail, with a breakdown of the dielectric inter-electrode space ensuing. Traces of three such breakdowns are shown in Fig. 6.

Fig. 6. Breakdown traces and general appearance of the cathode surface after a long period of work

The absence of saturation in the case of the contaminated cathode (Fig. 5, after a long period of work) suggests that there are structural changes on the cathode surface, as seen in Fig. 6. These appear in the course of operation under the action of plasma flow microrays, reproducing the contours of the anode holes. It has been known (Matare, 1974) that any disturbances of the crystalline lattice cause the interatomic bonds to be weakened. Such disturbances possess lower ionization potential due to ion bombardment compared with the core material, Thus as it would be expected, the potential of the high-voltage discharge ignition should be decreased, in accordance with the form of the curve in Fig. 5. In this case, the character of the curve is determined by the predominant emission of the cathode material, which begins at a lower pressure. Low pressure facilitates the elimination from the cathode surface of easily evaporated contamination particles, such as various atoms and molecules absorbed by the surface, leaving the ion-electron emission the only mechanism for maintaining the discharge.

Thus, for the cathode voltages in the range $3000 \leq U \leq 1000$ V the high-voltage discharge is mainly maintained with the α-process, whereas at $U \geq 1000$ V the discharge exists due to the γ-process.

4. Theoretical and experimental investigation of surface treatment mechanisms with the directed flows of the off-electrode plasma

In particular, (V.A. Kolpakov, 2002) has shown that high-voltage gas discharge is in principle suitable for plasma etching and reactive ion etching. At the same time, we are unaware of current reports in which the mechanism of surface treatment with the directed flows of the off-electrode plasma is explored in a practical context.

The aim of this part was to investigate surface treatment mechanisms with the directed flows of the off-electrode plasma. The process was applied to SiO_2 and also some other materials, widely used in micro-, nanoelectronics and diffractive optics.

4.1 Basic reactions in plasma etching and reactive ion etching by the off-electrode plasma

Kolpakov (V.A. Kolpakov, 2002) has shown that high-voltage gas discharge can provide plasma etching or reactive ion etching, depending on the applied voltage or the cathode–

wafer spacing. With plasma etching, the wafer is bombarded by normally incident ions. This feature enhances etching anisotropy and increases the etch rate, because the reactive species, such as atomic fluorine, are produced just on the wafer surface. The species are formed by interaction between negative ions and adsorbed neutral process-gas molecules.

Ion bombardment is the main source of reactive species in plasma etching. To show this, we examine plasma reactions in the case of CF_4. With radio-frequency or microwave discharge, reactive species, namely, F^* radicals, can be produced both in the bulk of the plasma and at the wafer surface by electron impact dissociation of neutral molecules (Flamm, 1979):

$$e^- + CF_4 \rightarrow CF_3^+ + F^* + 2e^-, \tag{4}$$

$$e^- + CF_4 \rightarrow CF_3^* + F^* + e^-, \tag{5}$$

$$e^- + CF_4 \rightarrow CF_3^* + F^-. \tag{6}$$

It appears reasonable to say that high-voltage gas discharge is an anomalous form of glow discharge. If the spacing between a solid anode and a cathode is reduced to the Aston dark space, the glow discharge will disappear, because $nG< 1$, where n and G are the respective densities of electrons (negative ions) and positive ions. If, however, an aperture is made in the anode, then we shall have $nG \geq 1$ near the aperture (Vagner et al., 1974). Gas discharge will thus arise at a certain distance from the anode. In high-voltage gas discharge, therefore, charged particles are strongly separated according to the sign of the charge: an as-produced negative ion or electron will move toward the wafer, while the corresponding positive ion will be heading toward the cathode. An interaction event may also yield two or more negatively charged particles (ions and/or electrons), but at the same time it must generate an appropriate number of positive ions in order to maintain charge equilibrium: $nG \geq 1$. If this condition is not fulfilled in a region, high-voltage gas discharge will cease to exist there. This occurs where the energy of negatively charged particles is too low to allow production of positive ions in collisions with process-gas molecules, as in regions outside the output stream of the plasma source (V.A. Kolpakov, 2002). In this respect, reaction (4) is the best, giving a ion. It has been emphasized that in the voltage range 0.5–2 kV electrons are lost mainly due to their capture by neutral atoms (V.A. Kolpakov, 2002). In particular, this is true of the plasma etching mode. The lifetime of reactive species is short at the voltages. The free radicals F^* decay as

$$F^* + e^- + e^- \rightarrow F^- + e^-. \tag{7}$$

Since high-voltage gas discharge produces a plasma stream, the particles rarely collide with the wall, so that wall recombination can be neglected when examining the plasma processes. Electron–ion recombination requires that, aside from an adequate density of free electrons, their energies be less than the ion ionization potential. As these conditions are not fulfilled in the plasma etching mode, charge neutralization is mainly by ion–ion recombination (Raizer, 1987). In addition to electron–ion recombination, we exclude electron-impact excitation and ionization of process gas molecules, because these effects can occur at a higher pressure (Chernyaev, 1987; Ivanovskii, 1986). Thus, the above considerations allow the following main reactions in the bulk of an high-voltage gas discharge plasma:

$$e^- + CF_4 \rightarrow CF_3^+ + F^- + e^- \tag{8}$$

$$F^- + CF_4 \rightarrow CF_3^+ + 2F^- \tag{9}$$

$$F^- + CF_3^+ \rightarrow CF_4 \,. \qquad 10)$$

Reaction (9) is possible because the energy E of F^- ions was found to exceed the ionization potential of CF_4 throughout their progress toward the wafer, as follows from the equation

$$E_n = E_{n-1}(1-\gamma) + \Delta U_n \,, \tag{11}$$

where ΔU_n is the accelerating potential difference after the corresponding collision and $\gamma = 4mM/(m+M)^2$, with m and M denoting the respective masses of an ion and a process-gas molecule (V.A. Kolpakov, 2002). We calculated that E should decrease from 400 eV just after a first collision to below 100 eV just before the collision with a CF_4 molecule adsorbed by the wafer. In the last collision a proportion of the ion energy (on the order of the ionization potential) is consumed by the ionization of the molecule, and the rest goes into the breakage or weakening of bonds between the atoms of SiO_2 molecules on the wafer surface. The collision produces free radicals by the equation

$$F^- + CF_4 + S_s \rightarrow CF_3^+ + 2F^* + 2e^- \,, \tag{12}$$

where S_s denotes a surface species. As-generated radicals react with SiO_2 to form volatile substances:

$$4F^* + SiO_2 \rightarrow SiF_4 \uparrow + O_2 \uparrow \,. \tag{13}$$

We see that every F^- ion generated in the bulk of the plasma creates a radical on the wafer surface, the reaction products being withdrawn from the work chamber. If a collision occurs between an F^- and a CF_3^+ ion such that the energy of the former is less than or equal to the ionization potential of the latter, the two ions recombine to produce a CF_4 molecule according to (10).

Thus, for high-voltage gas discharge (off-electrode) plasma etching, Eqs. (8)–(10), (12), and (13) imply the following advantages: (i) Reactive species are formed exactly on the wafer surface; therefore, they cannot decay by interaction with other plasma particles. (ii) F^- ions (due to ionization of CF_4) play the major part in the production of reactive species. (iii) The collision between an F^- ion and a process-gas molecule adsorbed on the SiO_2 surface yields two reactive species, the surface serving as a catalyst. (iiii) There is no carbon deposition on the wafer surface, because CF_3^+ ions are attracted by the cathode and so cannot produce $(C_xF_y)_n$ polymers on the surface (Fig. 1a).

In the reactive ion etching mode of treatment with CF_4 plasmas, the energy of F^- ions incident on the SiO_2 surface is so high (100–500 eV) as to strongly heat the surface. This impedes process-gas adsorption and hence virtually prevents reactive species from taking part in etching (Kireev et al., 1986; V.A. Kolpakov, 2002). Erosion is due to sputtering by F^-

ions and their reaction with the sputtered matter. Reactions in this case are similar to those in the plasma etching mode. Also note that the mechanism of reactive ion etching is extensively treated in the literature (Ivanovskii, 1986). We therefore shall not address reactive ion etching with pure CF_4 in the subsequent text.
Far more interesting is the high-voltage gas discharge etching in which CF_4 is mixed with O_2. Guided by the discussion above, we can reasonably expect that aside from reactions (8)–(10) the plasma will exhibit

$$e^- + O_2 \rightarrow O^+ + O^- + e^-, \tag{14}$$

$$e^- + OF \rightarrow O^+ + F^- + e^-, \tag{15}$$

$$F^- + O_2 \rightarrow O^+ + O^- + F^-, \tag{16}$$

$$F^- + OF \rightarrow O^+ + 2F^-, \tag{17}$$

$$O^- + CF_4 \rightarrow CF_3^+ + F^- + O^-, \tag{18}$$

$$O^+ + F^- \rightarrow OF, \tag{19}$$

$$CF_3^+ + O^- \rightarrow COF_2 \uparrow + F^* \tag{20}$$

Reactions (14)–(17) occur under ionization, while reactions (19) and (20) under recombination.
Furthermore, it has been noted that the volatile product, COF_2, decomposes to give free fluorine radicals (Gerlach-Meyer, 1981):

$$COF_2 \rightarrow CO \uparrow + 2F^* . \tag{21}$$

Finally, an F^* atom can capture an electron by Eq. (7) to become an F^- ion, and this in turn can take part in plasma etching, producing F^* according to Eq. (12). It appears reasonable to expect that O^- ions will undergo similar transformations, with the result that oxygen radicals will compete with F^* radicals for active sites on the SiO_2 surface. This factor is likely to reduce the rate of plasma etching at certain O_2 concentrations.

4.2 Results and discussion: etch rate in relation to oxygen percentage and other process parameters

To optimize the etch rate in CF_4–O_2 plasmas, it is important to know how it varies with oxygen percentage. Let us first consider the plasma etching mode of treatment. Figure 7a shows graphs of the dependence measured for different discharge currents. Notice that with increasing oxygen percentage the etch rate first rises and then falls to almost zero values. The graphs are similar in shape for all the discharge currents except the minimum one, 50 mA. For this current the insignificant variation in etch rate is attributable to a low density of charged particles in the plasma: with a low ionization rate of process-gas molecules by O^-

ions, these make a modest contribution to the production of F^- ions (see Eqs. (18), (20), and (21)). With pure CF_4, etching was not observed at the minimum discharge current.

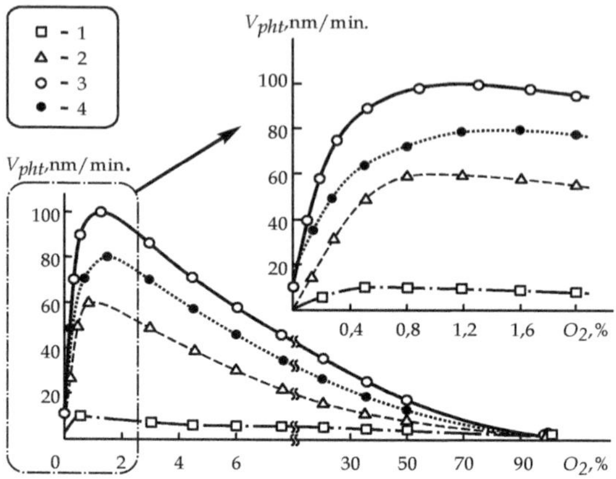

(a)

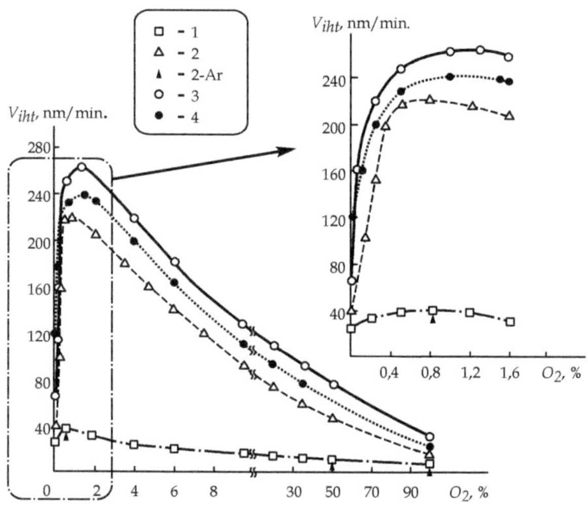

(b)

Fig. 7. Etch rate vs. oxygen percentage in (a) the plasma etching and (b) the reactive ion etching mode of treatment at discharge currents of (1) 50, (2) 80, (3) 120, and (4) 140 mA. The cathode voltage is (a) 0.8 or (b) 2 kV

The effect of discharge-current variation on the etch-rate pattern can be explained as follows. As the discharge current increases, so should do the density of charged particles in the plasma. This in turn should increase the ionization rate of CF_4 molecules by O^- ions and hence the density of F^- ions produced with the assistance of oxygen.

The steep, rising segments of curves in Fig. 7a should indicate deficiency in F^* radicals at the wafer surface, implying that etch rate is determined by the density of F^- ions. The pronounced peak, observed at each discharge current, should correspond to the situation in which all of the oxygen takes part in the production of F^- ions; at the same time, the oxygen does not compete with F^* radicals for active sites on the SiO_2 surface, nor does it passivate the surface. It is important to note that the etch rate peaks for an oxygen percentage as low as 0.5–1.5%. This finding must indicate high transverse uniformity of the plasma stream, its normal incidence on the wafer surface, and freedom from wall collisions. Also, every O^- ion produced in the bulk of the plasma by Eqs. (14) and (16) must be involved in the generation of an F^- ion, which in turn will create reactive species:

$$O_2 \xrightarrow{e, F^-} O^- \xrightarrow{CF_4} F^- \xrightarrow{S_S} F^* .$$

The falling segments of the etch-rate graphs should be due to occupation of vacant SiO_2 bonds by oxygen radicals, which thus compete with fluorine ones. Further, oxygen molecules excited at the SiO_2, surface should react with F^* radicals to convert them into F_2, a less reactive substance (Harsberger & Porter, 1979). The density of reactive species is thus reduced. When the plasma is generated in pure oxygen, the SiO_2 surface is fully passivated, so that the etch rate is close to zero; this conclusion is consistent with the established conception (Chernyaev, 1987; Ivanovskii, 1986; Kireyev & Danilin, 1983).

Let us now turn to the reactive ion etching mode of treatment. The corresponding etch-rate curves are shown in Fig. 7b. The etch rate also rises with oxygen percentage while the latter is not too high. However, such behavior in the reactive ion etching case is at variance with long-standing views (Horiike, 1983; Ivanovskii, 1986). To clarify the point, let us examine Fig. 7b. On the whole, the etch rate follows the same pattern as in the plasma etching case. This is obviously attributable to the fact that only neutral process-gas molecules and charged plasma particles are in the bulk of the plasma. Fluorocarbon and oxygen ions are unlikely to combine into stable molecules (CO, CO_2, and COF_2) on account of the above-mentioned separation of charged particles and the action of a strong, nonuniform electric field (Kolpakov & Rastegayev, 1979; V.A. Kolpakov, 2002). Consequently, high-energy O^- and F^- ions produced in the plasma stream (see Eqs. (14)–(18)) should not recombine as they travel toward the wafer. These ions will erode the material first by sputtering and then by chemical reactions. In the sputtering, highenergy ions penetrate a certain depth into the material and in doing so break interatomic bonds. Having lost energy, the ions can interact with the material only by chemical reactions. As with plasma etching, this stage of reactive ion etching is characterized by competition between reactive fluorine and oxygen species for active sites; however, these are now located in the bulk of SiO_2. This explains why the etch rate starts falling once the oxygen percentage has reached 1.5%. Also, the etch rate does not vanish, however high the oxygen percentage is, implying that pureoxygen etching occurs by sputtering with O^- ions. In fact, this mechanism starts acting at an oxygen percentage of 10%. It is manifested in characteristic dips in the etching profile (Orlikovskiy, 1999a), as shown in Fig. 8, which indicate that reevaporation rather than chemical erosion dominates the sputtering (Chernyaev, 1987).

Comparing Figs. 7a and 7b, we notice that the etch rate peaks for the same oxygen percentage. This fact is evidence that in plasma etching and reactive ion etching the same processes occur in the bulk of the plasma (or at least upstream of the wafer), thus supporting the mechanisms and equations proposed above. Otherwise, the etch rate would decrease at low oxygen percentages. The nonzero etch rate at zero oxygen percentage, observed even at a discharge current as low as 50 mA, signifies that the voltage between the electrodes is the major factor in the transport of reactive species to the wafer. The higher rate of change shown by the reactive ion etching curves should be due to sputtering.

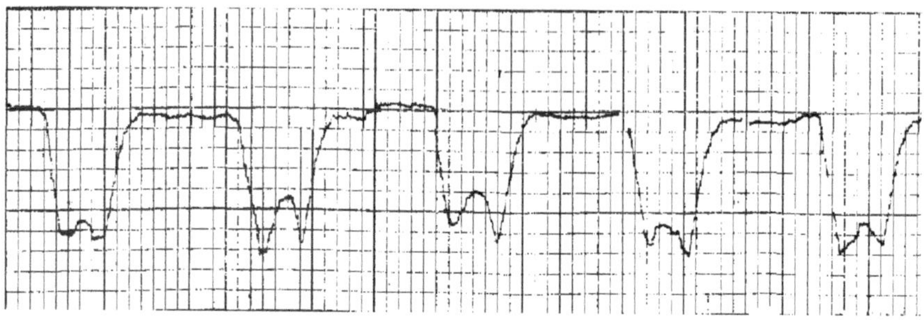

Fig. 8. Reactive ion etching trench profile obtained at an oxygen percentage above 10%. The horizontal and the vertical scale read to 2 and 0.2 μm, respectively

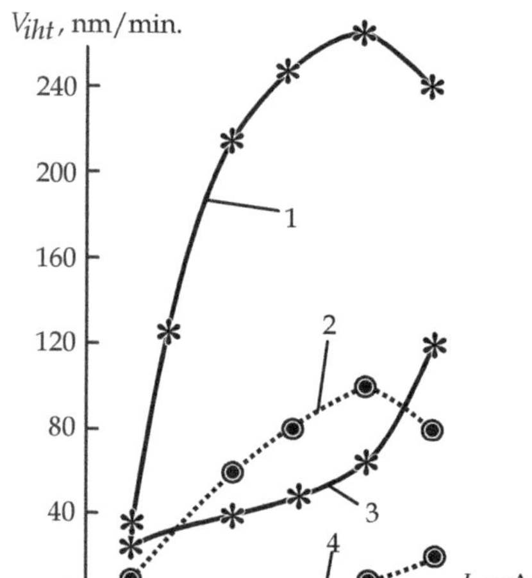

Fig. 9. Etch rate vs. discharge current for (1, 3) reactive ion etching or (2, 4) plasma etching in (1, 2) a CF_4–O_2 or (3, 4) a CF_4 plasma

It was found that addition of oxygen to CF_4 is most effective if the discharge current is in the range 80–120 mA, for both modes of etching (Fig. 7; Fig. 9, curves 1, 2). If the current is increased further, the etch rate falls because the large density of reactive species on the wafer surface makes it difficult to remove etch products. The removal is therefore the rate-determining factor. This conclusion is supported by etch-rate curves 3 and 4 of Fig. 9. These show consistent exponential growth, indicating deficiency of reactive species on the SiO_2 surface. Thus, the etch rate in a CF_4 plasma is determined by the density of F^- ions produced in the plasma, for both modes of etching. It was also observed that discharge currents above 140 mA cause high-temperature breakdown of the photoresist.

4.3 Effect of bulk modification of polymers in a directional off-electrode plasma flow

The treatment of polymers by low-temperature plasma is one of fundamental processes in preparing micro- and nanostructures. The regularities of this technological process have been studied for a long time (Moreau, 1988a; Sarychev, 1992; Valiev et al., 1985, 1987). However, in spite of the large number and apparent comprehensiveness of available experimental results, the mechanism of polymer etching is not completely clear in view of its complex multifactor dependence on the type of interaction of active particles in the plasma with the polymer matrix.

This part of chapter is devoted to experimental investigation of regularities of polymer etching in the plasma generated outside the electrode gap in oxygen. The experimental results are used for constructing a computational model of the etching process.

Figure 10 shows the experimental dependences of the thickness of etched polymer layer (h) on etching time (t) for two different values of the initial film thickness. Analysis of these dependences shows that both curve display identical behavior in the region $0 \leq t \leq 18$ s: the value of h increases for $0 \leq t \leq 6$ s and $15 \leq t \leq 18$ s ($15 \leq t \leq 21$ s for curve 1) and the rate of etching decreases for $6 \leq t \leq 15$ s. Both curves have regions of saturation for values of h equal to the corresponding values of the film thickness, which confirms the complete removal of polymer from the surface.

Let us use the experimental results for constructing the model of polymer etching in the oxygen plasma outside the electrode gap.

It should be noted that the most comprehensive mechanisms and models of polymer etching in the high-frequency and ultrahigh-frequency (microwave) plasma were proposed in (Sarychev, 1992; Valiev et al., 1985, 1987). It was assumed that a modified surface layer (K-layer) is formed during etching, which is more resistive to destruction than unmodified lower layers of the polymer structure.

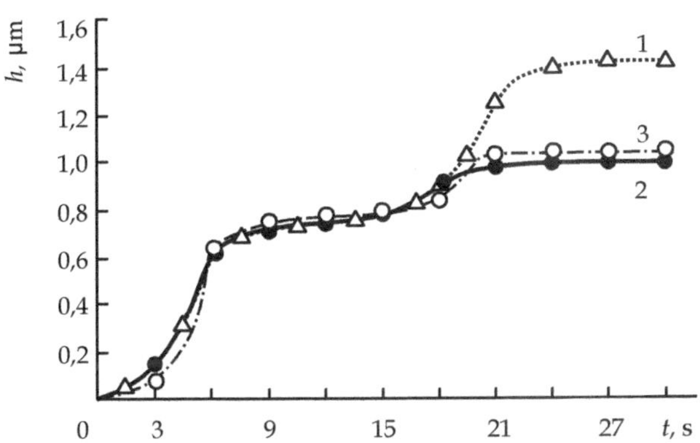

Fig. 10. Dependence of the thickness of the scoured polymer layer on the etching time for I = 100 mA and U = 2 kV: 1—initial thickness of polymer film is $1.4 \cdot 10^{-6}$ m; 2—$1 \cdot 10^{-6}$ m; 3—calculated dependence for an initial thickness of the polymer film of $1 \cdot 10^{-6}$ m

It should be noted, however, that the model of K-layer was developed on the basis of experiments on etching in the electrode plasma. Interpretation of our results on etching outside the electrode gap allows us to supplement this model by the idea that the modified layer in this case may lie in the bulk of the polymer.

In an oxygen plasma, atomic oxygen (O^{**}), negative oxygen ions (O^-), and excited molecular oxygen (O_2^*) with a low concentration on the order of 0.01% are active etching particles (Ivanovskii, 1986). Polymer etching may occur due to sputtering by high-energy O^- ions, as well as due to their chemical interaction with polymer molecules. In addition, atomic oxygen O^{**} present at the surface can also interact with these molecules. The reaction products form volatile compounds H_2O (water vapor), CO_2, and N_xO_y, which are removed from the working chamber by evacuation facilities.

The role of electrons in this process is controlled by the following circumstance. The electron mean free path in the gas and in the polymer is much larger than the mean free path of an ion due to smaller number of collisions with atoms and molecules of the medium. Electrons penetrate to the bulk of the polymer to a depth (Rykalin et al., 1978)

$$L = 10^{-5}\frac{U^{3/2}}{\rho}, \tag{22}$$

where ρ = 500 kg/m³ is the polymer density; U = 2 kV is the accelerating voltage; and $L = 0.57\cdot10^{-6}$ m, which is half the thickness of the polymer film and in good agreement with experimental curve 2 (see Fig. 10). Electrons are decelerated in the substance due to excitation of atoms in polymer molecules. In each collision, an electron spends for excitation an energy (Raizer, 1987)

$$\varepsilon = \frac{2m_e}{M}E_e, \tag{23}$$

where M is the mass of an atom in a polymer molecule and E_e is the initial energy of the electron. For E_e = 2000 eV, the value of $\varepsilon \approx 0.005$ eV, which is several orders of magnitude lower than the ionization loss. The electron energy loss distribution over the path depth in this case can be described by the Thomson—Widdington law (Popov, 1967). An electron experiences about 30 collisions over length L; in this case, it releases an energy of 1.9 keV at the end of its path, spending this energy for rupture of bonds between atoms in the polymer layer.

As a result of excitation, polymers may experience relaxation, which is observed at temperatures equal to or exceeding the glass-transition temperature T_s (Bartenev & Barteneva, 1992). For a DNQ protecting layer obtained from metacresol novolac, T_s = 423 K (Moreau, 1988a); consequently, relaxation does not take place. Hence, the increase in the dependences on segment $0 \le t \le 6$ s can be explained by the interaction of active plasma particles with excited polymer atoms, for which the number of active bonds N_a is determined by the flux of electrons, their energy E_e, and duration t of the process.

When the rupture of atomic bonds takes place, atoms containing a single uncompensated electron each on the outer orbital try to fill it. Bonds involving the collectivization of electron pairs are formed between adjacent carbon atoms.

Thus, a modified layer consisting predominantly of carbon atoms is formed at a depth L. This layer must possess an elevated density ρ_m (as compared to unmodified layers) and stability to destruction (Valiev et al., 1985). The degree of homogeneity of this layer depends on the uniformity of the distribution of charged particles over the plasma flow cross section, the dose and energy of electron irradiation recalculated for the number of carbon atoms in the layer with different numbers of ruptured (suppressed) bonds and, accordingly, with different degrees of modification (Fig. 11a).

Such a mechanism explains the existence of two first regions for $0 < t < 6$ s and $6 < t < 15$ s of curve 1 in Fig. 10.

For $15 \le t \le 21$ s, curve 1 (see Fig. 10) has a second segment in the dependence of $h = f(t)$, indicating the etching of a material with properties close to initial properties. Let us consider the mechanism of its formation.

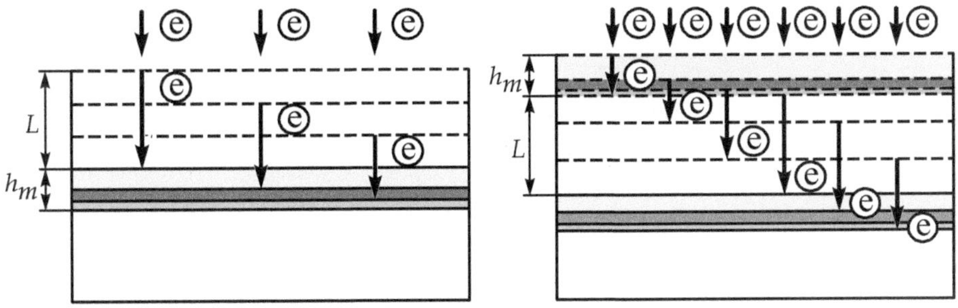

Fig. 11. Diagram illustrating the formation of a modified layer by electrons: (a) polymer etching stage with initial properties; (b) modified polymer layer etching

The motion of electrons in a denser medium is accompanied by their scattering, which is proportional to the mean free path. In the course of etching of a layer of modified polymer, the mean free path decreases, which increases the electron flux and energy (ΔE_e) carried by the electron flow to the lower (unmodified) region. This becomes possible if the etching rate V_m in the modified layer exceeds the rate V of its formation. In this case, if condition $\Delta E_e \ge E_{thr}$ is satisfied (E_{thr} is the threshold energy of delocalization, which is a part of the binding energy (Bechstedt & Enderlein, 1988), a new stage of formation of layers with different degrees of modification begins (it includes the stage of excitation of atoms) (Fig. 11b). The number of such layers is proportional to the thickness of the polymer film. The correctness of the above statements follows from experimental curve 1 (see Fig. 10). Indeed, this curve clearly displays the second peak corresponding to the stage of formation of the second modified layer.

Thus, the process of polymer removal consists of two stages: etching of unmodified and modified layers. The second stage for an individual region of the polymer lags behind the first stage by t_m, where t_m is the etching time for the unmodified polymer.

Let us estimate the height h of the etched layer as a function of parameters of the physical process (discharge current, accelerating voltage, and duration of etching) on the basis of the proposed mechanism and experimental results. The value of h is

$$h = \sum_{n=0}^{l-1} \left[\int_{nT}^{t_m+nT} V_0(t)\,dt + \int_{t_m+nT}^{(n+1)T} V_m(t)\,dt \right], \tag{24}$$

where $T = t_m + t_k$ (t_k is the time of etching of modified polymer); $n = 0, 1, 2, \ldots, l - 1$ (l is the number of modified layers); and t is the etching time. Considering that excitation of polymer atoms increases the etching rate, while the decrease in this rate is due to rupture (suppression) of bonds in the polymer at depth L, we can write

$$V_0(t) = V_0 \frac{N_a(t)}{N_{sn}} \; ; \; V_m(t) = V_m \left(1 - \frac{N(t)}{N_{sm}}\right), \tag{25}$$

where N_{sn} and N_{sm} are the total number of bonds in unmodified and modified layers of thickness L and h_m, N is the number of ruptures (suppressed) bonds, and V_0 and V_m are the etching rates for the polymer and the modified layer in the high-voltage gas discharge plasma flow (V.A. Kolpakov, 2002):

$$V_0 = \frac{BM}{\rho N_a} J_i^- \left| \exp\left(\frac{U - U_{gr}}{U}\right) - 1 \right| (k_1 + k_3), \tag{26}$$

$$V_m = \frac{BM}{\rho_m N_a} J_i^- \left| \exp\left(\frac{U - U_{gr}}{U}\right) - 1 \right| (k_1^m + k_3^m), \tag{27}$$

$$J_i^- = \left(1 - \frac{d}{d_{\max}}\right) \frac{I}{qeS_K} \left(1 - \frac{\gamma_e \eta}{(1 + \gamma_e)} \exp\left[(a - a_n) d_{\max}\right]\right), \tag{28}$$

where N_a is the Avogadro number, k_1 and k_1^m are the plasma-chemical etching coefficients, equal to the number of polymer atoms of unmodified and modified layers removed by a chemically active particle; k_3 and k_3^m are the physical sputtering coefficients equal to the number of atoms knocked from the surface of unmodified and modified layers by a bombarding particle; U_{gr} is the voltage across the gas discharge unit, for which the energy of an ion at the instant of its approach to the surface of treatment is at the threshold energies of plasma-chemical and ion-chemical etching; B is the (constant) value of the penalty function obtained from the natural experiment, which is $B \approx 0.6$ for $I = 100$ mA (V.A. Kolpakov, 2004); $d = 0.045$ m is the distance from the cathode to the sample surface; $d_{\max}$ is the maximal distance over which the plasma flow propagates for the given voltage across the electrodes; S_K is the surface area of the cathode; q is the geometrical transparency of the gauze anode; γ_e is the secondary emission coefficient; η is the electron beam focusing coefficient; a is the ionization factor; and a_n is the adhesion coefficient.

Having analyzed the nature of variation of experimental curves on segment $0 \le t \le 6$ s (see Fig. 10) and the above statement concerning the dependence of N_a on the electron flux and energy, as well as on the duration of the process, we approximate $N_a = f(J_e, E_e, t)$ by an exponential function of the form

$$N_a(t) = N_0 \exp\left(\frac{J_e S E_e}{E^*} t\right), \tag{29}$$

where N_0 is the number of bonds on the polymer surface (on the order of 10^{16}), $E^* = N_{sn} E_{thr}^*$ is the total energy required for exciting polymer atoms in a layer of thickness is the

threshold energy of excitation of a polymer atom, J_e is the electron flux, S is the area of interaction of the low-temperature plasma with the polymer, and E_e is the electron energy.
Analysis of the structure of the DNQ protecting material based on diazoquinone and metacreson novolac leads to the conclusion that carbon is the main bond-forming element. Knowing the number of carbon atoms n_C in a polymer molecule and its valence V_C, as well as the total number of atoms n_{at} in a polymer molecule, we can estimate quantities N_{sn} and N_{sm} from the formula

$$N_{sn} = \left(V_C n_C\right) \frac{\rho S}{M} \frac{N_A}{n_{at}} L ,$$

$$N_{sm} = \frac{N_{sn}}{L} h_m \; ; \; h_m = 10^{-5} \frac{U^{3/2}}{\rho_m} . \qquad (30)$$

Substituting the known values of n_C = 47, V_C = 4, and n_{at} = 108 into these formulas, we obtain $N_{sn} \approx 0.4 \cdot 10^{18}$ and $N_{sm} \approx 0.16 \cdot 10^{18}$. The energy released by an electron at the end on its path in the polymer for rupturing (suppression) of bonds is controlled by difference $E_e - E^*$, where E^* is the total energy spent by the electron for excitation of polymer atoms. In this case, quantity N can be represented analogously to relation (29) in the form

$$N(t) = N_0 \exp\left(\frac{J_e S}{N_{sm}} \frac{\left(E_e - E^*\right)}{E_{thr}} t \right) . \qquad (31)$$

In the time interval $0 \le t \le t_m$, etching of the polymer with initial properties takes place; as a result, k_1^m, $k_3^m = 0$ and the second term in relation (24) vanishes. The thickness of the scoured layer is proportional to the number of active bonds of excited polymer atoms. Ratio N_a/N_{sn} specifies the law of variation of the value of h on segment $0 \le t \le 6$ s in the dependence $h = f(t)$ depicted in Fig. 10. However, modified layers with various degrees of modification are formed at a depth $h \ge L$. By instant $t = t_m$ for which the number of active bonds becomes equal to the number of bonds in the unmodified layer ($N_a = N_{sn}$), etching of the unmodified polymer is completed, which leads to zero values of k_1 and k_3. For $6 \le t \le 15$ s, polymer layers with various degrees of modification experience etching; an increase in N slows down this process, which does not contradict the above mechanism. The law of variation of the value of h at a given segment of curve $h = f(t)$ specifies ratio N/N_{sm} subtracted from unity. The instant corresponding to completion of etching of the modified layer is determined by the equality $N = N_{sm}$, which leads to vanishing of the second term in relation (24). Alternation of the conditions for completion of etching of the modified and unmodified layers in time at the instant when these conditions hold makes it possible to use expression (24) for estimating the value of h for an arbitrary thickness of the polymer film for the given values of the discharge current, accelerating voltage, and t.
To obtain numerical values of N_a and N and, accordingly, the thickness of the scoured layer after the substitution of expressions (25) into (24), we must know the threshold values of excitation energy, delocalization energy, and the total energy spent by an electron for exciting polymer atoms. The results of computer and natural experiments lead to the

conclusion that the calculated dependence approximates experimental curve 2 (see Fig. 10) if $E^*_{thr} \approx 0.005$ eV, $E_{thr} \approx 0.015$ eV; in this case, the value of E^* varies from 10^3 eV to zero, which can be explained by the decrease in the electron mean free path in an unmodified polymer during its etching. The above values of E^*_{thr} and E_{thr} are two or three orders of magnitude lower than the ionization energy and satisfy the inequality $E^*_{thr} < E_{thr} < E_b$ (where E_b is the binding energy), which does not contradict the physical process and the firm opinion of the authors of (Bechstedt & Enderlein, 1988).
Our theoretical and experimental results which were obtained early allow to propose an analytical method of calculating the temperature of the surface exposed to an off-electrode plasma flux.

5. Temperature measurement of a surface exposed to an off-electrode plasma flux

At the present time, numerical methods of calculation are finding wide application in the theory of heat transfer. Indeed, the problem we are interested in can be viewed as the boundary-value problem inverse to the problem of heat conduction. In this case, taking measurements on one part of the surface, one can recover the heat load on other parts inaccessible to measurements. However, such an inverse problem of mathematical physics belongs to the class of ill-posed problems (Tikhonov & Arsenin, 1977); therefore, even an approximate solution can be obtained only with special numerical methods (Alifanov, 1983; Vabishchevich & Pulatov, 1986) providing its stability. At the same time, recent advances in the field of plasma physics make it possible to quantitatively evaluate the effect of the plasma in the form of a heat flux. Therefore, the problem of determining the sample temperature is suggested to be reduced to the analytical solution of the direct problem of heat conduction with mixed boundary conditions.
The charged particles of a plasma flux are uniformly distributed over its cross section in the region where they impinge upon the substrate surface (Kolpakov & V.A. Kolpakov, 1999). With this in mind and taking into account the geometry of the substrate and its single-crystal structure, one can use the heat conduction equation for the one-dimensional case (Samarskii & Vabishchevich, 1996) with the following boundary and initial conditions:

$$\begin{cases} T(x,0) = T_0 \\ T(b,t) = T_{low}(t). \\ q(0,t) = q_1 \end{cases} \tag{32}$$

As far as we know, analytical solutions to the heat conduction equation with conditions (32) are absent. Therefore, we take the most appropriate known solution to this equation with similar boundary and initial conditions (Alifanov, 1994),

$$T(x,t) = \int_0^t q_1(\varepsilon)\frac{\partial\theta(x,t-\varepsilon)}{\partial t}d\varepsilon + \int_0^t q_2(\varepsilon)\frac{\partial\theta(b-x,t-\varepsilon)}{\partial t}d\varepsilon + T_0, \tag{33}$$

where $q_1(\varepsilon)$ is the specific heat flux incident on the front surface $x = 0$ and $q_2(\varepsilon)$ is the specific heat flux carried away from the back surface of the sample. Function $\theta(x,t)$, the temperature response of the body to the unit heat flux incident on one of the boundaries, is given by (Carslaw & Jaeger, 1956)

$$\theta(x,t)=\frac{1}{\lambda}\left\{\frac{at}{b}+\frac{3(b-x)^2-b^2}{6b}+\frac{2b}{\pi^2}\sum_{k=1}^{\infty}\frac{(-1)^{k+1}}{k^2}\exp\left(-k^2\pi^2\frac{at}{b^2}\right)\cos\left(k\pi\frac{b-x}{b}\right)\right\}, \quad (34)$$

Where b is the sample thickness, $a=\lambda/C$ is the thermal diffusivity, λ is the thermal conductivity, and C is the heat capacity per unit volume.
The form of function $q_1(\varepsilon)$ can be found by measuring the temperature of the back (lower) surface of the sample, $T(b,t)=T_{low}(t)$, with incident heat flux $q_1(\varepsilon)$ known. In this case, temperature $T(0,t)$ of the upper (exposed) surface is a partial solution to initial equation (33) and depends on $q_1(\varepsilon)$ and calculated value of $q_2(\varepsilon)$.
Using the Laplace transformation,

$$F(p)=\int_0^{\infty} f(t)\exp(-pt)\,dt, \quad (35)$$

we represent the left-hand side of Eq. (33) as a function of complex variable p,

$$T_{low}(p)\leftarrow T(b,t)\ . \quad (36)$$

The convergence condition imposed on integral (35) implies the need for finite approximation of the sum in (34). Expression (34) is known to be a convergent alternate series (Tikhonov & Samarskii, 1964); therefore, the sum can be calculated with a desired accuracy by discarding the right-hand part, the approximation error being no more than the absolute value of the first of discarded terms.
Let us transform the right-hand side of (33) by applying the convolution theorem (Ditkin & Prudnikov, 1966), which allows one to determine the original of the product of images,

$$T_0+\int_0^t q_1(\varepsilon)\frac{\partial\theta(x,t-\varepsilon)}{\partial t}d\varepsilon+\int_0^t q_2(\varepsilon)\frac{\partial\theta(b-x,t-\varepsilon)}{\partial t}d\varepsilon\leftarrow T_0(p)+Q_1(p)K_1(p)+Q_2(p)K_2(p), \quad (37)$$

where $T_0(p)$, $Q_1(p)$, and $Q_2(p)$ are the images of initial temperature T_0, heat flux $q_1(\varepsilon)$, and heat flux $q_2(\varepsilon)$, respectively, and $K_1(p)$ and $K_2(p)$ are the images of the time derivatives of temperature responses $\theta(b,t)$ and $\theta(0,t)$, respectively,

$$\begin{cases} K_1(b,t)=\dfrac{a}{b\lambda}\left[1+2\sum\limits_{k=1}^{n}(-1)^k\exp\left(-k^2\pi^2\dfrac{at}{b^2}\right)\right] \\ K_2(0,t)=\dfrac{a}{b\lambda}\left[1+2\sum\limits_{k=1}^{n}(-1)^{k+1}\exp\left(-k^2\pi^2\dfrac{at}{b^2}\right)\right] \end{cases}. \quad (38)$$

With regard to formulas (36) and (37), the desired solution for $Q_2(p)$ takes the form

$$Q_2(p)=\frac{T_{low}(p)-T_0(p)-Q_1(p)K_1(p)}{K_2(p)}\ . \quad (39)$$

To find the surface temperature, it is necessary to substitute the known value of $q_1(\varepsilon)$ and original $Q_2(p)$ calculated by formula (39) into initial equation (33). Note that, in going from function $T(b, t)$ to desired function $T(0, t) = T_{surf}(t)$ (i. e., surface temperature), function $\theta(x, t)$ changes to $\theta(b - x, t)$, since coordinate $x = b$ is replaced by $x = 0$. With this in mind, we can write Eq. (33) in the complex form,

$$T_{surf}(p) = T_0(p) + Q_1(p)K_2(p) + Q_2(p)K_1(p) \; . \tag{40}$$

The temperature of the exposed surface can be found using formulas (39) and (40),

$$T_{surf}(p) = T_0(p) + Q_1(p)K_2(p) + \frac{K_1(p)}{K_2(p)}\left[T_{low}(p) - T_0(p) - Q_1(p)K_1(p)\right] \; . \tag{41}$$

Expression (41) shows that, if the sample is thin ($b \rightarrow 0$), T_{surf} approaches T_{low}. Indeed, as follows from (38), the expression under the summation sign is an infinitesimal; in this case, $K_1 = K_2$. Substitution of this equality into (32) gives a negligibly small temperature gradient in a plane sample.
Real function T_{surf} is found using the inverse Laplace transformation (Ditkin & Prudnikov, 1966),

$$\frac{1}{2\pi\imath}\int_{x-\imath\infty}^{x+\imath\infty} T_{surf}(p)\exp(pt)\,dp = \begin{cases} T_{surf}(t), if\ t > 0 \\ 0, if\ t < 0 \end{cases} \; . \tag{42}$$

Thus, using the integral transformations, we have derived the expression for the sample temperature in the region exposed to a directed flux of a low-temperature plasma as a function of known parameters. The disadvantage of our method is the difficulty of going to (41). However, this problem can be completely eliminated with program packages.
The model proposed was used to calculate the surface temperature of a silicon dioxide substrate exposed to plasma irradiation. The substrate (0.03 × 0.03 m in area and $b = 0.002$ m in thickness) was exposed to a plasma flux generated by a high-voltage gas discharge in a nonuniform electric field (Kazanskiy et al., 2004) with particle energies reaching 6 keV (Kazanskiy & V.A. Kolpakov, 2003). In air, the current was varied in the range 1–140 mA. In the near-surface layer, the plasma flux incident on the substrate surface produces heat flux q_1, which, having passed through the substrate, turns into flux q_2 (Fig. 12). The lower surface temperature was measured by a precision chromel–copel thermocouple.
The lower surface temperature was found not to exceed 700 K (its variation is presented in Fig. 13). The temperature curve was interpolated by a polynomial the order of which was determined by a given accuracy. In this temperature range, the mean values of the thermophysical parameters of the substrate were taken to be $a = 10^{-5}$ m^2/s and $\lambda = 10$ W/(m K) (Kikoin, 1976).
The cathode–anode distance in the gas discharge unit was close to the size of the Aston dark space. In this case, $q_1(\varepsilon)$ was calculated by the techniques described in (V.A. Kolpakov, 2002) as the product of the electron flux by the electron energy (the latter being specified by the accelerating voltage). Such an assumption is valid, since the ion component of the plasma has a low energy compared to the electron one. In addition, the ions rapidly lose energy in collisions with atoms of a working gas. On the other hand, at pressures of 0.1–1.0 Torr and

cathode–substrate distance $d = 0.05$ m, the number of elastic collisions of electrons with gas atoms is small and the energy loss is insignificant.

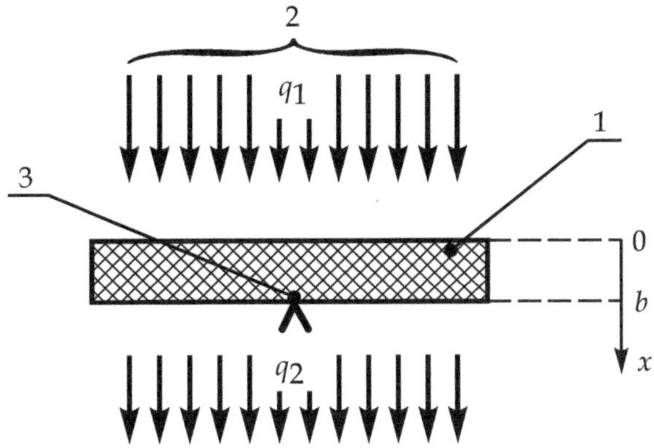

Fig. 12. Irradiation of the sample by the gas-discharge plasma flux: (1) insulating substrate, (2) directed flux of the low-temperature plasma, and (3) temperature sensor at the lower surface

It is known that whether series (34) converges or not depends on the value of at/b^2: the greater this parameter, the better the convergence. To find an exact solution at small at/b^2 (for example, at the initial stage of the process), it is necessary to leave 11–12 terms of the series (Malkovich, 2002). In this study, we took into account 12 terms of sum (34).

As was noted earlier, the boundary-value problem is rather difficult to solve analytically, because (41) contains the ratio of series K_1 and K_2 . Therefore, the proposed algorithm was implemented by applying the Maple 8 program package. Using (41), we constructed the dependences of temperature gradient ΔT in the substrate on the process time (Fig. 14).

As is seen from Fig. 14a, the curves first sharply ascend. This is because the substrate, being thin, heats up rapidly. In other words, incident flux $q_1(\varepsilon)$ passes through the sample almost instantly without noticeable energy losses and goes away from the lower surface, rapidly causing a temperature difference. When the irradiation time is long, the sample heats up at a constant temperature gradient (Fig. 14a).

It is this circumstance that may be responsible for the so-called "thermal shock" (Kartashov, 2001), when thin samples are almost instantly destroyed once the discharge power exceeds a critical value. Indeed, arising thermal stresses are determined by the temperature gradient, which rapidly runs through intermediate values and reaches a maximum virtually at the very beginning of the process (Fig. 14a). The simulation data suggest that the transient time increases as the thermal diffusivity of the sample decreases or it gets thicker, thermal action $q_1(\varepsilon)$ being the same. It is evident that this statement completely agrees with the theory of heat transfer: a more massive sample reaches the stationary state for a longer time. In addition, a material with a lower thermal conductivity will have a higher temperature gradient, which will be established for a longer time. The rigorous solution of this problem implies a combined consideration of the equations of heat transfer and thermoelasticity (Samarskii & Vabishchevich, 1996).

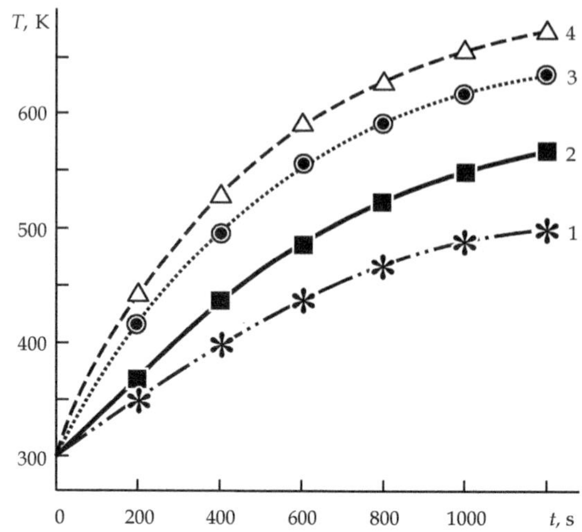

Fig. 13. Lower surface temperature vs. time: I = (1) 50, (2) 80, (3) 120, and (4) 140 mA. The voltage applied to the electrodes is 2 kV, the pressure is 1.5 Torr, and the working gas is air

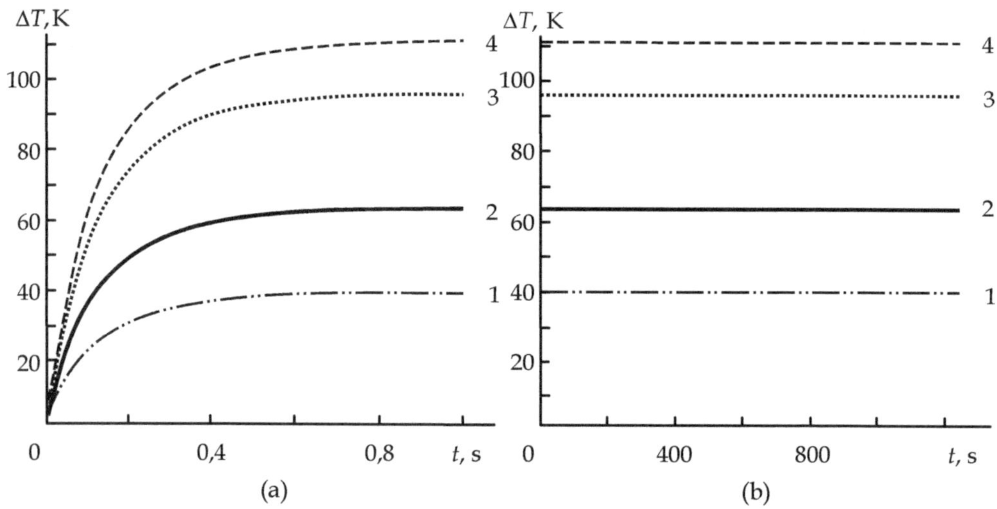

Fig. 14. Temperature difference between the upper and lower surfaces for an irradiation time of (a) 1 and (b) 1200 s. I = (1) 50, (2) 80, (3) 120, and (4) 140 mA

At high t, the temperature difference takes on a constant value (Fig. 14b). Therefore, failure of the sample at the final stage is unlikely. The model proposed was also experimentally verified using KÉF-32 silicon samples measuring 1×1×0.1 cm. The temperature of the sample was controlled by varying the plasma flux irradiation parameters: voltage from 2.6 to 5.2 kV and current from 24 to 80 mA. The irradiation duration was 10 min. The thermophysical parameters of the material were matched to the process conditions. The temperatures of the upper (exposed) and lower surface were measured by a Promin′ micropyrometer. The surface temperatures and temperature gradient are listed in the table.

The disagreement between the calculated and experimental values of the temperature difference does not exceed 12%, which confirms the adequacy of the estimation method.

The proposed method was applied for temperature measurement of a surface exposed to an off-electrode plasma flux during research of etch-rate-temperature characteristic. In the plasma etching mode of treatment the etch-rate–temperature characteristic is as shown in Fig. 15a. Notice that for every discharge current the etch rate is maximal at 360 K, the vaporization temperature of SiF_4. This point corresponds to the best conditions for etch-product removal. As the wafer temperature is raised further, the etch rate falls due to decrease in the amount of process gas adsorbed by SiO_2, in accord with earlier results (Ivanovskii, 1986; Kireyev & Danilin, 1983; Kireev et al., 1986).

In the reactive ion etching mode the temperature dependence is not so simple, as can be seen from Fig. 15b. At a discharge current as weak as 50 mA (Fig. 15b, curve 1), the etch rate is almost unaffected by wafer-temperature variation, because the etch rate in this case is determined by the density of F^- ions, as noted above. At 325–360 K, etching is possible because the SiO_2 surface is almost free from particles that could impede etch-product removal.

At stronger discharge currents, quite distinct behavior is observed (Fig. 15b, curves 2–4). The reason is that the removal of SiF_4 is impeded by the species (F^- ions, reactive species, and reaction products) that have accumulated on and underneath the SiO_2 surface, with the result that etching occurs only at wafer temperatures above 360 K, the vaporization temperature of SiF_4. As the wafer temperature increases from 360 K, the etch rate rises to a maximum. Notice that the temperature of maximum etch rate depends on the discharge current, being 390, 422, and 440 K for 80, 120, and 140 mA, respectively. An increase in wafer temperature weakens interatomic bonding in the SiO_2, making the material more susceptible to sputtering. Further, the higher the discharge current, the more ions penetrate the SiO_2 to enter into reactions there. As a result, the product species should migrate more slowly toward the surface with increasing discharge current at a fixed wafer temperature. Higher temperatures are therefore required to remove the products. The sharp fall in etch rate is attributable to increase in ion penetration depth; this factor seriously hinders removal of etch products (SiF_4) with growing wafer temperature. Plasma processing in this case is basically fluorine-ion doping of a SiO_2 surface layer and sputter etching. High temperature breakdown of the photoresist was found to occur at 440 K, showing up as a faster fall in etch rate with wafer temperature (etch rate should be the same in unmasked and opened areas). Breakdown starts from the edges of the mask and causes etch taper (Fig. 16a), which will guide ions just into trenches and so determine the trench profile (Fig. 16b). As the etch taper grows, so do its angles and the etch profile becomes a sinusoid (V.A. Kolpakov, 2002). This property is useful for making diffractive optical elements with a sinusoidal micropattern (Soifer, 2002).

6. Results and discussion: Quality of surface treatment

Figure 17 displays trench profiles obtained by off-electrode plasma etching at discharge currents of 50, 80, and 120 mA and oxygen percentages corresponding to maximum etch rates. Prior to photoresist stripping, processed wafers were examined and found to be free from etch undercut, an indicator of etching anisotropy. It can be seen from Fig. 17 that the profile approaches a vertical-walled pattern with growing discharge current, as predicted earlier. For example, a plasma with a current of 50 mA and a pressure of about 11 Pa is deficient in F^- ions, but these rarely collide with process-gas molecules and so have energies as high as 100–500 eV (see Eq. (11)). Favorable conditions thus arise for the reflection of F^- ions from trench sidewalls toward the center of the bottom. In this case the sidewalls may deviate from the normal by an angle as large as 70°–75° (Fig. 17a). At a higher density of F^- ions (current 80 mA, pressure 20 Pa), the ions strike the SiO_2 surface with a lower energy and so are more likely to enter surface reactions, mostly at the site of landing. Further, when isolated from other factors, the increase in reactive-species density is known to reduce the sidewall deviation to 10°–20° (Moreau, 1988b).

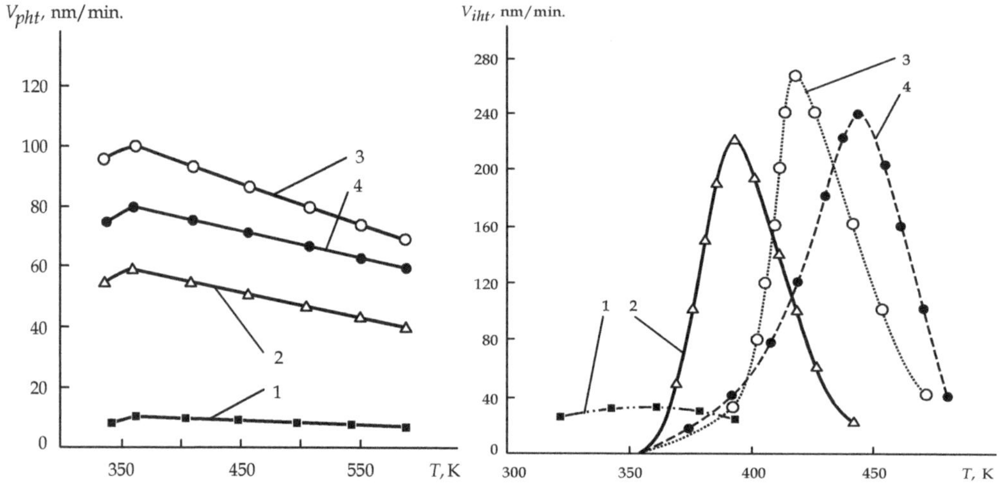

(a) (b)

Fig. 15. Etch rate vs. wafer temperature for (a) plasma etching or (b) reactive ion etching in a CF_4–O_2 plasma at discharge currents of (1) 50, (2) 80, (3) 120, and (4) 140 mA

Figure 17b,c shows that the trench bottoms meet the requirements of microelectronics manufacturing: they are smooth and free from acute angles. Moreover, etching at 120–140 mA and 25–33 Pa was found to produce trenches with vertical walls and a smooth bottom (Fig. 17d, e, f). Finally, the pressures employed satisfy the conditions given in (Orlikovskiy, 1999a). Thus, all the trench profiles presented could find use in microelectronics (Moreau, 1988b; Muller & Kamins, 1986) and diffractive optics (Soifer, 2002).

Off-electrode plasma etching in a CF_4–O_2 plasma was also applied to other materials used in microelectronics, as well as in diffractive optics. The respective etch rates are listed in the table. At the same time, it was observed that fairly thick deposit is formed on the cathode during etching (Fig. 18). Figure 19 is an x-ray diffraction pattern (Mirkin, 1961) from the deposit; it indicates elements and compounds present in the process gas (*C*), the etched material (SiO_2, *SiC*, *Si*, As_2S_3, and *C*), and the etch mask (Cr_2O_3, CrO_3, *C*, and H_2). Cathode deposit also includes large amounts of compounds containing the cathode material and different oxides. On the other hand, it is free from fluorine, a fact suggesting that fluorine is totally involved in etching (as part of reactive species). Moreover, the presence of the etched material in the deposit implies that the plasma ensures etch-product removal. It follows that the working plasma species (F^- ions) move toward the wafer, whereas the product ones toward the cathode. This result supports the mechanisms presented above. It is in accord with earlier research (V.A. Kolpakov, 2002).

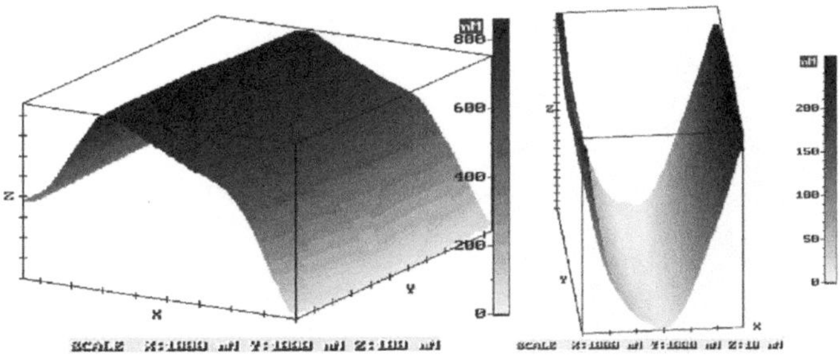

(a) (b)

Fig. 16. (a) Etch taper due to high-temperature photoresist breakdown and (b) the corresponding trench profile. Etching is carried out at a discharge current of 140 mA, a cathode voltage of 2 kV, and a wafer temperature of 440 K

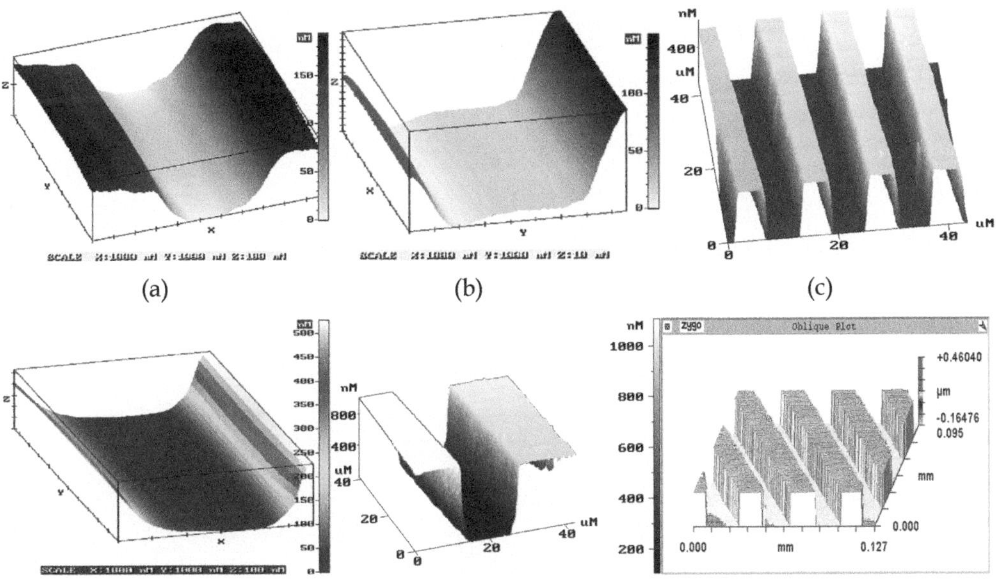

(a) (b) (c)

(d) (e) (f)

Fig. 17. Images of trenches obtained by etching in CF_4–O_2 plasma at different discharge currents, optimal oxygen percentages, and a cathode voltage of 2 kV. The discharge currents are (a) 50, (b, c) 80, and (d, e, f) 120 mA. The oxygen percentages are (a) 0.5, (b, c) 0.8, and (d, e, f) 1.3%

Thus, even with highly contaminated process gas and wafer surface, off-electrode plasma etching does not involve interactions other than a useful one (between reactive species and wafer-surface molecules), allowing one to take less expensive gases.

Etching uniformity is among major concerns in microfabrication, because etch rate can vary in a complicated manner over the wafer surface (Ivanovskii, 1986; Kovalevsky et al., 2002; Poulsen & Brochu, 1973). In essence, all the recent improvements in plasma etching technology aim to give high etching uniformity and rate; hence the high complexity and cost of the equipment.

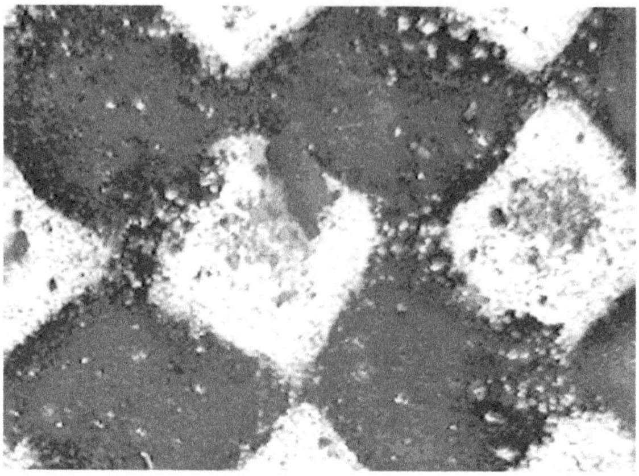

Fig. 18. Cathode surface after etching (magnification ×36)

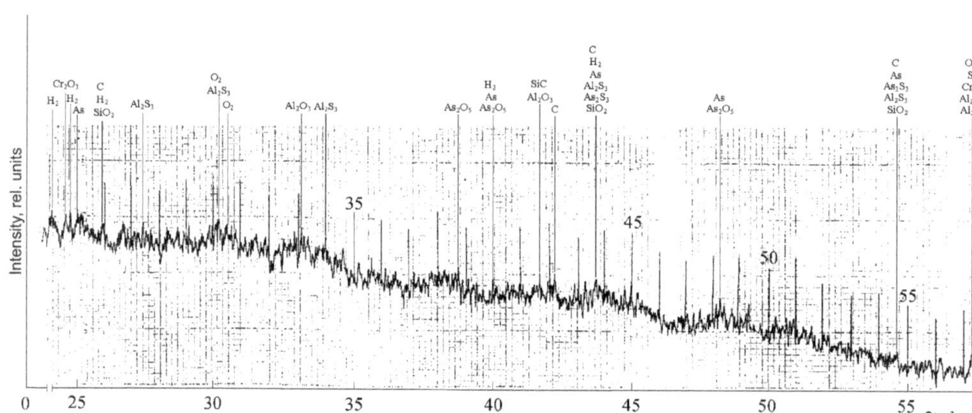

Fig. 19. X-ray diffraction pattern (wavelength 0.154 nm) from cathode deposit, with t denoting the x-ray reflection angle from the atomic planes

Our evaluation of off-electrode plasma etching in terms of uniformity, for a wafer of diameter 100 mm, showed that both the plasma etching and the reactive ion etching mode are uniform within 1% over the whole wafer. Etch profile was measured in different areas on the wafer, and etch depth was found to be almost the same. The minor variations in etch depth are in all likelihood linked with surface imperfections (lattice defects, contamination, etc.) rather than the plasma conditions.

7. Conclusion

It has been shown that a major feature that distinguishes the high-voltage gas discharge from the existing discharges is that the former can be induced in the dark Aston space, provided an anode hole. This feature allows to generate a low-temperature plasma flux outside the electrode gap.
Based on our experiments, a method for estimating the surface temperature of a sample irradiated by a low-temperature plasma flux is produced. The relationships obtained in this paper make it possible to evaluate the surface temperature directly at the site exposed to the plasma flux. A slight excess of the theoretical estimate seems to be associated with the fact that the plasma flux is incompletely absorbed by the solid: part of the flux is reflected from the surface, decreasing the gradient. During ion–plasma processing, the temperature gradient in the sample may become very high according to the geometry and material of the sample, as well as to the amount of the thermal action.
The method makes it possible to trace the surface temperature of a sample being etched by directed low-temperature plasma fluxes in a vacuum. This opens the way of improving the quality of micro- and nanostructures by stabilizing the process temperature and optimizing the rate of etching in the low-temperature plasma.
The phenomenon of thermal shock taking place at ion–plasma processing of flat surfaces is theoretically explained. It is shown that the failure probability of thin samples is the highest early in irradiation under the action of rapidly increasing thermal stresses. To determine the critical power of the discharge, it is necessary to jointly solve the equations of heat conduction and thermoelasticity.
Among disadvantages of the method is the neglect of the temperature dependence of thermophysical parameters. This point becomes critical for semiconductors operating in a wide temperature range. As a result, the temperature gradient versus process time dependence becomes ambiguous. A more rigorous solution can be obtained by applying numerical methods to the direct problem of heat conduction with mixed boundary conditions. This would be a logical extension of this investigation.

8. Acknowledgment

The work was financially supported by the RF Presidential grant # NSH-7414.2010.9, the Program of the President of the Russian Federation for Supporting Young Russian Scientists (grant no. MD-1041.2011.2) and the Carl Zeiss grant # SPBGU 7/11 KTS.

9. References

Orlikovskiy, A.A. (1999a). Plasma Processes in Micro- and Nanoelectronics, Part 1: Reactive Ion Etching. *Mikroelektronika*, Vol. 28, No. 5, pp. 344–362 (In Russian)

Alifanov, O. M. (1983). *Inzhen.Fizich. Zhurnal*. Vol. 45, No. 5, p.p. 742-752 (In Russian)

Alifanov, O. M. (1994). *Inverse Heat Transfer Problem*, Springer, New York

Bartenev, G. M. & Barteneva, A. G. (1992). *Relaxation Properties of Polymers*, Khimiya, Moscow (In Russian)

Bechstedt, F. & Enderlein, R. (1988). *Semiconductor Surfaces and Interfaces*, Akademie-Verlag, Berlin

Carslaw, H. S. & Jaeger, J. C. (1956). *Conduction of Heat in Solids*, Clarendon Press, Oxford

Chernetsky, A. V. (1969). *Introduction into Plasma Physics*, Atomizdat Publishers, Moscow (In Russian)

Chernyaev, V.N. (1987). *Fiziko-khimicheskie protsessy v tekhnologii REA (Physical and Chemical Processes in Electronics Manufacture)*, Vysshaya Shkola, Moscow (In Russian)

Ditkin, V. A. & Prudnikov, A. P. (1966). *Integral Transforms and Operational Calculus*, Pergamon, Oxford

Doh Hyun-Ho et al. (1997). Effects of bias frequency on reactive ion etching lag in an electron cyclotron resonance plasma etching system. *J.Vac. Sci. and Technol. A., Pt 1*, Vol.15, No. 3, p.p. 664-667

Flamm, D.L. (1979). Measurements and Mechanisms of Etchant Production During the Plasma Oxidation of CF_4 and C_2F_6. *Solid State Technol.*, Vol. 22, No. 4, pp. 109–116

Gerlach-Meyer, V. (1981). Ion Enhanced Gas-Surface Reactions: A Kinetic Model for the Etching Mechanism. *Surface Sci.*, Vol. 103, No. 213, pp. 524–534

Harsberger, W.R. & Porter, R.A. (1979). Spectroscopic Analysis of RF Plasmas. *Solid State Technol.*, Vol. 22, No. 4, pp. 90–103

Hebner, G.A. et al. (1999). Influence of surface material on the boron chloride density in inductively coupled discharges. *J.Vac. Sci. and Technol. A.*, Vol.17, No. 6, p.p. 3218-3224

Horiike, Y. (1983). Dry Etching: An Overview. *Jap. Annual Revue in Electronics, Computers and Telecommunicated Semiconductor Technologies*, Vol. 8, pp. 55–72

Ivanovskii, G.F. (1986). *Ionno-plazmennaya obrabotka materialov (Plasma and Ion Surface Engineering)*, Radio i Svyaz', Moscow (In Russian)

Izmailov, S. V. (1939). On the thermal theory of electron emission under the impact of fast ions. *Russian Journal of Experimental and Theoretical Physics*, Vol.9, No. 12, p.p. 1473 - 1483 (In Russian)

Kartashov, E. M. (2001). *Analytical Methods in the Theory of Heat Conduction in Solids*, Vysshaya Shkola, Moscow (In Russian)

Kazanskiy, N.L. & Kolpakov, V.A. (2003). Studies into mechanisms of generating a low-temperature plasma in high-voltage gas discharge. *Computer Optics*, No. 25, p.p. 112-117 (In Russian)

Kazanskiy, N. L. et al. (2004). Anisotropic Etching of SiO_2 in High-Voltage Gas-Discharge Plasmas. *Russian Microelectronics*, Vol. 33, No. 3, p.p. 169-182

Kikoin, I. K. (Ed.). (1976). *Tables of Physical Quantities*, Atomizdat, Moscow (In Russian)

Kireyev, V. Yu. & Danilin, B. S. (1983). *Plasmo-chemical and ion-chemical etching of microstructures*, Radio i Svyaz (Radio and Communications) Publishers, Moscow (In Russian)

Kireev, V.Yu. et al., (1986). Ion-Enhanced Dry Etching. *Elektron. Obrab. Mater. (Electron Treatment of Materials)*, No. 67, pp. 40–43 (In Russian)

Kolpakov, A.I. & Kolpakov, V.A. (1999). Dragging of Silicon Atoms by Vacancies Created in Molten Aluminum under Ion–Electron Irradiation. *Technical Physics Letters*, Vol. 25, No. 15, p. 618

Kolpakov, V.A. (2002). Modeling the High-Voltage Gas-Discharge Plasma Etching of SiO_2. *Mikroelektronika*, Vol. 31, No. 6, pp. 431–440 (In Russian)

Kolpakov, A. I. et al. (1996). Ion-plasma cleaning of low-power relay contacts. *Electronics Industry*, No. 5, p.p. 41-44 (In Russian)

Kolpakov, A. I. & Rastegayev, V. P. (1979). *Calculating the electric field of a high-voltage gas discharge gun*, VINITI, Moscow (In Russian)

Kolpakov, V. A. (2004). *Candidate's Dissertation*, SGAU & ISOI RAN, Samara (In Russian)

Kolpakov, V. A. (2006). Studying an Adhesion Mechanism in Metal - Dielectric Structures Following the Surface Ion-Electron Bombardment. Part 1. Modeling an Adhesion Enhancement Mechanism. *Phys. and Chem. of Mat. Proc.*, No. 5, pp. 41-48 (In Russian)

Komine Kenji et al. (1996). Residuals caused by the CF_4 gas plasma etching process. *Jap. J. Appl. Phys. Pt.1*, Vol. 35, No. 5b, p.p. 3010-3014

Komov, A.N. et al. (1984). Electron-Beam Soldering Machine for Semiconductor Devices. *Prib. Tekh. Eksp. (Scientific Instruments and Methods)*, No. 5, pp. 218–220 (In Russian)

Kovalevsky, A. A. et al. (2002). Studies into the process of isotropic plasmo-chemical etching of silicon dioxide films. *Mikroelektronika*, Vol. 31, No. 5, p.p. 344-349 (In Russian)

Malkovich, R. Sh. (2002). *Technical Physics Letters*, Vol. 28, No. 21, p. 923

Matare, G. (1974). *Electronics of semiconductor defects*, Mir Publishers, Moscow (In Russian)

McLane, G.F. et al. (1997). Dry etching of germanium in magnetron enhanced SF_6 plasmas. *J.Vac. Sci. and Technol. B.*, Vol. 15, No. 4, p.p. 990-992

Mirkin, L.I. (1961). *Spravochnik po rentgenostrukturnomu analizu polikristallov (Handbook of X-ray Crystallography for Polycrystalline Materials)*, Gosudarstvennoe Izdatel'stvo Fiziko-Matematicheskoi Literatury Publisher, Moscow (In Russian)

Miyata Koji et al. (1996). CF_x radical generation by plasma interaction with fluorocarbon films on the reactor wall. *J.Vac. Sci. and Technol. A.*, Vol. 14, No. 4, p.p. 2083-2087

Molokovsky, S. I. & Sushkov, A. D. (1991). *High-intensity Electron and Ion Beams*, Energoatomizdat Publishers, Moscow (In Russian)

Moreau, W. M. (1988a). *Semiconductor Lithography: Principles, Practices and Materials. Chap. 1*, Plenum, New York

Moreau, W.M. (1988b). *Semiconductor Lithography: Principles, Practices, and Materials. Chap. 2*, Plenum, New York

Muller, R.S. & Kamins, T.I. (1986). *Device Electronics for Integrated Circuits*, Wiley, New York

Orlikovskiy, A.A. (1999b). Plasma Processes in Micro- and Nanoelectronics, Part 2: New-Generation Plasmochemical Reactors in Microelectronics. *Mikroelektronika*, Vol. 28, No. 6, pp. 415–426 (In Russian)

Popov, V. K. (1967). *Fiz. Khim. Obrab. Mater. (Physics and Chemistry of Materials Processing)*, No. 4, p.p. 11-24 (In Russian)

Poulsen, R.G. & Brochu, M. (1973). *Importance of Temperature and Temperature Control in Plasma Etching*, Si Bricond Silicon, New-Jersey

Raizer, Yu.P. (1987). *Fizika gazovogo razryada (Gas-Discharge Physics)*, Nauka, Moscow (In Russian)

Rykalin, N. N. et al. (1978). *Principles of Electron-beam Material Processing*, Mashinostroyenie (Mechanical Engineering) Publishers, Moscow (In Russian)

Samarskii, A. A. & Vabishchevich, P. N. (1996). *Computational Heat Transfer*, Wiley, Chichester

Sarychev, M. E. (1992). Non-liner Diffusion Model of Polymer Resist Plasma-chemical Etching Process. Simulation of Technological Processes of Microelectronics. *Tr. FTIAN (FTIAN Annals)*, Vol. 3, p.p. 74-84 (In Russian)

Soifer, V.A. (Ed.). (2002). *Methods for Computer Design of Diffractive Optical Elements*, Wiley, New York

Tikhonov, A. N. & Samarskii, A. A. (1964). *Equations of Mathematical Physics*, Pergamon, Oxford

Tikhonov, A. N. & Arsenin, V. Ya. (1977). *Solutions of Ill-Posed Problems*, Halsted, New York

Vabishchevich, P. N. & Pulatov, P. A. (1986). *Inzhen.Fizich. Zhurnal*. Vol. 51, No. 3, p.p. 470-474 (In Russian)

Vagner, I.V. et al. (1974). Simple Beam-Forming Arrangement for Generating Arbitrarily Shaped Electron Beams under High-Voltage Gas Discharge. *Zh. Tekh. Fiz.(Russian Journal of Technical Physics)*, Vol. 44, No. 8, pp. 1669–1674 (In Russian)

Valiev, K. A. et al. (1985). Polymer Plasma-chemical Etching Mechanism. *Dokl. Akad. Nauk SSSR (Sov. Phys. Dokl)*, Vol. 30, p. 609 (In Russian)

Valiev, K. A. et al. (1987). Investigation of Etching Kinetics of Polymetilmetacrelat in Low-temperature Plasma. *Poverkhnost' (Surface)*, No. 1, p.p. 53-57 (In Russian)

Woodworth, J.R. et al. (1997). Effect of bumps on the wafer on ion distribution functions in high-density argon and argon-chlorine discharges. *Appl. Phys. Lett.*, Vol. 70, No. 15, p.p. 1947-1949

Part 2

Heat Conduction – Engineering Applications

6

A Prediction Model for Rubber Curing Process

Shigeru Nozu[1], Hiroaki Tsuji[1] and Kenji Onishi[2]
[1]*Okayama Prefectural University*
[2]*Chugoku Rubber Industry Co. Ltd.*
Japan

1. Introduction

A prediction method for rubber curing process has historically received considerable attention in manufacturing process for rubber article with relatively large size. In recent years, there exists increasing demand for simulation driven design which will cut down the cost and time required for product development. In case of the rubber with relatively large dimensions, low thermal conductivity of the rubber leads to non-uniform distributions of the temperature history, which results in non-uniform cure state in the rubber. Since rubber curing process is an exothermic reaction, both heat conduction equation and expressions for the curing kinetics must be solved simultaneously.

1.1 Summary of previous works

In general, three steps exist for rubber curing process, namely, induction, crosslinking and post-crosslinking (e.g. Ghoreishy 2009). In many previous works, interests are attracted in the former two, and a sampling of the relevant literature shows two types of the prediction methods for the curing kinetics.

First type of the method consists of a set of rate equations describing chemical kinetics. Rubber curing includes many complicated chemical reactions that might delay the modelling for practical use. Coran (1964) proposed a simplified model which includes the acceleration, crosslinking and scorch-delay. After the model was proposed, some improvements have been performed (e.g. Ding et al, 1996). Onishi and Fukutani (2003a,2003b) performed experiments on the sulfur curing process of styrene butadiene rubber with nine sets of sulfur/CBS concentrations and peroxide curing process for several kinds of rubbers. Based on their results, they proposed rate equation sets by analyzing the data obtained using the oscillating rheometer operated in the range 403 K to 483 K at an interval of 10 K. Likozar and Krajnc (2007) proposed a kinetic model for various blends of natural and polybutadiene rubbers with sulphur curing. Their model includes post-crosslinking chemistry as well as induction and crosslinking chemistries. Abhilash et al. (2010) simulated curing process for a 20 mm thick rubber slab, assuming one-dimensional heat conduction model. Likazor and Krajnc (2008, 2011) studied temperature dependencies of relevant thermophysical properties and simulated curing process for a 50 mm thick rubber sheet heated below, and good agreements of temperature and degree of cure have been obtained between the predicted and measured values.

The second type prediction method combines the induction and crosslinking steps in series. The latter step is usually expressed by an equation of a form $d\varepsilon/d\tau = f(\varepsilon,T)$, where ε is the degree of cure, τ is the elapsed time and T is the temperature. Ghoreishy (2009) and Rafei et al. (2009) reviewed recent studies on kinetic models and showed a computer simulation technique, in which the equation of the form $d\varepsilon/d\tau= f(\varepsilon,T)$ is adopted. The form was developed by Kamal and Sourour (1973) then improved by many researchers (e.g. Isayev and Deng, 1987) and recently the power law type models are used for non-isothermal, three-dimensional design problems (e.g. Ghoreishy and Naderi, 2005).

Temperature field is governed by transient, heat conduction equation with internal heat generation due to the curing reaction. Parameters affecting the temperature history are dimensions, shape and thermophysical properties of rubbers. Also initial and boundary conditions are important factors. Temperature dependencies of relevant thermophysical properties are, for example, discussed in Likozar and Krajnc (2008) and Goyanes et al.(2008). Few studies have been done accounting for the relation between curing characteristics and swelling behaviour (e.g. Ismail and Suzaimah, 2000). Most up-to-date literature may be Marzocca et al. (2010), which describes the relation between the diffusion characteristics of toluene in polybutadiene rubber and the crosslinking characteristics. Effects of sulphur solubility on rubber curing process are not fully clarified (e.g. Guo et al., 2008).

Since the mechanical properties of rubbers strongly depend on the degree of cure, new attempts can be found for making a controlled gradient of the degree of cure in a thick rubber part (e.g. Labban et al., 2007). To challenge the demand, more precise considerations for the curing kinetics and process controls are required.

1.2 Objective of the present chapter

As reviewed in the above subsection, many magnificent experimental and theoretical studies have been conducted from various points of view. However, few fundamental studies with relatively large rubber size have been done to develop a computer simulation technique. Nozu et al. (2008), Tsuji et al. (2008) and Baba et al. (2008) have conducted experimental and theoretical studies on the curing process of rubbers with relatibely large size. Rubbers tested were styrene butadiene rubbers with different sulphur concentration, and a blend of styrene butadiene rubber and natural rubber. Present chapter is directed toward developing a prediction method for curing process of rubbers with relatively large size. Features of the chapter can be summarized as follows.

1. Experiments with one-dimensional heat conduction in the rubber were planned to consider the rubber curing process again from the beginning. Thick rubber samples were tested in order to clarify the relation between the slow heat penetration in the rubber and the onset and progress of the curing reaction.
2. The rate equation sets derived by Onishi and Fukutani (2003a,2003b) were adopted for describing the curing kinetics.
3. Progress of the curing reaction in the cooling process was studied.
4. Distributions of the crosslink density in the rubber were determined from the equation developed by Flory and Rehner (1943a, 1943b) using the experimental swelling data.
5. Comparisons of the distributions of the temperature history and the degree of cure between the model calculated values and the measurements were performed.

2. Experimental methods

The most typical curing agent is sulfur, and another type of the agent is peroxide (e.g. Hamed, 2001). In this section, summary of our experimental studies are described. Two types of curing systems were examined. One is the styrene butadiene rubber with sulfur/CBS system (Nozu et al., 2008). The other is the blend of styrene butadiene rubber and natural rubber with peroxide system (Baba et al., 2008).

2.1 Styrene Butadiene Rubber (SBR)

Figure 1 illustrates the mold and the positions of the thermocouples for measuring the rubber temperatures (rubber thermocouples). A steel pipe with inner diameter of 74.6 mm was used as the mold in which rubber sample was packed. On the outer surface of the mold, a spiral semi-circular groove with diameter 3.2 mm was machined with 9 mm pitch, and four sheathed-heaters with 3.2 mm diameter, a ~ d, were embedded in the groove. On the outer surface of the mold, silicon coating layer was formed and a grasswool insulating material was rolled. The method described here provides one-dimensional radial heat conduction excepting for the upper and lower ends of the rubber.

Four 1-mm-dia type-E sheathed thermocouples, A ~ D, were located in the mold as the wall thermocouples. Four 1-mm-dia Type-K sheathed thermocouples were equipped with the mold to control the heating wall temperatures. The top and bottom surfaces were the composite walls consisting of a Teflon sheet, a wood plate and a steel plate to which an auxiliary heater is embedded.

To measure the radial temperature profile in the rubber, eight type-J thermocouples were located from the central axis to the heating wall at an interval of 5 mm. At the central axis just below 60 mm from the mid-plane of the rubber, a type-J thermocouple was also located to measure the temperature variation along the axis. All the thermocouples were led out through the mold and connected to a data logger, and all the temperature outputs were subsequently recorded to 0.1 K.

Styrene butadiene rubber (SBR) was used as the polymer. Key ingredients include sulfur as the curing agent, carbon blacks as the reinforced agent. Ingredients of the compounded rubber are listed in Table 1, where sulfur concentrations of 1 wt% and 5wt% were prepared. To locate the rubber thermocouples at the prescribed positions, rubber sheets with 1 and 2 mm thick were rolled up with rubber thermocouples and packed in the mold.

Two curing methods, Method A and Method B, were adopted. Method A is that the heating wall temperature was maintained at 414K during the curing process. Method B is that the first 45 minutes, the wall temperature was maintained at 414K, then the electrical inputs to the heaters were switched off and the rubber was left in the mold from 0 to 75 minutes at an interval of 15 minutes to observe the progress of curing without wall heating. By adopting the Method B, six kinds of experimental data with different cooling time were obtained. The heater inputs were ac 200 volt at the beginning of the experiment to attain the quick rise of to the prescribed heating wall temperature.

After each experiment was terminated, the rubber sample was brought out quickly from the mold then immersed in ice water, and a thin rubber sheet with 5 mm thick was sliced just below the rubber thermocouples to perform the swelling test. As shown in Fig.2, eight test pieces at an interval of 5 mm were cut out from the sliced sheet. Each test piece has dimensions of 3mm×3mm×5mm and swelling test with toluene was conducted.

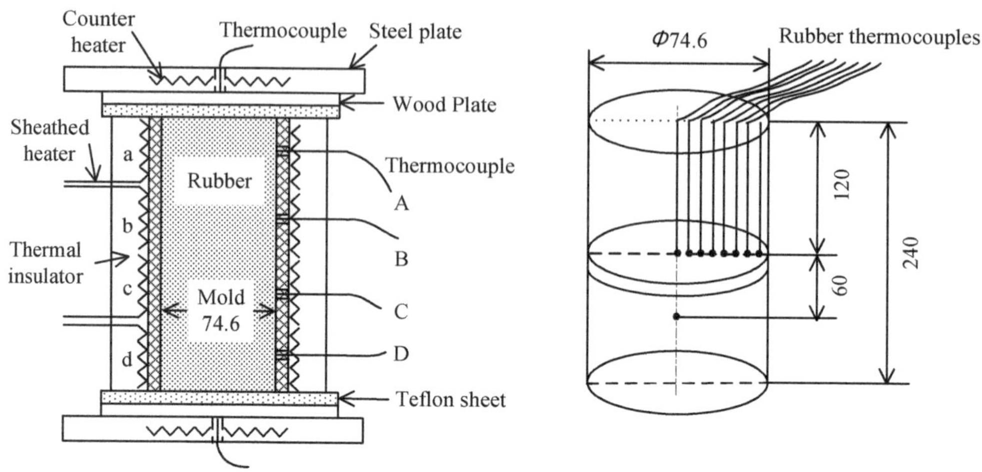

Fig. 1. Experimental mold and positions of rubber thermocouples for SBR

Ingredients	wt%	wt%
Polymer (SBR)	53.8	51.6
Cure agent (Sulfur)	1.0	5.0
Vulcanization accelerator	0.9	0.9
Reinforcing agent (Carbon black)	31.9	30.6
Softner	8.0	7.6
Activator (1)	2.6	2.5
Activator (2)	0.5	0.5
Antioxidant (1)	0.5	0.5
Antioxidant (2)	0.3	0.3
Antideteriorant	0.3	0.5

Table 1. Ingredients of compounded SBR

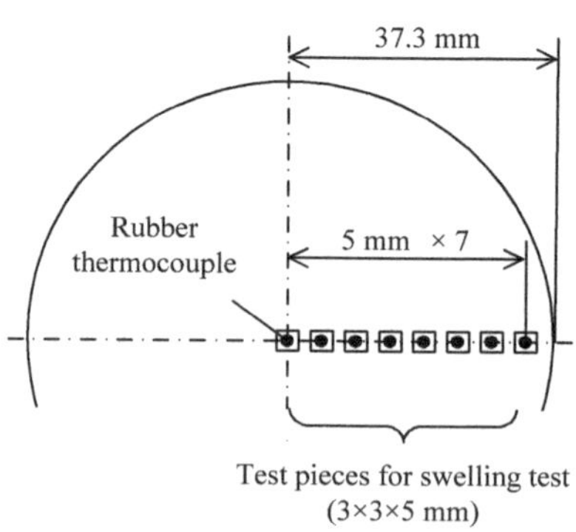

Fig. 2. Cross sectional view at mid-plane of rubber sample

The crosslink density was evaluated from the equation proposed by Flory and Rehner (1943a,1943b) using the measured results of the swelling test. In the present study, the degree of cure ε is defined by

$$\varepsilon = [RX]/[RX]_0 \tag{1}$$

where [RX] is the crosslink density at an arbitrary condition and $[RX]_0$ is that for the fully cured condition obtained from our preliminary experiment.

2.2 Styrene Butadiene Rubber and Natural Rubber blend (SBR/NR)

Figure 3 illustrates cross-section of the mold which consists of a rectangular mold with inner dimensions of 100mm×100mm×30 mm and upper and lower aluminum-alloy hot plates heated by steam. Rubber sample was packed in the cavity.

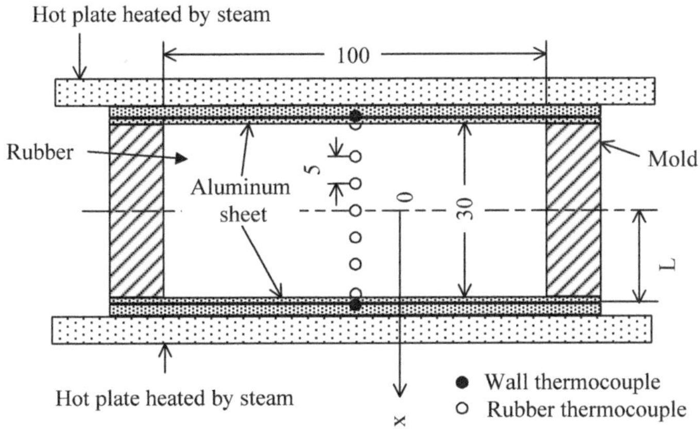

Fig. 3. Experimental mold and hot plates for SBR/NR

Energy transfer in the rubber is predominantly one-dimensional, transient heat conduction from the top and bottom plates to the rubber. To measure the through-the-thickness temperature profile along the central axis in the rubber, type-J thermocouples were located at an interval of 5 mm. Two wall thermocouples were packed between the hotplates and the rubber. All the thermocouples were led out through the mold and connected to the data logger, and the temperature outputs were subsequently recorded to 0.1K. The blend prepared includes 70 wt% styrene butadiene rubber (SBR) and 30 wt% natural rubber (NR). The peroxide was used as the curing agent. Ingredients are listed in Table 2.

To locate the rubber thermocouples at the prescribed positions rubber sheets with 5 mm thick were superposed appropriately.

Experiments were conducted under the condition of the heating wall temperature 433 K by changing the heating time in several steps from 50 to 120 minutes in order to study the dependencies of the degree of cure on the heating time. After the heating was terminated, the rubber was led out from the mold then immersed in ice water. The rubber was sliced 3 mm thick × 30 mm long in the vicinity of the central axis. Test pieces were prepared with dimensions of 3mm×3mm×3mm at x= -10, -5, 0, 5 and 10 mm, where the coordinate x is defined in Fig.3. The crosslink density was evaluated from the Flory-Rehner equation using the measured swelling data.

Ingredients	wt%
Polymer (SBR/NR) 70wt%SBR, 30wt%NR	86.2
Cure agent (Peroxide)	0.4
Reinforcing agent (Silica)	8.6
Processing aid	0.3
Activator	1.7
Antioxidant (1)	0.9
Antioxidant (2)	0.9
Coloring agent (1)	0.2
Coloring agent (2)	0.8

Table 2. Ingredients of compounded SBR/NR

3. Numerical prediction

Rubber curing processes such as press curing in a mold and injection curing are usually operated under unsteady state conditions. In case of the rubber with relatively large dimensions, low thermal conductivity of the rubber leads to non-uniform thermal history, which results to non-uniform degree of cure.

The present section describes theoretical models for predicting the degree of cure for the SBR and SBR/NR systems shown in the previous section. The model consists of solving one-dimensional, transient heat conduction equation with internal heat generation due to cureing reaction.

3.1 Heat conduction

Heat conduction equation with constant physical properties in cylindrical coordinates is

$$c\rho\frac{\partial T}{\partial \tau}=\frac{\lambda}{r}\frac{\partial}{\partial r}\left(r\frac{\partial T}{\partial r}\right)+\frac{dQ}{d\tau} \tag{2}$$

subject to

$$T = T_{init} \text{ for } \tau = 0 \tag{3a}$$

$$T = T_w(\tau) \text{ for } \tau > 0 \text{ and } r = r_M \tag{3b}$$

$$\partial T/\partial r = 0 \text{ for } \tau > 0 \text{ and } r = 0 \tag{3c}$$

where r is the radial coordinate, τ is the time , ρ is the density, c is the specific heat, λ is the thermal conductivity, $T_M(\tau)$ is the heating wall temperature, T_{init} is the initial temperature in the rubber and r_M is the inner radius of the mold.

Heat conduction equation in rectangular coordinates is

$$c\rho\frac{\partial T}{\partial \tau}=\lambda\frac{d^2T}{dx^2}+\frac{dQ}{d\tau} \tag{4}$$

subject to

$$T = T_{init} \text{ for } \tau = 0 \quad (5a)$$

$$T = T_w(t) \text{ for } \tau > 0 \text{ and } x = \pm L \quad (5b)$$

where x is the coordinate defined as shown in Fig. 3. The second term of the right hand sides of equations (2) and (4), $dQ/d\tau$, show the effect of internal heat generation expressed as

$$dQ/d\tau = \rho \Delta H \, d\varepsilon/d\tau \quad (6)$$

where ΔH is the heat of curing reaction and ε is the degree of cure.

3.2 Curing reaction kinetics

Prediction methods for the degree of cure ε in equations (2) and (4) have been derived by Onishi and Fukutani(2003a,2003b) and the models are adopted in this chapter.

3.2.1 Styrene Butadiene Rubber (SBR)

Curing process of SBR with sulfur has been analyzed and modeled by Onishi and Fukutani (2003a). A set of reactions is treated as the chain one which includes CBS thermal decomposition.

Simplified reaction model is shown in Fig. 4, where a is the effective accelerator, N is the mercapt of accelerator, M is the polysulfide, RN is the polysulfide of rubber, R^* is the active point of rubber, and RX is the crosslink site.

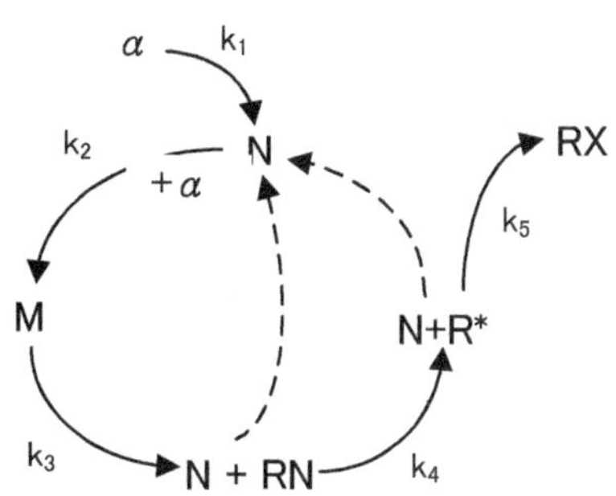

Fig. 4. Simplified curing model for SBR

The model can be expressed by a set of the following five chemical reactions.

$$\left.\begin{array}{ll} k_1 & a \rightarrow N \\ k_2 & N + a \rightarrow M \\ k_3 & M \rightarrow RN + N \\ k_4 & RN \rightarrow R^* + N \\ k_5 & R^* \rightarrow RX \end{array}\right\} \quad (7)$$

which leads the following rate equation set

$$\left.\begin{aligned} d[a]/d\tau &= -k_1[a] - k_2[N][a] \\ d[N]/d\tau &= k_1[a] - k_2[N][a] + k_3[M] + k_4[RN] \\ d[M]/d\tau &= k_2[N][a] - k_3[M] \\ d[RN]/d\tau &= k_3[M] - k_4[RN] \\ d[R^*]/d\tau &= k_4[RN] - k_5[R^*] \\ d[RX]/d\tau &= k_5[R^*] \end{aligned}\right\} \quad (8)$$

where $[a]$,$[N]$,$[M]$,$[RN]$,$[R^*]$ and $[RX]$ are the molar densities of appropriate species. Initial conditions of equation (8) are $[a] = 1$ and zero conditions for the rest of species. Rate constants k_i ($i = 1 \sim 5$) in the set were expressed using the Arrhenius form as

$$k_i = A_i \exp(-E_i/RT) \quad (9)$$

where A_i is the frequency factor of reaction i, E_i is the activation energy of reaction i, R is the universal gas constant, T is the absolute temperature. Values of A_i and E_i are shown in Table 3, where these values were derived from the analysis of the isothermal curing data using the oscillating rheometer in the range 403 K to 483K at an interval of 10K (Onishi and Fukutani, 2003a).

Sulfur concentration	1 wt %		5 wt %	
	A_i (1/s)	E_i/R (K)	A_i (1/s)	E_i/R (K)
k_1	1.034×10^{7}	1.166×10^{4}	1.387×10^{-1}	3.827
k_2	3.159×10^{13}	1.466×10^{4}	5.492×10^{8}	9.973
k_3	2.182×10^{7}	8.401×10^{3}	1.880×10^{9}	9.965
k_4	1.089×10^{7}	8.438×10^{3}	1.160×10^{9}	9.863
k_5	1.523×10^{9}	1.119×10^{4}	1.281×10^{9}	1.135×10

Table 3. Frequency factor and activation energy for SBR

3.2.2 Styrene butadiene rubber and natural rubber blend (SBR/NR)

Peroxide curing process for rubbers has been analyzed and modeled by Onishi and Fukutani (2003b). Simplified reaction model is shown in Fig. 5, where R is possible crosslink site of polymer, R^* is active cure site, PR the polymer radical, RX^* is the polymer radical with crosslinks and RX is the crosslink site.

	A_i (1/s)	E_i/R (K)
k_1	1.243×10^{8}	1.095×10^{4}
k_2	1.007×10^{15}	1.826×10^{4}
k_3	9.004×10^{2}	6.768×10^{3}
k_4	2.004×10^{6}	8.860×10^{3}
k_5	1.000×10^{-6}	-3.171×10^{3}

Table 4. Frequency factor and activation energy for SBR/NR

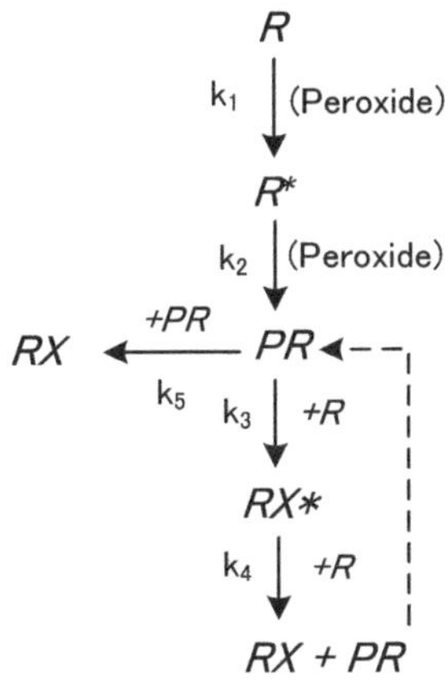

Fig. 5. Simplified curing model for SBR/NR

The model can be expressed by a set of the following five chemical reactions.

$$\left.\begin{array}{ll} k_1 & R \rightarrow R^* \\ k_2 & R^* \rightarrow PR \\ k_3 & PR + R \rightarrow RX^* \\ k_4 & RX^* + R \rightarrow RX + PR \\ k_5 & PR + R^* \rightarrow RX \end{array}\right\} \quad (10)$$

which leads the following rate equation set

$$\left.\begin{array}{l} d[R]/d\tau = - k_1[R] - k_3[PR][R] - k_4[RX^*][R] \\ d[R^*]/d\tau = k_1[R] - k_2[R^*] \\ d[PR]/d\tau = k_2[R^*] - k_3[PR][R] + k_4[RX^*][R] - 2k_5[PR]^2 \\ d[RX^*]/d\tau = k_3[PR][R] - k_4[RX^*][R] \\ d[RX]/d\tau = k_4[RX^*][R] + k_5[PR]^2 \end{array}\right\} \quad (11)$$

where [R], [R^*], [PR], [RX^*] and [RX] are the molar densities of appropriate species. Initial conditions of equation (11) are [R] = 2 and zero conditions for the rest of species.
Rate constants k_i (i = 1∼5) are listed in Table 4, where the values were obtained from the similar method conducted by Onishi and Fukutani (2003b).

3.3 Usage of the equations

For the SBR with sulfur curing system described in the previous section, we need to solve heat conduction equation (2) together with rate equation set (8) to obtain cure state distributions. Initial and boundary conditions for the temperatures were given by equation (3). Initial concentration conditions are described below equation set (8). Similar method can be adopted for estimating the SBR/NR system.

	SBR		SBR/NR
	Sulfur 1 wt%	Sulfur 5wt%	
Density ρ (kg/m³)	1.165×10^3		1.024×10^3
Thermal conductivity l (W/mK)	0.33		0.20
Specific heat capacity c (J/kgK)	1.84×10^3		1.95×10^3
Heat of reaction ΔH (J/kg)	1.23×10^4	3.99×10^4	2.78×10^4

Table 5. Physical properties used for prediction

The density ρ was determind using the mixing-rule. The thermal conductivity λ was measured using the cured rubber at 293K. DSC measurements of the specific heat capacity c and that of the heat of curing reaction ΔH for the rubber compounds were performed in the range 293 K to 453 K. The theromophysical properties used for the prediction are tabulated in Table 5. For the case of SBR with 5 wt% sulfur, a small correction of the specific heat capacity was made in the range 385.9 K to 392.9K to account for the effect of the fusion heat of crystallized sulfer. The solubility of sulfur in the SBR was assumed to be 0.8 wt% fom the literature (Synthetic Rubber Divison of JSR, 1989). Heat conduction equations (2) and (4) were respectively reduced to systems of simultaneous algebraic equations by a control-volume-based, finite difference procedure. Number of control volumes were 37 for SBR with 37.3 mm radius and 30 for SBR/NR with 30 mm thick. Time step of 0.5 sec was chosen after some trails.

4. Comparison with experimental data

4.1 Styrene Butadiene Rubber (SBR)

Figure 6 shows the temperature profile for the cured rubber with Method A, where solid and dashed lines respectively show the numerical results and the measured heating wall temperature. Symbols present measured rubber temperatures. In the figure for the measured temperatures, typical one-dimensional transient temperature field can be observed and it takes about 180 minutes to reach T_R to the final temperature T_w. Comparisons of the measured and predicted temperatures show good agreements between them. Also, the measured temperature difference along the axis between the positions at mid-cross section and that at 60 mm downward was less than 0.5 K. Since the difference is considerably smaller as compared to the radial one, one-dimensional transient heat conduction field is well established in the present experimental mold.

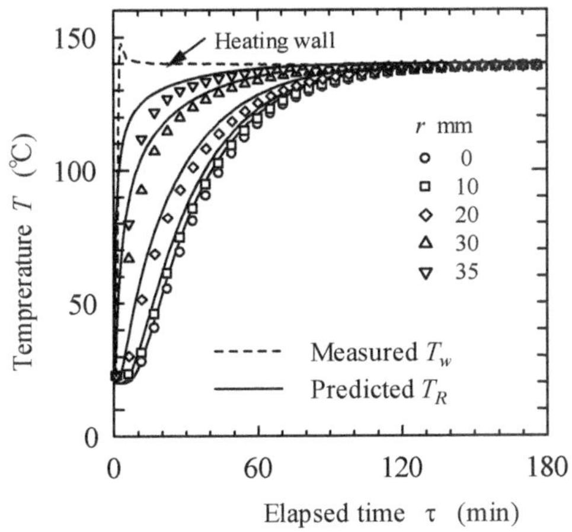

Fig. 6. Temperature profile for cured SBR, Method A

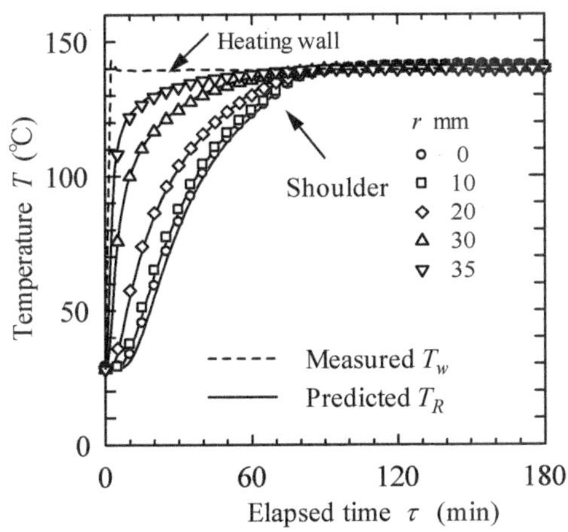

Fig. 7. Temperature profile for SBR with 1 wt% sulfur, Method A

Figure 7 is the result of the compounded rubber with Method A. The temperature rise is faster for the compounded rubber than for the cured one, and a uniform temperature field is observed at about τ = 95 minutes. The former may be caused by the internal heat generation due to curing reaction. The numerical results well follow the measured temperature history.

Figure 8 shows the numerical results of the internal heat generation rate $dQ/d\tau$ and the degree of cure ε corresponding to the condition of Fig.7. The $dQ/d\tau$ at each radial position r

shows a sharp increase and takes a maximum then decreases moderately. It can also be seen that the onset of the heat generation takes place, for example, at τ = 15 minutes for r = 35 mm, and at τ = 65 minutes for r = 0mm. This means that the induction time is shorter for nearer the heating wall due to slow heat penetration. Another point to note here is that the symmetry condition at r = 0, equation (3c), leads to the rapid increase of T_R near $r = 0$ after τ = 60 minutes is reached as shown in Fig.7. The degree of cure ε increases rapidly just after the onset of curing, then approaches gradually to 1 as shown in the lower part of Fig.8.
Figure 9 shows the profiles of rubber temperature and that of degree of cure, both are model calculated results. An overall comparison of the Figs. 9(a) and 9(b) indicates that the progress of the curing is much slower than the heat penetration. The phenomenon is pronounced in the central region of the rubber. Temperature profiles at τ = 90 and 105 minutes were almost unchanged, thus the two profiles can not be distinguished in the figure.

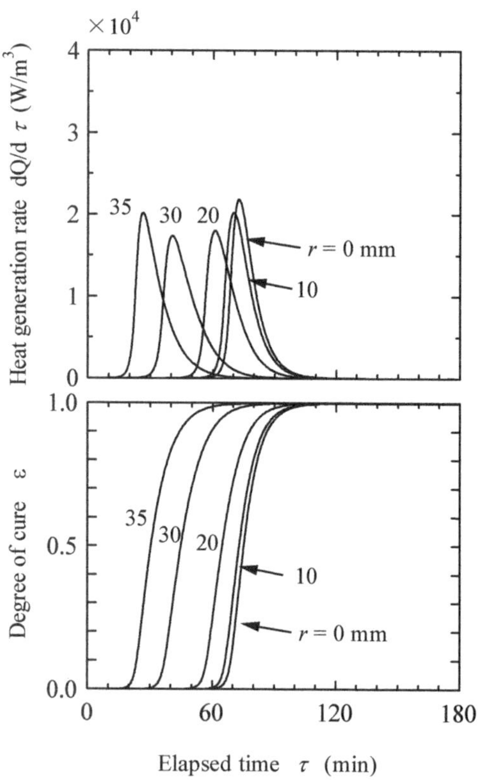

Fig. 8. Heat generation rate dQ/dτ and degree of cure ε for SBR with 1 wt% sulphur, Method A, corresponding to Fig.7

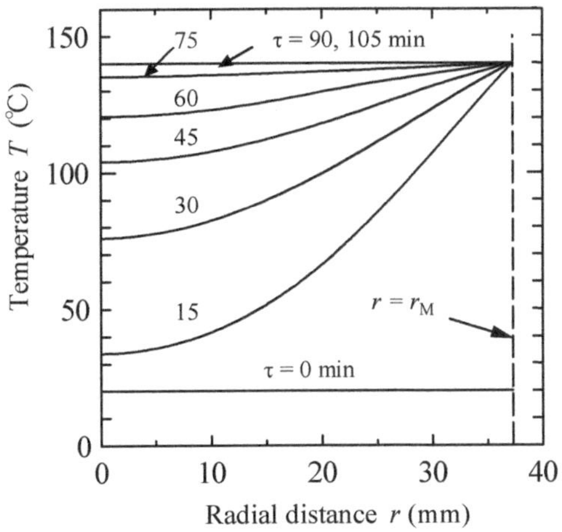

(a) Rubber temperature

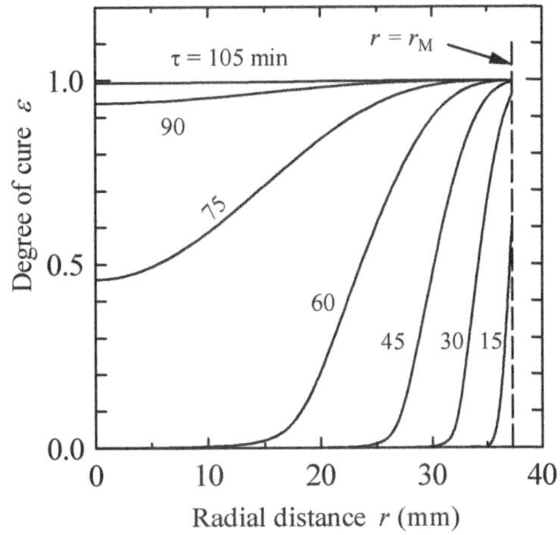

(b) Degree of cure

Fig. 9. Profiles of rubber temperature and degree of cure, SBR with 1wt% sulphur, Method A, corresponding to Figs.7 and 8

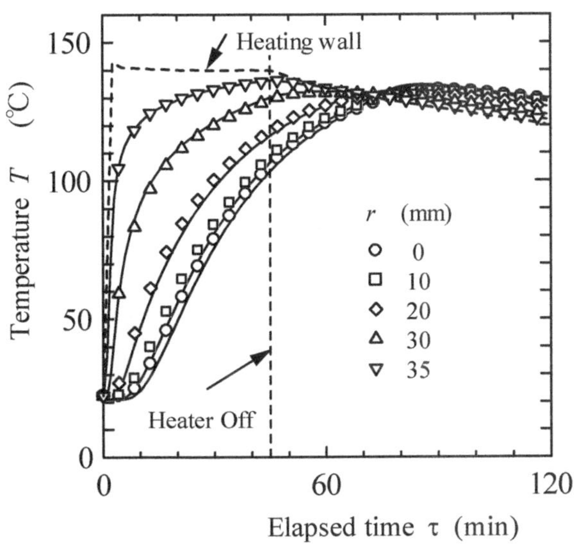

Fig. 10. Profile of rubber temperature for SBR with 1wt% sulfur, Method B

Figures 10 and 11 show the results of the SBR with 1 wt% sulfur with Method B. In Fig.10 for after τ = 45 minutes, T_R at r = 35 mm decreases monotonically, while that at r = 0 mm increases and takes maximum at τ = 90 minutes, then decreases. After τ = 75 minutes, negative temperature gradient to the heating wall can be observed in the rubber. This implies that the outward heat flow to the heating wall exists in the rubber.

In Fig.11 at τ = 45 minutes, the reaction only proceeds in the region $r > 30$ mm, and after τ = 45 minutes, the reaction proceeds without wall heating. Especially at r = 0 mm, degree of cure ε increases after τ = 75 minutes, where the negative temperature gradient to the heating wall is established as shown in Fig.10. The predicted ε well follow the measurements after τ = 45 minutes. These results indicate that the curing reaction proceeds without wall heating after receiving a certain amount of heat.

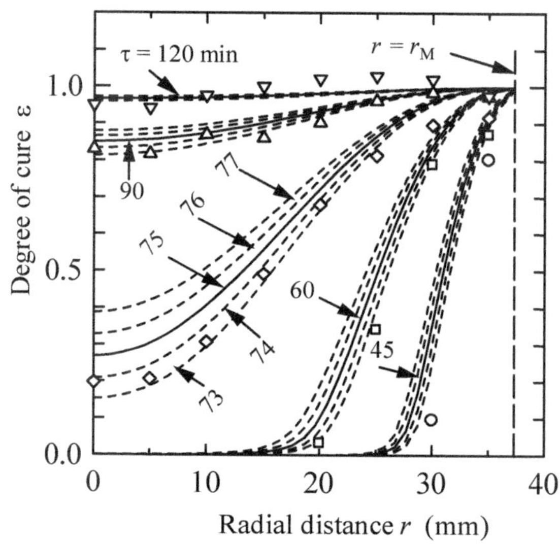

Fig. 11. Profile of degree of cure for SBR with 1wt% sulfur, Method B, corresponding to Fig.10

The results in Fig. 10 and 11 reveal that the cooling process plays an important role in the curing process, and wall heating time may be reduced by making a precise modelling for the curing process. Also good agreements of T_R and ε between the predictions and the measurements conclude that the present prediction method is applicable for practical use.

Figure 12 plots the numerical results of the internal heat generation rate $dQ/d\tau$ and the degree of cure ε corresponding to the condition of Fig.10. Comparison of $dQ/d\tau$ between Fig.8 for Method A and Fig.12 for Method B shows that the effects of wall heating after τ = 45 minutes is only a little, namely, the onset of the curing for Method B is a little slower than that for Method A. It takes 91 and 100 minutes for Method A and for Method B respectively to arrive at the condition of ε = 0.95 at r = 0mm.

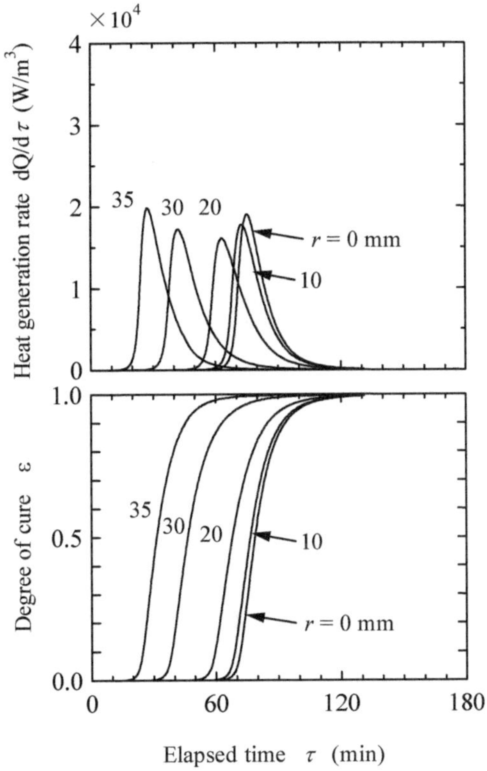

Fig. 12. Heat generation rate $dQ/d\tau$ and degree of cure ε for SBR with 1 wt% sulphur, Method B, corresponding to Figs. 10 and 11

Figures 13 and 14 show the results of SRB with 5wt% sulfur. Model calculated temperatures in Fig.13 well follow the measurements before τ = 80 minutes. An overall inspection of $dQ/d\tau$ in Figs. 8 and 14 shows that the induction time for 5wt% is shorter than for 1 wt%, whereas the $d\varepsilon/d\tau$ for 5wt% is smaller than for 1 wt%. Also the amount of heat generation rate for each r is more remarkable for 5wt% than for 1 wt%. It takes 89 minutes to arrive at the condition of $\varepsilon = 0.95$ at $r = 0$ mm.

Again in Fig.13, in the region after τ = 80 minutes, the rubber temperature less than r = 20 mm increases and takes maximum of 429 K then gradually decrease. The model can not predict the experimental results. It is not possible to make conclusive comments, but the experimental results may be caused by the crosslink decomposition reaction, complex behaviour of free sulfur, etc. not taken into consideration in the present prediction model. Effects of sulphur on the curing kinetics, especially on the post-crosslinking chemistry have not been well solved. For reference, recently Miliani,G. and Miliani,F (2011) reviewed relevant literature and proposed a macroscopic analysis.

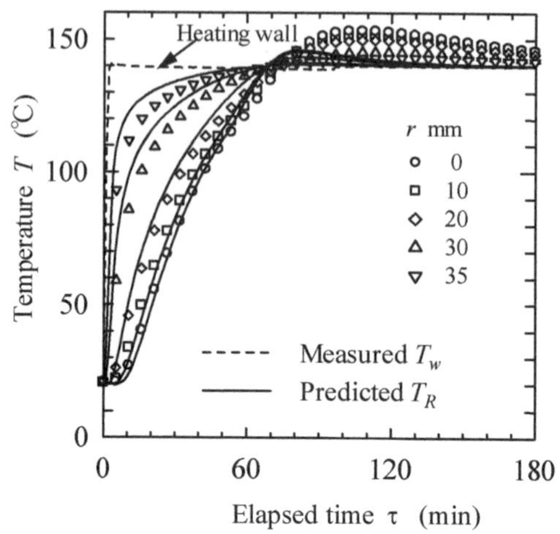

Fig. 13. Temperature profile for SBR with 5wt% sulphur

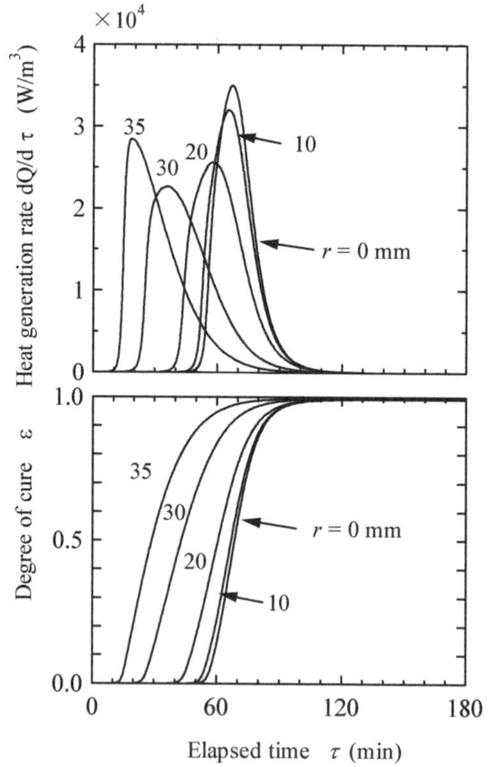

Fig. 14. Heat generation rate dQ/dτ and degree of cure ε for SBR with 5 wt% sulphur, corresponding to Fig.13

4.2 Styrene Butadiene Rubber and Natural Rubber blend (SBR/NR)

Figures 15 and 16 show the SBR/NR results. As shown in the figures, the experimental results of rubber temperature and degree of cure almost follow the predictions.

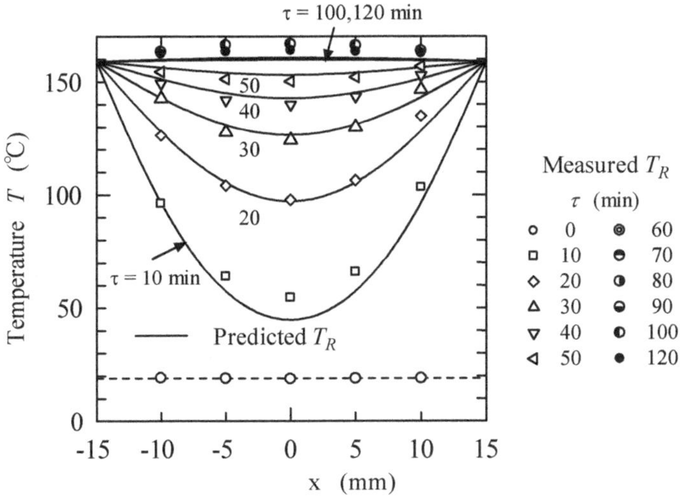

Fig. 15. Cross-sectional view of temperature history for SBR/NR

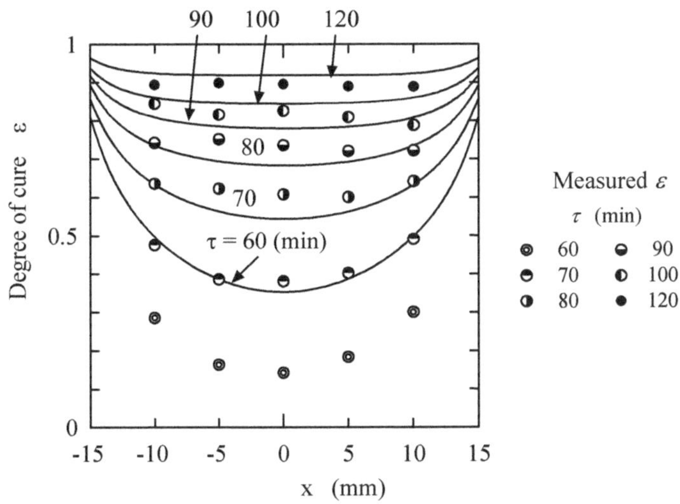

Fig. 16. Cross-sectional view of degree of cure for SBR/NR, corresponding to Fig.15

It is clear that the temperature at smaller x after τ = 60 minutes is higher than the heating wall value. This is due to the effect of internal heat generation.

Figure 17 plots the model calculated values of $dQ/d\tau$ and ε, where values at the heating wall (x = 15 mm) are also shown for reference. A quick comparison of Fig.8 for SBR and Fig.17 for SBR/NR apparently indicates that the gradient of $d\varepsilon/d\tau$ for SBR is larger than that for

SBR/NR. This means that more precious expression for curing kinetics may be required for making a controlled gradient of the degree of cure in a thick rubber part, because thermal history influences the curing process.

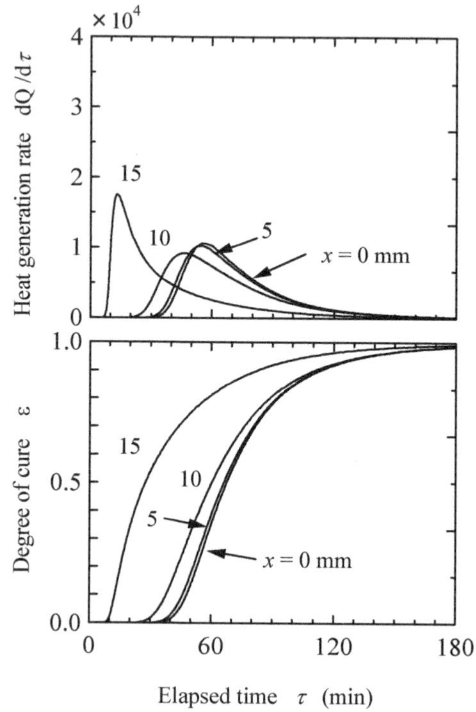

Fig. 17. Heat generation rate $dQ/d\tau$ and degree of cure ε for SBR/NR, corresponding to Figs. 15 and 16

5. Concluding remarks

A Prediction method for rubber curing process has been proposed. The method is derived from our experimental and numerical studies on the styrene butadiene rubber with sulphur curing and the blend of styrene butadiene rubber and natural rubber with peroxide curing systems. Following concluding remarks can be derived.

1. Rate equation sets (8) and (11), both obtained from isothermal oscillating rheometer studies, are applicable for simulation driven design of rubber article with relatively large size. The former set is applicable to SBR with sulphur/CBR curing and the latter for rubbers with peroxide curing.
2. Rafei et al. (2009) has pointed out that no experimental verification on the accuracy of the predicted degree of cure comparing with directly measured data. It is important to solve this problem as quickly as possible. When the problem is solved, it can determine whether or not to take into account the effects of temperature dependencies of thermophysical properties.

3. Curing reaction under the temperature decreasing stage can also be evaluated by the present prediction method.
4. Extension of the present prediction methods to realistic three-dimensional problems may be relatively easy, since we have various experiences in the fields of numerical simulation and manufacturing technology.

6. References

Abhilash, P.M. et al., (2010). Simulation of Curing of a Slab of Rubber, *Materials Science and Engineering B,* Vol.168, pp.237-241, ISSN 0921-5107

Baba T. et al., (2008). A Prediction Method of SBR/NR Cure Process, *Preprint of the Japan Society of Mechanical Engineers, Chugoku-Shikoku Branch,* No.085-1, pp.217-218, Hiroshima, March, 2009

Coran, A.Y. (1964). Vulcanization. Part VI. A Model and Treatment for Scorch Delay Kinetics, *Rubber Chemistry and Technology,* Vol.37, pp. 689-697, ISSN= 0035-9475

Ding, R. et al., (1996). A Study of the Vulcanization Kinetics of an Accelerated-Sulfur SBR Compound, *Rubber Chemistry and Technology,* Vol.69, pp. 81-91, ISSN= 0035-9475

Flory, P.J and Rehner,J (1943a). Statistical Mechanics of Cross-Linked Polymer Networks I. Rubberlike Elasticity, *Journal of Chemical Physics,* Vol.11, pp.512- ,ISSN= 0021-9606

Flory, P.J and Rehner,J (1943b). Statistical Mechanics of Cross-Linked Polymer Networks II. Swelling, *Journal of Chemical Physics,* Vol.11, pp.521- ,ISSN=0021-9606

Guo,R., et al., (2008). Solubility Study of Curatives in Various Rubbers, *European Polymer Journal,* Vol.44, pp.3890-3893, ISSN=0014-3057

Ghoreishy, M.H.R. and Naderi, G. (2005). Three-dimensional Finite Element Modeling of Rubber Curing Process, *Journal of Elastomers and Plastics,* Vol.37, pp.37-53, ISSN 0095-2443

Ghoreishy M.H.R. (2009). Numerical Simulation of the Curing Process of Rubber Articles, In : *Computational Materials,* W. U. Oster (Ed.) , pp.445-478, Nova Science Publishers, Inc., ISBN= 9781604568967, New York

Goyanes, S. et al., (2008). Thermal Properties in Cured Natural Rubber/Styrene Butadiene Rubber Blends, *European Polymer Journal,* Vol.44, pp.1525-1534, ISSN= 0014-3057

Hamed,G.R. (2001). *Engineering with Rubber; How to Design Rubber Components* (*2nd edition*), Hanser Publishers, ISBN=1-56990-299-2, Munich

Ismail,H. and Suzaimah,S. (2000). Styrene-Butadiene Rubber/Epoxidized Natural Rubber Blends: Dynamic Properties, Curing Characteristics and Swelling Studies, *Polymer Testing,* Vol.19, pp.879-888, ISSN=01420418

Isayev, A.I. and Deng, J.S. (1987). Nonisothermal Vulcanization of Rubber Compounds, *Rubber Chemistry and Technology,* Vol.61, pp.340-361, ISSN 0035-9475

Kamal, M.R. and Sourour,S., (1973). Kinetics and Thermal Characterization of Thermoset Cure, *Polymer Engineering and Science,* Vol.13, pp.59-64, ISSN= 0032-3888

Labban A. EI. et al., (2007). Numerical Natural Rubber Curing Simulation, Obtaining a Controlled Gradient of the State of Cure in a Thick-section Part, In:*10th ESAFORM Conference on Material Forming (AIP Conference Proceedings),* pp.921-926, ISBN= 9780735404144

Likozar,B. and Krajnc,M. (2007). Kinetic and Heat Transfer Modeling of Rubber Blends' Sulfur Vulcanization with N-t-Butylbenzothiazole-sulfenamide and N,N-Di-t-butylbenzothiazole-sulfenamide, *Journal of Applied Polymer Science*, Vol.103, pp.293-307. ISSN=0021-8995

Likozar, B. and Krajnc, M. (2008). A Study of Heat Transfer during Modeling of Elastomers, *Chemical Engineering Science*, Vol.63, pp.3181-3192, ISSN 0009-2509

Likozar,B. and Krajnc,M. (2011). Cross-Linking of Polymers: Kinetics and Transport Phenomena, *Industrial & Engineering Chemistry Research*, Vol.50, pp.1558-1570. ISSN= 0888-5885

Marzocca,A.J. et al., (2010). Cure Kinetics and Swelling Behaviour in Polybutadiene Rubber, *Polymer Testing*, Vol.29, pp.477-482, ISSN= 0142-9418

Milani,G and Milani,F. (2011). A Three-Function Numerical Model for the Prediction of Vulcanization-Reversion of Rubber During Sulfur Curing, *Journal of Applied Polymer Science*, Vol.119, pp.419-437, ISSN= 0021-8995

Nozu,Sh. et al., (2008). Study of Cure Process of Thick Solid Rubber, *Journal of Materials Processing Technology*, Vol.201, pp.720-724 , ISSN=0924-0136

Onishi,K and Fukutani,S. (2003a). Analyses of Curing Process of Rubbers Using Oscillating Rheometer, Part 1. Kinetic Study of Curing Process of Rubbers with Sulfur/CBS, *Journal of the Society of Rubber Industry, Japan*, Vol.76, pp.3-8, ISSN= 0029-022X

Onishi,K and Fukutani,S. (2003b). Analysis of Curing Process of Rubbers Using Oscillating Rheometer, Part 2. Kinetic Study of Peroxide Curing Process of Rubbers, *Journal of the Society of Rubber Industry, Japan*, Vol.76, pp.160-166, ISSN= 0029-022X

Rafei, M et al., (2009). Development of an Advanced Computer Simulation Technique for the Modeling of Rubber Curing Process. *Computational Materials Science*, Vol.47, pp. 539-547, ISSN 1729-8806

Synthetic Rubber Division of JSR Corporation, (1989). *JSR HANDBOOK*, JSR Corporation, Tokyo

Tsuji, H. et al., (2008). A Prediction Method for Curing Process of Styrene-butadien Rubber, *Transactions of the Japan Society of Mechanical Engineers, Ser.B*, Vol.74, pp.177-182, ISSN=0387-5016

7

Experimental and Numerical Evaluation of Thermal Performance of Steered Fibre Composite Laminates

Z. Gürdal[1], G. Abdelal[2] and K.C. Wu[3]
[1]*Delft University of Technology*
[2]*Virtual Engineering Centre, University of Liverpool*
[3]*Structural Mechanics and Concepts, NASA Langley Research Center*
[1]*The Netherlands*
[2]*UK*
[3]*USA*

1. Introduction

For Variable Stiffness (VS) composites with steered curvilinear tow paths, the fiber orientation angle varies continuously throughout the laminate, and is not required to be straight, parallel and uniform within each ply as in conventional composite laminates. Hence, the thermal properties (conduction), as well as the structural stiffness and strength, vary as functions of location in the laminate, and the associated composite structure is often called a "variable stiffness" composite structure. The steered fibers lead not only to the alteration of mechanical load paths, but also to the alteration of thermal paths that may result in favorable temperature distributions within the laminate and improve the laminate performance. Evaluation of VS laminate performance under thermal loading is the focus of this chapter. Thermal performance evaluations require experimental and numerical analysis of VS laminates under different processing and loading conditions. One of the advantages of using composite materials in many applications is the tailoring capability of the laminate, not only during the design phase but also for manufacturing. Heat transfer through variable conduction and chemical reaction (degree of cure) occurring during manufacturing (curing) plays an important role in the final thermal and mechanical performance, and shape of composite structures.

Three case studies are presented in this chapter to evaluate the thermal performance of VS laminates prior to and after manufacturing. The first case study is a numerical analysis that investigates the effect of variable conductivity within the VS laminate on the temperature and degree of cure distribution during its cure cycle. The second case study is a numerical analysis that investigates the transient thermal performance of a rectangular VS composite laminate. Variable thermal conductivity will affect the temperature profile and results are compared to a unidirectional composite. The effect of fiber steering on transient time and steady state solutions is compared to the effect of unidirectional fibers. The third case study is an experimental one that was conducted to evaluate the thermal performance of two variable stiffness panels fabricated using an Advanced Fiber Placement (AFP) Machine.

These variable stiffness panels have the same layup, but one panel has overlapping bands of unidirectional tows (which lead to thickness variations) and the other panel does not. Results of thermal tests of the variable stiffness panels are presented and compared to results for a baseline cross-ply panel. These case studies will show the impact of the steering parameters of variable stiffness laminates during the manufacturing and design phases.

2. Preliminaries

Laminated composite plates structures, which have high strength-to-weight and stiffness-to-weight ratios, are extensively used in aerospace and automotive applications that are exposed to elevated temperatures. Accurate knowledge of the thermal response of these materials is essential for the optimum design of thermal protection systems. In some circumstances, high thermally induced compressive stresses may be developed in the constrained plates and can therefore lead to buckling failures. In addition, conductivity of the fiber-reinforced layers is direction dependent, and therefore the degree of anisotropy of the laminate can substantially influence the conductivity of the laminate in different directions.

Variable stiffness laminates with steered fiber paths offer stiffness tailoring possibilities that can lead to alteration of load paths, resulting in favorable temperature distributions within the laminate and improved laminate structural performance. A further generalization of this idea was to allow the direction of the fiber orientation angle variation to be rotated with respect to the coordinate direction x, rather than limiting it to be only along the x-axis or the y-axis. As shown in Figure 1, a fiber orientation angle T_0 is defined at an arbitrary reference point A with respect to direction x' that is rotated by an angle ϕ from the coordinate axis x. The fiber orientation angle is then assumed to reach a value T_1 at point B located a characteristic distance d from point A. With the linear variation of the fiber orientation angle between the points A and B, the equation for the fiber orientation angle along this reference path takes the form,

$$\theta(x') = \phi + (T_1 - T_0)\frac{|x'|}{d} + T_0 \tag{1}$$

A variable stiffness layer can be represented with three angles and a characteristic distance to represent a single layer. Assuming the characteristic distance to be associated with a geometric property of the part, representation of a single curvilinear layer may be specified by $\phi<T_0|T_1>$. A conventional representation, a $\pm$ sign in front of either ϕ or $<T_0|T_1>$ means that there are two adjacent layers with equal and opposite variation of the fiber orientation angle. A laminate with $\pm\phi \pm<T_0|T_1>$ designation will have four curvilinear layers: a $\pm\phi$ $<T_0|T_1>$ pair along the $+\phi$ direction, and a $\pm\phi$ $<T_0|T_1>$ pair along the $-\phi$ direction.

Variable stiffness (VS) laminates were introduced by (Gürdal and Olmedo, 1993; Olmedo and Gürdal, 1993). Examples of fiber orientation angle tailoring include theoretical and numerical studies by (Banichuk, 1981; Banichuk and Sarin, 1995), (Pedersen, 1991, 1993)], and (Duvaut et al., 2000). In a design study by (Gürdal et al., 2008), analyses of variable stiffness panels for in-plane and buckling responses are developed and demonstrated for two distinct cases of stiffness variation. Later optimization studies (Setoodeh et al., 2006, 2007, 2009) were carried out demonstrating the theoretical benefits of variables stiffness laminates in improving structural performance. For variable stiffness laminates, which also have spatially variable coefficients of thermal expansion along with the stiffness properties, spatial variation of residual stresses are induced. More recent studies investigated the effect of thermal residual stresses on the mechanical buckling performance of variable stiffness laminates (Abdalla et al., 2009).

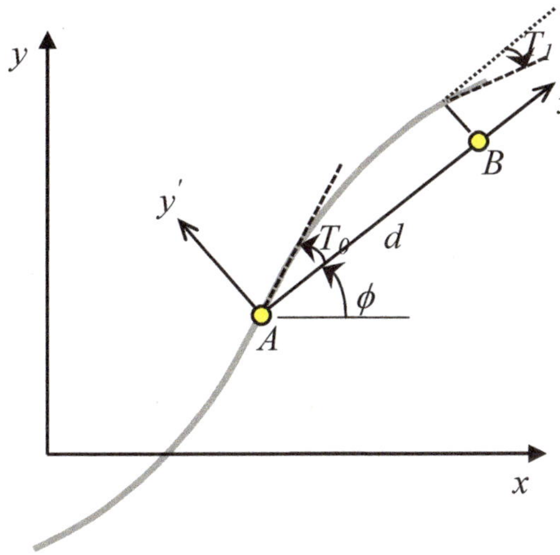

Fig. 1. Reference path definition of a variable layer

The degree of damage and strength degradation of the VS laminates subjected to severe thermal environments is a major limiting factor in relation to service requirements and lifetime performance. In order to predict these thermally induced stresses, a detailed understanding of the transient temperature distributions is essential. A number of studies of isotropic materials and composites have been carried out to explore the potentially complicated time dependence of the temperature field (Hetnarski, 1996). There have also been numerous analytical models developed over the years to describe the transient behaviors of commonly encountered geometries (Mittler et al., 2003; Obata and Noda, 1993).

Alternatively, in order to endure the severe thermal loads, many structural components are made such that they are non-homogeneous. Thermal barrier coatings of super alloys on ceramics used in jet engines, stainless steel cladding of nuclear pressure vessels, and a great variety of diffusion-bonded materials used in microelectronics may be mentioned as some examples. Typically, these non-homogeneous materials and structures are subjected to severe residual stresses upon cooling from their processing temperatures. They may also undergo thermal cycling during operations. Depending on the temperature gradients, the underlying thermal stress problem may be treated as a thermal shock problem or as a quasi-static isothermal problem in the sense that the problem may still be time-dependent but with no variation of temperature within the composite solid.

Clearly, there is an actual need for investigating the thermal transients developed within VS composites. However, because of the inherent mathematical difficulties, thermal analyses of non-homogeneous structural materials are considerably more complex than in the corresponding homogeneous case. Numerical techniques such as the finite element method are of great importance for solving for temperature profile and stress distribution in VS laminates.

3. Numerical heat conduction analysis

The main purpose of this section is to study a finite-element approximation to the solution of the transient and steady-state heat conduction with different boundary conditions in a two-dimensional rectangular region. A Galerkin finite element formulation is applied to the 2-D heat conduction equation using four planar finite elements. The composite conduction and capacitance matrices are derived as functions of the steered fiber orientation angles.

3.1 Two-dimensional heat conduction in Cartesian coordinates

The governing differential equation for a two dimensional heat conduction problem is given by,

$$k_x\, T_{,xx} + k_y\, T_{,yy} + Q = \rho\, c_p\, T_{,t} \tag{2}$$

where T is the temperature, k_x and k_y are material conductivity along the x- and y-directions, ρ is material density, Q is the inner heat-generation rate per unit volume, and c_p is material heat capacity.
Boundary conditions are described as,

$$\begin{aligned} &T = T_0\big|_{S=S1} \quad \text{(specified temperature)} \\ &k_x\, T_{,x} + k_y\, T_{,y} = q\big|_{S=S2} \quad \text{(specified heat flow)} \\ &k_x\, T_{,x} + k_y\, T_{,y} = h(T_s - T_e)\big|_{S=S3} \quad \text{(convection boundary condition)} \end{aligned} \tag{3}$$

where h is the convection coefficient, Ts is an unknown surface temperature, Te is a convective exchange temperature, and q is the incident heat flow per unit surface area.

3.2 Heat conduction simulation using finite element

Using a typical Galerkin finite element approach to Eqn. (2) the residual equation for a plate with unit thickness assumes the form,

$$\iint [N]\left[k_x\, T_{,xx} + k_y\, T_{,yy} + Q - \rho\, c_p\, T_{,t}\right] dx\, dy = 0 \tag{4}$$

where the temperature field function is expressed in terms of the interpolation functions as,

$$[T] = [N][T]^e \tag{5}$$

Integration of Eqn. (4) by parts yields,

$$\begin{aligned} &\iint \left[[N_{,x}]^T k_x [N_{,x}] + [N_{,y}]^T k_y [N_{,y}] \right] dx\, dy\, \{T\}^e + \\ &\rho\, c_p \iint [N]^T [N] dx\, dy\, \{T_{,t}\} + \int_V Q\, N_i\, dV = \int [N]^T q\, dS \end{aligned} \tag{6}$$

where q is the heat flow through unit area. Finally the governing equation takes the form,

$$[K_T + K_h]\ \{T\}\ + [C]\ \{\dot{T}\}\ =\ \{F_T\}\ + \{R_Q\} + \{R_h\} \tag{7}$$

where,

$[K_C]$ = composite conduction matrix, $Watt / C^\circ$ = $\int_V N^T [k] N\, dV$

$[K_h]$ = composite convection matrix, $Watt / C^\circ$ = $\int_{S3} h\, N^T\, N\, dS$

$\{T\}$ = unknown nodal temperature vector, C°.

$[C_T]$ = capacitance matrix, J/C° = $\int_V \rho\, c_p\, N^T\, N\, dV$

$\{\dot{T}\}$ = $\partial\{T\}^e / \partial t$

$\{F_T\}$ = composite element nodal force vector, Watt = $\int_V N^T\, q_s\, dS$

$\{R_Q\}$ = internal-heat vector, $Watt / m^3\, C^\circ$ = $\int_V Q\, N^T\, dV$

$\{R_h\}$ = convection heat vector, $Watt / m^2\, C^\circ$ = $\int_{S3} h\, T_e\, N^T\, dS$

N = interpolation function, $\{T\} = [N]\{T\}^e$

Introducing the normalized lamination parameters [Gürdal et al., 1999],

$$\begin{aligned} \{V_1, V_2, V_3, V_4\} &= \int_{-1/2}^{1/2} \{\cos 2\theta, \sin 2\theta, \cos 4\theta, \sin 4\theta\} d\bar{z} \\ \{W_1, W_2, W_3, W_4\} &= 12 \int_{-1/2}^{1/2} \bar{z}^2\, \{\cos 2\theta, \sin 2\theta, \cos 4\theta, \sin 4\theta\} d\bar{z} \end{aligned} \tag{8}$$

the conductivity matrix [k] can be expressed as a function of the lamination parameters,

$$[k] = K_0 + K_1\, V_1 + K_2\, V_2 \tag{9}$$

$$K_0 = \begin{bmatrix} 0.5(k_{11}+k_{22}) & 0 \\ 0 & 0.5(k_{11}+k_{22}) \end{bmatrix} \quad K_1 = \begin{bmatrix} 0.5(k_{11}-k_{22}) & 0 \\ 0 & 0.5(k_{11}-k_{22}) \end{bmatrix}$$
$$K_2 = \begin{bmatrix} 0 & 0.5(k_{22}-k_{11}) \\ 0.5(k_{22}-k_{11}) & 0 \end{bmatrix} \tag{10}$$

where k_{11} and k_{22} are the conductivity of the lamina along and perpendicular to the fibers directions respectively.

The finite element solution of time-dependent field problems produces a system of linear first-order differential equations in the time domain. These equations must be solved before the variation of the temperature T in space and time is known. There are several procedures for numerically solving Eqn. (7). Finite difference approximation (central difference) in the time domain is applied to generate a numerical solution,

$$\left([C_T] + \frac{\Delta t}{2}[K_T]\right)\{T\}_{i+1} = \left([C_T] - \frac{\Delta t}{2}[K_T]\right)\{T\}_i + \frac{\Delta t}{2}\left(\{F_T\}_i + \{F_T\}_{i+1}\right) \tag{11}$$

If Δt (time increment) and the material properties are independent of time, Eqn. (11) can be formulated as,

$$[A]\{T\}_{i+1} = [P]\{T\}_i + \{F^*\} \quad (12)$$

where [A] and [P] are combinations of [C] and [K], and $\{F^*\} = \frac{\Delta t}{2}\left(\{F_T\}_i + \{F_T\}_{i+1}\right)$.

3.3 Numerical stability techniques

Transient heat conduction problems can be solved by first discretizing the spatial dimensions using the finite element method, then transforming the space-time partial differential equation (PDE) into an ordinary differential equation (ODE) in time. The time integration of the discrete Eqn. (7) is then performed using ODE integrators. These integrators replace the time derivative by a finite difference approximation (forward, central, or backward differences). This integration method is inexpensive per step, but numerical stability requires using small time-steps. The numerical oscillations in the values of {T} from one time-step to the next are related to $[A]^{-1}[P]$, and can be avoided by applying the following criteria (Segerlind, 1984),

$$\begin{aligned} &\det\left([C_T]^e - \delta [K_T]^e\right) = 0 \\ &\Delta t < 2\delta \end{aligned} \quad (13)$$

(Trujillo, 1977) has proposed an explicit algorithm that has a time-step fifteen times greater than the conventional method used. (Hughes et al., 1982) proposed an element-by-element implicit algorithm to solve Eqn. (7). (Zienkiwicz et al., 1980) proposed a procedure based on systemic partitioning of the discrete Eqn. (7) and involving extrapolation.

4. Curing simulation of variable stiffness laminate

Thermoset polymers often release a significant heat of reaction during processing. The chemical reaction that occurs during the curing of thermoset polymers plays an important role in the process modeling of thermoset composites. The exothermic heat released during the curing process can cause excessive temperatures in the interior of composites. Cure kinetics that provide information on the curing rate and the amount of exothermic heat release during the chemical reaction are important in the process simulation of composite materials with thermoset polymers. Therefore, the inclusion of an accurate cure kinetics model is essential for the processing simulation of thermoset composites. Several studies have shown that amine-cured epoxy resins are governed by an autocatalytic reaction (Johnston, 1997; Scott, 1991). The epoxy group that reacts with a primary amine produces a secondary amine and then forms a tertiary amine. These reactions are also accelerated by the catalytic action of the hydroxyl group that is formed as a by-product of the amine–epoxy reaction.

The majority of heat transfer models for composites processing consider heat flow in the through-thickness direction only (a one-dimensional model) or make the even simpler assumption of a uniform laminate temperature (White and Hahn, 1992). More sophisticated models examining heat transfer in two and three dimensions have also been developed (Bogetti and Gillespie, 1991). The resin chemical reaction is generally indicated by a time-dependent measure known as the degree of cure, α, as in Equation (14) which is usually defined based on a measure of the heat given off by bond formation as follows:

$$\alpha = \frac{1}{H_R}\int_0^t \left(\frac{dq}{dt}\right) dt \tag{14}$$

where (dq/dt) is the rate of heat generation and H_R is the total amount of heat generated per unit volume during a full reaction. The value of α can be easily determined using standard differential scanning calorimetry (DSC) heat flow measurements. The heat generated due to cure of the resin Q^* is expressed as,

$$Q^{\bullet} = \rho\, H_R \frac{d\alpha}{dt} \tag{15}$$

where $d\alpha/dt$ is designated as the cure rate and ρ is the composite material density.

A number of different models have been proposed to simulate the cure kinetics of various resin systems. These models can be divided into two general types: mechanistic and empirical. Mechanistic models use as their basis a detailed understanding of the chemical reactions that take place throughout the cure process. Such models are not always practical due to the complexity of the chemical reactions that can vary greatly from one resin system to another depending on the combination of base resin, hardeners and catalysts. Furthermore, most commercial resin systems are proprietary and the user is often prohibited from even trying to determine their exact composition. Empirical models are expressions derived to provide a fit to experimentally determined cure rates. Most cure models employ Arrhenius-type equations, with the rate of reaction expressed as some function of degree of cure and temperature. An example of a commonly used semi-empirical expression for cured epoxy systems is (Scott, 1991):

$$\begin{aligned} \frac{d\alpha}{dt} &= \left(Y_1 + Y_2\,\alpha^m\right)(1-\alpha)^n \\ Y_i &= A_i \exp\left(-\Delta E_i / RT\right) \end{aligned} \tag{16}$$

where ΔE_i are activation energies, R is the gas constant, and A_i, m, n are experimentally-determined constants. To account for the effect of glass transition on reaction rate, a combination of expressions of the form of Eqns. (16) (Lee et al., 1982; White and Hahn, 1992) and modified expressions of similar type have been used (Dusi et al., 1987; Cole et al., 1991).

$$\frac{d\beta}{dt} = \left(Y_1 + Y_2\,\beta^m\right)(1-\beta)^n \tag{17}$$

where Y_i are defined as in Eqn. (16) and m and n are other experimentally determined constants. The parameter β is the "isothermal degree of cure", defined as the limiting degree of cure at a given temperature. Its relation to the degree of cure is given by:

$$\alpha = \beta\,\frac{H_T(T)}{H_R} \tag{18}$$

where $H_T(T)$ is the total amount of heat that would be given off by isothermally curing the resin at temperature T for infinite time.

4.1 Thermal-chemical model

A thermo-chemical model simulation requires the determination of the reaction kinetics of each resin and thermal transport of the heat of reaction across the laminate panel to calculate changes in the laminate temperature that affect thermal strains, and degree of cure that affects resin modulus. The following mathematical equation is modeled applying transient thermal analysis.

$$\rho c_p \frac{\partial T}{\partial t} = \frac{\partial}{\partial x_i}\left(k_{ij} \frac{\partial T}{\partial x_j} \right) + \rho H_R \frac{\partial c}{\partial t} + Q_v \qquad (i,j=1,2) \tag{19}$$

where ρ denotes the composite density, c_p the specific heat, T the temperature, t the time, x_i the spatial coordinates, and k_{ij} the components of the thermal conductivity tensor. The degree of cure c is defined as the ratio of the heat released by the reaction to the ultimate heat of reaction H_R, and Q_v is the heat convection to the surrounding air in the autoclave. Orthotropic conductivity is assumed for all materials so that values of k_{11}, k_{22}, and k_{33} are required. Assuming resin conductivity is isotropic and fibre conductivity is transversely isotropic, a rule of mixture is used (Twardowski et al., 1993) to evaluate lamina thermal conductivity:

$$\begin{aligned} K_{11} &= V_f . K_{f11} + \left(1 - V_f\right) K_r \\ K_{22} &= K_r . \left[\left(1 - 2.\sqrt{\frac{V_f}{\pi}} \right) + \left(\frac{1}{B} \right) . \left[\pi - \left(\frac{4}{\sqrt{1 - B^2 . \frac{V_f}{\pi}}} \right) . a\tan \left(\frac{\sqrt{1 - B^2 . \frac{V_f}{\pi}}}{1 + B . \sqrt{\frac{V_f}{\pi}}} \right) \right] \right] \\ B &= 2. \left[\left(\frac{K_r}{Kf_{22}} \right) - 1 \right] \end{aligned} \tag{20}$$

where K_{f11}, K_{f22} are the longitudinal and transverse conductivities of the fibers, and K_r is the isotropic conductivity of the resin. These values are assumed as a function of temperature. Laminate global conductivity is calculated as described in eqn. (9) and eqn. (10).

The lamina specific heat capacity is calculated using the following equation (Gürdal et al., 1999),

$$C_p = \frac{V_f C_{pf} \rho_f + \left(1 - V_f\right) C_{pr} \rho_r}{V_f \rho_f + \left(1 - V_f\right) C_{pr} \rho_r} \tag{21}$$

where C_{pf} and C_{pr} are the respective specific heats of the fibers and resin, and are modelled as a function of temperature. The internal heat generated due to the exothermic cure reaction is described as in (White and Hahn, 1992),

$$\frac{dc}{dt} = Y\alpha^{m}(1-\alpha)^{n}$$
$$Y = Ae^{-\Delta E/RT} \tag{22}$$

where Y is the Arrhenius rate, R=8.31 J/Mol.K is the universal gas constant, A the frequency factor, (m, and n) are experimentally-determined constants for a given material, and ΔE the activation energy. The heat transfer coefficient for an autoclave of size (1.8 m x 1.5 m) is measured and can be expressed as (Johnston, 1997),

$$h = 20.1 + 9.3\ (10^{-5})\ P \quad (W/m^2\ K) \tag{23}$$

where P is the autoclave pressure.

4.2 Cure simulation using finite elements

A thermochemical model is used for calculation of temperature and the degree of cure in composite components. Accurate prediction of these parameters is potentially useful to the composites modeling in a number of ways. First of all, one of the main objectives of processing thermoset composite materials is to achieve uniform cure of the matrix resin so that a structure can attain its maximum stiffness, static and fatigue strength and resistance to moisture and chemical degradation. Achieving maximum degree of cure in minimum time would seem to indicate that a high temperature cure cycle be used. However, this approach has other, potentially negative, implications for the cure process. For example, the rapid heat evolution of the resin's exothermic reaction at high temperatures can lead to a 'runaway' reaction as heat is generated more quickly than it can be removed, potentially resulting in resin thermal degradation. It has also been found that the large spatial and temporal gradients in resin degree of cure and temperature induced by rapid cure can be an important source of process-induced stress (Levitsky and Shaffer, 1975; Bogetti, 1989), especially in thick-section composites. Thermochemical model predictions are also important to the simulation of other processing phenomena such as resin flow and the generation of residual stress and deformation. One reason this is so is that temperature and degree of cure are two of the most important 'state' variables used to predict composite material properties during processing. Thus, prediction of everything from resin viscosity to thermal expansion and resin cure shrinkage strains are dependent on thermochemical model predictions.

The thermochemical model consists of a combination of analyses for heat transfer and resin reaction kinetics. The model of (Bogetti and Gillespie, 1992) can be applied to general two-dimensional composite modeling. It has the added capability of incorporating multiple composite and non-composite materials as well as process tooling. Other important features include the potential for improved boundary condition modeling using the autoclave simulation and consideration of material property variation during processing. Also, by integrating another model that considers resin flow, the effect of fiber volume fraction variation during processing can also be considered.

The governing equation of the thermochemical model is the unsteady-state two-dimensional anisotropic heat conduction equation with an internal heat generation term from the resin's exothermic curing reaction is modeled using Eqn. (19). The governing equations of the heat transfer portion of the problem are solved employing the finite element approximation, and time integration is performed as described in section 3.3 using a central finite difference approximation.

The heat transfer equation, Eqn. (19), and the cure kinetics equations outlined in section 4.1 are coupled. Ideally, therefore, a coupled solution technique would be employed in which both temperature and degree of cure would be solved in a single calculation. For the current model, for each time increment, the temperature will be solved first at the discretized locations (nodes) using the previous increment degree of cure (internal heat) and applied boundary conditions. Then using the calculated temperature, the degree of cure and the new internal heat generated are updated using Eqn. (22). Composite thermal properties that are used in our simulation are listed in Table 1.

Specific heat capacity (J/Kg.K)	$C_{pf} = 800$
Thermal conductivity (W/m.K)	$k_{fl} = 6.5$ $k_r = 0.65$
Fibre volume fraction	$V_f = 0.6$
Cure kinetic model	$H_R = 590(10^3)$, $\Delta E = 66.9$ KJ/gmole, A =5.333E+5 /s, m=0.79, n=2.16, α_0=0.01

Table 1. Thermal material properties.

Boundary conditions considered on the two-dimensional simulation analysis are:

- *Convective heat transfer (applied on the top surface of the laminate)* ($q = h\ (T_e - T)$, where h is the heat transfer coefficient (see Eqn. (23)) and Te and T are the air and boundary temperatures, respectively). Convective heat transfer is usually the dominant heat transfer mechanism in an autoclave at temperatures normally encountered in thermoset processing.
- Adiabatic boundary (along the laminate's edges) ($q = 0$ or $\partial T/\partial n = 0$, where n is the surface normal vector).

4.3 Numerical results

The composite plate considered in this study is a mid-plane symmetric laminate with dimensions (0.36m x 0.36m x 0.003m). It consists of 6 sub-laminates, each of which consists of 4 layers [$\pm\phi\pm\langle T_0 | T_1 \rangle$], for a total of 24 plies. The laminate is made from carbon/epoxy (AS4/3501) with material properties defined in Table 1. The analysis is performed by discretizing the panel using a uniform mesh of rectangular four-noded Kirchhoff plate elements. The transient response of the VS laminate and the different shape of the heat transfer channels from the straight-fiber ones (to be discussed in the next section), could affect both the temperature and degree of cure response. However, performance of the simulation as a function of the autoclave time cycle showed that there is no difference in thermal performance "in autoclave" between VS and straight fiber composites.

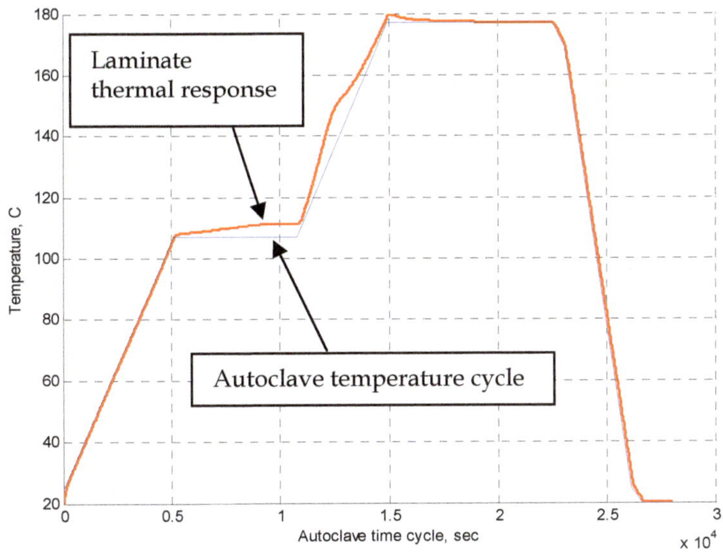

Fig. 2. Autoclave temperature cycle and temperature response of composite.

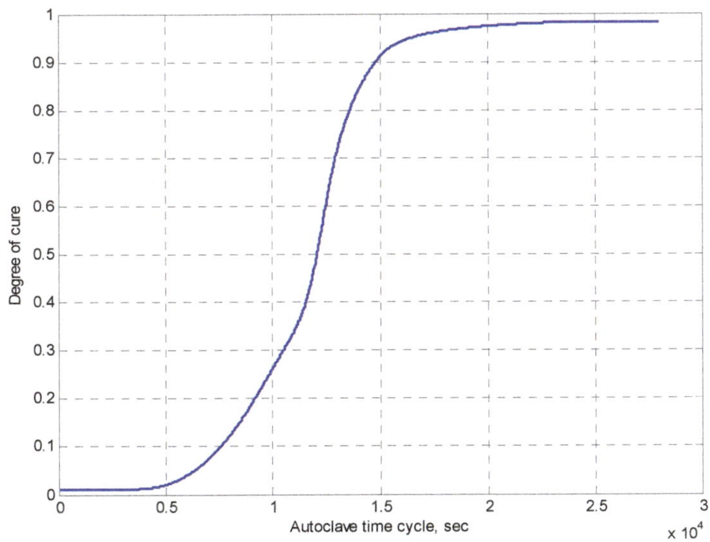

Fig. 3. Degree of cure of VS and straight-fiber composite.

The autoclave temperature cycle and the laminate thermal response at the composite plate center are shown in Figure 2. The autoclave thermal cycle can be divided to three phases. The first phase involves raising the autoclave air temperature from room temperature (25 °C) to 107 °C, and then holding this temperature for 1 hour. The second phase raises the autoclave air temperature to 177 °C and holds this temperature for 2 hours. The heat that is generated by the cure kinetics during the autoclave thermal cycle starts when the composite panel temperature is close to 107 °C. This internal heat leads to an increase in the composite panel temperature, then the extra heat is transferred back to the autoclave air. Once the epoxy is fully cured, there is no exothermic curing heat reaction. The degree of cure is shown in Figure 3 as a function of the cure cycle time. The change in the degree of cure is low during the initial phase of the autoclave thermal cycle, once the exothermic curing reaction started, we can see faster curing until the autoclave air temperature is 177 °C, then curing slows down until it is fully cured. The main factor behind the similarity of curing of VS and straight fibers is the composite panel is exchanging heat through the transverse direction and not through the in-plane direction (as boundaries are isolated). Heat exchange through boundaries will be affected by the in-plane fiber layout and is discussed in next section.

5. Transient heat response of variable stiffness laminate

Clearly, there is an actual need for investigating the thermal transients developed within VS composites. However, because of the inherent mathematical difficulties, thermal analysis of non-homogeneous structural materials is considerably more complex than in the corresponding homogeneous case. Numerical techniques such as finite element methods are of great importance for solving for temperature profile and stress distribution.
This section investigates thermal transient analysis of rectangular variable stiffness composite under uniform partial heat flux using finite element analysis. Variable thermal conductivity will affect the temperature profile results compared to unidirectional composite. The effect of the extra design parameter of variable stiffness laminates (steering of fibers) on transient response is investigated. The need for cooling of a structural panel is determined by transient time and boundary conditions. The transient solution of variable stiffness laminate is compared to the unidirectional one. Later, the steady state temperature profile is used to determine thermal stresses of the variable stiffness laminate compared to a unidirectional one applying forced straight edge boundary conditions. The mechanics of variable-stiffness laminates are discussed in (Abdalla et al., 2009).

5.1 Transient response

A two-dimensional finite element (FE) model describing heat transfer analysis of a variable-stiffness composite laminate exposed to partial surface heating is illustrated in Figure 4. The effect of fiber steering (T_0, T_1) on transient time and steady state solution is compared to straight fibers. A Matlab computer program was written for the FE procedure discussed in previous sections. The program has been used to evaluate the transient response of different fiber steering. Material properties of the laminate are listed in Table 1. The composite wall is assumed to have a uniform initial temperature; T_i = 0 °C. The results presented here are for a partially heated plate at input heat energy of h_{Input} = 100 W/(m^2). The accuracy of the results was verified by comparing the straight-fiber results to the ABAQUS transient heat transfer analysis. Boundary conditions for laminate edges act as a heat sink and are fixed at T = 0 °C.
The transient response of a heat conduction problem can be characterized by two parameters. One parameter is the time to reach steady state. A longer time to reach steady state is better for thermal protection panels as it allows more time for the structure to absorb heat, which reduces thermal stresses and delays the ablation process that degrades material properties. The second parameter is the maximum temperature reached at the center of the

panel. A lower maximum temperature is desirable for the same reasons to reduce thermal stresses, which improves buckling response and delays the ablation process. A variable stiffness laminate adds more three design parameters (ϕ, T_0, T_1) that can be used to achieve the optimal thermal performance. For example, Figure 5 compares the performance of variable stiffness panels while varying T_0 and T_1 versus the straight-fiber laminate performance. The transient time is normalized by the maximum transient time that can be achieved using a straight-fiber laminate, and the maximum temperature at plate center is normalized by the minimum temperature achieved using a straight-fiber laminate. The straight-fiber laminate is represented by the heavy line, and is a symmetric angle-ply layup that has a linear variation of a pair of fiber angles $[\pm\theta]_s$, so the 0° and 90° laminates have the same maximum temperature and transient time. The performance changes only in the range between 0° and ±45° degrees, and appears to be nearly linear with constant transient time, and independent of the orientation angle. The maximum temperature for the 0° and 90° laminates is about 1.3 times that of the minimum temperature, which is achieved for the ±45° laminate.

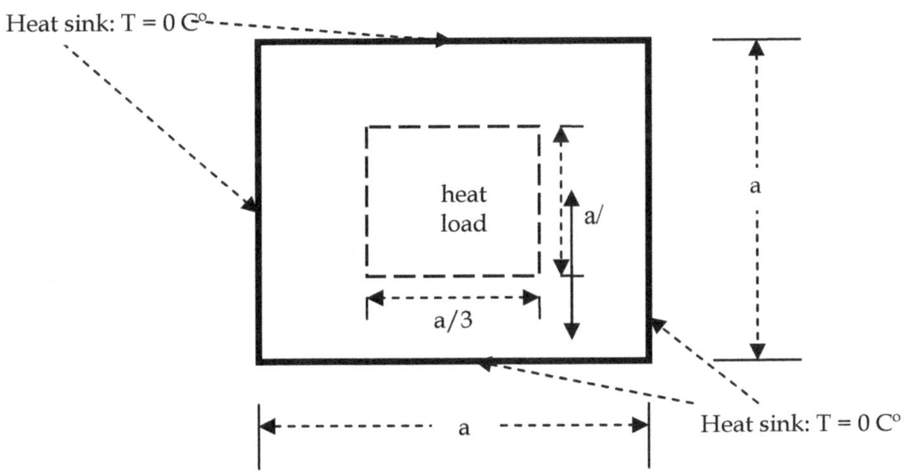

Fig. 4. Square variable stiffness panel under thermal load.

Two types of variable stiffness laminates are studied. A type-1 laminate, which is represented by $[0\pm\langle T_0/T_1\rangle]_{2S}$, is constructed of layers with steered fiber orientation angles θ which are a function of x-coordinate only. Type-2, which is represented by $[0\pm\langle T_0/T_1\rangle/90\pm\langle T_0/T_1\rangle]_S$, is constructed with half of the layers having fiber orientations which are functions of the x-coordinate, and the other half functions of the y-coordinate. The type-1 laminate performance is presented by incrementally changing T_1 in the range of 0° to 90°. For each increment of T_1, T_0 is varied between 0° and 90°, and then the normalized maximum temperature is plotted versus the normalized transient time in Figure 5. The minimum temperature at the plate center is achieved at [T_1=0, T_0=60]. The longest transient time is achieved for the VS lamina with [T_1=90, T_0=0]. Incrementing T_1 of the VS laminate leads to higher temperatures at the plate center and higher transient times. This result can be explained by the fact that the heat is easily channeled when the fiber angle at the plate edges (T_1) is perpendicular to the plate boundaries. The lowest heat flux is achieved at T_1=0°, where fibers at plate edges are tangent to boundaries. Changing the stacking sequence of the variable stiffness laminate to type-2, where $[0 \pm <T_0/T_1>/90\pm <T_0/T_1>]_S$ leads to a minimum temperature at the plate center at $[T_0=75°, T_1=0°]$.

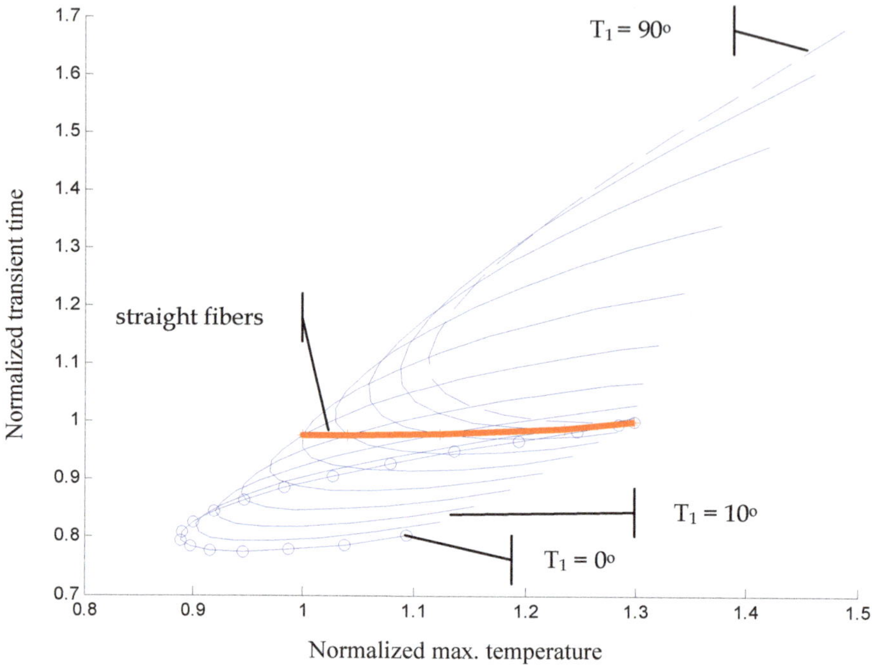

Fig. 5. Transient time and maximum temperature for straight-fibers, type-1 $[0\pm<T_0 \mid T_1>]_{2s}$

The steady-state heat flux is studied to investigate the effect of varying fiber paths for selected design configurations from Figure 5. Steady state temperature and transient time are controlled by heat dissipation along two channels, along the longitudinal and transverse directions of the composite plate. The heat dissipation profile is dependent on the fiber orientation angle in the laminate, as the fibers are more conductive than the matrix. A comparison of the heat flux profile for two straight-fiber laminates $[0]_{2S}$ and $[\pm45]_{2S}$ is shown in Figures 6 and 7. For the laminate with fiber angle θ=0°, the heat is dissipated only along x-direction, while the $[\pm45]_{2S}$ laminate dissipates heat in two directions, which promotes lower temperatures and smaller transient times. As shown on Figure 6, the normalized heat channel along the x-direction where θ=0° is broader from the one where θ=±45°, which has an extra heat channel along the y-direction. The heat flux for the $[0\pm<60/0>]_{2S}$ laminate is shown in Figure 8. Clearly, the laminate has more heat channels along the longitudinal and transverse direction than the θ=±45° laminate, which leads to 20% less transient time and 12% less maximum temperature at plate center, as shown on Figure 5.

Next, plots of heat flux for one quarter of the plate for the θ=0° straight-fiber laminate, $[0\pm<45/0>]_{2S}$, and $[0\pm<60/0>]_{2S}$ laminates are shown in Figure 9 and Figure 10. These figures indicate that the straight fibers at θ=0° have the highest heat flux along the x-direction, and the lowest along the y-direction. The heat flux for the $[0\pm<60/0>]_{2S}$ laminate has two heat channels along the x- and y-directions, and both dissipate heat higher than the $\theta=\pm45^{\circ}$ laminate. The success of the variable stiffness laminate in dissipating heat faster and reaching a lower temperature compared to the straight-fiber laminate prompted the idea of introducing the second laminate type in which the fiber orientation is steered in both the x- and y-directions. The heat flux plots indicate that the variable stiffness laminate has the best characteristics in the x- and y-directions.

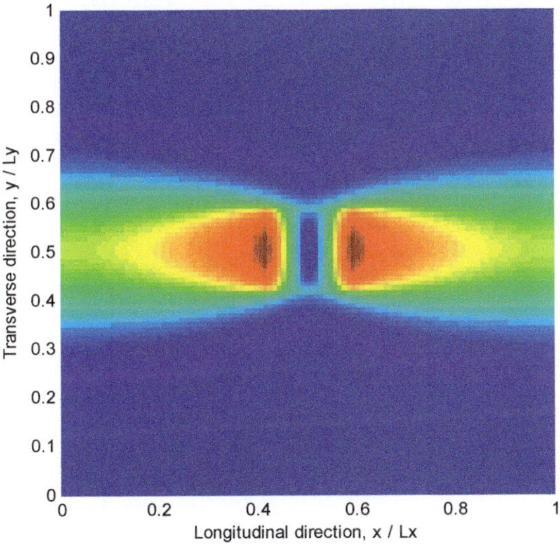

Fig. 6. Heat Flux for 0° straight angle laminate.

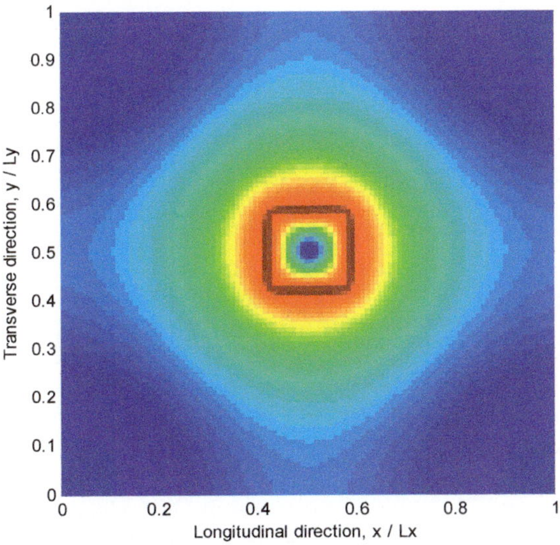

Fig. 7. Heat Flux for ±45° straight angle laminate.

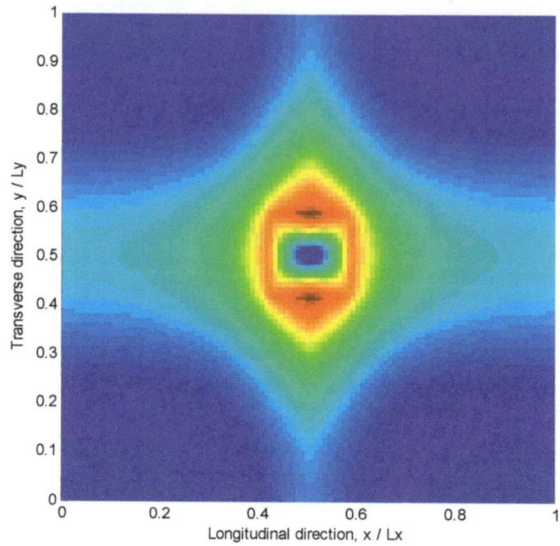

Fig. 8. Heat Flux for [$0 \pm <60/0>$]$_{2S}$ VS laminate.

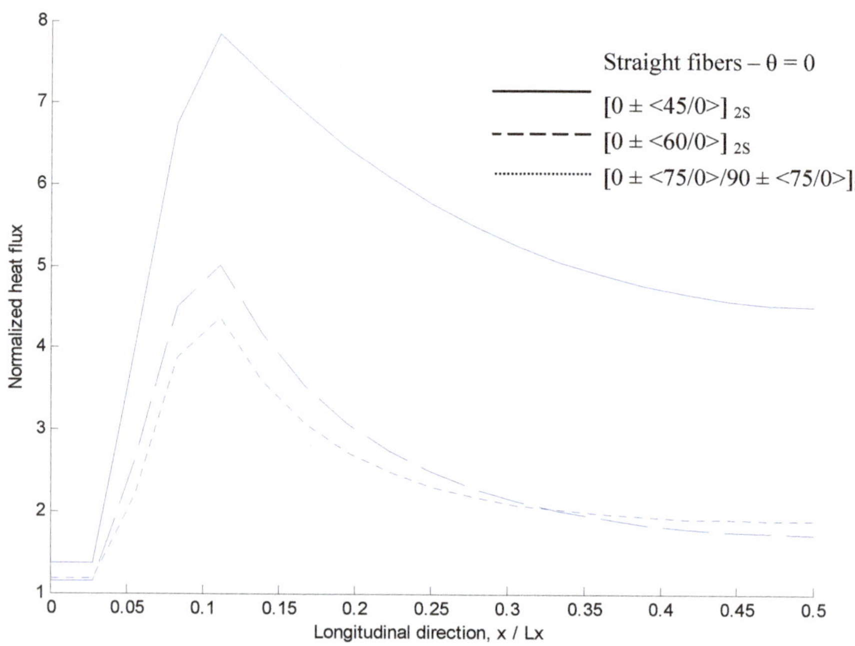

Fig. 9. Heat Flux along longitudinal direction for one-quarter of plate

Steering the fibers provides an additional design parameter to control both the transient time and temperature profile. Incrementing T_1 of the steered fibers leads to a higher transient time and temperature at plate center, while incrementing T_0 has opposite effect in the range of T_0=0° to 60°. Next, the stress resultants and plane deformations are examined for two types of stacking sequences under forced straight edge boundary conditions (where nodes that are located on the laminate edge are coupled with one selected node to force equal displacements).

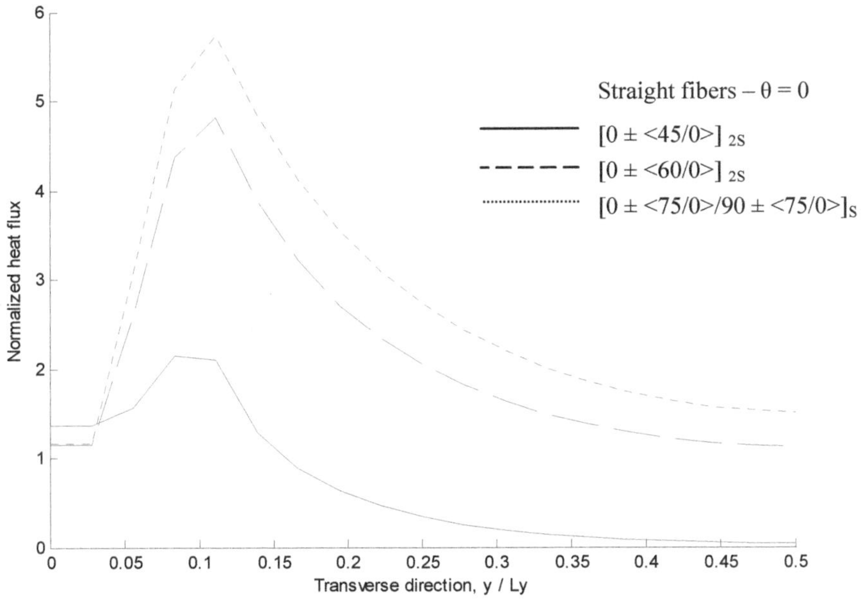

Fig. 10. Heat Flux along transverse direction for one-quarter of plate

5.2 Steady state stress results

The in-plane response of the two types of laminates, $[0\pm\langle T_0 | T_1\rangle]_{2S}$ and $[0\pm\langle T_0/T_1\rangle/90\pm\langle T_0/T_1\rangle]_S$ (type 1 and type 2, respectively), are considered under a steady-state temperature input. Under an applied non-uniform temperature profile, one or both of the in-plane stress and displacement distributions can be highly non-uniform, which results from the coupling of the in-plane equilibrium equations. An in-plane analysis of a variable stiffness laminate is important because one has to determine the location of the largest stress in a laminate to implement a failure constraint. Deformation and stress variations in the panel will depend on the in-plane boundary conditions applied on the longitudinal and transverse edges, $y = \pm b/2$ and x= $\pm a/2$. Here, a straight-edge boundary that can expand or contract in the transverse and longitudinal directions is enforced.

For the type-1 laminate; the fiber orientation angle θ is a function of the x-coordinate only $[0\pm\langle T_0 | T_1\rangle]_{2S}$, and is subjected to a non-uniform temperature profile. In Figure 11, the maximum axial stress resultant of the panels is normalized by the maximum axial stress resultant for a straight-fiber θ=0° laminate, and are shown as a function of the right edge displacement in the longitudinal direction normalized by the edge width. The heavy green line in the figure is the locus of results for constant-stiffness straight-fiber format panels, and shows the variation of both the maximum stress resultant and plate right edge displacement as the fiber orientation angle is changed from 0° to 90°. The lowest normalized stress resultant for this square panel with straight fibers and forced straight edges is 0.468 and corresponds to a ±45° laminate with normalized longitudinal deflection value of about 0.084. For variable-stiffness panels a family of curves corresponding to various values of T_1 (from 0° to 90° with increments of 15°) is shown in Figure 11. Each curve is generated by varying the value of T_0 between 0° and 90° for a given value of T_1 as labeled in the figure. Intersection of these curves with the curve for the straight-fiber panel is a constant-fiber angle panel where T_1 is equal to T_0. The lowest normalized value of the stress resultant is 0.126, and is obtained for a variable stiffness configuration of T_0 = 0° and T_1 = 45°. The corresponding normalized longitudinal deflection of this panel is about 0.084, which is 73% lower than the lowest value of 0.468 obtained with a straight-fiber ±45° configuration. Most variable stiffness panels with T_0 = 0° have a higher stress resultant than straight-fiber configurations for T_1 in the range of 0° to 45°.

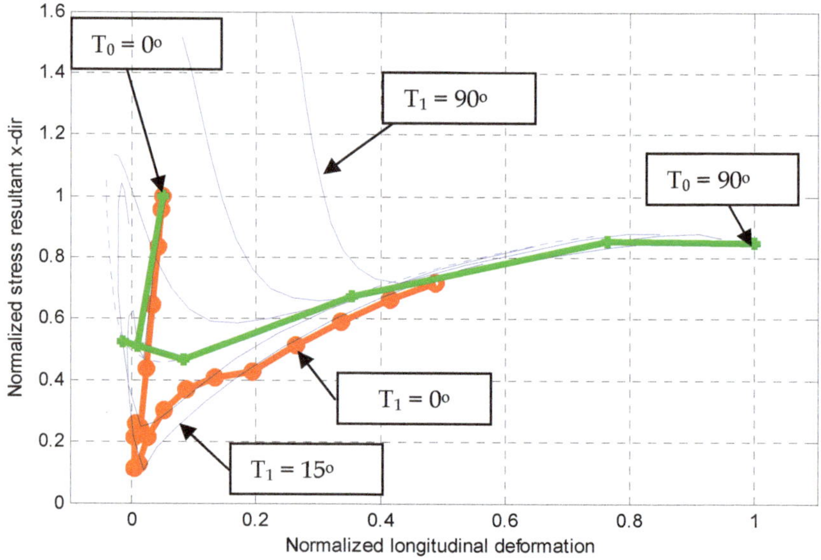

Fig. 11. Normalized longitudinal stress resultant for $[0\pm<T_0\,|\,T_1>]_{2S}$

For the type-2 laminate; the fiber orientation θ is a function of both the x- and y-coordinates, where $\theta=[0\ \pm<T_0/T_1>/90\pm\ <T_0/T_1>]_S$. A non-uniform temperature profile is again applied to the panels. The maximum axial stress resultant of panels normalized by maximum axial stress resultant for straight-fiber laminate $\theta=0^\circ$ are shown in Figure 12 as a function of the right edge displacement in the longitudinal direction normalized by the edge width. The heavy green line in the figure is for constant-stiffness, straight-fiber format panels, and shows the variation of both the maximum stress resultant and plate right edge displacement as the fiber orientation angle ranges from 0° to 90°. The lowest normalized stress resultant for this square panel with straight fibers and forced straight edges is 0.573, and corresponds to a ±45° laminate. The normalized longitudinal deflection value is constant at a value of 1 for all straight-fiber configurations.

For variable-stiffness panels a family of curves corresponding to various values of T_1 (from 0° to 90° in increments of 15°) is plotted in Figure 12. The lowest normalized value of stress-resultant is 0.185, and is obtained for a variable stiffness configuration of $T_0 = 85°$ and $T_1 = 0°$, with normalized longitudinal deflection value of about 1.127. This value is 68% lower than the lowest value of 0.577 obtained with a straight-fiber configuration, but with 12% increase of normalized longitudinal deformation. Most variable stiffness panels with $T_0 = 0°$ and T_1 in the range of 0° to 45° have a higher stress resultant than the corresponding straight-fiber configurations.

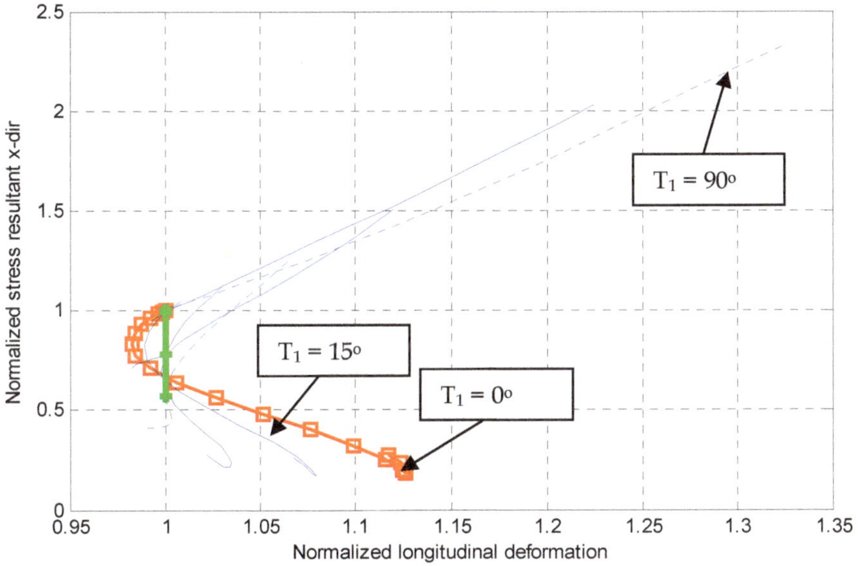

Fig. 12. Normalized longitudinal stress resultant for $[0 \pm\langle T_0/T_1\rangle/90\pm \langle T_0/T_1\rangle]_S$

6. Thermal testing of variable stiffness laminates

The thermal-structural responses of two variable stiffness panels and a third cross-ply panel are evaluated under thermal loads. A brief description of the variable stiffness panels and their fiber orientation angles is given, along with an overview of the thermal test setup and instrumentation. Results of these tests are presented and discussed, and include measured thermal strains and calculated coefficients of thermal expansion.

6.1 Fiber tow path definition

The layups of the three composite panels tested in this study are described herein. The two variable stiffness panel layups are $[\pm45/(\pm\theta)_4]_s$, where the steered fiber orientation angle θ varies linearly from ±60° on the panel axial centerline, to ±30° near the panel vertical edges 30.5 cm away. The curvilinear tow paths that the fiber placement machine followed during fabrication of these variable stiffness panels are shown in Figure 13. One panel has all 24, 0.32-cm-wide tows placed during fabrication. This results in significant tow overlaps and thickness buildups on one side of the panel, and therefore it is designated as the panel with overlaps. The fiber placement system's capability to drop and add individual tows during fabrication is used to minimize the tow overlaps of the second variable stiffness panel, which is designated as the panel without overlaps. The third panel has a straight-fiber $[\pm 45]_{5s}$ layup and provides a baseline for comparison with the two variable stiffness panels. The overall panel dimensions are 66.0 cm in the axial direction, and 62.2 cm in the transverse dimension, as indicated by the dashed lines in the figure. Further details of the panel construction are given in (Wu, 2006).

6.2 Test setup and instrumentation

The thermal test was performed in an insulated oven with feedback temperature control. Electrical resistance heaters and a forced-air heater unit were used to heat the enclosure. Perforated metal baffles were used to evenly distribute hot air over the back surface of the panel. The oven's front was glass to allow observation of the panel using shadow moiré interferometry. The panel was supported inside the oven with fixtures that restricted its rigid-body motion but allowed free thermal expansion. The panel was placed on two small quartz rods that prevented direct contact with the lower heated platen. The panel surfaces were supported between quartz cones and spring-loaded steel probes with low axial stiffnesses.

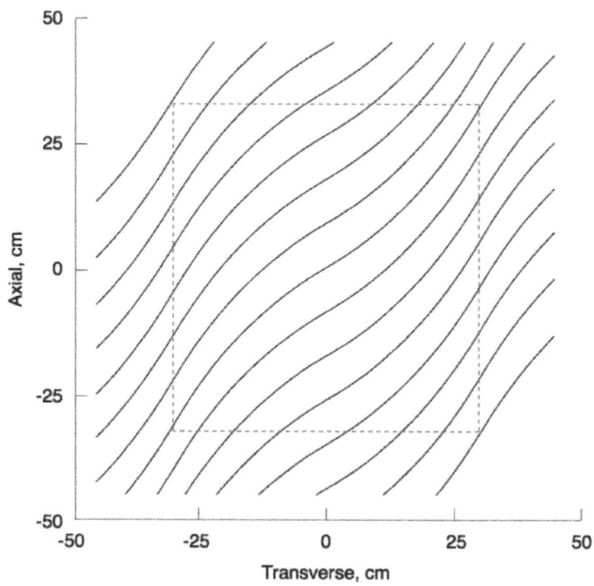

Fig. 13. Variable stiffness panel tow paths.

Each composite panel was gradually heated from room temperature up to approximately 65 °C. A feedback control system provided closed-loop, real-time thermal control based on readings from five K-type thermocouples on the heated platens and air inlet surrounding the panel. These separate data were then averaged into a single temperature provided to the control system. The thermocouples used in this study have a measurement uncertainty of ±1 °C. For a thermal test, the control temperature inside the oven was first raised to 32 °C and held there for 5 minutes. After the hold period, the control temperature was raised at 1 °C/min. to a maximum of 65 °C and held there for 20 minutes before the test was ended. The solid line in Figure 14 shows the average of the five control thermocouples plotted against time for a typical test.

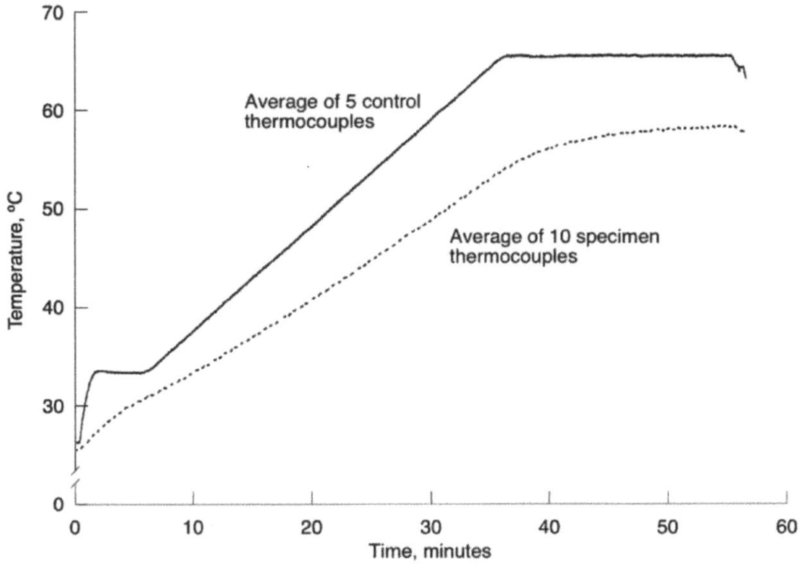

Fig. 14. Temperature profiles for thermal tests.

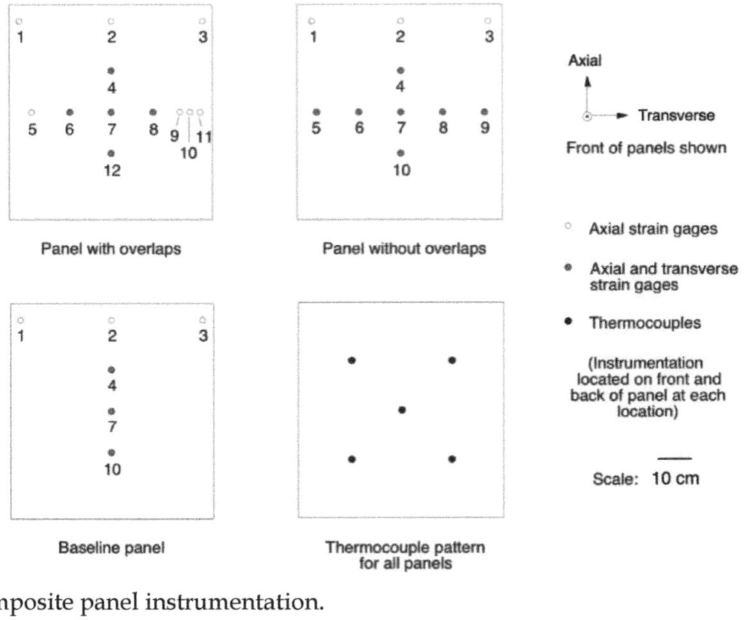

Fig. 15. Composite panel instrumentation.

The panel response was measured during the thermal test with thermocouples and strain gages, and these data were collected using a personal computer-based system. Panel front and back surface temperatures were measured with five pairs of K-type thermocouples. The average panel temperature is shown as a function of time as the dashed line in Figure 14. The thermocouples, denoted as black-filled circles, are located at the corners and center of a 30.5-cm square centered on the panel, as shown in Figure 15.

Back-to-back pairs of electrical-resistance strain gages (each with a nominal ±1 percent measurement error) are bonded to the panel surfaces using the procedures described in (Moore, 1997). The locations of the strain gage pairs on each panel are also shown in Figure 15. The strain gages measure either axial strains (the open circles in the figure), or both axial and transverse strains (the gray filled circles), and are deployed along the top edge, and axial and transverse centerlines of the panels. The closely spaced axial gage pairs (locations 9, 10 and 11) on the panel with overlaps span a region of varying laminate thickness along the transverse centerline. In addition to the axial gage pairs along the upper edge of the baseline panel, biaxial gages are fitted at locations 4, 7 and 10 along the axial centerline.

6.3 Test results

The heating profile shown in Figure 14 is applied to the panels, and the resulting panel thermal response is measured. An initial thermal cycle is performed for each panel to fully cure the adhesives used to attach the strain gages to the panels. Since the strain gage response is dependent on both its operating temperature and the motion of the surface to which it is bonded, the thermal output of the strain gages themselves (Anon., 1993; Kowalkowski et al., 1998) must first be determined. Strain data are recorded for gages bonded to Corning ultralow-expansion titanium silicate (coefficient of thermal expansion 0 ± 3.06 x 10-8 cm/cm/°C) blocks that are subjected to the same thermal loading. After completion of each thermal test, this thermal output measurement is then subtracted from the total (apparent) strain of each strain gage recorded during the test to obtain the actual mechanical strains presented below.

6.3.1 Variable stiffness panels

Measured axial and transverse strains at the center (gage location 7) of the panel with overlaps are plotted against the panel temperature in Figure 16 for a representative thermal test. The plotted strains on the front and back panel surfaces are proportional to the temperature, and are qualitatively similar to the responses at the other panel gage locations. The membrane strain at the laminate mid-plane is defined as the average strain from a back-to-back gage pair. The panel's local coefficient of thermal expansion (CTE) at that gage location is then defined as the linear best-fit slope of the membrane strain as a function of temperature. Using the panel center strains shown in Figure 16, the measured axial CTE there is 9.11 x 10-6 cm/cm/°C, and the transverse CTE is 0.11 x 10-6 cm/cm/°C (units of 1 x 10-6 cm/cm are denoted as με or microstrain). Note that these local CTEs for the variable stiffness panels are dependent on the non-uniform fiber orientation angles, and may not be equal to straight-fiber CTEs calculated using classical lamination theory.

The maximum measured strains at each of the 12 gage locations on the panel with overlaps are plotted in Figure 17, with the corresponding axial CTEs shown in Figure 18. The axial CTEs increase from -3.98 με/°C near the edges (θ=±30°) to 10.67 με/°C along the axial centerline (θ=±60°). Transverse CTEs are also plotted in the figure and range from -0.94 to 1.35 με/°C. In general, the fiber-dominated ±30° layups near the panel edges have low axial CTEs and high transverse CTEs. The opposite is true for the matrix-dominated ±60° laminates on the panel axial centerline, which have high axial CTEs and low transverse CTEs.

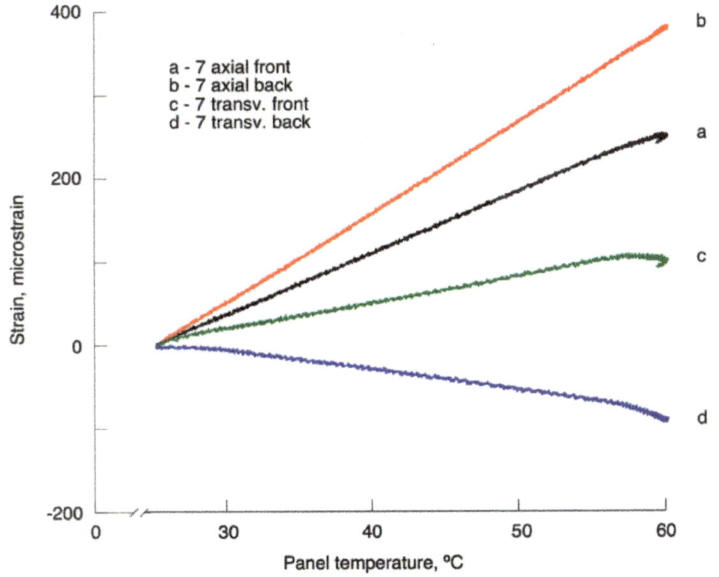

Fig. 16. Strain vs. temperature at center of panel with overlaps.

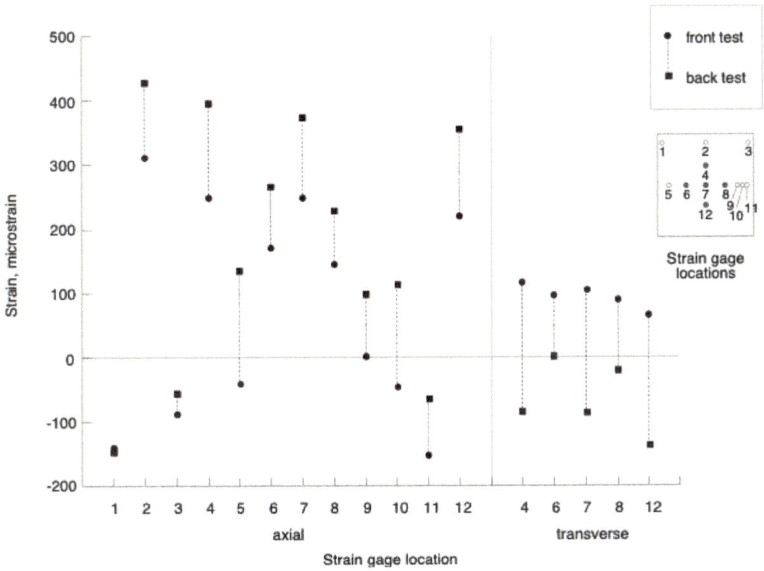

Fig. 17. Maximum strains for panel with overlaps.

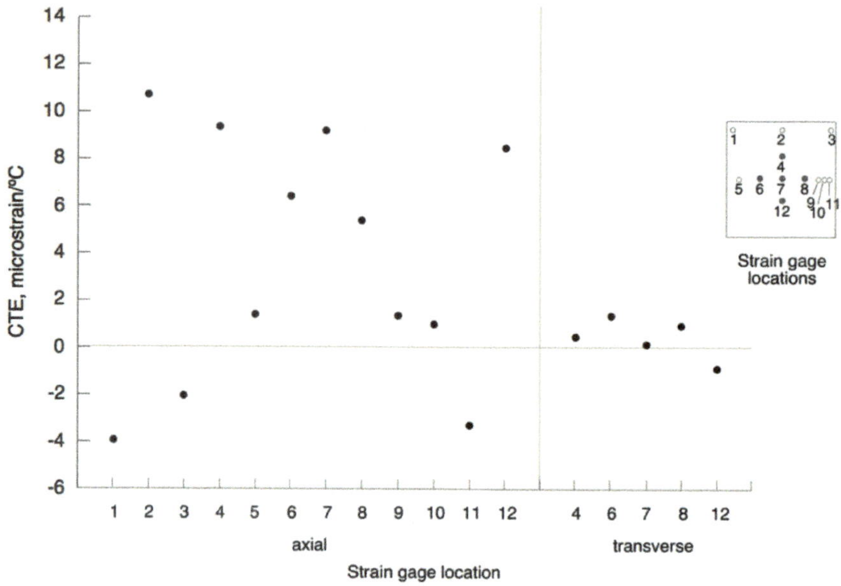

Fig. 18. CTEs for panel with overlaps.

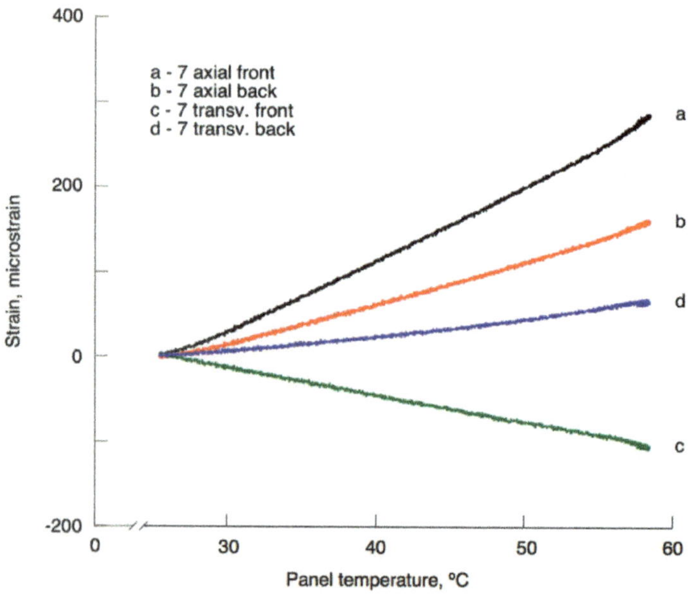

Fig. 19. Strain vs. temperature at center of panel without overlaps.

The 20-ply laminate on the transverse centerline 12.7 cm on either side of the panel center has a [±45/(±48)4]s layup. However, the measured axial CTEs (6.35 and 5.33 με/°C) at gage locations 6 and 8 there are much higher than the corresponding transverse CTEs (1.35 and 0.92 με/°C). Since the CTEs of an [±45]5s orthotropic cross-ply laminate should all be equal, the observed differences strongly suggest that the variable stiffness laminate CTEs can be highly sensitive to relatively small changes in the fiber orientation angles.

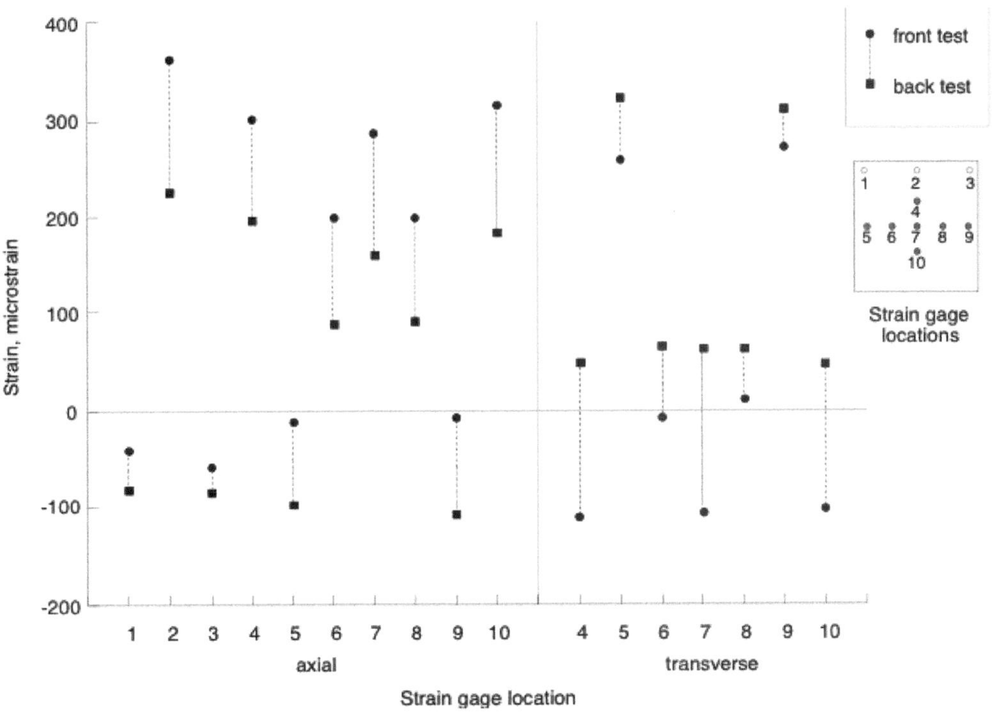

Fig. 20. Maximum strains for panel without overlaps.

Measured axial and transverse strains on the front and back surfaces of the center of the panel without overlaps are shown plotted against the corresponding panel temperature in Figure 19. The axial and transverse strains at the maximum test temperature at each of the 10 strain gage locations on this panel are shown in Figure 20. The axial and transverse CTEs plotted in Figure 21 are then calculated from the membrane strains. Axial CTEs for the panel without overlaps range from -2.14 με/°C near the panel edges to 9.16 με/°C along the axial centerline, with transverse CTEs ranging from -0.79 με/°C on the axial centerline to 9.07 με/°C on the transverse centerline near the panel edge. The CTEs for the panel without overlaps are much more symmetric with respect to the panel axial and transverse centerlines than those described previously for the panel with overlaps. However, similar qualitative trends are observed in the plotted CTEs for both panels.

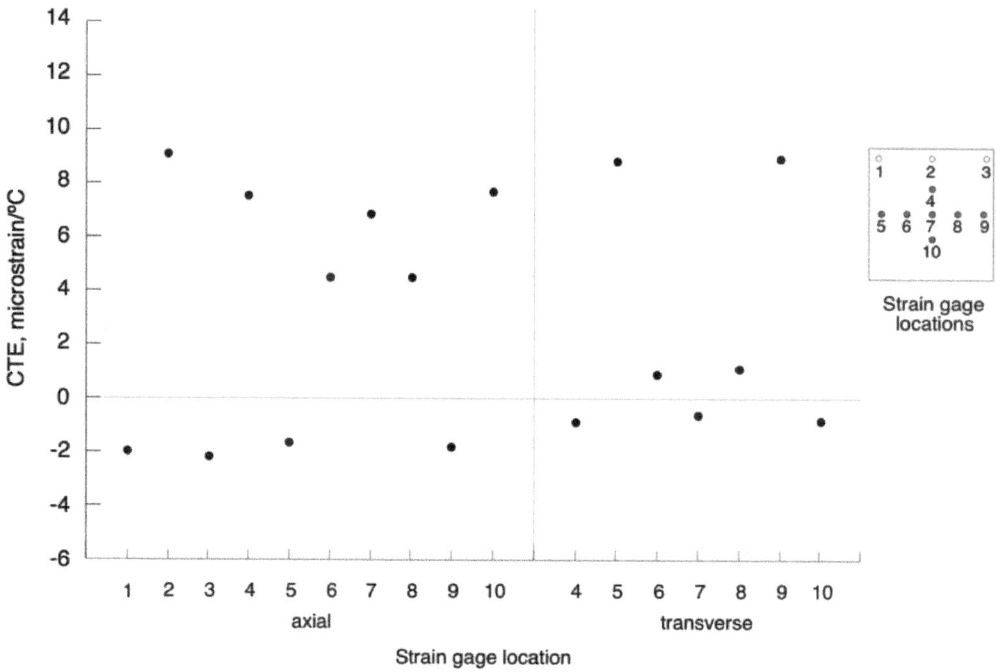

Fig. 21. CTEs for panel without overlaps.

6.3.2 Baseline panel

Front and back surface axial and transverse strains at the baseline panel center are plotted as functions of the panel temperature in Figure 22. The measured strains are linear and very nearly equal, which is to be expected since the [±45]5s layup has the same response in both the axial and transverse directions. The range of measured CTEs for the baseline panel is from 2.34 to 3.40 $\mu\varepsilon/°C$, with an average CTE of 2.92 $\mu\varepsilon/°C$. The corresponding standard deviation is 0.32 $\mu\varepsilon/°C$, resulting in an 11 percent coefficient of variation. The maximum temperature for the baseline panel thermal test is about 3.9 °C lower than the maximum temperature for the variable stiffness panels because the heating profile was terminated when the temperature reached 65 °C.

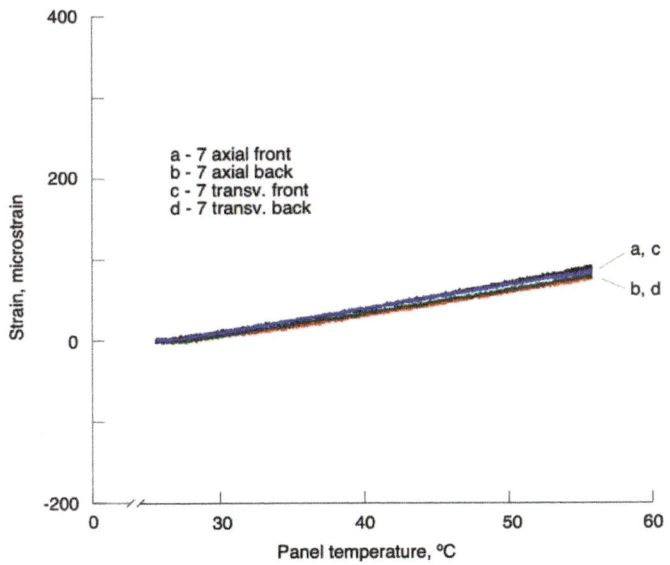

Fig. 22. Strain vs. temperature at center of baseline panel.

6.4 Summary

The measured strain response at each gage location on each of the composite panels is generally linear with increasing temperature. The membrane strain at each gage location is defined and used to compute the laminate CTE at that location. The measured axial CTEs for both variable stiffness panels are lowest near the panel edges and increase to their maximum values along the axial centerline, while the transverse CTEs show the opposite behavior. This corresponds to the fiber-dominated ±30° layup towards the panel edges and a matrix-dominated ±60° layup on the axial centerline. For a given orientation, the measured CTEs along the panel axial centerlines are all fairly close to one another. This is as expected, since the fiber orientation angle varies along the panel transverse axis, with only the ply shifts contributing to any axial fiber orientation angle variation.

7. References

Abdalla, M, Gürdal, Z and Abdelal, G. (2009). Thermomechanical response of variable stiffness composite panels. *Journal of Thermal Stresses*, Vol. 32, No. 1, pp. (187 - 208).

Anon. (1993). Strain Gage Thermal Output and Gage Factor Variation with Temperature. *TN-504-1, Measurements Group*, Inc., Raleigh, North Carolina.

Banichuk, NV. (1981). Optimization Problems for Elastic Anisotropic Bodies. *Archive of Mechanics*, 33, 1981, pp. (347-363).

Banichuk, NV and Sarin, V. (1995). Optimal Orientation of Orthotropic Materials for Plates Designed Against Buckling. *Structural and Multidisciplinary Optimization*, Vol. 10, No. 3-4, 1995, pp. (191-196).

Bogetti, T. (1989). Process-Induced Stress and Deformation in Thick-Section Thermosetting Composites. *Technical Report CCM-89-32*, Center for Composite Materials, University of Delaware, Newark, Delaware, 1989.

Bogetti, T and Gillespie, J. (1992). Process-Induced Stress and Deformation in Thick-Section Thermoset composite Laminates", *Journal of Composite Materials, Vol.* 26, No. 5, pp. (626-660).

Cole, K, Hechler, J and Noël, D. (1991). A New Approach to Modelling the Cure Kinetics of Epoxy Amine Thermosetting Resin. 2. Application to a Typical System Based on Bis[4-diglycidylamino)phenyl]methane and Bis(4-aminophenyl) Sulphone", *Macromolecules.* Vol. 24, No. 11, pp. (3098-3110).

Dusi, M, Lee, W, Ciriscioli, P and Springer, G. (1987). Cure Kinetics and Viscosity of Fiberite 976 Resin," *Journal of Composite Materials.* Vol. 21, No. 3, pp. (243-261).

Duvaut, G, Terrel, G, Léné, F and Verijenko, V. (2000). Optimization of Fiber Reinforced Composites. *Composite Structures*, Vol. 48, 2000, pp. (83-89).

Gürdal, Z and Olmedo, R. (1993). In-Plane Response of Laminates with Spatially Varying Fiber Orientations: Variable Stiffness Concept. *AIAA Journal*, Vol. 31, (4), pp. (751-758), 0001-1452.

Gürdal, Z, Haftka, RT and Hajela, P. (1999). *Design and Optimization of Laminated Composite Materials.* John Wiley & Sons, Inc., New York, NY.

Gürdal, Z, Tatting, BF and Wu, KC. (2008). Variable stiffness composite panels: Effects of stiffness variation on the in-plane and buckling response. *Composite: Part A*, Vol. 39, 2008, pp. (911-922).

Hetnarski, RB. (1996). *Thermal stresses (I–IV).* Amsterdam: Elsevier Science Pub. Co.

Hughes, T, Levit, I and Winget, J. (1982). Unconditionally stable element-by-element implicit algorithm for heat conduction analysis. U.S. Applied Mechanics Conference, Cornell University, Ithaca, USA.

Johnston, A. (1997). *An Integrated Model of the Development of Process-Induced Deformation in Autoclave Processing of Composite Structures.* PhD dissertation, University of British Columbia.

Levitsky, M and Shaffer, B. (1975). Residual Thermal Stresses in a Solid Sphere Cast From a Thermosetting Material. *Journal of Applied Mechanics*, pp. (651-655).

Kowalkowski, M, Rivers, HK and Smith, RW. (1998). Thermal Output of WK-Type Strain Gauges on Various Materials at Elevated and Cryogenic Temperatures. *NASA TM-1998-208739*, October 1998.

Lee, W, Loos, A and Springer, S. (1982). Heat of Reaction, Degree of Cure, and Viscosity of Hercules 3501-6 Resin", *Journal of Composite Materials.* Vol. 16, pp. (510-520).

Mittler, G, Klima, R, Alapin, B, et al. (2003). Determination and application of thermo-mechanical characteristics for the optimization of refractory linings. *STAHL UND EISEN* Vol. 123, No. 11 pp. (109–12).

Moore, T. (1997). Recommended Strain Gage Application Procedures for Various Langley Research Center Balances and Test Articles. *NASA TM-110327*, March 1997.

Obata, Y and Noda, N. (1993). Unsteady thermal stresses in a functionally gradient material plate - Analysis of one-dimensional unsteady heat transfer problem. Japan Society of Mechanical Engineers, Transactions A (ISSN 0387-5008), Vol. 59, No. 560, pp. (1090-1096).

Olmedo, R and Gürdal, Z. (1993). Buckling Response of Laminates with Spatially Varying Fiber Orientations, Proceedings of the 34th AIAA/ASME/ASCE/AHS/ASC Structures, Structural Dynamics and Materials (SDM) Conference, La Jolla, CA, April 1993.

Pedersen, P. (1991). On Thickness and Orientation Design with Orthotropic Materials. *Structural Optimization*, Vol. 3, 1991, pp. (69-78).

Pedersen, P. (1993). Optimal Orientation of Anisotropic Materials, Optimal Distribution of Anisotropic Materials, Optimal Shape Design with Anisotropic Materials, Optimal Design for a Class of Non-linear Elasticity. *Optimization of Large Structural Systems*, Ed. Rozvany, G. I. N., Vol. 2, 1993, pp. (649-681).

Scott, P. (1991). Determination of Kinetic Parameters Associated with the Curing of Thermoset Resins Using Dielectric and DSC Data", *Composites: Design, Manufacture, and Application, ICCM/VIII*, Honolulu, 1991.

Segerlind, LJ. (1984). *Applied Finite Element Analysis*. John Wiley & Sons.

Setoodeh, S, Abdalla, M and Gürdal, Z. (2006). Design of variable–stiffness laminates using lamination parameters. *Composites Part B: Engineering*. Vol. 37, No. 4-5, pp. (301-309).

Setoodeh, S, Abdalla, M and Gürdal, Z. (2007). Design of variable stiffness composite panels for maximum fundamental frequency using lamination parameters. *Composite Structures*. Vol. 81, No. 2, pp. (283-291).

Setoodeh, S, Abdalla, M, IJsselmuiden, S and Gürdal, Z. (2009). Design of variable-stiffness composite panels for maximum buckling load. *Composite Structures*. Vol. 87, No. , pp. (109-117).

Thornton, EA. (1992). *Thermal structures and materials for high-speed flight*. American Institute of Aeronautics and Astronautics.

Trujillo, D. (1977). An unconditionally stable explicit algorithm for structural dynamics. *International Journal of Numerical Methods Engineering*, Vol. 1, pp. (1579-1592).

Twardowski, T, Lin, S and Geil, P. (1993). Curing in Thick Composite Laminates: Experiments and Simulation", *Journal of Composite Materials*. Vol. 27, No. 3, pp. (216-250).

White, S and Hahn, H. (1992). Process Modelling of Composite Materials: Residual Stress Development during Cure. Part I. Model Formulation", *J. of Composite Materials*. Vol. 26, No. 16,pp. (2402-2422).

Wu, KC. (2006). Thermal and Structural Performance of Tow-Placed, Variable Stiffness Panels. ISBN 1-58603-681-5, *Delft University Press/IOS Press*, Amsterdam, The Netherlands, 2006.

Zienkiewicz, O, Hinton, E, Leung, K and Taylor, R. (1980). Staggered time marching schemes in dynamic soil analysis and a selective explicit extrapolation algorithm. Second International Symposium on Innovative Numerical Analysis in Applied Engineering Sciences, Canada.

8

Thermal Transport in Metallic Porous Media

Z.G. Qu[1], H.J. Xu[1], T.S. Wang[1], W.Q. Tao[1] and T.J. Lu[2]
[1]*Key Laboratory of Thermal Fluid Science and Engineering, MOE*
[2]*Key Laboratory of Strength and Vibration, MOE of Xi'an Jiaotong University in Xi'an, China*

1. Introduction

Using porous media to extend the heat transfer area, improve effective thermal conductivity, mix fluid flow and thus enhance heat transfer is an enduring theme in the field of thermal fluid science. According to the internal connection of neighbouring pore elements, porous media can be classified as the consolidated and the unconsolidated. For thermal purposes, the consolidated porous medium is more attractive as its thermal contact resistance is considerably lower. Especially with the development of co-sintering technique, the consolidated porous medium made of metal, particularly the metallic porous medium, gradually exhibits excellent thermal performance because of many unique advantages such as low relative density, high strength, high surface area per unit volume, high solid thermal conductivity, and good flow-mixing capability (Xu et al., 2011b). It may be used in many practical applications for heat transfer enhancement, such as catalyst supports, filters, bio-medical implants, heat shield devices for space vehicles, novel compact heat exchangers, and heat sinks, et al. (Banhart, 2011; Xu et al., 2011a, 2011b, 2011c).
The metallic porous medium to be introduced in this chapter is metallic foam with cellular micro-structure (porosity greater than 85%). It shows great potential in the areas of acoustics, mechanics, electricity, fluid dynamics and thermal science, especially as an important porous material for thermal aspect. Principally, metallic foam is classified into open-cell foam and close-cell foam according to the morphology of pore element. Close-cell metallic foams are suitable for thermal insulation, whereas open-cell metallic foams are often used for heat transfer enhancement. Open-cell metallic foam is only discussed for thermal performance. Figure 1(a) and 1(b) show the real structure of copper metallic foam

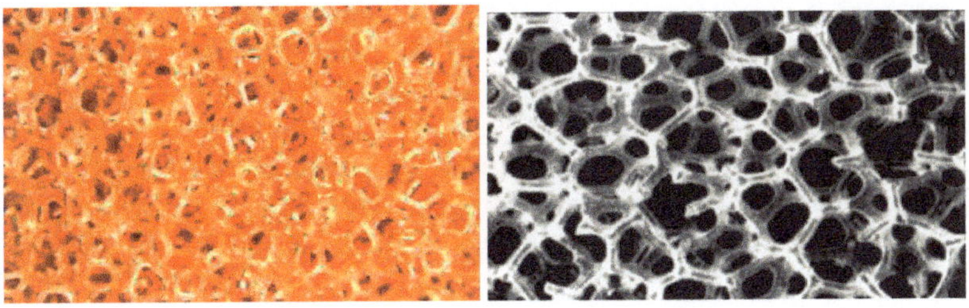

(a) (b)

Fig. 1. Metallic foams picture: (a) sample; (b) SEM (scanning electron microscope) and its SEM image respectively It can be noted that metallic foams own three-dimensional space structures with interconnection between neighbouring pore elements (cell). The morphology structure is defined as porosity (ε) and pore density (ω), wherein pore density is the pore number in a unit length or pores per inch (PPI).

In the last two decades, there have been continuous concerns on the flow and heat transfer properties of metallic foam. Lu et al. (Lu et al., 1998) performed a comprehensive investigation of flow and heat transfer in metallic foam filled parallel-plate channel using the fin-analysis method. Calmidi and Mahajan (Calmidi & Mahajan, 2000) conducted experiments and numerical studies on forced convection in a rectangular duct filled with metallic foams to analyze the effects of thermal dispersion and local non-thermal equilibrium with quantified thermal dispersion conductivity, k_d, and interstitial heat transfer coefficient, h_{sf}. Lu and Zhao et al. (Lu et al., 2006; Zhao et al., 2006) performed analytical solution for fully developed forced convective heat transfer in metallic foam fully filled inner-pipe and annulus of tube-in-tube heat exchangers. They found that the existence of metallic foams can significantly improve the heat transfer coefficient, but at the expense of large pressure drop. Zhao et al. (Zhao et al., 2005) conducted experiments and numerical studies on natural convection in a vertical cylindrical enclosure filled with metallic foams; they found favourable correlation between numerical and experimental results. Zhao et al. (Zhao & Lu et al., 2004) experimented on and analyzed thermal radiation in highly porous metallic foams and gained favourable results between the analytical prediction and experimental data. Zhao et al. (Zhao & Kim et al., 2004) performed numerical simulation and experimental study on forced convection in metallic foam fully filled parallel-plate channel and obtained good results. Boomsma and Poulikakos (Boomsma & Poulikakos, 2011) proposed a three-dimensional structure for metallic foam and obtained the empirical correlation of effective thermal conductivity based on experimental data. Calmidi (Calmidi, 1998) performed an experiment on flow and thermal transport phenomena in metallic foams and proposed a series of empirical correlations of fibre diameter d_f, pore diameter d_p, specific surface area a_{sf}, permeability K, inertia coefficient C_I, and effective thermal conductivity k_e. Simultaneously, a numerical simulation was conducted based on the correlations developed and compared with the experiment with reasonable results. Overall, metallic foam continues to be a good candidate for heat transfer enhancement due to its excellent thermal performance despite its high manufacturing cost.

For thermal modeling in metallic foams with high solid thermal conductivities, the local thermal equilibrium model, specifically the one-energy equation model, no longer satisfies the modelling requirements. Lee and Vafai (Lee & Vafai, 1999) addressed the viewpoint that for solid and fluid temperature differentials in porous media, the local thermal non-equilibrium model (two-energy equation model) is more accurate than the one-equation model when the difference between thermal conductivities of solid and fluid is significant, as is the case for metallic foams. Similar conclusions can be found in Zhao (Zhao et al., 2005) and Phanikumar and Mahajan (Phanikumar & Mahajan, 2002). Therefore, majority of published works concerning thermal modelling of porous foam are performed with two equation models.

In this chapter, we report the recent progress on natural convection on metallic foam sintered surface, forced convection in ducts fully/partially filled with metallic foams, and modelling of film condensation heat transfer on a vertical plate embedded in infinite metallic foams. Effects of morphology and geometric parameters on transport performance are discussed, and a number of useful suggestions are presented as well in response to engineering demand.

2. Natural convection on surface sintered with metallic porous media

Due to the use of co-sintering technique, effective thermal resistance of metallic porous media is very high, which satisfies the heat transfer demand of many engineering applications such as cooling of electronic devices. Natural convection on surface sintered with metallic porous media has not been investigated elsewhere. Natural convection in an enclosure filled with metallic foams or free convection on a surface sintered with metallic foams has been studied to a certain extent (Zhao et al., 2005; Phanikumar & Mahajan, 2002; Jamin & Mohamad, 2008).

The test rig of natural convection on inclined surface is shown in Fig. 2. The experiment system is composed of plexiglass house, stainless steel holder, tripod, insulation material, electro-heating system, data acquisition system, and test samples. The dashed line in Fig. 2 represents the plexiglass frame. This experiment system is prepared for metallic foam sintered plates. The intersection angle of the plate surface and the gravity force is set as the inclination angle θ. The Nusselt number due to convective heat transfer (with subscript 'conv') can be calculated as:

$$Nu_{\mathrm{conv}} = h_{\mathrm{conv}} \frac{L}{k} = \frac{\phi_{\mathrm{conv}}}{A(T_{\mathrm{w}} - T_{\infty})} \frac{L}{k} = \frac{\phi - \mathrm{E}\sigma\left(T_{\mathrm{rad}}{}^{4} - T_{\infty}{}^{4}\right) A_{\mathrm{rad}}}{A\left(T_{\mathrm{w}} - T_{\infty}\right)} \frac{L}{k} . \tag{1}$$

where h, L, k, ϕ, A, Tw, T_∞, E and σ respectively denotes heat transfer coefficient, length, thermal conductivity, heat, area, wall temperature, surrounding temperature, emissive power and Boltzmann constant. The subscript 'rad' refers to 'radiation'.

Meanwhile, the average Nusselt number due to the combined convective and radiative heat transfer can be expressed as follows:

$$Nu_{\mathrm{av}} = h_{\mathrm{av}} \frac{L}{k} = \frac{\phi}{A(T_{\mathrm{w}} - T_{\infty})} \frac{L}{k} . \tag{2}$$

In Eq.(2), the subscript 'av' denotes 'average'.

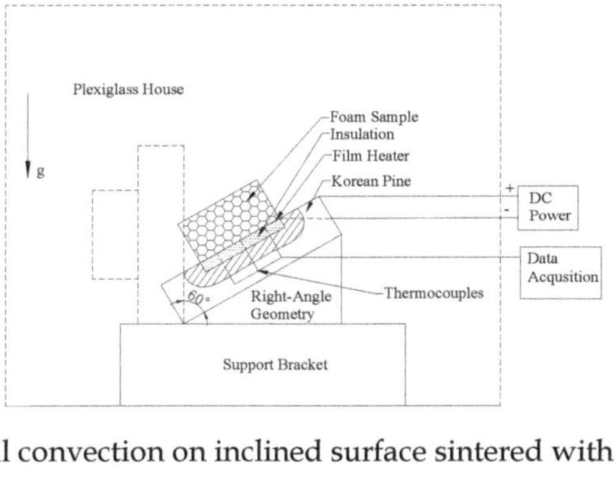

Fig. 2. Test rig of natural convection on inclined surface sintered with metallic foams

Experiment results of the conjugated radiation and natural convective heat transfer on wall surface sintered with open-celled metallic foams at different inclination angles are presented. The metal foam test samples have the same length and width of 100 mm, but different height of 10 mm and 40 mm. To investigate the coupled radiation and natural convection on the metal foam surface, a black paint layer with thickness of 0.5 mm and emissivity 0.96 is painted on the surface of the metal foam surface for the temperature testing with infrared camera. Porosity is 0.95, while pore density is 10 PPI.

Figure 3(a) shows the comparison between different experimental data, several of which were obtained from existing literature. The present result without paint agrees well with existing experimental data (Sparrow & Gregg, 1956; Fujii T. & Fujii M., 1976; Churchil & Ozoe, 1973). However, experimental result with paint is higher than unpainted metal foam block. This is attributed to the improved emissivity of black paint layer of painted surface.

Figure 3(b) presents effect of inclination angle on the average Nusselt number for two thicknesses of metallic foams (δ / L=0.1 and 0.4). As inclination angle increases from 0° (vertical position) to 90° (horizontal position), heat transferred in convective model initially increases and subsequently decreases. The maximum value is between 60° and 80°. Hence, overall heat transfer increases initially and remains constant as inclination angle increases.

To investigate the effect of radiation on total heat transfer, a ratio of the total heat transfer occupied by the radiation is introduced in this chapter, as shown below:

$$R = \frac{\phi_{\mathrm{rad}}}{\phi}. \tag{3}$$

Figure 3(c) provides the effect inclination angle on R for different foam samples (δ / L=0.1 and 0.4). In the experiment scope, the fraction of radiation in the total heat transfer is in the range of 33.8%-41.2%. For the metal foam sample with thickness of 10 mm, R is decreased as the inclination angle increases. However, with a thickness of 40 mm, R decreases initially and eventually increases as the inclination angle increases, reaching the minimum value of approximately 75°.

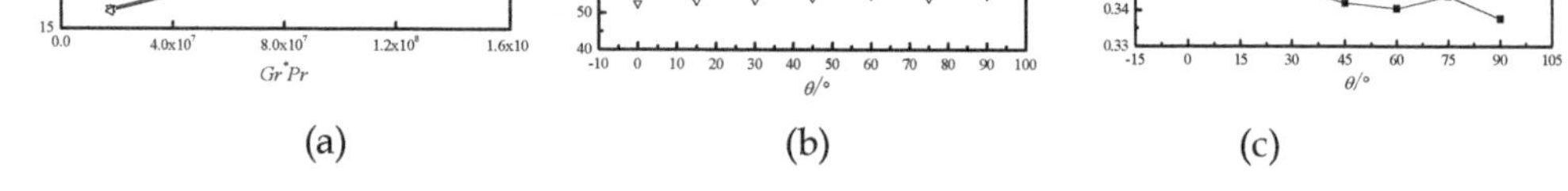

(a) (b) (c)

Fig. 3. Experimental results: (a) comparison with existing data; (b) effect of inclination angle on heat transfer; (c) effect of inclination angle on R

Figure 4 shows the infrared result of temperature distribution on the metallic foam surface with different foam thickness. It can be seen that the foam block with larger thickness has less homogeneous temperature distribution.

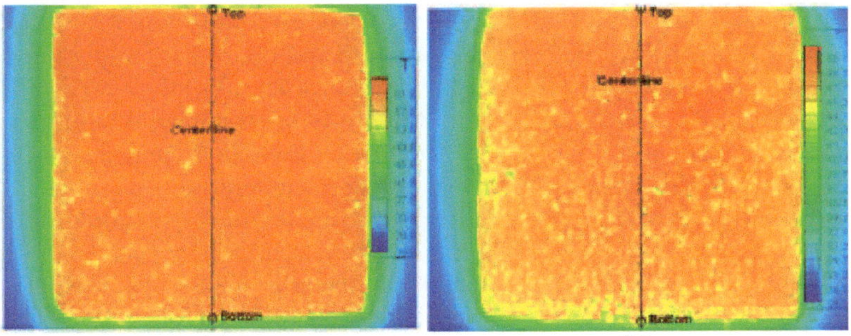

(a) (b)

Fig. 4. Temperature distribution of metallic foam surface predicted by infrared rays: (a) δ/L=0.1; (b) δ/L=0.4

3. Forced convection modelling in metallic foams

Research on thermal modeling of internal forced convective heat transfer enhancement using metallic foams is presented here. Several analytical solutions are shown below as benchmark for the improvement of numerical techniques. The Forchheimer model is commonly used for establishing momentum equations of flow in porous media. After introducing several empirical parameters of metallic foams, it is expressed for steady flow as:

$$\frac{\rho_{\mathrm{f}}}{\varepsilon^2}\left\langle\left(\vec{U}\cdot\nabla\right)\vec{U}\right\rangle = -\nabla p_{\mathrm{f}} + \frac{\mu_{\mathrm{f}}}{\varepsilon}\nabla^2\left\langle\vec{U}\right\rangle - \frac{\mu_{\mathrm{f}}}{K}\left\langle\vec{U}\right\rangle - \frac{\rho_{\mathrm{f}}C_{\mathrm{I}}}{\sqrt{K}}\left[\left\langle\vec{U}\right\rangle\cdot\left\langle\vec{U}\right\rangle\right]J\,. \tag{4}$$

where ρ, p, μ, K, C_{I}, $\vec{U}$ is density, pressure, kinematic viscousity, permeability, inertial coefficient and velocity vector, respectively. And J is the unit vector along pore velocity vector $J=\overrightarrow{U_{\mathrm{P}}}/\left|\overrightarrow{U_{\mathrm{P}}}\right|$. The angle bracket means the volume averaged value. The term in the left-hand side of Eq. (4) is the advective term. The terms in the right-hand side are pore pressure gradient, viscous term (i.e., Brinkman term), Darcy term (microscopic viscous shear stress), and micro-flow development term (inertial term), respectively. When porosity approaches 1, permeability becomes very large and Eq. (4) is converted to the classical Navier-Stokes equation.

Thermal transport in porous media owns two basic models: local thermal equilibrium model (LTE) and local thermal non-equilibrium model (LNTE). The former with one-energy equation treats the local temperature of solid and fluid as the same value while the latter has two-energy equations taking into account the difference between the temperatures of solid and fluid. They take the following forms [Eq. (5) for LTE and Eqs. (6a-6b) for LNTE]:

$$\langle\rho_{\mathrm{f}}\rangle c_{\mathrm{f}}\left\langle\vec{U}\right\rangle\nabla\langle T\rangle = \nabla\left[\left(k_{\mathrm{fe}}+k_{\mathrm{d}}\right)\nabla\langle T\rangle\right]. \tag{5}$$

$$0 = \nabla\left[k_{\mathrm{se}}\nabla\langle T_{\mathrm{s}}\rangle\right] - h_{\mathrm{sf}}a_{\mathrm{sf}}\left(T_{\mathrm{s}}-T_{\mathrm{f}}\right). \tag{6a}$$

$$\langle\rho_{\mathrm{f}}\rangle c_{\mathrm{f}}\left\langle\vec{U}\right\rangle\nabla\langle T_{\mathrm{f}}\rangle = \nabla\left[k_{\mathrm{fe}}\nabla\langle T_{\mathrm{f}}\rangle\right] + h_{\mathrm{sf}}a_{\mathrm{sf}}\left(T_{\mathrm{s}}-T_{\mathrm{f}}\right). \tag{6b}$$

Subscripts 'f', 's', 'fe', 'se', 'd' and 'sf' respectively denotes 'fluid', 'solid', 'effective value of fluid', 'effective value of solid', 'dispersion' and 'solid and fluid'. ***T*** is temperature variable.

As stated above, Lee and Vafai (Lee & Vafai, 1999) indicated that the LNTE model is more accurate than the LTE model when the difference between solid and fluid thermal conductivities is significant. This is true in the case of metallic foams, in which difference between solid and fluid phases is often two orders of magnitudes or more. Thus, LNTE model with two-energy equations (also called two-equation model) is employed throughout this chapter.

For modeling forced convective heat transfer in metallic foams, the metallic foams are assumed to be isotropic and homogeneous. For analytical simplification, the flow and temperature fields of impressible fluid are fully developed, with thermal radiation and natural convection ignored. Simultaneously, thermal dispersion is negligible due to high solid thermal conductivity of metallic foams (Calmidi & Mahajan, 2000; Lu et al., 2006; Dukhan, 2009). As a matter of convenience, the angle brackets representing the volume-averaging qualities for porous medium are dropped hereinafter.

3.1 Fin analysis model

As fin analysis model is a very simple and useful method to obtain temperature distribution, fin theory-based heat transfer analysis is discussed here and a modified fin analysis method of present authors (Xu et al., 2011a) for metallic foam filled channel is introduced. A comparison between results of present and conventional models is presented.

Fin analysis method for heat transfer is originally adopted for heat dissipation body with extended fins. It is a very simple and efficient method for predicting the temperature distribution in these fins. It was first introduced to solve heat transfer problems in porous media by Lu et al. in 1998 (Lu et al., 1998). As presented, the heat transfer results with fin analysis exhibit good trends with variations of foam morphology parameters. However, it has been pointed out that this method may overpredict the heat transfer performance. This fin analysis method treats the velocity and temperature of fluid flowing through the porous foam as uniform, significantly overestimating the heat transfer result. With the assumption of cubic structure composed of cylinders, fin analysis formula of Lu et al. (Lu et al., 1998) is expressed as:

$$\frac{\partial^2 T_s(x,y)}{\partial y^2} - \frac{4h_{sf}}{k_s d_f}\left[T_s(x,y) - T_{f,b}(x)\right] = 0. \tag{7}$$

where (x,y) is the Cartesian coordinates and d_f is the fibre diameter. The subscript 'f,b' denotes 'bulk mean value of fluid'.

In the previous model (Lu et al., 1998), heat conduction in the cylinder cell is only considered and the surface area is taken as outside surface area of cylinders with thermal conductivity k_s. Based on the assumption, thermal resistance in the fin is artificially reduced, leading to the previous fin method that overestimates heat transfer. Fluid with temperature $T_{f,b}(x)$ flows through the porous channel. The fluid heat conduction in the foam is considered together with the solid heat conduction. The effective thermal conductivity k_e and extended surface area density of porous foam a_{sf} instead of k_s and surface area of solid cylinders are applied to gain the governing equation. The modified heat conduction equation proposed by present authors (Xu et al., 2011a) is as follows:

$$\frac{\partial^2 T_{e,f}(x,y)}{\partial y^2} - \frac{h_{sf}a_{sf}}{k_e}\left[T_{e,f}(x,y) - T_{f,b}(x)\right] = 0. \tag{8}$$

Temperature $T_{e,f}(x)$ in Eq. (8) representing the temperature of porous foam is defined as the equivalent foam temperature. With the constant heat flux condition, equivalent foam temperature, and Nusselt number are obtained in Eq. (9) and Eq. (10):

$$T_{e,f}(x,y) = T_{f,b}(x) + q_w \cosh(my) / \left[mk_e \sinh(mH)\right]. \tag{9}$$

$$Nu = \frac{q_w}{T_w(x) - T_{f,b}(x)}\frac{4H}{k_f} = \frac{q_w}{T_{e,f}(x,0) - T_{f,b}(x)}\frac{4H}{k_f} = 4mH \cdot \frac{k_e}{k_f} \cdot \tanh(mH). \tag{10}$$

where q_w is the wall heat flux and H is the half width of the parallel-plate channel. The fin efficiency m is calculated with $m=h_{sf}a_{sf}/k_e$.

To verify the improvement of the present modified fin analysis model for heat transfer in metallic foams, the comparison among the present fin model, previous fin model (Lu et al., 1998), and the analytical solution presented in Section 3.2 is shown in Fig. 5. Figure 5(a) presents the comparison between the Nusselt number results predicted by present modified fin model, previous fin model (Lu et al., 1998), and analytical solution in Section 3.2. Evidently, the present modified fin model is closer to the analytical solution. It can replace the previous fin model (Lu et al., 1998) to estimate heat transfer in porous media with improved accuracy. Only the heat transfer results of the present modified fin model and analytical solution in Section 3.2 are compared in Fig. 5(b). It is noted that when k_f/k_s is sufficiently small, the present modified fin model can coincide with the analytical solution. The difference between the two gradually increases as k_f/k_s increases.

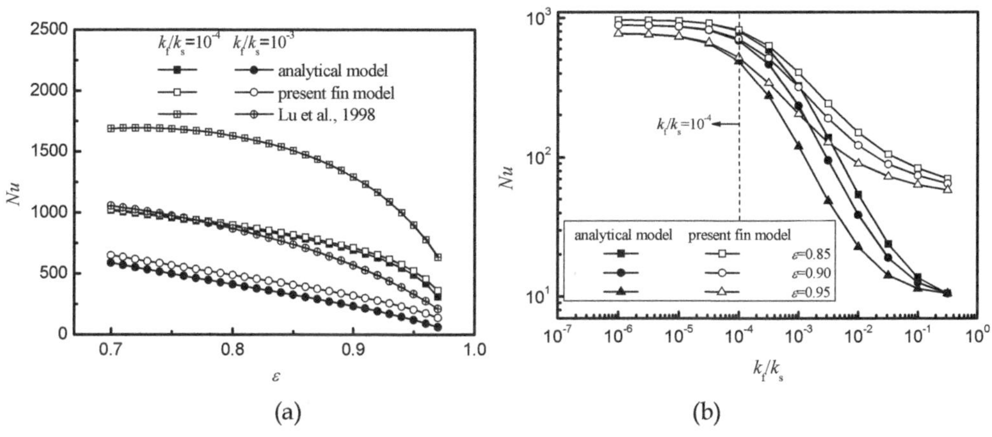

(a) (b)

Fig. 5. Comparisons of Nu (a) among present modified fin model, previous fin model (Lu et al., 1998), and analytical solution in Section 3.2.1 (ω=10 PPI, H=0.005 m, u_m=1 m·s^{-1}); (b) between present modified fin model and analytical solution in Section 3.2.1 (ω=10 PPI, H=0.005 m, u_m=1 m·s^{-1})

3.2 Analytical modeling

3.2.1 Metallic foam fully filled duct

In this part, fully developed forced convective heat transfer in a parallel-plate channel filled with highly porous, open-celled metallic foams is analytically modeled using the Brinkman-Darcy and two-equation models and the analytical results of the present authors (Xu et al., 2011a) are presented in the following. Closed-form solutions for fully developed fluid flow and heat transfer are proposed.

Figure 6 shows the configuration of a parallel-plate channel filled with metallic foams. Two infinite plates are subjected to constant heat flux q_w with height $2H$. Incompressible fluid flows through the channel with mean velocity u_m and absorbs heat imposed on the parallel plates.

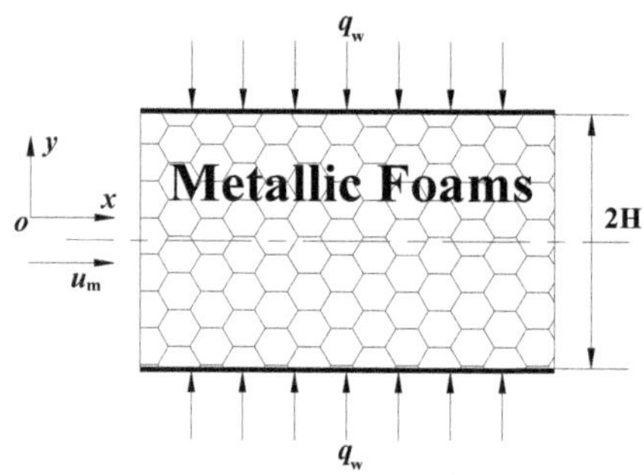

Fig. 6. Schematic diagram of metallic foam fully filled parallel-plate channel

For simplification, the angle brackets representing volume-averaged variables are dropped from Eqs. (4), (6a), and (6b). The governing equations and closure conditions are normalized with the following qualities:

$$Y=\frac{y}{H}, U=\frac{u}{u_m}, P=\frac{K}{\mu_f u_m}\frac{dp}{dx}, \theta_s=\frac{T_s-T_w}{q_w H/k_{se}}, \theta_f=\frac{T_f-T_w}{q_w H/k_{se}}, \tag{11a}$$

$$Da=\frac{K}{H^2}, B=\frac{k_f}{k_{se}}, C=\frac{k_{fe}}{k_{se}}, D=\frac{h_{sf}a_{sf}H^2}{k_{se}}, s=\sqrt{\varepsilon/Da}, t=\sqrt{D(C+1)/C}. \tag{11b}$$

Empirical correlations for these parameters are listed in Table 1.

After neglecting the inertial term in Eq. (4), governing equations for problem shown in Fig. 6 can be normalized as:

$$\frac{\partial^2 U}{\partial Y^2}-s^2(U+P)=0. \tag{12a}$$

$$\frac{\partial^2 \theta_s}{\partial Y^2}-D(\theta_s-\theta_f)=0. \tag{12b}$$

$$C\frac{\partial^2 \theta_f}{\partial Y^2}+D(\theta_s-\theta_f)=U. \tag{12c}$$

Parameter	Correlation	Reference
Pore diameter d_p	$d_p = 0.0254 / \omega$	Calmidi, 1998
Fibre diameter d_f	$d_f = d_p \cdot 1.18\sqrt{(1-\varepsilon)/(3\pi)}\left[1-\exp((\varepsilon-1)/0.04)\right]^{-1}$	Calmidi, 1998
Specific surface area a_{sf}	$a_{sf} = 3\pi d_f\left[1-e^{-((1-\varepsilon)/0.04)}\right] / \left(0.59d_p\right)^2$	Zhao et al., 2001
Permeability K	$K = 0.00073(1-\varepsilon)^{-0.224}\left(d_f / d_p\right)^{-1.11} d_p^2$	Calmidi, 1998
Local heat transfer coefficient h_{sf}	$h_{sf} = \begin{cases} 0.76Re_d^{0.4}Pr^{0.37}k_f / d, & (1 \le Re_d \le 40) \\ 0.52Re_d^{0.5}Pr^{0.37}k_f / d, & \left(40 \le Re_d \le 10^3\right) \\ 0.26Re_d^{0.6}Pr^{0.37}k_f / d, & \left(10^3 \le Re_d \le 2\times10^5\right) \end{cases}$ $Re_d = \rho_f u d / \mu_f$	Lu et al., 2006
Effective thermal conductivity k_e	$R_A = \dfrac{4\lambda}{\left(2e^2+\pi\lambda(1-e)\right)k_s+\left(4-2e^2-\pi\lambda(1-e)\right)k_f}$ $R_B = \dfrac{(e-2\lambda)^2}{(e-2\lambda)e^2k_s+\left(2e-4\lambda-(e-2\lambda)e^2\right)k_f}$ $R_C = \dfrac{\left(\sqrt{2}-2e\right)^2}{2\pi\lambda^2\left(1-2\sqrt{2}e\right)k_s+2\left(\sqrt{2}-2e-\pi\lambda^2\left(1-2\sqrt{2}e\right)\right)k_f}$ $R_D = \dfrac{2e}{e^2k_s+\left(4-e^2\right)k_f}$ $\lambda = \sqrt{\dfrac{\sqrt{2}\left(2-(5/8)e^3\sqrt{2}-2\varepsilon\right)}{\pi\left(3-4\sqrt{2}e-e\right)}}, \ e = 0.339$ $k_e = \dfrac{1}{\sqrt{2}\left(R_A+R_B+R_C+R_D\right)}$ $k_{se} = k_e\big\|_{k_f=0}, \ k_{fe} = k_e\big\|_{k_s=0}$	Boomsma & Poulikakos, 2001

Table 1. Semi-empirical correlations of parameters for metallic foams

The dimensionless closure conditions can likewise be derived as follows:

$$Y=0, \frac{\partial U}{\partial Y}=\frac{\partial \theta_s}{\partial Y}=\frac{\partial \theta_f}{\partial Y}=0; Y=1, U=0, \theta_s=\theta_f=0. \tag{13}$$

Thus, the dimensionless velocity profile is expressed by the hyperbolic functions:

$$U = P\left[\cosh(sY)/\cosh(s)-1\right]. \tag{14a}$$

$$P = \frac{1}{\tanh(s)/s-1}. \tag{14b}$$

Meanwhile, dimensionless fluid and solid temperatures can be derived as follows:

$$\theta_s + C\theta_f = P\left[\frac{\cosh(sY)}{s^2\cosh(s)} - \frac{1}{2}Y^2 + \frac{1}{2} - \frac{1}{s^2}\right]. \tag{15a}$$

$$\theta_f = P\left[\frac{C\cdot s^2}{D(C+1)^2\left[C\cdot s^2 - D(C+1)\right]}\frac{\cosh(tY)}{\cosh(t)} + \frac{\left(1-D/s^2\right)}{\left[C\cdot s^2 - D(C+1)\right]}\frac{\cosh(sY)}{\cosh(s)} - \frac{1}{2(C+1)}Y^2 + \frac{D\left(1/2-1/s^2\right)-1/(C+1)}{D(C+1)}\right]. \tag{15b}$$

$$\theta_s = P\left[-\frac{C^2\cdot s^2}{D(C+1)^2\left[C\cdot s^2 - D(C+1)\right]}\frac{\cosh(tY)}{\cosh(t)} - \frac{D/s^2}{\left[C\cdot s^2 - D(C+1)\right]}\frac{\cosh(sY)}{\cosh(s)} - \frac{1}{2(C+1)}Y^2 + \frac{D\left(1/2-1/s^2\right)+C/(C+1)}{D(C+1)}\right]. \tag{15c}$$

The dimensionless numbers, friction factor f, and Nusselt number Nu are shown below.

$$f = \frac{\mathrm{d}p/\mathrm{d}x\cdot 4H}{\rho_f u_m^2/2} = \frac{32P}{Re\cdot Da}. \tag{16a}$$

$$Nu = \frac{q_w}{T_w - T_{f,b}}\frac{4H}{k_f} = -\frac{4}{B\cdot\theta_{f,b}}. \tag{16b}$$

In Eq. (20), the dimensionless bulk fluid temperature can be obtained using Eq. (19).

$$\begin{aligned}\theta_{f,b} = \frac{\frac{1}{A}\int_A U\theta_f \mathrm{d}A}{\frac{1}{A}\int_A U\mathrm{d}A} = \int_0^1 U\theta_f \mathrm{d}Y = P^2\Bigg\{ & \frac{\left[\frac{\cosh(s+t)}{s+t} - \frac{\cosh(s-t)}{s-t}\right]C\cdot s^2}{2D(C+1)^2\left[C\cdot s^2 - D(C+1)\right]\cosh(s)\cosh(t)} \\ & + \frac{\left(1-\frac{D}{s^2}\right)\cosh(2s)}{4\left[C\cdot s^2 - D(C+1)\right]s\cdot\cosh^2(s)} + \frac{\left(1-\frac{D}{s^2}\right)\left[e^s\left(1-\frac{2}{s}+\frac{2}{s^2}\right) - e^{-s}\left(1+\frac{2}{s}+\frac{2}{s^2}\right)\right]}{4\left[C\cdot s^2 - D(C+1)\right]s\cdot\cosh^2(s)} \\ & + \frac{1}{s}\left(\frac{C\cdot s^2}{D(C+1)^2\left[C\cdot s^2 - D(C+1)\right]} + \frac{1}{2(C+1)}\right) - \frac{C\cdot s^2}{D(C+1)^2\left[C\cdot s^2 - D(C+1)\right]t} \\ & + \frac{1}{6(C+1)} + \frac{1-\frac{D}{s^2}}{\left[C\cdot s^2 - D(C+1)\right]\cosh^2(s)} - \frac{D\left(\frac{1}{2}-\frac{1}{s^2}\right) - \frac{1}{C+1}}{D(C+1)}\Bigg\}.\end{aligned} \tag{17}$$

The Nusselt number of the present analytical solution is compared with experiment data (Zhao & Kim, 2001) for forced convection in rectangular metallic foam filled duct [Fig. 7(a)]. As illustrated, the difference between the analytical and previous experiment results is attributed to the experimental error and omission of the dispersion effect and quadratic term in the velocity equation. Overall, the analytical and experimental results are correlated with each other, with reasonable similarities. To examine the effect of the Brinkman term, comparison between temperature profiles of the present solution and that of Lee and Vafai (Lee & Vafai, 1999) is presented in Fig. 7(b). It can be observed that temperature profiles of the two solutions are similar. In particular, the predicted temperatures for solid and fluid of Lee and Vafai (Lee & Vafai, 1999) are closer to the wall temperature than the present solution. This is due to the completely uniform cross-sectional velocity assumed for the Darcy model by Lee and Vafai (Lee & Vafai, 1999). The comparison provides another evidence for the feasibility of the present solution.

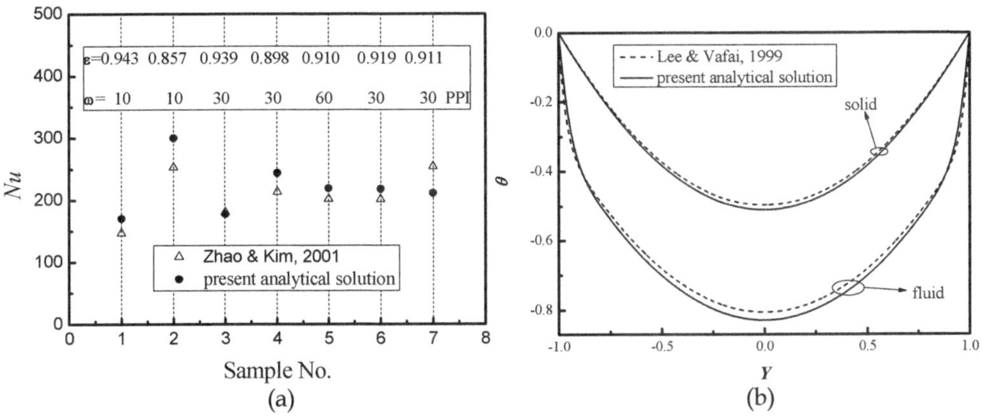

Fig. 7. Validation of present solution: (a) compared with the experiment; (b) compared with Lee and Vafai (Lee & Vafai, 1999) (ε=0.9, ω=10 PPI, H=0.01 m, u_m=1 m/s, k_f/k_s=10^{-4})

Figure 8(a) displays velocity profiles for smooth and metallic foam channels with different porosities and pore densities from the analytical solution for velocity in Eq. (14a). The existence of metallic foams can dramatically homogenize the flow field and velocity gradient near the impermeable wall because the metallic foam channel is considerably higher than that for the smooth channel. With decreasing porosity and increasing pore density, the flow field becomes more uniform and the boundary layer becomes thinner. This implies that high near wall velocity gradient, created by the ability to homogenize flow field for metallic foams, is an important reason for heat transfer enhancement.

Figure 8(b) shows the effect of porosity on the temperature profiles of solid and fluid phases. Porosity has significant influence on both solid and fluid temperatures. When porosity increases, temperature difference between the solid matrix and channel wall is improved. This is because the solid ligament diameter is reduced with increasing porosity under the same pore density, which results in an increased thermal resistance of heat conduction in solid matrix. In addition, the temperature difference between the fluid and channel wall has a similar trend. This is attributed to the reduction of the specific surface area caused by increasing porosity to create higher heat transfer temperature difference under the same heat transfer rate.

Figure 8(c) illustrates the effect of pore density on temperature profile. It is found that θ_s / k_{se} almost remains unchanged with increasing pore density since effective thermal conductivity and thermal resistance in solid matrix are affected not by pore density but by porosity. While θ_f / k_{se} increases with an increase in pore density, it shows that temperature difference between fluid and solid wall is reduced since the convective thermal resistance is reduced due to the extended surface area.

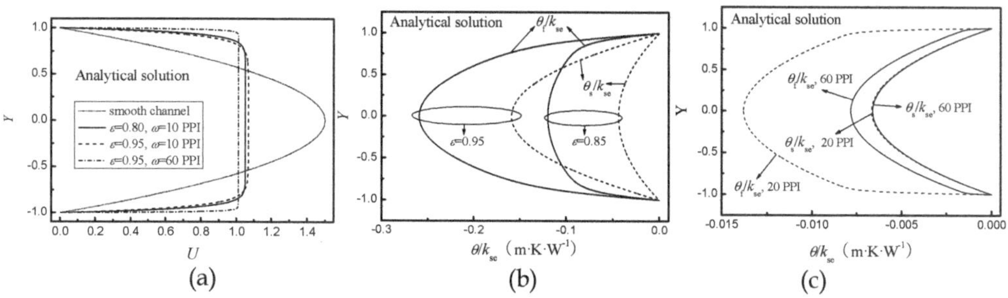

Fig. 8. Effect of metallic foam morphology parameters on velocity and temperature profiles: (a) velocity profile (H=0.005 m); (b) temperature profile affected by porosity (ω=10 PPI, H=0.01 m, u_m=5 m s^{-1}, k_f/k_s=10^{-4}); (c) temperature profile affected by pore density (ε=0.9, H=0.01 m, u_m=5 m s^{-1}, k_f/k_s=10^{-5})

3.2.2 Metallic foam partially filled duct

In the second part, fully developed forced convective heat transfer in a parallel-plate channel partially filled with highly porous, open-celled metallic foam is analytically investigated and results proposed by the present author (Xu et al., 2011c) is presented in this section. The Navier-Stokes equation for the hollow region is connected with the Brinkman-Darcy equation in the foam region by the flow coupling conditions at the porous-fluid interface. The energy equation for the hollow region and the two energy equations of solid and fluid for the foam region are linked by the heat transfer coupling conditions. The schematic diagram for the corresponding configuration is shown in Fig. 9. Two isotropic

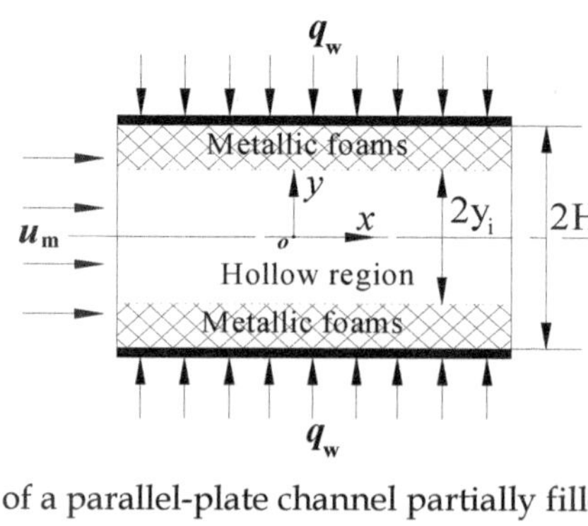

Fig. 9. Schematic diagram of a parallel-plate channel partially filled with metallic foam

and homogeneous metallic foam layers are symmetrically sintered on upper and bottom plates subjected to uniform heat flux. Fluid is assumed to possess constant thermal-physical properties. Thermal dispersion effect is neglected for metallic foams with high solid thermal conductivity (Calmidi & Mahajan 2000). To obtain analytical solution, the inertial term in Eq.(4) is neglected as well.

For the partly porous duct, coupling conditions of flow and heat transfer at the porous-fluid interface can be divided into two types: slip and no-slip conditions. Alazmi and Vafai (Alazmi & Vafai, 2001) reviewed different kinds of interfacial conditions related to velocity and temperature using the one-equation model. It was indicated that the difference between different interfacial conditions for both flow and heat transfer is minimal. However, for heat transfer, since the LTNE model is more suitable for highly conductive metallic foams rather than the LTE model, the implementation of LTNE model on thermal coupling conditions at the foam-fluid interface is considerably more complex in terms of number of temperature variables and number of interfacial conditions compared with LTE model. The momentum equation in the fluid region belongs to the Navier-Stokes equation while that in the foam region is the Brinkman-extended-Darcy equation, which is easily coupled by the flow conditions at foam-fluid interface.

For heat transfer, Ochoa-Tapia and Whitaker (Ochoa-Tapia & Whitaker, 1995) have proposed a series of interface conditions for non-equilibrium conjugate heat transfer at the porous-fluid interface, which can be used for thermal coupling at the foam-fluid interface of a domain party filled with metallic foams. Given that the two sets of governing equations are coupled at the porous-fluid interface, the interfacial coupling conditions must be determined to close the governing equations. Continuities of velocity, shear stress, fluid temperature, and heat flux at the porous-fluid interface should be guaranteed for meaningful physics. The corresponding expressions are shown in Eqs. (18)-(21):

$$u\big|_{y_i^-} = u\big|_{y_i^+} \,. \tag{18}$$

$$\mu_f \frac{du}{dy}\bigg|_{y_i^-} = \frac{\mu_f}{\varepsilon}\frac{du}{dy}\bigg|_{y_i^+} \,. \tag{19}$$

$$T_f\big|_{y_i^-} = T_f\big|_{y_i^+} \,. \tag{20}$$

$$k_f \frac{\partial T_f}{\partial y}\bigg|_{y_i^-} = \left(k_{fe}\frac{\partial T_f}{\partial y} + k_{se}\frac{\partial T_s}{\partial y} \right)\bigg|_{y_i^+} \,. \tag{21}$$

There are three variables for two-energy equations: the fluid and solid temperatures of the foam region and the temperature of the open region. Therefore, another coupling condition is required to obtain the temperature of solid and fluid in the foam region (Ochoa-Tapia & Whitaker, 1995), as:

$$k_{se}\nabla T_s\big|_{y_i^+} = h_{sf}\left(T_s\big|_{y_i^+} - T_f\big|_{y_i^-} \right) . \tag{22}$$

For the reason that axial heat conduction can be neglected, Eq. (22) is simplified as follows:

$$k_{\mathrm{se}} \frac{\partial T_{\mathrm{s}}}{\partial y}\bigg|_{y_{\mathrm{i}}^{+}} = h_{\mathrm{sf}}\left(T_{\mathrm{s}}\big|_{y_{\mathrm{i}}^{+}} - T_{\mathrm{f}}\big|_{y_{\mathrm{i}}^{-}}\right). \tag{23}$$

Due to the fact that the solid ligaments are discontinuous at the foam-fluid interface, heat conduction through the solid phase is totally transferred to the fluid in the manner of convective heat transfer across the foam-fluid interface. Thus, the physical meaning of Eq.(22) stands for the convective heat transfer at the foam-fluid interface from the solid ligament to the fluid nearby. Thus, the governing equations in the foam region and that in the fluid region are linked together via these interfacial coupling conditions.
With the dimensionless qualities in Eqs.(11a)–(11b) and the following variables,

$$Y_i = \frac{y_i}{H}, A = \frac{h_{\mathrm{sf}} H}{k_{\mathrm{se}}}. \tag{24}$$

governing equations for the fluid region and the foam region are normalized. Dimensionless governing equations for the hollow region ($0 \le Y \le Y_{\mathrm{i}}$) are as follows:

$$\frac{\partial^2 U}{\partial Y^2} - \frac{P}{Da} = 0. \tag{25}$$

$$\frac{\partial^2 \theta_{\mathrm{f}}}{\partial Y^2} = \frac{1}{B} U. \tag{26}$$

Dimensionless governing equations for the foam region ($Y_{\mathrm{i}} \le Y \le 1$) are as follows:

$$\frac{\partial^2 U}{\partial Y^2} - s^2 (U + P) = 0. \tag{27}$$

$$\frac{\partial^2 \theta_{\mathrm{s}}}{\partial Y^2} - D(\theta_{\mathrm{s}} - \theta_{\mathrm{f}}) = 0. \tag{28}$$

$$C\frac{\partial^2 \theta_{\mathrm{f}}}{\partial Y^2} + D(\theta_{\mathrm{s}} - \theta_{\mathrm{f}}) = U. \tag{29}$$

Corresponding dimensionless closure conditions are as follows:

$$Y = 0 : \frac{\partial U}{\partial Y} = \frac{\partial \theta_{\mathrm{f}}}{\partial Y} = 0. \tag{30a}$$

$$Y = 1 : U = 0, \theta_{\mathrm{s}} = \theta_{\mathrm{f}} = 0. \tag{30b}$$

When $Y = Y_{\mathrm{i}}$, the dimensionless coupling conditions are as follows:

$$U\big|_{Y_{\mathrm{i}}^{-}} = U\big|_{Y_{\mathrm{i}}^{+}}. \tag{31a}$$

$$\left.\frac{dU}{dY}\right|_{Y_i^-} = \frac{1}{\varepsilon}\left.\frac{dU}{dY}\right|_{Y_i^+}. \tag{31b}$$

$$\left.\theta_f\right|_{Y_i^-} = \left.\theta_f\right|_{Y_i^+}. \tag{31c}$$

$$B\left.\frac{\partial\theta_f}{\partial Y}\right|_{Y_i^-} = \left.\left(C\frac{\partial\theta_f}{\partial Y} + \frac{\partial\theta_s}{\partial Y}\right)\right|_{Y_i^+}. \tag{31d}$$

$$\left.\frac{\partial\theta_s}{\partial Y}\right|_{Y_i^+} = A\left(\left.\theta_s\right|_{Y_i^+} - \left.\theta_f\right|_{Y_i^-}\right). \tag{31e}$$

The velocity field is typically obtained ahead of the temperature field for the uncoupled relationship between the momentum and energy equations. With closure conditions for flow in Eqs. (30a), (30b), (31a), and (31b), the dimensionless velocity equation of Eqs. (25) and (27) can be solved and the solution for velocity is as follows:

$$U = \begin{cases} P\left(\frac{1}{2Da}Y^2 + C_0\right), & 0 \le Y \le Y_i \\ P\left(C_1 e^{sY} + C_2 e^{-sY} - 1\right), & Y_i \le Y \le 1 \end{cases}. \tag{32}$$

where dimensionless pressure drop P, constants $\mathbf{C_0}$, $\mathbf{C_1}$, and $\mathbf{C_2}$ are as follows:

$$P = \frac{1}{\frac{1}{s}\left(C_1 e^s - C_2 e^{-s}\right) + \frac{Y_i^3}{6Da} + C_0 Y_i - 1}. \tag{33a}$$

$$C_0 = C_1 e^{sY_i} + C_2 e^{-sY_i} - Y_i^2/(2Da) - 1. \tag{33b}$$

$$C_1 = \frac{e^{-sY_i} + sY_i e^{sY_i}}{e^{s(1-Y_i)} + e^{-s(1-Y_i)}}. \tag{33c}$$

$$C_2 = \frac{e^{sY_i} - sY_i e^{s}}{e^{s(1-Y_i)} + e^{-s(1-Y_i)}}. \tag{33d}$$

The solution to the energy equations is as follows:

$$\theta_f = \begin{cases} P\cdot\frac{1}{B}\left(\frac{1}{24Da}Y^4 + \frac{C_0}{2}Y^2 + C_3\right), & 0 \le Y \le Y_i \\ P\begin{Bmatrix} C_6 e^{tY} + C_7 e^{-tY} + \frac{1 - D/s^2}{C\cdot s^2 - D(C+1)}\left(C_1 e^{sY} + C_2 e^{-sY}\right) \\ -\frac{1}{2(C+1)}Y^2 + \frac{C_4}{C+1}Y + \frac{C_5}{C+1} + \frac{1}{D(C+1)^2} \end{Bmatrix}, & Y_i \le Y \le 1 \end{cases}. \tag{34a}$$

$$\theta_s = P\left\{\begin{array}{l} -\dfrac{1}{C}\left(C_6 e^{tY} + C_7 e^{-tY}\right) - \dfrac{D/s^2}{C\cdot s^2 - D(C+1)}\left(C_1 e^{sY} + C_2 e^{-sY}\right) \\ -\dfrac{1}{2(C+1)}Y^2 + \dfrac{C_4}{C+1}Y + \dfrac{C_5}{C+1} - \dfrac{C}{D(C+1)^2} \end{array}\right\}, \quad Y_i \le Y \le 1 \tag{34b}$$

The constants in the above equation are defined as follows:

$$C_3 = B\left[\begin{array}{l} C_6 e^{tY_i} + C_7 e^{-tY_i} + \dfrac{1 - D/s^2}{C\cdot s^2 - D(C+1)}\left(C_1 e^{sY_i} + C_2 e^{-sY_i}\right) \\ -\dfrac{1}{2(C+1)}Y_i^2 + \dfrac{C_4}{C+1}Y_i + \dfrac{C_5}{C+1} + \dfrac{1}{D(C+1)^2} \end{array}\right] - \frac{1}{24Da}Y_i^4 - \frac{C_0}{2}Y_i^2. \tag{35a}$$

$$C_4 = \frac{1}{6Da}Y_i^3 + C_0 Y_i. \tag{35b}$$

$$C_5 = \frac{1}{2} - \frac{1}{s^2} - C_4. \tag{35c}$$

$$C_6 = \frac{\left[A(C+1) + Ct\right]e^{-tY_i}C_8 - e^{-t}C_9}{\left[A(C+1) + Ct\right]e^{t(1-Y_i)} - \left[A(C+1) - Ct\right]e^{-t(1-Y_i)}}. \tag{35d}$$

$$C_7 = \frac{e^{t}C_9 - \left[A(C+1) - Ct\right]e^{tY_i}C_8}{\left[A(C+1) + Ct\right]e^{t(1-Y_i)} - \left[A(C+1) - Ct\right]e^{-t(1-Y_i)}}. \tag{35e}$$

$$C_8 = -\frac{C\cdot s^2}{D(C+1)^2\left[C\cdot s^2 - D(C+1)\right]}. \tag{35f}$$

$$C_9 = A\left[-\frac{C_1 e^{sY_i} + C_2 e^{-sY_i}}{C\cdot s^2 - D(C+1)} - \frac{1}{D(C+1)}\right] - \frac{C\cdot s^2\cdot Y_i}{(C+1)\left[C\cdot s^2 - D(C+1)\right]} - \frac{C_4}{C+1}. \tag{35g}$$

The bulk dimensionless fluid temperature is expressed as follows:

$$\begin{aligned}
\theta_{f,b} = \frac{\frac{1}{A}\int_A U\theta_f \mathrm{d}A}{\frac{1}{A}\int_A U\mathrm{d}A} &= \int_0^{Y_i} U\theta_f \mathrm{d}Y + \int_{Y_i}^1 U\theta_f dY = \frac{P^2}{B}\left[\frac{1}{336Da^2}Y_i^7 + \frac{7C_0}{120Da}Y_i^5 + \left(\frac{C_3}{Da} + C_0^2\right)\frac{Y_i^2}{6} + C_0 C_3 Y_i\right] \\
&+ P^2\left\{\frac{C_1 C_6}{s+t}\left(e^{(s+t)} - e^{(s+t)Y_i}\right) + \frac{C_1 C_7}{s-t}\left(e^{(s-t)} - e^{(s-t)Y_i}\right) - \frac{C_2 C_6}{(s-t)}\left(e^{-(s-t)} - e^{-(s-t)Y_i}\right)\right. \\
&- \frac{C_2 C_7}{(s+t)}\left(e^{-(s+t)} - e^{-(s+t)Y_i}\right) + \frac{N_1 C_1^2}{2s}\left(e^{2s} - e^{2sY_i}\right) - \frac{N_1 C_2^2}{2s}\left(e^{-2s} - e^{-2sY_i}\right) + \\
&\frac{C_1}{s}\left[e^{s}\left(N_2 + N_3 - \frac{2N_2}{s} - N_1 + \frac{2N_2}{s^2} - \frac{N_3}{s} + N_4\right) - e^{sY_i}\left(N_2 Y_i^2 + \left(N_3 - \frac{2N_2}{s}\right)Y_i - N_1 + \frac{2N_2}{s^2} - \frac{N_3}{s} + N_4\right)\right] - \\
&\frac{C_2}{s}\left[e^{-s}\left(N_2 + N_3 + \frac{2N_2}{s} - N_1 + \frac{2N_2}{s^2} + \frac{N_3}{s} + N_4\right) - e^{-sY_i}\left(N_2 Y_i^2 + \left(N_3 + \frac{2N_2}{s}\right)Y_i - N_1 + \frac{2N_2}{s^2} + \frac{N_3}{s} + N_4\right)\right] \\
&\left. - \frac{C_6}{t}\left(e^{t} - e^{tY_i}\right) + \frac{C_7}{t}\left(e^{-t} - e^{-tY_i}\right) - \frac{N_3}{3}\left(1 - Y_i^3\right) - \frac{N_3}{2}\left(1 - Y_i^2\right) + (2N_1 C_1 C_2 - N_4)(1 - Y_i)\right\}
\end{aligned} \tag{36}$$

In Eq. (36), the constants N_1, N_2, N_3, and N_4 are as follows:

$$N_1 = \frac{1 - D / s^2}{C \cdot s^2 - D(C+1)}, N_2 = -\frac{1}{2(C+1)}, N_3 = \frac{C_4}{C+1}, N_4 = \frac{1}{D(C+1)^2}. \quad (37)$$

Friction factor f and Nusselt number Nu are the same with Eqs. (16a) and (16b).
Velocity profiles predicted by the analytical solutions shown in Eq. (32) at three combinations of porosity and pore density are presented in Fig. 10(a). It is found that velocity in the hollow space is considerably higher than that in the foam region with obstructing foam ligaments. Increasing porosity or decreasing pore density can both increase velocity in foam region and decrease velocity in open region simultaneously since flow resistance in the foam region can be reduced by increasing porosity and decreasing pore density of metallic foams. Figure 10(b) illustrates the comparison of velocity profiles for four hollow ratios: 0 (foam fully filled channel), 0.2, 0.5, 0.8, and 1.0 (smooth channel). The velocity profile is a parabolic distribution for the smooth channel and the profile for fully filled channel is similar, except that the distribution is more uniform since no sudden change of permeability is in the cross section. Comparatively for the foam partially filled channel, sudden change in permeability occurs at the porous-fluid interface; the average velocity in the central fluid region for foam partially filled channel (Y_i=0.2, 0.5, 0.8) is higher than those of the foam fully filled channel and empty channel. Moreover, maximum velocity in the central region decreases with the increase in hollow ratio due to the sizeable difference in permeability between porous and clear fluid regions. Velocity in the foam region of foam partially filled channel is considerably lower than those of foam fully filled channel and smooth channel.

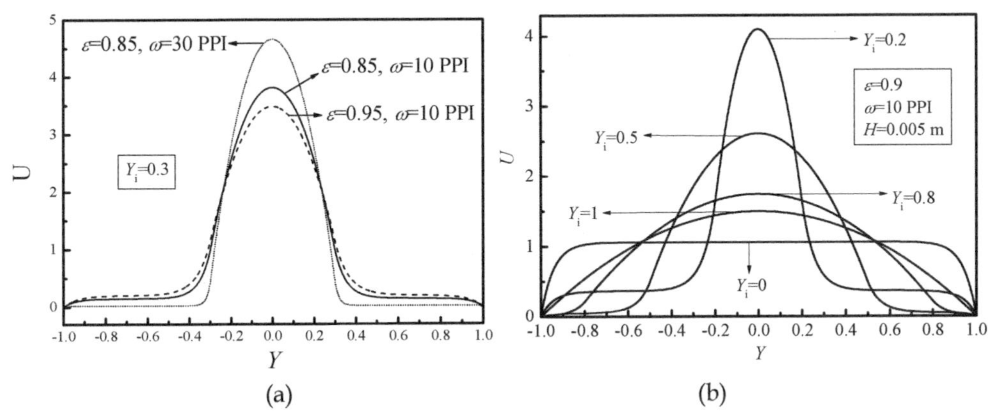

Fig. 10. Effect of key parameters on velocity profiles: (a) metal foam morphology parameters; (b) hollow ratio

Figure 11(a) presents the influence of porosity on temperature profile. Evidently, the nominal excess temperature of fluid in the hollow region decreases in the y direction and the decreasing trend becomes less obvious in the foam region. This is due to the total thermal resistance in the foam region, which is lower than that in the hollow region for the significant heat transfer surface extension. The nominal excess temperatures of fluid and solid in the foam region for ε =0.9 were lower than that for ε =0.95 because decreased porosity leads to the increase in both the effective thermal conductivity and the foam surface area to improve the corresponding heat transfer with the same heat flux. Effect of pore density on temperature profile is shown in Fig. 11(b). The solid excess temperature is almost the same in the foam region for different pore densities. This is attributed mainly to the effect of porosity on heat conduction thermal resistance of the foam, as shown in Table 1. On the other hand, the nominal excess fluid temperature of 5 PPI is significantly smaller than that of 30 PPI. The trend inconsistency of the fluid and solid temperatures in the two regions for the two pore densities is caused by mass flow fraction in the foam region. The local convective heat transfer coefficient for 5 PPI was higher than 30 PPI due to the relative higher mass flow fraction in the foam region.

However, the heat transfer surface area inside the foam of 5 PPI is lower than that of 30 PPI. The two opposite effects competed with each other, resulting in the identical temperature difference between wall and fluid in the foam region. However, in the hollow region, the porous-fluid interface area becomes the only surface area where porosity for the two pore densities is the same. Hence, the temperature difference between wall and fluid for 5 PPI is reduced and obviously lower than that for 30 PPI.

Figure 11(c) presents the comparison between fluid and solid temperature distribution for different channels, including empty channel (Y_i=1), foam partially filled channel (Y_i=0.5), and foam fully filled channel (Y_i=0). The nominal excess temperature becomes dependent on the heat transfer area and local heat transfer coefficient. In the hollow region, the heat transfer surface area is reduced to the interface area for the foam partially filled channel (Y_i=0.5), which was considerably smaller than the volume surface area of the fully filled channel (Y_i=0). The nominal fluid excess temperature increases in the order of Y_i=1, Y_i=0.5, and Y_i=0 since the total convective thermal resistance $1/(h_{sf}a_{sf})$ decreases in the order Y_i=0, Y_i=0.5, Y_i=1 in the clear fluid region. In the near-wall foam region, local heat transfer coefficient for Y_i=0.5 is reduced compared with that for Y_i=0. However, the effect of fluid heat conduction dominates in the near-wall area, resulting in a lower nominal fluid excess temperature for Y_i=0.5 compared with that for the foam fully filled channel (Y_i=0). Thus, an intersection point occurs in the curve of the fluid excess temperature distribution.

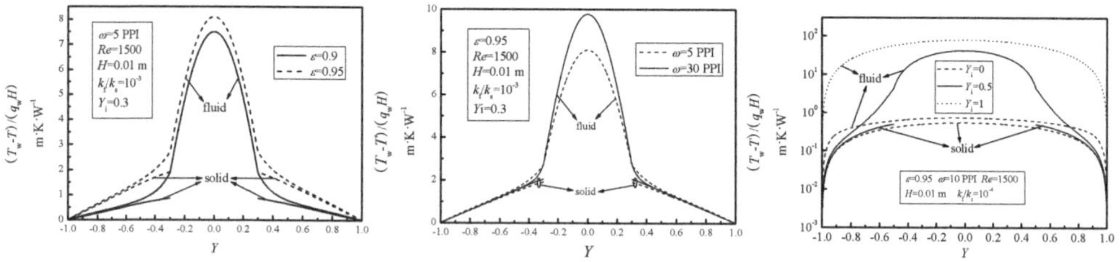

Fig. 11. Effects of key parameters on temperature profiles: (a) porosity; (b) pore density; (c) hollow ratio

Figure 12(a) presents the effect of porosity on *Nu* for four different metal materials: steel, nickel, aluminum, and copper with air as working fluid. The Nusselt number does not monotonically increase with porosity increase and a maximum value of Nu exists at a critical porosity. This can be attributed to the fact that increasing porosity will lead to a decrease in effective thermal conductivity and an increase in mass flow rate in the foam region. Below the critical porosity, the increase in the mass flow region prevails and Nu increases to the maximum value. When porosity is higher than the critical value, which approaches 1, the decrease in thermal conductivity prevails. The Nusselt number sharply reduces and approaches the value of the smooth channel.

It is observed that the increase in the solid thermal conductivity can result in an increase in Nu. The critical porosity likewise increases with the solid thermal conductivity. It is implied that porosity should be maintained at an optimal value in the design of related heat transfer devices to maximize the heat transfer coefficient. Figure 12(b) shows the effect of pore density on Nu for different hollow ratios in which the two limiting cases for Y_i =1 (J.H. Lienhard IV & J.H. Lienhard V, 2006) and Y_i=0 (Mahjoob & Vafai, 2009) are compared as references. As Y_i approaches 1, the predicted Nu of the present analytical solution, with a value of 8.235, coincides accurately with that of the smooth channel (J.H. Lienhard IV & J.H. Lienhard V, 2006). As Y_i approaches 0, the difference between the present analytical result and that of Mahjoob and Vafai (Mahjoob & Vafai, 2009) is very mild since the effect of viscous force of impermeable wall is considered in the present work and not considered in the research of Mahjoob and Vafai (Mahjoob & Vafai, 2009) with the Darcy model. This provides another evidence for feasibility of present analytical solution.

Moreover, it is found that the Nusselt number gradually decreases to a constant value as pore density increases. Increasing pore density can improve the heat transfer surface area but lead to drastic reduction in mass flow rate in the foam region. Hence, small pore density is recommended to maintain heat transfer performance and to reduce pressure drop for thermal design of related applications. The effect of hollow ratio on Nu under various k_f/k_s is shown in Fig. 12(c). At high k_f/k_s (1, 10^{-1}), a minimized Nu exists as Y_i varies from 0 to 1, which is in accordance with the thermal equilibrium result of Poulikakos and Kazmierczak (Poulikakos & Kazmierczak, 1987). However, Nu monotonically decreases as Y_i varies from 0 to 1 for low k_f/k_s, as in the case of metallic foams with high solid thermal conductivities. This is attributed to the fact that both the mass flow rate and foam surface area in the foam region are reduced as Y_i increases. As Y_i approaches 1, the Nusselt number gradually converges to the value 8.235, which is the exact value of forced convective heat transfer in the smooth channel (Y_i=1).

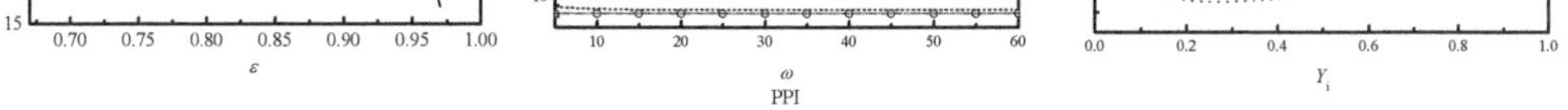

(a) (b) (c)

Fig. 12. Effects of key parameters on Nu: (a) porosity; (b) pore density; (c) hollow ratio

3.3 Numerical modeling for double-pipe heat exchangers

In this section, the two-energy-equation numerical model has been applied to parallel flow double-pipe heat exchanger filled with open-cell metallic foams. The numerical results of the present authors (Du et al., 2010) are introduced in this section. In the model, the solid-fluid conjugated heat transfer process with coupling heat conduction and convection in the open-celled metallic foam, interface wall, and clear fluid in both inner and annular space in heat exchanger is fully considered. The non-Darcy effect, thermal dispersion (Zhao et al., 2001), and wall thickness are taken into account as well.

Figure 13 shows the schematic diagram of metal foam filled double-pipe heat exchanger with parallel flow, in which the cylindrical coordinate system and adiabatic boundary condition in the outer surface of the annular duct are shown. The interface-wall with thickness δ is treated as conductive solid block in the internal part of metallic foams. R_1, R_2, and R_3 represent the inner radius of the inner pipe, outer radius of the inner pipe, and inner radius of the outer pipe, respectively. Fully developed conditions of the velocity and temperature at the exit are adopted. For simplification, incompressible fluids with constant physical properties are considered. Metallic foams are isotropic, possessing no contact resistance on the interface wall.

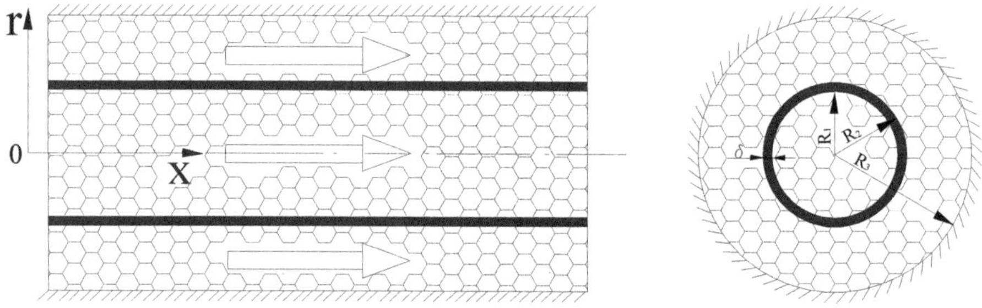

Fig. 13. Schematic diagram of double-pipe heat exchanger with parallel flow

With the Forchheimer flow model for momentum equation and two-equation model for energy equations, the flow and heat transfer problem shown in Fig. 13 is described with the following governing equations:

Continuity equation:

$$\frac{\partial(\rho_f u)}{\partial x}+\frac{1}{r}\frac{\partial(r\rho_f v)}{\partial r}=0\,. \tag{38}$$

Momentum equations:

$$\frac{\partial(\rho_f u^2)}{\partial x}+\frac{1}{r}\frac{\partial(r\rho_f uv)}{\partial r}=-\varepsilon^2\frac{\partial p}{\partial x}+\frac{\partial}{\partial x}(\mu_f\varepsilon\frac{\partial u}{\partial x})+\frac{1}{r}\frac{\partial}{\partial r}(r\mu_f\varepsilon\frac{\partial u}{\partial r})-\frac{\mu_f\varepsilon^2}{K}u-\frac{\rho_f\varepsilon^2 C_I u^2}{\sqrt{K}}\,. \tag{39a}$$

$$\frac{\partial(\rho_f uv)}{\partial x}+\frac{1}{r}\frac{\partial(r\rho_f v^2)}{\partial r}=-\varepsilon^2\frac{\partial p}{\partial r}+\frac{\partial}{\partial x}(\mu_f\varepsilon\frac{\partial v}{\partial x})+\frac{1}{r}\frac{\partial}{\partial r}(r\mu_f\varepsilon\frac{\partial v}{\partial r})-\frac{\mu_f\varepsilon^2}{K}v-\frac{\rho_f\varepsilon^2 C_I v^2}{\sqrt{K}}\,. \tag{39b}$$

Two energy equations:

$$\frac{\partial}{\partial x}\left(k_{\mathrm{se}}\frac{\partial T_{\mathrm{s}}}{\partial x}\right)+\frac{1}{r}\frac{\partial}{\partial r}\left(rk_{\mathrm{se}}\frac{\partial T_{\mathrm{s}}}{\partial r}\right)-h_{\mathrm{sf}}a_{\mathrm{sf}}(T_{\mathrm{s}}-T_{\mathrm{f}})=0\,. \tag{40a}$$

$$\frac{\partial(\rho_{\mathrm{f}}uT_{\mathrm{f}})}{\partial x}+\frac{1}{r}\frac{\partial(r\rho_{\mathrm{f}}vT_{\mathrm{f}})}{\partial r}=\frac{\partial}{\partial x}\left(\frac{k_{\mathrm{fe}}+k_{\mathrm{d}}}{c_{\mathrm{f}}}\frac{\partial T_{\mathrm{f}}}{\partial x}\right)+\frac{1}{r}\frac{\partial}{\partial r}\left(r\frac{k_{\mathrm{fe}}+k_{\mathrm{d}}}{c_{\mathrm{f}}}\frac{\partial T_{\mathrm{f}}}{\partial r}\right)+\frac{h_{\mathrm{sf}}a_{\mathrm{sf}}}{c_{\mathrm{f}}}(T_{\mathrm{s}}-T_{\mathrm{f}})\,. \tag{40b}$$

where x and r pertain to cylindrical coordinate system. The tortuous characteristic of fluids in metallic foams enhances heat transfer coefficients between fluid and solid, influence of which is considered by introducing dispersion conductivity k_{d} (Zhao et al., 2001).

In the interfacial wall domain, the conventional two-equation model cannot be directly used since no fluid can flow through the wall. Hence, particular treatments are proposed to take into account the interface wall. Special fluids can be assumed to exist in the interface wall such that the fluid-phase equation in Eq. (40b) can be applied. However, the dispersion conductivity k_{d} is considered to be zero and the viscosity of fluid can be considered to be infinite, thus leading to zero fluid velocity. As such, the temperature and efficient thermal conductivity of the special fluid and solid are the same, that is to say, $T_{\mathrm{s}}=T_{\mathrm{f}}$, $k_{\mathrm{fe}}=k_{\mathrm{se}}$. In this condition, Eqs. (40a) and (40b) are unified into one for the interface wall, as seen in Eq. (41):

$$\frac{\partial}{\partial x}\left(k_{\mathrm{se}}\frac{\partial T_{\mathrm{s}}}{\partial x}\right)+\frac{1}{r}\frac{\partial}{\partial r}\left(rk_{\mathrm{se}}\frac{\partial T_{\mathrm{s}}}{\partial r}\right)=0\,. \tag{41}$$

According to this extension, the temperature and wall heat flux distribution can be determined by Eq. (43a) during the iteration process, instead of being described as a constant value in advance. Due to the continuity of the interfacial wall heat flux, the inner side wall heat flux q_{inner} and the annular side wall heat flux q_{annular} are formulated in Eq. (42).

$$q_{\mathrm{inner}}=-q_{\mathrm{annular}}\frac{R_2}{R_1}\neq 0\,. \tag{42}$$

where q is the heat flux and the subscripts 'inner' and 'annular' respectively denotes physical qualities relevant to the inner-pipe and the annular space. Simultaneously heat transfer through solid and fluid at the wall can be obtained using the method formulated by Lu and Zhao et al. (Lu et al., 2006; Zhao et al., 2006), which is frequently used and validated in relevant research. The two heat fluxes are expressed as:

$$q_{\mathrm{inner}}=\left.\left(k_{\mathrm{se}}\frac{\partial T_{\mathrm{s}}}{\partial r}+k_{\mathrm{fe}}\frac{\partial T_{\mathrm{f}}}{\partial r}\right)\right|_{r=R_1}\quad\text{(inner side)} \tag{43a}$$

$$q_{\mathrm{annular}}=-\left.\left(k_{\mathrm{se}}\frac{\partial T_{\mathrm{s}}}{\partial r}+k_{\mathrm{fe}}\frac{\partial T_{\mathrm{f}}}{\partial r}\right)\right|_{r=R_2}\quad\text{(annular side)} \tag{43b}$$

Conjugated heat transfer between heat conduction in the interfacial wall and metal ligament, as well as the convection in the fluid, are solved within the entire computational domain.

On the center line of the double-pipe heat exchanger, symmetric conditions are adopted. No-slip velocity and adiabatic thermal boundary conditions for the outside wall of the heat exchanger are conducted as well. Since the wall is adiabatic, the heat flux equals to zero when temperatures of the fluid and solid matrices are equivalent. At the entrance of the double pipe, velocities and fluid temperatures in both flow passages are given as uniform distribution profiles, while the gradient of the metallic foam temperature is equal to zero, according to Lu and Zhao et al. (Lu et al., 2006; Zhao et al., 2006). Fully developed conditions are adopted at the outlet. For the interface wall, velocity of the fluid is zero, as previously determined, which indicates that dynamic viscosity is infinite. The specifications of the boundary conditions are shown in Table 2. Both governing equations are described using the volume-averaging method. During code development, the two equations are unified over the entire computational domain. The above special numerical treatment is implemented in the wall domain.

	x-velocity $\boldsymbol{u}$	y-velocity $\boldsymbol{v}$	Fluid temperature T_{f}	Solid temperature T_{s}
$x=0$	$u=u_{\mathrm{in}}$	$v=0$	$T_{\mathrm{f}}=T_{\mathrm{f,in}}$	$\frac{\partial T_{\mathrm{s}}}{\partial x}=0$
$x=L$	$\frac{\partial u}{\partial x}=0$	$\frac{\partial v}{\partial x}=0$	$\frac{\partial T_{\mathrm{f}}}{\partial x}=0$	$\frac{\partial T_{\mathrm{s}}}{\partial x}=0$
$r=0$	$\frac{\partial u}{\partial r}=0$	$v=0$	$\frac{\partial T_{\mathrm{f}}}{\partial r}=0$	$\frac{\partial T_{\mathrm{s}}}{\partial r}=0$
$R_1 \le r \le R_2$	$u=0$	$v=0$	$T_{\mathrm{s}}=T_{\mathrm{f}}$	$T_{\mathrm{s}}=T_{\mathrm{f}}$
$r=R_3$	$u=0$	$v=0$	$q=k_{\mathrm{fe}}\frac{\partial T_{\mathrm{f}}}{\partial r}+k_{\mathrm{se}}\frac{\partial T_{\mathrm{s}}}{\partial r}=0$, $T_{\mathrm{s}}=T_{\mathrm{f}}$	

Table 2. Boundary conditions for numerical simulation of double-pipe heat exchanger

The simulation is performed according to the volume-averaging method, based on the geometrical model of open-cell metallic foams provided by Lu et al. (Lu et al., 2006). The codes are validated by comparison with Lu and Zhao et al. (Lu et al., 2006; Zhao et al., 2006). The criterion for ceasing iterations is a relative error of temperatures less than 10^{-5}. The thermo-physical properties of fluid and important parameters in the numerical simulation are shown in Table 3.

To monitor vividly the temperature distribution along the flow direction, a dimensionless temperature is defined as follows:

$$\theta_{\mathrm{s}}=\frac{T_{\mathrm{s}}-T_{\mathrm{s}}\big|_{r=R_2}}{T_{\mathrm{s,b}}-T_{\mathrm{s}}\big|_{r=R_2}}. \tag{44a}$$

$$\theta_{\mathrm{f}}=\frac{T_{\mathrm{f}}-T_{\mathrm{s}}\big|_{r=R_2}}{T_{\mathrm{f,b}}-T_{\mathrm{s}}\big|_{r=R_2}}. \tag{44b}$$

where$T_{\mathrm{s,b}}$ and $T_{\mathrm{f,b}}$ represent the cross-sectional bulk mean temperature of solid matrix and fluid phase, respectively, defined as follows:

Parameter	Unit	Value
Reynolds number Re	1	3329
Prandtl number Pr	1	0.73
Density of inner fluid ρ_{inner}	$\text{kg}\cdot\text{m}^{-3}$	1.13
Density of annular fluid ρ_{annular}	$\text{kg}\cdot\text{m}^{-3}$	1.00
Thermal conductivity of inner fluid $k_{\text{f,inner}}$	$\text{W}\cdot\text{m}^{-1}\cdot\text{K}^{-1}$	0.0276
Thermal conductivity of annular fluid $k_{f,annular}$	$\text{W}\cdot\text{m}^{-1}\cdot\text{K}^{-1}$	0.0305
Kinematic viscousity of inner fluid $\mu_{\text{f,inner}}$	$\text{Pa}\cdot\text{S}$	1.91×10^{-5}
Kinematic viscousity of annular fluid $\mu_{\text{f,annular}}$	$\text{Pa}\cdot\text{S}$	2.11×10^{-5}
Solid thermal conductivity k_s	$\text{W}\cdot\text{m}^{-1}\cdot\text{K}^{-1}$	100
Heat capacity at constant pressure of inner fluid $c_{\text{p,inner}}$	$\text{J}\cdot\text{kg}^{-1}\cdot\text{K}^{-1}$	1005
Inlet temperature of inner fluid $T_{\text{in,1}}$	$^{\circ}\text{C}$	38
Inlet temperature of annular fluid $T_{\text{in,2}}$	$^{\circ}\text{C}$	78

Table 3. Fluid flow and metal foam parameters of the double-pipe heat exchanger

$$T_{\text{f,b}} = \frac{\int_0^{R_1} u\cdot T_{\text{f}}\cdot r\cdot \text{d}r}{\int_0^{R_1} u\cdot r\cdot \text{d}r}, \quad T_{\text{s,b}} = \frac{2}{R_1^{\,2}}\int_0^{R_1} T_{\text{s}}\cdot r\cdot \text{d}r \text{ (inner side)} \tag{45a}$$

$$T_{\text{f,b}} = \frac{\int_{R_2}^{R_3} u\cdot T_{\text{f}}\cdot r\cdot \text{d}r}{\int_{R_2}^{R_3} u\cdot r\cdot \text{d}r}, \quad T_{\text{s,b}} = \frac{2}{R_3^{\,2}-R_2^{\,2}}\int_{R_2}^{R_3} T_{\text{s}}\cdot r\cdot \text{d}r \text{ (annular side)} \tag{45b}$$

The Reynolds number for the inner and annular sides is defined as follows:

$$Re = \frac{\rho u_{\text{m}} D_{\text{h}}}{\mu_{\text{f}}}. \tag{46}$$

where D_h is the hydraulic diameter equaling $2R_1$ for the inner side and $2(R_3\text{-}R_2)$ for the annular side. The Nusselt number at the inner and annular sides is defined as follows:

$$Nu = \frac{hD_{\text{h}}}{k_{\text{f}}}. \tag{47}$$

where h is the average convective heat transfer coefficient defined in Eq. (21) for the entire double-pipe heat exchanger for each space.

$$h = \frac{\int_0^L h_{\text{x}}(T_{\text{w,x}} - T_{\text{b,x}})\pi 2R_1 \text{d}x}{A_{\text{inner}}\left(T_{\text{w,av}} - T_{\text{f,b}}\right)} = \frac{\int_0^L q_{\text{x}}\text{d}x}{L\left(T_{\text{w,av}} - T_{\text{f,b}}\right)}. \tag{48a}$$

$$h_{\mathrm{x}} = \frac{q_{\mathrm{x}}}{T_{\mathrm{w,x}} - T_{\mathrm{f,b,x}}}. \tag{48b}$$

where $T_{\mathrm{w,av}}$ is the average interface wall temperature and $T_{\mathrm{f,b}}$ is the mean fluid temperature, which are defined in Eqs.(22) and (23) for both sides.

$$T_{\mathrm{f,inner}} = \frac{T_{\mathrm{in,1}} + T_{\mathrm{out,1}}}{2} \text{ (for inner space)} \tag{49a}$$

$$T_{\mathrm{f,annular}} = \frac{T_{\mathrm{in,2}} + T_{\mathrm{out,2}}}{2} \text{ (for annular side space)} \tag{49b}$$

The overall heat transfer coefficient $\boldsymbol{U}$ is defined as follows:

$$U = \frac{c_{\mathrm{inner}} \rho_{\mathrm{inner}} u_{\mathrm{inner}} R_1 \left(T_{\mathrm{out,1}} - T_{\mathrm{in,1}}\right)}{8L(T_{\mathrm{f,annlular}} - T_{\mathrm{f,inner}})}. \tag{50}$$

Figure 14 shows the fluid temperature distribution along the axial direction in the inner-pipe and the annulus for smooth double-pipe HEX (heat exchanger) and metal foam fully filled HEX in condition of ε =0.9, 20 PPI, $Re_{\mathrm{inner}}=Re_{\mathrm{annular}}=3329$. It is found that $\partial\theta_{\mathrm{f}} / \partial x = 0$ holds in hollow heat exchanger when $L/D>125$, which has critical value $L/2R_1 \approx 0.05\, RePr=121$ for fully developed flow region of empty channel. However, the ratio of length to inner diameter for the entrance zone for foam filled heat exchanger is extended to approximately 180 because the foam creates redistribution of the flow profile and enlarges the length of entrance region. Hence, predicted results at axial location ($L/2R_1=200$) could be compared with previous results (Lu et al., 2006; Zhao et al., 2006), which are suitable for fully developed regions.

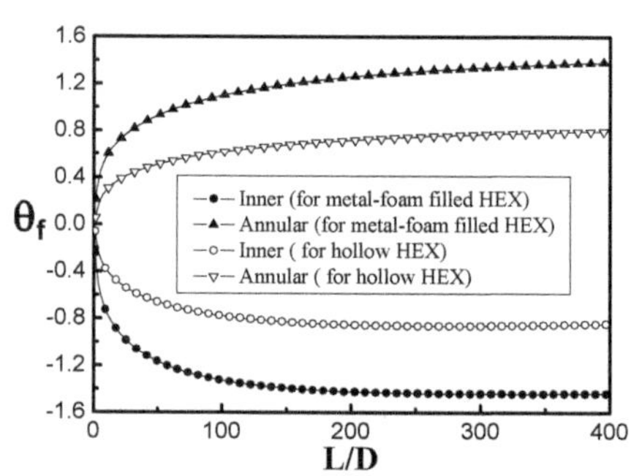

Fig. 14. Predicted dimensionless temperature distribution along the axial direction (for metallic foams ε =0.9, 20PPI, $Re_{\mathrm{inner}}=Re_{\mathrm{annular}}=3329$)

Interface wall heat flux distribution along axial direction could be further predicted by the coupling interface wall and fluids in two sides. Interface wall heat flux is gained with

$$q_{\mathrm{x}} = k_{\mathrm{s}} \frac{T_{r=R_2} - T_{r=R_2}}{\delta}. \tag{51}$$

The interface wall heat flux distributions and local heat transfer coefficients for metal foam filled double-pipe and hollow double-pipe heat exchangers are shown in Fig. 15. The interface wall heat flux decreased significantly from the inlet to the outlet, as shown in Fig. 15(a). The heat flux at the inlet location is approximately 4.5 times that at the outlet location, while previous models neglect this variation trend. The predicted average interface wall heat flux in the case of exchangers filled with metallic foams is a little lower than the case without foams due to the significant extension of surface area. The local heat transfer coefficient for metallic foam filled double-pipe exchangers is considerably higher than the air flow across the smooth plate from the analytical solution (Incropera et al., 1985), which is attributed to the increased specific surface area by metallic foams, as seen in Fig. 15(b).

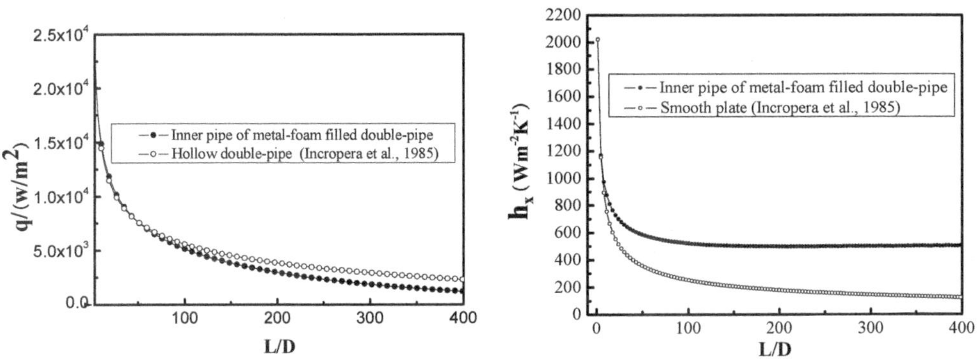

(a) (b)

Fig. 15. Predicted interface wall heat flux and local heat transfer coefficient along axial direction (ε =0.9, 20PPI, Re_{inner}=Re_{annular}=3329): (a) Local heat flux; (b) Local heat transfer coefficient

Figure 16 displays the influence of metallic foam porosity on the dimensionless fluid and solid temperatures in the radial direction under the same pore density (20 PPI). Figure 16(a)

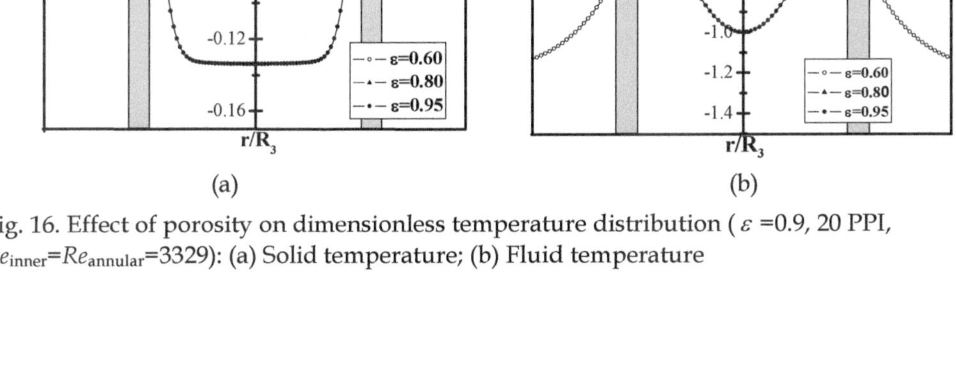

(a) (b)

Fig. 16. Effect of porosity on dimensionless temperature distribution (ε =0.9, 20 PPI, Re_{inner}=Re_{annular}=3329): (a) Solid temperature; (b) Fluid temperature shows that with increasing porosity, the dimensionless solid foam matrix temperature increases, and the temperature difference between the matrix and interface wall is improved. This is attributed to reductions in the porous ligament diameter with increasing porosity under the same pore density, leading to increased thermal resistance of heat conduction. The influence of porosity on dimensionless fluid temperatures is more complicated due to conjugated heat transfer, as shown in Fig. 16(b). With increase in porosity, the dimensionless fluid temperature increases in the inner tube, but decreases in the annular space.

Figure 17 displays the effects of pore density and porosity of metallic foams on overall heat-transfer performance. The predicted value is lower than that in the conventional model, shown in Figs. 17(a) and 17(b) since the conventional uniform heat flux model is an ideal case for amplifying the thermal performance of the heat exchanger to over-evaluate the overall heat transfer coefficient. It is indicated that the heat transfer performance is enhanced by increasing the pore density due to increasing heat transfer surface area; it is weakened by increasing porosity due to decreasing diameter of cell ligament.

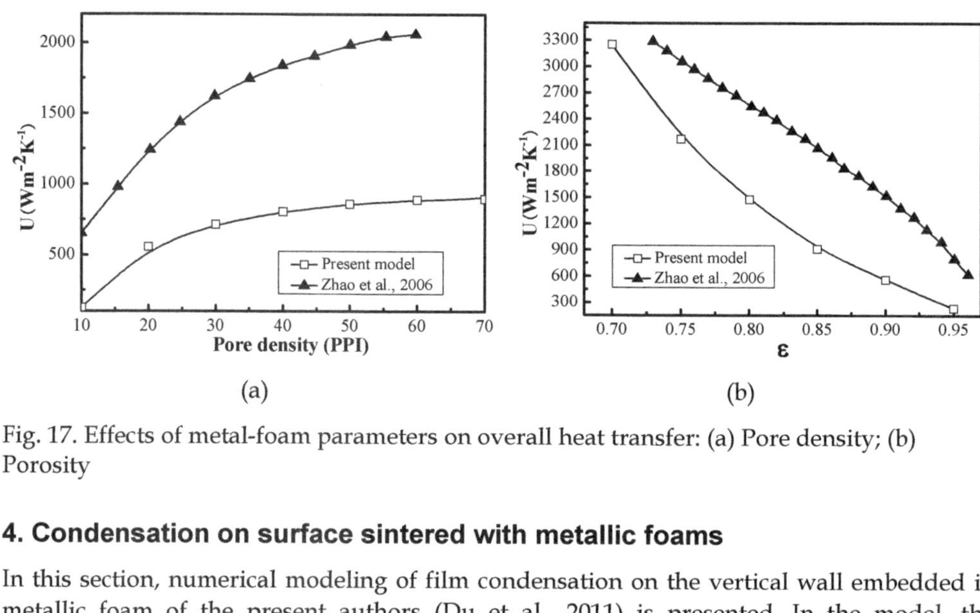

Fig. 17. Effects of metal-foam parameters on overall heat transfer: (a) Pore density; (b) Porosity

4. Condensation on surface sintered with metallic foams

In this section, numerical modeling of film condensation on the vertical wall embedded in metallic foam of the present authors (Du et al., 2011) is presented. In the model, the advection and inertial force in the condensate film are thoroughly considered, from which the non-linear effect of cross-sectional temperature distribution on the condensation heat transfer is involved as well.

Due to substantial heat transferred from saturated vapor to super-cold solid wall, the condensation phenomenon extensively exists in nature, human activities, and industrial applications, such as dew formation in the morning, frost formation on the window in the cold winter, condensation from refrigerant vapor in air-conditioning, water-cooled wall in boiler, and so on. In principle, condensation can be classified into film-wise condensation and drop-wise condensation. However, most condensation phenomena are film-wise form for the reason that drop-wise condensation is difficult to produce and that it will not last in the face of generation of liquid drops adhered on the solid surface.

Film-wise condensation, likewise called film condensation, was first analytically treated by Nusselt as early as 1916 by introducing the Nusselt theory (Nusselt, 1916). Similar extensive works or experimental studies of film condensation were performed extensively (Dhir & Lienhard, 1971; Popiel & Boguslawski, 1975; Sukhatme et al., 1990; Cheng & Tao, 1994). The use of porous structure for extensive surface of condensation was not a new idea and numerous related studies were found in open literature (Jain & Bankoff, 1964; Cheng & Chui, 1984; Masoud et al., 2000; Wang & Chen et al., 2003; Wang & Yang et al., 2003; Chang, 2008). Metallic foams possess a three-dimensional basic cell geometry of tetrakaidecahedron with specific surface area as high as 10^3-10^4 m^2 m^{-3}. This provides a kind of ideal extended surface for condensation heat transfer. This attractive metallic porous structure motivates the present authors to perform related research on condensation heat transfer, which is shown below.

Overall, the physical process for film condensation is a thickening process of liquid film. The present authors aim to explore condensation mechanisms on surfaces covered with metallic foams and establish a numerical model for film condensation heat transfer within metallic foams, in which the non-linear temperature distribution in the condensation layer was solved. The characteristics of condensation on the vertical plate embedded in metallic foams were discussed and compared with those on the smooth plate.

The schematic diagram for the problem discussed is shown in Fig. 18 with the coordinate system and boundary conditions for the computational domain denoted. Dry saturated vapor of water at atmospheric pressure is static near the vertical plate covered with metallic foams. Due to gravity, condensate fluid flows downward along the plate. Width and length of the computational domain are 5×10^{-4} m and 1 m, respectively. Flow direction of condensing water is denoted by x while thickness direction is denoted by y. The positive x direction is the direction of gravity as well. The temperature of the plate is below the saturated temperature of the vapor, allowing condensation to take place favorably.

Effects of wall conduction and forced convection of the saturated gas are neglected due to the condition of unchanged temperature and zero velocity, respectively. Because of the complexity of the tortuous characteristic of fluid flow in metallic foams, laminar film condensation is assumed for simplification. The metallic foams are assumed to be both isotropic and homogeneous. The fluid is considered to be incompressible with constant properties. Thermal radiation is ignored. Thermal dispersion effect is neglected. The quadratic term for high-speed flow in porous medium are neglected as well for momentum equation in the present study.

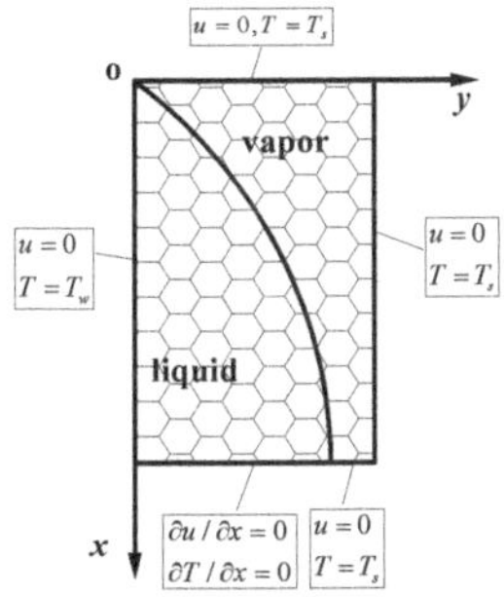

Fig. 18. Computational domain and boundary conditions

Based on the Nusselt theory, vapor density, shear stress on the interface, convection, and inertial force in the condensate layer are all considered. The Darcy model with local thermal equilibrium is adopted to establish the momentum equation. To condensate film of water within the range of $x \geq 0$, $0 \leq y \leq \delta(x)$, the momentum equation is as follows:

$$\frac{\mu}{\varepsilon}\frac{\partial^2 u_1}{\partial y^2} - \frac{\mu}{K}u_1 + g(\rho_1 - \rho_v) = 0 . \tag{52}$$

where K, g pertain to the permeability of metallic foams and acceleration of gravity, respectively, and u_l, ρ_1, ρ_v denote liquid velocity, liquid density, and vapor density, respectively. Based on the liquid velocity distribution, liquid mass flux at a certain location is shown as follows:

$$q_m = \int_0^{\delta(x)} \rho u_1 dy . \tag{53}$$

For the energy equation, the convective term is partially considered since the span wise is assumed to be zero based on the Nusselt theory. Hence, the energy equation is expressed as:

$$u_1\frac{\partial T_1}{\partial x} = \frac{k_{e,1}}{\rho_1 c_1}\frac{\partial^2 T_1}{\partial y^2} . \tag{54}$$

where $k_{e,l}$ is the effective thermal conductivity of metallic foams saturated with liquid and referred from Boomsma and Poulikakos (Boomsma & Poulikakos, 2002), while u_l, T_l pertain to vapor velocity and liquid temperature, respectively. In the vertical wall for liquid film, heat transfers through phase change of saturated vapor are equal to the heat condition through the vertical wall along the thickness direction of the condensate layer.

$$r \cdot \mathrm{d}q_m = k_1 \cdot \left.\frac{\partial T_1}{\partial y}\right|_{y=0} \cdot \mathrm{d}x . \tag{55}$$

where r, k_l denote the latent heat of saturated water, conductivity of liquid water, and thickness of the condensate layer, respectively. T_s, T_w pertain to saturated temperature and wall temperature, respectively. $\mathrm{d}q_m$ is the differential form of liquid film mass flux.
For convenience of numerical implementation, a square domain is selected as computational domain. Boundary conditions of governing equations in the whole computational domain for Eqs. (52) and (54) are shown in Fig. 18. By iterative methods, the thickness of the condensate layer is obtained with Eq. (52)-(55). The inner conditions at the liquid-vapor interface are set by considering the shear stress continuity as follows:

$$y = 0, \quad u = 0 \quad T = T_w . \tag{56a}$$

$$y = \delta(x), \quad \frac{\partial u_1}{\partial y} = 0 \quad T = T_s . \tag{56b}$$

Certain parameters used are defined below. The Jacobi number is defined as follows:

$$Ja = \frac{r}{c_l \cdot (T_s - T_w)} . \tag{57}$$

The local heat transfer coefficient and Nusselt number along the x direction can be obtained in Eqs. (59) and (60):

$$h(x) = \frac{k_e}{T_w - T_s} \frac{\partial T_l}{\partial y}\bigg|_{y=0} = \frac{1}{\delta(x)} \frac{k_e}{T_w - T_s} \frac{\partial T_l}{\partial [y / \delta(x)]}\bigg|_{y=0} . \tag{58}$$

$$Nu(x) = h(x)\frac{x}{k_e} . \tag{59}$$

The condensate film Reynolds number is expressed as follows:

$$Re(x) = \frac{4h(x)}{c_l \cdot Ja \cdot \mu_l} . \tag{60}$$

In the region outside condensation layer, the domain extension method is employed, where special numerical treatment is implemented during the inner iteration to ensure that velocity and temperature in this extra region are set to be zero and T_s, and that these values cannot affect the solution of velocity and temperature field inside condensation layer.

The governing equations in Eqs. (52)-(54) are solved with using SIMPLE algorithm (Tao, 2005). The convective terms are discritized using the power law scheme. A 200×20 grid system has been checked to gain a grid independent solution. The velocity field is solved ahead of the temperature field and energy balance equation. By coupling Eqs. (52)-(55), the non-linear temperature field can be obtained. The thermal-physical properties in the numerical simulation, involving the fluid thermal conductivity, fluid viscosity, fluid specific heat, fluid density, fluid saturation temperature, fluid latent heat of vaporization, and gravity acceleration are presented in Table 4.

Parameter	Unit	Value
Liquid density ρ_l	$kg \cdot m^{-3}$	977.8
Vapor density ρ_v	$kg \cdot m^{-3}$	0.58
Liquid kinematic viscousity μ_l	$Pa \cdot s$	2.825×10^{-4}
Liquid thermal conductivity k_l	$W \cdot m^{-1} \cdot K^{-1}$	0.683
Liquid heat capacity at constant pressure c_l	$J \cdot kg^{-1} \cdot K^{-1}$	4200
Saturation temperature T_s	°C	100
Latent heat r	$J \cdot kg^{-1}$	297030
Gravity acceleration g	$m \cdot s^{-2}$	9.8

Table 4. Constant parameters in numerical procedure of film condensation

For a limited case of porosity being equal to 1, the present numerical model can predict film condensation on the vertical smooth plate for reference case validation. The distribution of film condensate thickness and local heat transfer coefficient on the smooth plate predicted by the present numerical model with those of Nusselt (Nusselt, 1916) and Al-Nimer and Al-Kam (Al-Nimer and Al-Kam, 1997) are shown in Fig. 19. It can be seen that the numerical solution is approximately consistent with either Nusselt (Nusselt, 1916) or Al-Nimer and Al-Kam (Al-Nimer and Al-Kam, 1997). The maximum deviation for condensate thickness and local heat transfer coefficient is 14.5% and 12.1%, respectively.

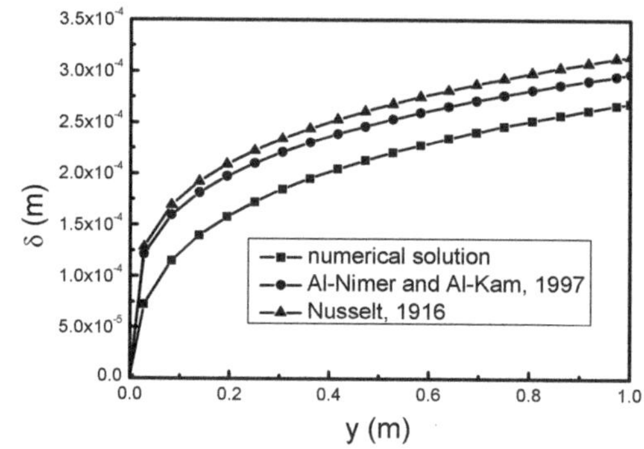

Fig. 19. Distribution of condensate thickness for the smooth plate (ε =0.9, 10 PPI)

Figure 20(a) exhibits the temperature distribution in condensate layer for three locations in the vertical direction (x/L=0.25, 0.5, and 0.75) with porosity and pore density being 0.9 and 10 PPI, respectively. Evidently, the temperature profile is nonlinear. The non-linear characteristic is more significant, or the defined temperature gradient $\partial T_1 / \partial\left[y / \delta(x)\right]$ is higher in the downstream of condensate layer since the effect of heat conduction thermal resistance of the foam matrix in horizontal direction becomes more obvious.

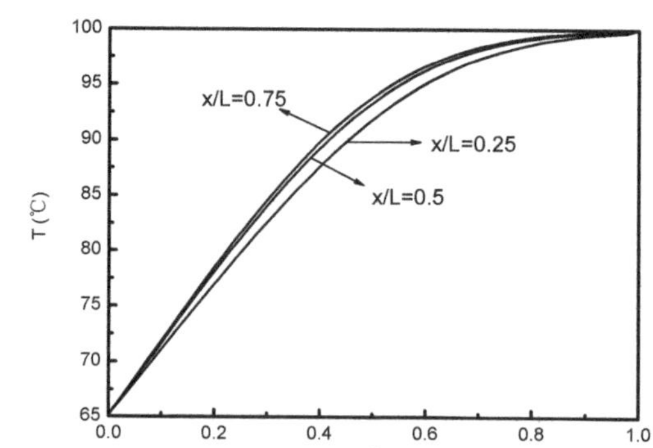

Fig. 20. Temperature distribution in condensate layer for different x (ε =0.9, 10 PPI)

Effects of parameters involving Jacobi number, porosity, and pore density are discussed in this section. Super cooling degree can be controlled by changing the value of ***Ja***. The effect of ***Ja*** on the condensate layer thickness is shown in Fig. 21(a). It can be seen that condensate layer thickness decreases as the Jacobi number increases. This can be attributed to the fact that the super cooling degree, which is the key factor driving the condensation process, is reduced as the ***Ja*** number increases, leading to a thinner liquid condensate layer. For a limited case of zero super cooling degree, condensation cannot occur and the condensate layer does not exist.

The effect of porosity on the condensate film thickness is shown in Fig. 21(b). It is found that in a fixed position, increase in porosity can lead to the decrease in the condensate film thickness, which is helpful for film condensation. This can be attributed to the fact that the increase in porosity can make the permeability of the metallic foams increase, decreasing the flow resistance of liquid flowing downwards. The effect of pore density on the condensate film thickness is shown in Fig. 21(c). It can be seen that for a fixed x position, the increase in pore density can make the condensate film thickness increase greatly, which enlarges the thermal resistance of the condensation heat transfer process. The reason for the above result is that the increasing pore density can significantly reduce metal foam permeability and substantially increase the flow resistance of the flowing-down condensate. Thus, with either an increase in porosity or a decrease in pore density, condensate layer thickness is reduced for condensation heat transfer coefficient.

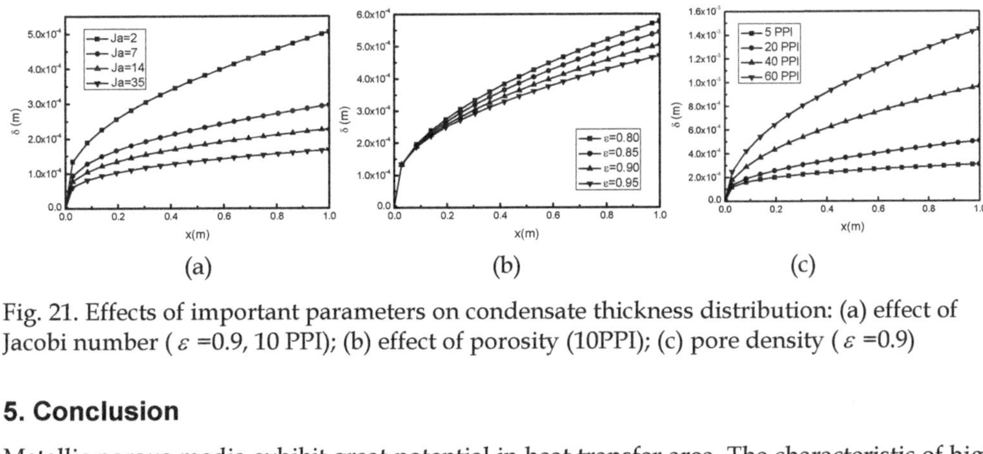

(a) (b) (c)

Fig. 21. Effects of important parameters on condensate thickness distribution: (a) effect of Jacobi number (ε =0.9, 10 PPI); (b) effect of porosity (10PPI); (c) pore density (ε =0.9)

5. Conclusion

Metallic porous media exhibit great potential in heat transfer area. The characteristic of high pressure drop renders those with high porosity and low pore density considerably more attractive in view of pressure loss reduction. For forced convective heat transfer, another way to lower pressure drop is to fill the duct partially with metallic porous media.

In this chapter, natural convection in metallic foams is firstly presented. Their enhancement effects on heat transfer are moderate. Next, we exhibit theoretical modeling on thermal performance of metallic foam fully/partially filled duct for internal flow with the two-equation model for high solid thermal conductivity foams. Subsequently, a numerical model for film condensation on a vertical plate embedded in metallic foams is presented and the effects of advection and inertial force are considered, which are responsible for the non-linear effect of cross-sectional temperature distribution. Future research should be focused on following areas with metallic porous media: implementation of computation and parameter optimization for practical design of thermal application, phase change process, turbulent flow and heat transfer, non-equilibrium conjugate heat transfer at porous-fluid interface, thermal radiation, experimental data/theoretical model/flow regimes for two-phase/multiphase flow and heat transfer, and so on.

6. Acknowledgment

This work is supported by the National Natural Science Foundation of China (No. 50806057), the National Key Projects of Fundamental R/D of China (973 Project: 2011CB610306), the Ph.D. Programs Foundation of the Ministry of Education of China (200806981013) and the Fundamental Research Funds for the Central Universities.

7. References

Alazmi, B. & Vafai, K. (2001). Analysis of Fluid Flow and Heat Transfer Interfacial Conditions Between a Porous Medium and a Fluid Layer. *International Journal of Heat and Mass Transfer*, Vol.44, No.9, (May 2001), pp. 1735-1749, ISSN 0017-9310

Al-Nimer, M.A. & Al-Kam, M.K. (1997). Film Condensation on a Vertical Plate Imbedded in a Porous Medium. *Applied Energy*, Vol. 56, No.1, (January 1997), pp. 47-57, ISSN 0306-2619

Banhart, J. (2001). Manufacture, characterisation and application of cellular metals and metal foams, *Progress in Materials Science*, Vol.46, No.6, (2001), pp. 559-632, ISSN 0079-6425

Boomsma, K. &Poulikakos, D. (2001). On the Effective Thermal Conductivity of a Three-Dimensionally Structured Fluid-Saturated Metal Foam. *International Journal of Heat and Mass Transfer*, Vol.44, No.4, (February 2001), pp. 827-836, ISSN 0017-9310

Calmidi, V.V. (1998). *Transport phenomena in high porosity fibrous metal foams*. Ph.D. thesis, University of Colorado.

Calmidi, V.V. & Mahajan, R.L. (2000). Forced convection in high porosity metal foams. *Journal of Heat Transfer*, Vol.122, No.3, (August 2000), pp. 557-565, ISSN 0022-1481

Chang, T.B. (2008). Laminar Film Condensation on a Horizontal Wavy Plate Embedded in a Porous Medium. *International Journal of Thermal Sciences*, Vol. 47, No.4, (January 2008), pp. 35–42, ISSN 1290-0729

Cheng, B. & Tao, W.Q. (1994). Experimental Study on R-152a Film Condensation on Single Horizontal Smooth Tube and Enhanced Tubes. *Journal of Heat Transfer*, Vol.116, No.1, (February 1994), pp. 266-270, ISSN 0022-1481

Cheng, P. & Chui, D.K. (1984). Transient Film Condensation on a Vertical Surface in a Porous Medium. *International Journal of Heat and Mass Transfer*, Vol.27, No.5, (May 1984), pp. 795–798, ISSN 0017-9310

Churchil S.W. & Ozoe H. (1973). A Correlation for Laminar Free Convection from a Vertical Plate. *Journal of Heat Transfer*, Vol.95, No.4, (November 1973), pp. 540-541, ISSN 0022-1481

Dhir, V.K. & Lienhard, J.H. (1971). Laminar Film Condensation on Plane and Axisymmetric Bodies in Nonuniform Gravity. *Journal of Heat Transfer*, Vol.93, No.1, (February 1971), pp. 97-100, ISSN 0022-1481

Du, Y.P.; Qu, Z.G.; Zhao, C.Y. &Tao, W.Q. (2010). Numerical Study of Conjugated Heat Transfer in Metal Foam Filled Double-Pipe. *International Journal of Heat and Mass Transfer*, Vol.53, No.21, (October 2010), pp. 4899-4907, ISSN 0017-9310

Du, Y.P.; Qu, Z.G.; Xu, H.J.; Li, Z.Y.; Zhao, C.Y. &Tao, W.Q. (2011). Numerical Simulation of Film Condensation on Vertical Plate Embedded in Metallic Foams. *Progress in Computational Fluid Dynamics*, Vol.11, No.3-4, (June 2011), pp. 261-267, ISSN 1468-4349

Dukhan, N. (2009). Developing Nonthermal-Equilibrium Convection in Porous Media with Negligible Fluid Conduction. *Journal of Heat Transfer*, Vol.131, No.1, (January 2009), pp. 014501.1-01450.3, ISSN 0022-1481

Fujii T. & Fujii M. (1976). The Dependence of Local Nusselt Number on Prandtl Number in Case of Free Convection Along a Vertical Surface with Uniform Heat-Flux. *International Journal of Heat and Mass Transfer*, Vol. 19, No.1, (January 1976), pp. 121-122, ISSN 0017-9310

Incropera, F.P.; Dewitt, D.P. & Bergman, T.L. (1985). *Fundamentals of heat and mass transfer* (2nd Edition), ISBN 3540295267, Springer, New York, USA

Jain, K.C. & Bankoff, S.G. (1964). Laminar Film Condensation on a Porous Vertical Wall with Uniform Suction Velocity. *Journal of Heat Transfer*, Vol. 86, (1964), pp. 481-489, ISSN 0022-1481

Jamin Y.L. & Mohamad A.A. (2008). Natural Convection Heat Transfer Enhancements From a Cylinder Using Porous Carbon Foam: Experimental Study. *Journal of Heat Transfer*, Vol.130, No.12, (December 2008), pp. 122502.1-122502.6, ISSN 0022-1481

Lee, D.Y. & Vafai, K. (1999). Analytical characterization and conceptual assessment of solid and fluid temperature differentials in porous media. *International Journal of Heat and Mass Transfer*, Vol.42, No.3, (February 1999), pp. 423-435, ISSN 0017-9310

Lienhard, J.H. IV & Lienhard J.H.V (2006). *A heat transfer textbook* (3rd Edition), Phlogiston, ISBN 0-15-748821-1, Cambridge in Massachusetts, USA

Lu, T.J.; Stone, H.A. & Ashby, M.F. (1998). Heat transfer in open-cell metal foams. *Acta Materialia*, Vol.46, No.10, (June 1998), pp. 3619-3635, ISSN 1359-6454

Lu, W.; Zhao, C.Y. & Tassou, S.A. (2006). Thermal analysis on metal-foam filled heat exchangers, Part I: Metal-foam filled pipes. *International Journal of Heat and Mass Transfer*, Vol.49, No.15-16, (July 2006), pp. 2751-2761, ISSN 0017-9310

Mahjoob, S. & Vafai, K. (2009). Analytical Characterization of Heat Transport through Biological Media Incorporating Hyperthermia Treatment. *International Journal of Heat and Mass Transfer*, Vol.52, No.5-6, (February 2009), pp. 1608–1618, ISSN 0017-9310

Masoud, S.; Al-Nimr, M.A. & Alkam, M. (2000). Transient Film Condensation on a Vertical Plate Imbedded in Porous Medium. *Transport in Porous Media,* Vol. 40, No.3, (September 2000), pp. 345–354, ISSN 0169-3913

Nusslet, W. (1916). Die Oberflachenkondensation des Wasserdampfes. *Zeitschrift des Vereines Deutscher Ingenieure*, Vol. 60, (1916), pp. 541-569, ISSN 0341-7255

Ochoa-Tapia, J.A. & Whitaker, S. (1995). Momentum Transfer at the Boundary Between a Porous Medium and a Homogeneous Fluid-I: Theoretical Development. *International Journal of Heat and Mass Transfer*, Vol.38, No.14, (September 1995), pp. 2635-2646, ISSN 0017-9310

Phanikumar, M.S. & Mahajan, R.L. (2002). Non-Darcy Natural Convection in High Porosity Metal Foams. *International Journal of Heat and Mass Transfer*, Vol.45, No.18, (August 2002), pp. 3781–3793, ISSN 0017-9310

Popiel, C.O. & Boguslawski, L. (1975). Heat transfer by laminar film condensation on sphere surfaces. *International Journal of Heat and Mass Transfer*, Vol.18, No.12, (December 1975), pp. 1486-1488, ISSN 0017-9310

Poulikakos, D. & Kazmierczak, M. (1987). Forced Convection in Duct Partially Filled with a Porous Material. *Journal of Heat Transfer*, Vol.109, No.3, (August 1987), pp. 653-662, ISSN 0022-1481

Sparrow E.M. & Gregg, J.L. (1956). Laminar free convection from a vertical plate with uniform surface heat flux. *Transactions of ASME*, Vol. 78, (1956), pp. 435-440

Sukhatme, S.P.; Jagadish, B.S. & Prabhakaran P. (1990). Film Condensation of R-11 Vapor on Single Horizontal Enhanced Condenser Tubes. *Journal of Heat Transfer*, Vol. 112, No.1, (February 1990), pp. 229-234, ISSN 0022-1481

Tao, W.Q. (2005). *Numerical Heat Transfer* (2nd Edition), Xi'an Jiaotong University Press, ISBN 7-5605-0183-4, Xi'an, China.

Wang, S.C.; Chen, C.K. & Yang, Y.T. (2006). Steady Filmwise Condensation with Suction on a Finite-Size Horizontal Plate Embedded in a Porous Medium Based on Brinkman and Darcy models. *International Journal of Thermal Science*, Vol.45, No.4, (April 2006), pp. 367–377, ISSN 1290-0729

Wang, S.C.; Yang, Y.T. & Chen, C.K. (2003). Effect of Uniform Suction on Laminar Film-Wise Condensation on a Finite-Size Horizontal Flat Surface in a Porous Medium. *International Journal of Heat and Mass Transfer*, Vol.46, No.21, (October 2003), pp. 4003-4011, ISSN 0017-9310

Xu, H.J.; Qu, Z.G. & Tao, W.Q. (2011a). Analytical Solution of Forced Convective Heat Transfer in Tubes Partially Filled with Metallic Foam Using the Two-equation Model. *International Journal of Heat and Mass Transfer*, Vol. 54, No.17-18, (May 2011), pp. 3846–3855, ISSN 0017-9310

Xu, H.J.; Qu, Z.G. & Tao, W.Q. (2011b). Thermal Transport Analysis in Parallel-plate Channel Filled with Open-celled Metallic Foams. *International Communications in Heat and Mass Transfer*, Vol.38, No.7, (August 2011), pp. 868-873, ISSN 0735-1933

Xu, H.J.; Qu, Z.G.; Lu, T.J.; He, Y.L. & Tao, W.Q. (2011c). Thermal Modeling of Forced Convection in a Parallel Plate Channel Partially Filled with Metallic Foams. *Journal of Heat Transfer*, Vol.133, No.9, (September 2011), pp. 092603.1-092603.9, ISSN 0022-1481

Zhao, C.Y.; Kim, T.; Lu, T.J. & Hodson, H.P. (2001). *Thermal Transport Phenomena in Porvair Metal Foams and Sintered Beds*. Technical report, University of Cambridge.

Zhao, C.Y.; Kim, T.; Lu, T.J. & Hodson, H.P. (2004). Thermal Transport in High Porosity Cellular Metal Foams. *Journal of Thermophysics and Heat Transfer*, Vol.18, No.3, (2004), pp. 309-317, ISSN 0887-8722

Zhao, C.Y.; Lu, T.J. & Hodson, H.P. (2004). Thermal radiation in ultralight metal foams with open cells. *International Journal of Heat and Mass Transfer*, Vol. 47, No.14-16, (July 2004), pp. 2927–2939, ISSN 0017-9310

Zhao, C.Y.; Lu, T.J. & Hodson, H.P. (2005). Natural Convection in Metal Foams with Open Cells. *International Journal of Heat and Mass Transfer*, Vol.48, No.12, (June 2005), pp. 2452–2463, ISSN 0017-9310

Zhao, C.Y.; Lu, W. & Tassou, S.A. (2006). Thermal analysis on metal-foam filled heat exchangers, Part II: Tube heat exchangers. *International Journal of Heat and Mass Transfer*, Vol.49, No.15-16, (July 2006), pp. 2762-2770, ISSN 0017-9310

9

Coupled Electrical and Thermal Analysis of Power Cables Using Finite Element Method

Murat Karahan[1] and Özcan Kalenderli[2]
[1]*Dumlupinar University, Simav Technical Education Faculty,*
[2]*Istanbul Technical University, Electrical-Electronics Faculty,*
Turkey

1. Introduction

Power cables are widely used in power transmission and distribution networks. Although overhead lines are often preferred for power transmission lines, power cables are preferred for ensuring safety of life, aesthetic appearance and secure operation in intense settlement areas. The simple structure of power cables turn to quite complex structure by increased heat, environmental and mechanical strains when voltage and transmitted power levels are increased. In addition, operation of existing systems at the highest capacity is of great importance. This requires identification of exact current carrying capacity of power cables.

Analytical and numerical approaches are available for defining current carrying capacity of power cables. Analytical approaches are based on IEC 60287 standard and there can only be applied in homogeneous ambient conditions and on simple geometries. For example, formation of surrounding environment of a cable with several materials having different thermal properties, heat sources in the vicinity of the cable, non constant temperature limit values make the analytical solution difficult. In this case, only numerical approaches can be used. Based on the general structure of power cables, especially the most preferred numerical approach among the other numerical approaches is the finite element method (Hwang et al., 2003), (Kocar et al., 2004), (IEC TR 62095).

There is a strong link between current carrying capacity and temperature distributions of power cables. Losses produced by voltage applied to a cable and current flowing through its conductor, generate heat in that cable. The current carrying capacity of a cable depends on effective distribution of produced heat from the cable to the surrounding environment. Insulating materials in cables and surrounding environment make this distribution difficult due to existence of high thermal resistances.

The current carrying capacity of power cables is defined as the maximum current value that the cable conductor can carry continuously without exceeding the limit temperature values of the cable components, in particular not exceeding that of insulating material. Therefore, the temperature values of the cable components during continuous operation should be determined. Numerical methods are used for calculation of temperature distribution in a cable and in its surrounding environment, based on generated heat inside the cable. For this purpose, the conductor temperature is calculated for a given conductor current. Then, new calculations are carried out by adjusting the current value.

Calculations in thermal analysis are made usually by using only boundary temperature conditions, geometry, and material information. Because of difficulty in identification and implementation of the problem, analyses taking into account the effects of electrical parameters on temperature or the effects of temperature on electrical parameters are performed very rare (Kovac et al., 2006). In this section, loss and heating mechanisms were evaluated together and current carrying capacity was defined based on this relationship. In numerical methods and especially in singular analyses by using the finite element method, heat sources of cables are entered to the analysis as fixed values. After defining the region and boundary conditions, temperature distribution is calculated. However, these losses are not constant in reality. Evaluation of loss and heating factors simultaneously allows the modeling of power cables closer to the reality.

In this section, use of electric-thermal combined model to determine temperature distribution and consequently current carrying capacity of cables and the solution with the finite element method is given. Later, environmental factors affecting the temperature distribution has been included in the model and the effect of these factors to current carrying capacity of the cables has been studied.

2. Modelling of power cables

Modelling means reducing the concerning parameters' number in a problem. Reducing the number of parameters enable to describe physical phenomena mathematically and this helps to find a solution. Complexity of a problem is reduced by simplifying it. The problem is solved by assuming that some of the parameters are unchangeable in a specific time. On the other hand, when dealing with the problems involving more than one branch of physics, the interaction among those have to be known in order to achieve the right solution. In the future, single-physics analysis for fast and accurate solving of simple problems and multi-physics applications for understanding and solving complex problems will continue to be used together (Dehning et al., 2006), (Zimmerman, 2006).

In this section, theoretical fundamentals to calculate temperature distribution in and around a power cable are given. The goal is to obtain the heat distribution by considering voltage applied to the power cable, current passing through the power cable, and electrical parameters of that power cable. Therefore, theoretical knowledge of electrical-thermal combined model, that is, common solution of electrical and thermal effects is given and current carrying capacity of the power cable is determined from the obtained heat distribution.

2.1 Electrical-thermal combined model for power cables

Power cables are produced in wide variety of types and named with various properties such as voltage level, type of conductor and dielectric materials, number of cores. Basic components of the power cables are conductor, insulator, shield, and protective layers (armour). Conductive material of a cable is usually copper. Ohmic losses occur due to current passing through the conductor material. Insulating materials are exposed to an electric field depending on applied voltage level. Therefore, there will be dielectric losses in that section of the cable. Eddy currents can develop on grounded shield of the cables. If the protective layer is made of magnetic materials, hysteresis and eddy current losses are seen in this section.

Main source of warming on the power cable is the electrical power loss ($R \cdot I^2$) generated by flowing current (I) through its conductor having resistance (R). The electrical power (loss) during time (t) spends electrical energy ($R \cdot I^2 \cdot t$), and this electric energy loss turns into heat energy. This heat spreads to the environment from the cable conductor. In this case, differential heat transfer equation is given in (1) (Lienhard, 2003).

$$\nabla \cdot (k \nabla \theta) + W = \rho c \frac{\partial \theta}{\partial t} \tag{1}$$

Where;

θ : temperature as the independent variable (°K),

k : thermal conductivity of the environment surrounding heat source (W/Km),

ρ : density of the medium as a substance (kg/m^3),

c : thermal capacity of the medium that transmits heat (J/kg°K),

W : volumetric heat source intensity (W/m^3).

Since there is a close relation between heat energy and electrical energy (power loss), heat source intensity (W) due to electrical current can be expressed similar to electrical power.

$$P = J \cdot E \,\mathrm{dxdydz} \tag{2}$$

Where $\boldsymbol{J}$ is current density, $\boldsymbol{E}$ is electrical field intensity; dx.dy.dz is the volume of material in the unit. As current density is $\boldsymbol{J} = \sigma \cdot \boldsymbol{E}$ and electrical field intensity is $\boldsymbol{E} = \boldsymbol{J}/\sigma$, ohmic losses in cable can be written as;

$$P = \frac{1}{\sigma} J^2 \,\mathrm{dxdydz} \tag{3}$$

Where σ is electrical conductivity of the cable conductor and it is temperature dependent. In this study, this feature has been used to make thermal analysis by establishing a link between electrical conductivity and heat transfer. In equation (4), relation between electrical conductivity and temperature of the cable conductor is given as;

$$\sigma = \frac{1}{\rho_0 \cdot (1 + \alpha(\theta - \theta_0))} \tag{4}$$

In the above equation ρ_0 is the specific resistivity at reference temperature value θ_0 ($\Omega \cdot$m); α is temperature coefficient of specific resistivity that describes the variation of specific resistivity with temperature.

Electrical loss produced on the conducting materials of the power cables depends on current density and conductivity of the materials. Ohmic losses on each conductor of a cable increases temperature of the power cable. Electrical conductivity of the cable conductor decreases with increasing temperature. During this phenomenon, ohmic losses increases and conductor gets more heat. This situation has been considered as electrical-thermal combined model (Karahan et al., 2009).

In the next section, examples of the use of electric-thermal model are presented. In this section, 10 kV, XLPE insulated medium voltage power cable and 0.6 / 1 kV, four-core PVC insulated low voltage power cable are modeled by considering only the ohmic losses. However, a model with dielectric losses is given at (Karahan et al., 2009).

2.2 Life estimation for power cables

Power cables are exposed to electrical, thermal, and mechanical stresses simultaneously depending on applied voltage and current passing through. In addition, chemical changes occur in the structure of dielectric material. In order to define the dielectric material life of power cables accelerated aging tests, which depends on voltage, frequency, and temperature are applied. Partial discharges and electrical treeing significantly reduce the life of a cable. Deterioration of dielectric material formed by partial discharges particularly depends on voltage and frequency. Increasing the temperature of the dielectric material leads to faster deterioration and reduced cable lifetime. Since power cables operate at high temperatures, it is very important to consider the effects of thermal stresses on aging of the cables (Malik et al., 1998).

Thermal degradation of organic and inorganic materials used as insulation in electrical service occurs due to the increase in temperature above the nominal value. Life span can be obtained using the Arrhenius equation (Pacheco et al., 2000).

$$\frac{dp}{dt} = A \cdot e^{\frac{-E_a}{k_B \theta}} \tag{5}$$

Where;

dp/dt : Change in life expectancy over time
A : Material constant
k_B : Boltzmann constant [eV/K]
θ : Absolute temperature [°K]
E_a : Excitation (activation) energy [eV]

Depending on the temperature, equation (6) can be used to estimate the approximate life of the cable (Pacheco et al., 2000).

$$p = p_i \cdot e^{\left(\frac{-E_a}{k_B}\right) \cdot \frac{\Delta\theta}{\theta_i \cdot (\theta_i + \Delta\theta)}} \tag{6}$$

In this equation, p is life [days] at $\Delta\theta$ temperature increment; p_i is life [days] at θ_i temperature; $\Delta\theta$ is the amount of temperature increment [°K]; and θ_i is operating temperature of the cable [°K].

In this study, temperature distributions of the power cables were obtained under electrical, thermal and environmental stresses (humidity), and life span of the power cables was evaluated by using the above equations and obtained temperature variations.

3. Applications

3.1 5.8/10 kV XLPE cable model

In this study, the first electrical-thermal combined analysis were made for 5.8/10 kV, XLPE insulated, single core underground cable. All parameters of this cable were taken from (Anders, 1997).

The cable has a conductor of 300 mm² cross-sectional area and braided copper conductor with a diameter of 20.5 mm. In Table 1, thicknesses of the layers of the model cable are given in order.

Layer	Thickness (mm)
Inner semiconductor	0.6
XLPE insulation	3.4
Outer semiconductor	0.6
Copper wire shield	0.7
PVC outer sheath	2.3

Table 1. Layer thicknesses of the power cable.

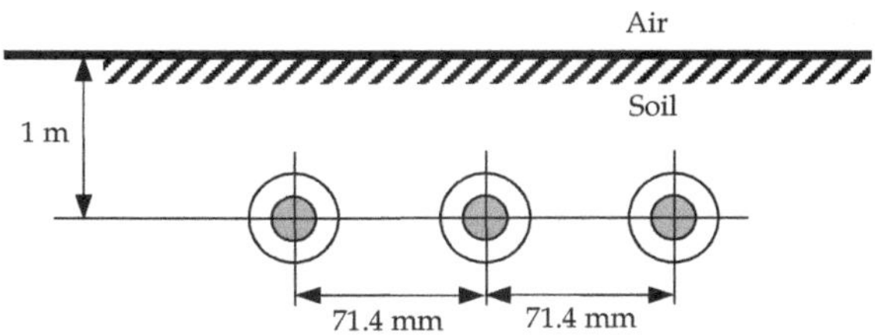

Fig. 1. Laying conditions of the cables.

Figure 1 shows the laying conditions taken into account for the cable. Here, it has been accepted that three exactly same cables having the above given properties are laid side by side at a depth of 1 m underground and they are parallel to the surface of the soil. The distance between the cables is left up to a cable diameter. Thermal resistivity of soil surrounding cables was taken as the reference value of 1 K·m/W. The temperature at far away boundaries is considered as 15°C.

3.1.1 Numerical analysis

For thermal analysis of the power cable, finite element method was used as a numerical method. The first step of the solution by this method is to define the problem with geometry, material and boundary conditions in a closed area. Accordingly the problem has been described in a rectangle solution region having a width of 10 m and length of 5 m, where three cables with the specifications given above are located. Description and consequently solution of the problem are made in two-dimensional Cartesian coordinates. In this case the third coordinate of the Cartesian coordinate system is the direction perpendicular to the solution plane. Accordingly, in the solution region, the axes of the cables defined as the two-dimensional cross-section will be parallel to the third coordinate axis. In the solution, the third coordinate, and therefore the cables are assumed to be infinite length cables.

Thermal conductivity (k) and thermal capacity (c) values of both cable components and soil that were taken into account in analysis are given in Table 2. The table also shows the ρ density values considered for the materials. These parameters are the parameters used in the heat transfer equation (1). Heat sources are defined according to the equation (3).

After geometrical and physical descriptions of the problem, the boundary conditions are defined. The temperature on bottom and side boundaries of the region is assumed as fixed (15°C), and the upper boundary is accepted as the convection boundary. Heat transfer coefficient h is computed from the following empirical equation (Thue, 1999).

Material	Thermal Conductivity k (W/K.m)	Thermal Capacity c (J/kg.K)	Density ρ (kg/m³)
Copper conductor	400	385	8700
XLPE insulation	1/3.5	385	1380
Copper wire screen	400	385	8700
PVC outer sheath	0.1	385	1760
Soil	1	890	1600

Table 2. Thermal properties of materials in the model.

$$h = 7.371 + 6.43 \cdot u^{0.75} \tag{7}$$

Where u is wind velocity in m/s at ground surface on buried cable. In the analysis, wind velocity is assumed to be zero, and the convection is the result of the temperature difference.

Second basic step of the finite element method is to discrete finite elements for solution region. Precision of computation increases with increasing number of finite elements. Therefore, mesh of solution region is divided 8519 triangle finite elements. This process is applied automatically and adaptively by used program.

Changing of cable losses with increasing cable temperature requires studying loss and warm-up mechanisms together. Ampacity of the power cable is determined depending on the temperature of the cable. The generated electrical-thermal combined model shows a non-linear behavior due to temperature-dependent electrical conductivity of the material.

Fig. 2 shows distribution of equi-temperature curve (line) obtained from performed analysis using the finite element method. According to the obtained distribution, the most heated cable is the one in the middle, as a result of the heat effect of cables on each side. The current value that makes the cable's insulation temperature 90°C is calculated as 626.214 A. This current value is calculated by multiplying the current density corresponding to the temperature of 90°C with the cross-sectional area of the conductor. This current value is the current carrying capacity of the cable, and it is close to result of the analytical solution of the same problem (Anders, 1997), which is 629 A.

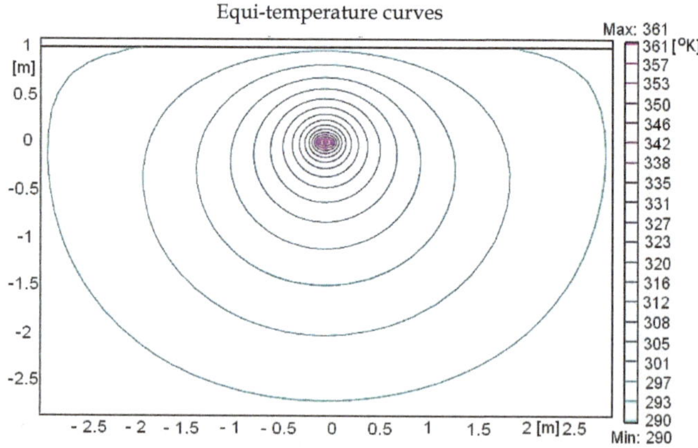

Fig. 2. Distribution of equi-temperature curves.

In Fig. 3, variation of temperature distribution depending on burial depth of the cable in the soil is shown. As shown in Fig. 3, the temperature of the cable with the convection effect shows a rapid decline towards the soil surface. This is not the case in the soil. It can be said that burial depth of the cables has a significant impact on cooling of the cables.

3.1.2 Effect of thermal conductivity of the soil on temperature distribution

Thermal conductivity or thermal resistance of the soil is seasons and climate-changing parameter. When the cable is laid in the soil with moisture more than normal, it is easier to disperse the heat generated by the cable. If the heat produced remains the same, according to the principle of conservation of energy, increase in dispersed heat will result in decrease in the heat amount kept by cable, therefore cable temperature drops and cable can carry more current. Thermal conductivity of the soil can drop up to 0.4 W/K·m value in areas where light rainfall occurs and high soil temperature and drying event in soil are possible. In this case, it will be difficult to disperse the heat generated by the cable; the cable current carrying capacity will drop. The variation of the soil thermal resistivity (conductivity) depending on soil and weather conditions is given in Table 3 (Tedas, 2005).

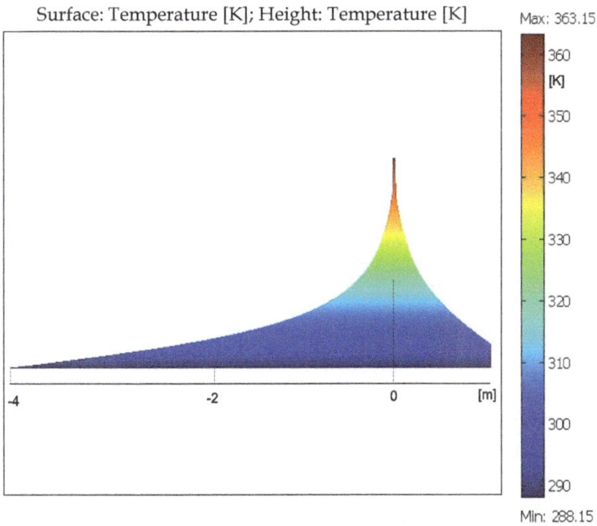

Fig. 3. Variation of temperature distribution with buried depth of the cable in soil.

Thermal Resistivity (K.m/W)	Thermal Conductivity (W/K.m)	Soil Conditions	Weather Conditions
0.7	1.4	Very moist	Continuous moist
1	1	Moist	Regular rain
2	0.5	Dry	Sparse rain
3	0.3	Very dry	too little rain or drought

Table 3. Variation of the soil thermal resistivity and conductivity with soil and weather conditions.

As can be seen from Table 3, at the continuous rainfall areas, soil moisture, and the value of thermal conductivity consequently increases.

While all the other circuit parameters and cable load are fixed, effect of the thermal conductivity of the surrounding environment on the cable temperature was studied. Therefore, by changing the soil thermal conductivity, which is normally encountered in the range of between 0.4 and 1.4 W/K·m, the effect on temperature and current carrying capacity of the cable is issued and results are given in Fig. 4. As shown in Fig. 4, the temperature of the cable increases remarkably with decreasing thermal conductivity of the soil or surrounding environment of the cable. This situation requires a reduction in the cable load.

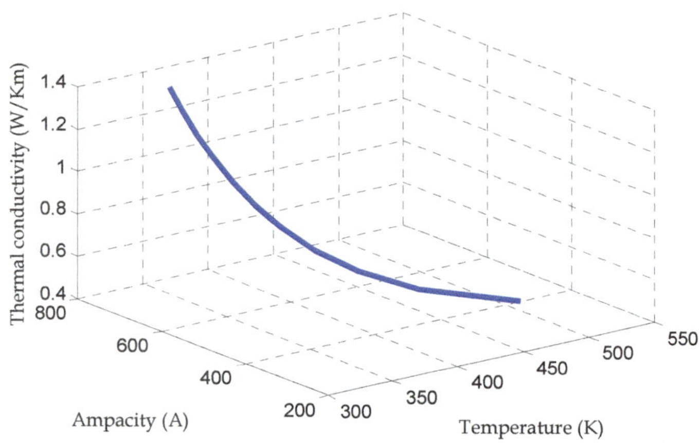

Fig. 4. Effect of variation in thermal conductivity of the soil on temperature and current carrying capacity (ampacity) of the cable.

When the cable load is 626.214 A and thermal conductivity of the soil is 1 W/Km, the temperature of the middle cable that would most heat up was found to be 90°C. For the thermal conductivity of 0.4 W/Km, this temperature increases up to 238°C (511.15°K). In this case, load of the cables should be reduced by 36%, and the current should to be reduced to 399.4 A. In the case of thermal conductivity of 1.4 W/K·m, the temperature of the cable decreases to 70.7°C (343.85°K). This value means that the cable can be loaded %15 more (720.23 A) compared to the case which the thermal conductivity of soil is 1 W/K·m.

3.1.3 Effect of drying of the soil on temperature distribution and current carrying capacity

In the numerical calculations, the value of thermal conductivity of the soil is usually assumed to be constant (Nguyen et al., 2010) (Jiankang et al., 2010). However, if the soil surrounding cable heats up, thermal conductivity varies. This leads to form a dry region around the cable. In this section, effect of the dry region around the cable on temperature distribution and current carrying capacity of the cable was studied.

In the previous section, in the case of the soil thermal conductivity is 1.4 W/K·m, current carrying capacity of the cable was found to be 720.23 A. In that calculation, the thermal conductivity of the soil was assumed that the value did not change depending on temperature value. In the experimental studies, critical temperature for drying of wet soil was determined as about 60°C (Gouda et al., 2011). Analyses were repeated by taking into account the effect of drying of the soil and laying conditions. When the temperature for the surrounding soil exceeds 60°C, which is the critical temperature, this part of the soil was accepted as the dry soil and its thermal conductivity was included in the calculation with the value of 0.6 W/K·m.

The temperature distribution obtained from the numerical calculation using 720.23 A cable current, 1.4 W/K·m initial thermal conductivity of soil, as well as taking into account the effect of drying in soil is given in Fig. 5. As shown in Fig. 5, considering the effect of soil drying, temperature increased to 118.6°C (391.749°K). The cable heats up 28.6°C more compared to the case where the thermal conductivity of the soil was taken as a constant value of 1.4. The boundary of the dried soil, which means the temperature is higher than critical value of 60°C (333.15°K), is also shown in the figure. Then, how much cable current should be reduced was calculated depending on the effect of drying in the soil, and this value was calculated as 672.9 A.

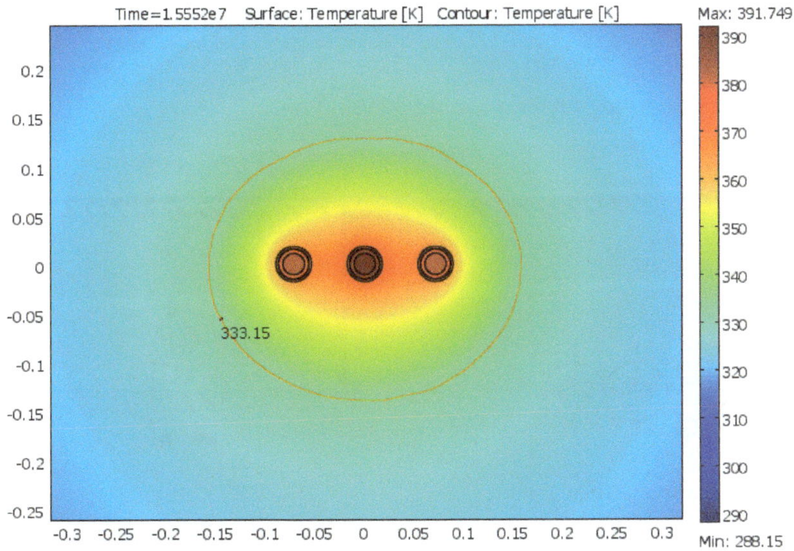

Fig. 5. Effect of drying in the soil on temperature distribution.

The new temperature distribution depending on this current value is given in Fig. 6. As a result of drying effect in soil, the current carrying capacity of the cable was reduced by about 7 %.

3.1.4 Effect of cable position on temperature distribution

In the calculations, the distance between the cables has been accepted that it is up to a cable diameter. If the distances among the three cables laid side by side are reduced, the cable in the middle is expected to heat up more because of two adjacent cables at both sides, as shown in Fig. 7(a). In this case, current carrying capacity of the middle cable will be reduced. Table 4 indicates the change in temperature of the middle cable depending on the distance between cables and corresponding current carrying capacity, obtained from the numerical solution.

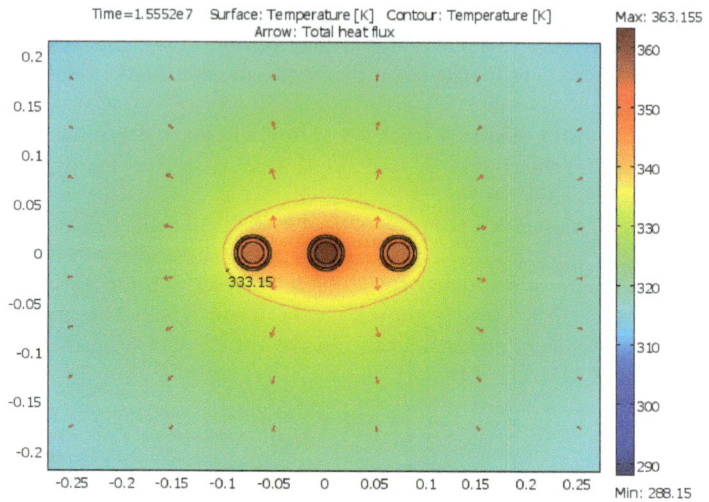

Fig. 6. Effect of the soil drying on temperature distribution.

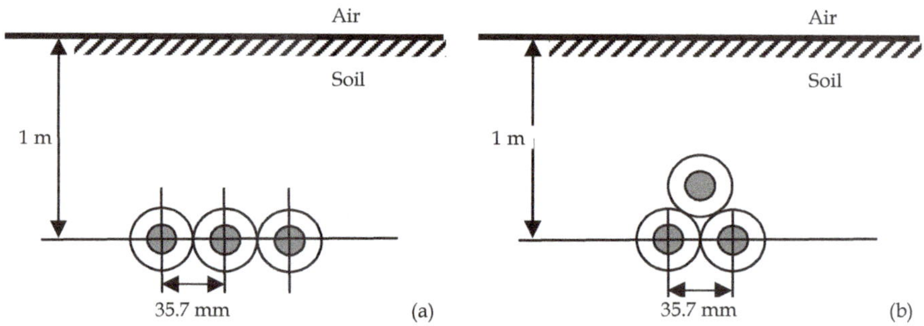

Fig. 7. Laying conditions of the cables.

As shown in Table 4, if there is no distance between the cables, temperature of the cable in the middle increases 10ºC. This situation requires about 6% reduction in the cable load. The case where the distance between the cables is a diameter of a cable is the most appropriate case for the current carrying capacity of the cable.

Distance between the cables (mm)	Cable temperature (ºC)	Current carrying capacity (A)
0	100.03	591.51
10	96.14	604.16
20	93.35	613.85
30	91.12	622.00
36	90.00	626.21

Table 4. Variation of temperature and current carrying capacity of the cable in middle with changing distance between the cables.

Triangle shaped another type of set-up, in which the cables contact to each other, is shown in Fig. 7(b). Each cable heats up more by the effect of two adjacent cables in this placement. The temperature distribution of a section with a height of 0.20 m and width of 0.16 m, which was obtained from the numerical analysis by using the same material and environmental properties given in section 3.1.1, is shown in Fig 8. As a result of this analysis, cables at the bottom heated up more when compared with the cable at the top but, the difference has been found to be fairly low. The current value that increases the temperature value of the bottom cables to 90 ºC was found to be 590.63 A. This value is the current carrying capacity for the cables laid in the triangle shaped set-up.

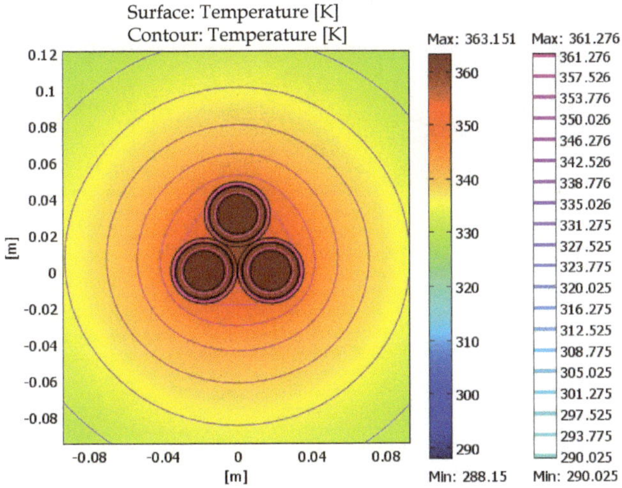

Fig. 8. Temperature distribution for the triangle-shaped set-up.

3.1.5 Single-cable status

In the studies conducted so far, the temperature distribution and current carrying capacity of 10 kV XLPE insulated cables having the triangle shaped and flat shaped set-up with a cable diameter distance have been determined. Other cables lay around or heat sources in the vicinity of the cable reduce the current carrying capacity remarkably. In case of using a single cable, the possible thermal effect of other cables will be eliminated and cable will carry more current. In this section, as shown in Fig. 9, the current carrying capacity of a

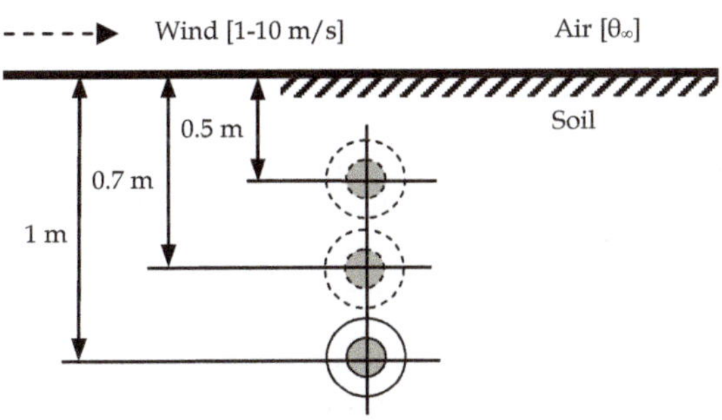

Fig. 9. A power cable buried in different depths.

single cable was calculated for different burial depths and then the impact of wind on the current carrying capacity of the cable has been examined.

In the created model, it is assumed that one 10 kV, XLPE insulated power cable is buried in soil and burial depth is 1 m. Physical descriptions and boundary conditions are the same as the values specified in section 3.1.1. The temperature distribution obtained by numerical analysis is shown in Fig. 10.

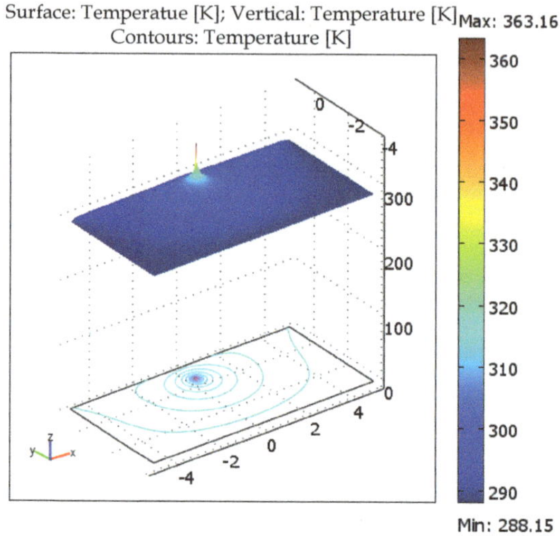

Fig. 10. Three-dimensional temperature distribution in the cable.

According to this distribution, current value which makes the temperature of the cable insulation is 90°C is calculated as 890.97 A. This value is the current carrying capacity for the configuration of stand-alone buried cables and it is 264 A more than that of the side by side configuration and 300 A more than that of triangle shaped set-up. The laying of the cable as closer to the ground surface changes the temperature distribution in and around the cable. For example, at 0.5, 0.7, and 1 m deep-buried case for the cable, the temperatures of the cable insulation depending on the current passing through the cable are shown in Fig. 11.

From the Fig. 11, it is shown that the current carrying capacity increases with the laying of the cable closer to the ground surface. When the cable was laid at a depth of 0.7 m, the current value that makes insulation temperature 90°C was found to be 906.45 A. Current value for a depth of 0.5 m is 922.63 A. Current carrying capacity of the buried cable to a depth of 0.5 m is about 32 A more than current carrying capacity of the buried cable to a depth of 1 m.

So far, it was assumed in the calculations that the wind speed was zero and the convection is the result of the temperature difference. In this section, the effect of change in wind speed on temperature distribution of buried cables has been investigated. Insulator temperatures have been calculated for the different wind speeds changing in the range of 1-10 m/s at each of burial depth by considering the current values that make the insulator temperature 90°C as constant value. As shown in Fig. 12, the increasing wind speed contributes to the cooling of the cables. In this case, the cable temperature will drop and small increase will be seen in the current carrying capacity.

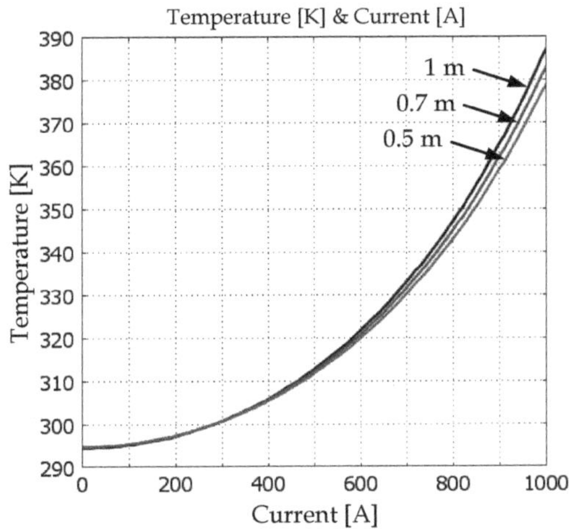

Fig. 11. Variation of temperature as a function of current in different buried depth.

The average wind speed for Istanbul is 3.2 m/s. (Internet, 2007). By taking into account this value, the temperature of the cable buried at 1 m depth will decrease about 0.8°C, while the temperature of the cable buried at 0.5 m depth will decrease about 2°C. This decrease for the cable buried at a depth of 0.5 m means the cable can be loaded 11 A more.

3.1.6 Relationship between cable temperature and cable life

In this section, the life of three exactly same cables laid side by side at a depth of 1 m has been calculated by using the temperature values determined in section 3.1.2 and 3.1.3. Decrease in the value of thermal conductivity of the soil and distance between the cables results in significant increase in temperature of the cables and consequently significant decrease in their current carrying capacities. This condition also reduces the life of the cable.

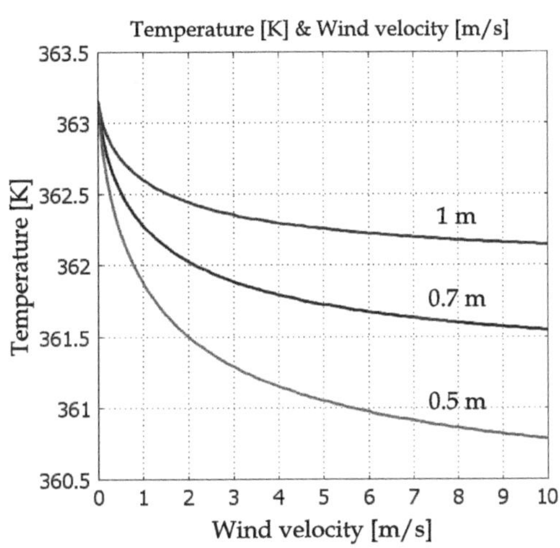

Fig. 12. Variation of temperature of the cable insulation with wind velocity.

In order to see the borders of this effect, cable life has been calculated for both cases by using the equation (6) and the results are indicated in Fig. 13 and Fig. 14. Activation energy of 1.1 eV for XPLE material, Boltzmann constant of $8.617 \cdot 10^{-5}$ eV/K was taken for the calculations and it is assumed that the life of XPLE insulated power cable at 90°C is 30 years. The relationship between the cable distances and life of cables for three different soil thermal conductivities has been shown in Fig. 13. As it is seen from the figure, cable life decreases

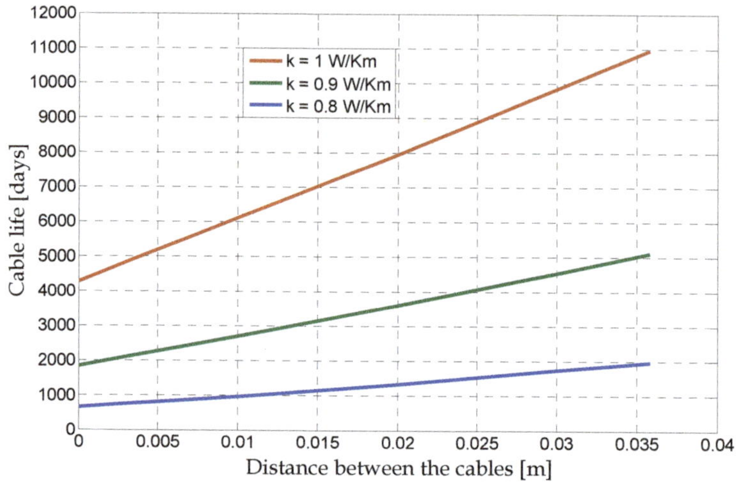

Fig. 13. Variation of the cable life as a function of distance between the cables in different thermal conductivities of the soil.

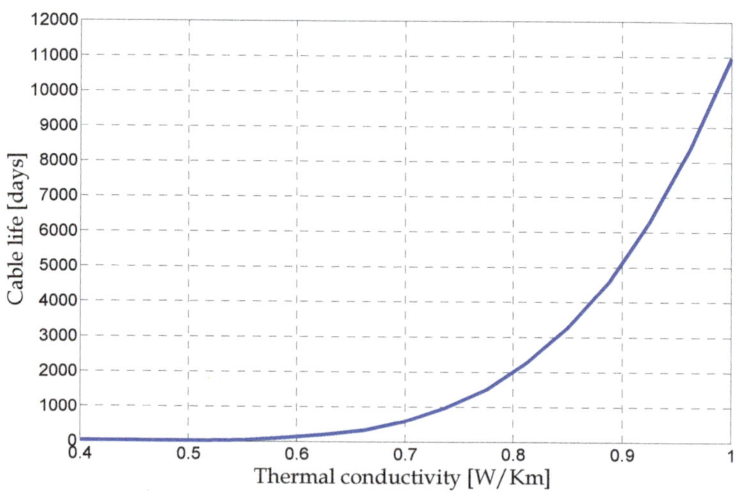

Fig. 14. Variation of the cable life as a function of thermal conductivity of the soil.

linearly depending on decrease in distance between the cables. In this analysis, the currents in the cables were assumed to be constant values and the cable temperatures (changing with

the cable distances) obtained from the numerical solution, were used for the calculation of cable's life. A decrease of 0.5 cm at cable distances leads to loss of 1000 days in the cable life when the thermal conductivity is 1 W/Km, as it is seen from the figure.
The life of cables laid side by side with a one diameter distance has been calculated in another analysis, depending on the change in the thermal conductivity of soil and given in Fig. 14. As a result of this analysis, in which the current values were assumed to be constant, it was seen that cable life increases logarithmically depending on the increase in the thermal conductivity of the soil. As it is seen from the figure, 10% decrease in the thermal conductivity of the soil results in 50% reduction in the cable life unless load conditions are adjusted.

3.2 0.6/1 kV PVC cable model

3.2.1 Experimental studies

This section covers the experimental studies performed in order to examine the relationship between current and temperature in power cables. For this purpose, current and also conductor and sheath temperatures were recorded for a current carrying low voltage power cable in an experiment at laboratory conditions and the obtained experimental data was used in numerical modeling of that cable.
The first cable used in the experiment is a low voltage power cable having the properties of 0.6/1/1.2 (U0/Un/Um) kV, 3 x 35/16 mm^2, 3$^{1/2}$ core (3 phase, 1 neutral), PVC insulated, armored with galvanized flat steel wire, cross-hold steel band, PVC inner and outer sheaths. The catalog information of this PVC insulated cable having 29.1 mm outer diameter specifies that DC resistance at 20°C is 0.524 Ω/km and the maximum operating temperature is 70°C (Turkish Prysmian Cable and Systems Inc.).
In order to examine the relationship between current and temperature in case of the power cable in water and air, a polyester test container was used. During measurements, the cable was placed in the middle and at a 15 cm distance from the bottom of the container. In the first stage, current-temperature relation of the power cable placed in air was studied. The experimental set-up prepared for this purpose is shown in Fig. 15.

Fig. 15. Experimental set-up for 0.6/1 kV cable.

The required current for the power cable has been supplied from alternating current output ends of a 10 kW welding machine. Its the highest output current is 300 A. Current flowing through the cable is monitored by two ammeters which are iron-core, 1.5 classes, and 150 A. Output current is adjusted by use of a variac on the welding machine.

A digital thermometer having the properties of double input, ability to measure temperatures between -200 and 1370 °C, and ± (%0.1 rdg + 0,7°C) precision was used during the measurements. Two K-type thermocouples can be used with the thermometer and this enables to monitor the temperatures of different points simultaneously. These thermocouples were used to measure the conductor and sheath temperatures of the cable.

Conductor and sheath temperatures were measured on cable components at a 50 cm distance from the current source's both ends in accordance with the defined temperature measurement conditions in the Turkish Standard (TS EN 50393, 2006).

During the experiment phase conductors of the cable were connected to each other in serial order and alternative current was applied. Throughout the experiments, cable conductor and sheath temperatures at the point where the current source is connected to the cable and also ambient temperature were recorded with an interval of 10 min. Fig. 16 indicates the variations of current applied to the cable; the cable and ambient temperatures with time.

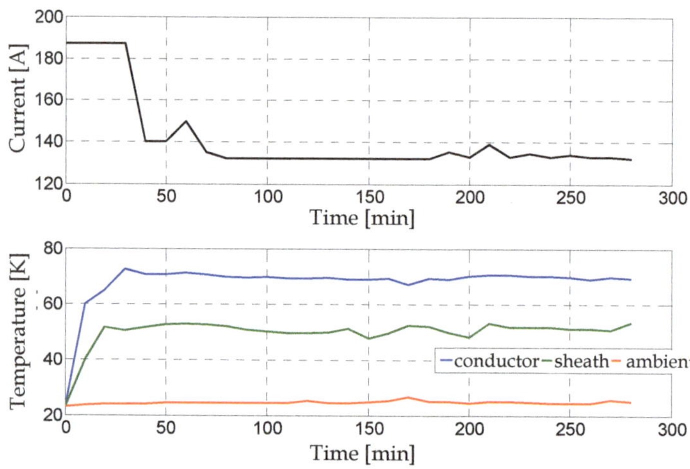

Fig. 16. Variations of current applied to the cable; the cable components and ambient temperatures with time

Conductor and sheath temperatures in the figure are the average of the values obtained from the both measurement points. In order to find the current carrying capacity of the cable it was starded with a high current value and then current was adjusted so that the conductor temperature can be kept constant at 70°C. After almost 3 hours later the current and cable temperatures were stabilized. In that case, the cable was continued to be energized for another 2 hours. The highest current value that cable can carry in steady state operation was found to be 132 A, as it was in agreement with the defined value in the catalog of that cable.

As a second stage, first of all it was waited almost 3 hours for cooling of the cable warmed up during the measurements and then it was started to study the current-temperature relation of the cable that is under water. At this stage, test container was completely filled with water and 2.5 m of 4 m cable was immersed in water placing it at a distance of 35 cm from the water surface.

As it was performed earlier in the case where the power cable was in air, the current value that makes the conductor temperature 70°C was tried to be found and the cable was run at that current value for a certain time. The conductor and sheath temperatures were measured from the sections which are out of water, as it was explained above; at a 50 cm distance from the current source's both ends. Water temperature was also monitored to see the effect of current passing through to cable on the surrounding environment. Fig. 17 indicates the variations of current applied to the cable; the cable components, the ambient, and the water temperatures with time.

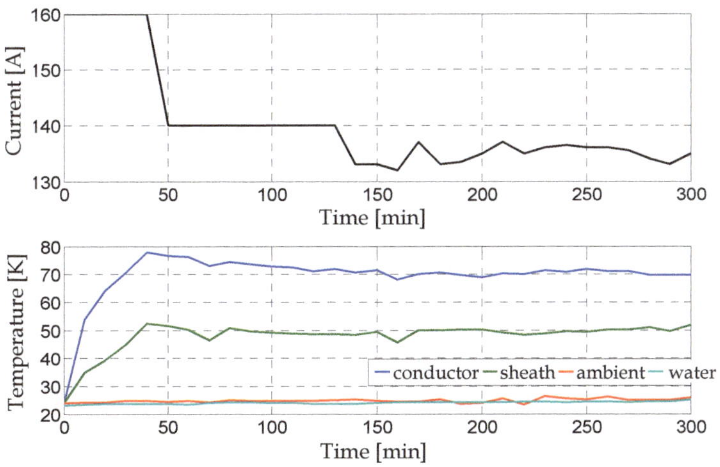

Fig. 17. Variations of current applied to the cable; the cable components, the ambient, and the water temperatures with time

As shown in Fig. 17, the conductor and the sheath temperatures have reached steady state values at the end of nearly two-hour work period. The average current value for the stable operation state is approximately 135 A. The current value that was obtained in the case where substantial portion of cable was immersed in water is a few amps higher than that of air environment.

3.2.2 Numerical solution

Cross section of 0.6 / 1 kV power cable is shown in Fig. 18. In the figure, O shows the center of the cable, O_1 and O_2 indicate the centers of the phase and neutral conductor, respectively. The radiuses of the other cable components are given in Table 5.

Numerical solution of the problem has two-stages. The numerical model of the power cable was created firstly for the air configuration, secondly for the water configuration and the steady-state temperature distributions were determined.

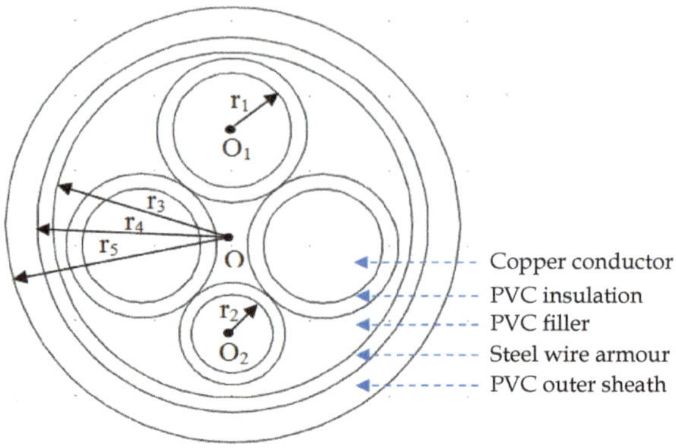

Fig. 18. View of 0.6/1 kV, 3 x 35/16 mm², PVC insulated power cable.

Cable Components	Radius (mm)
Phase conductors (r1)	3.8
Neutral conductor (r2)	2.6
Filling material (r3)	11.5
Armour (r4)	12.5
Outer sheath (r5)	14.5

Table 5. Radiuses of the cable components.

The first step of finding the temperature distrubiton of a power cable in air is to create the geometry of the problem. The problem was defined at 2 x 2 m solution region, where the cable with the given properties above was located. After creating the geometry of the problem, thermal parameters of the cable components and the surrounding environment are defined as given in Table 6.

Cable Material	Density ρ (kg/m³)	Thermal Capacity c (J/kg·K)	Thermal Conductivity k (W/K·m)
Conductor (copper)	8700	385	400
Insulator (PVC)	1760	385	0.1
Armour (steel)	7850	475	44.5
Air	1.205	1005	k_air(θ)

Table 6. Thermal parameters of the cable components.

Thermal conductivity of air varies with temperature. As shown in Fig. 19, the thermal conductivity of air increases depending on the increasing temperature of the air (Remsburg, 2001).

This case, which depends on increased temperature of power cables, provides better distribution of heat to the surrounding environment. By including the values given in Table 7 in the cable model, intermediate values corresponding to change in the air temperature have been found.

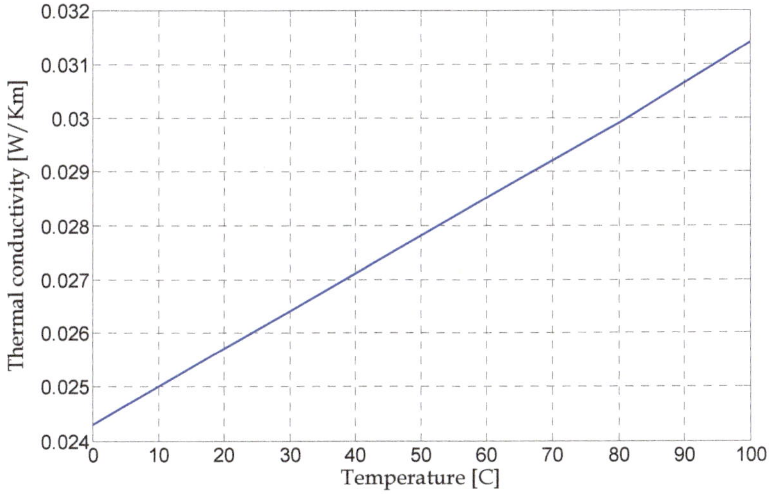

Fig. 19. Variation of thermal conductivity of air with temperature.

Temperature (°C)	Thermal Conductivity (W/mK)
0	0.0243
20	0.0257
40	0.0271
60	0.0285
80	0.0299
100	0.0314

Table 7. Variation in thermal conductivity of air with temperature.

The most important heat source for the existing cable is the ohmic losses formed by current flowing through the cable conductors. The equation of $P = J^2 / \sigma$ is used to calculate these losses. Ohmic losses in the conductor are described as "$(132/(pi * 0.0038^2))^2$ /condCu" (W/m^3)$(132/(pi * 0.0038^2))^2$ / condCu" (W/m^3). In this equation, condCu expression is the value of the electrical conductivity of the material, and it is a temperature-dependent parameter as shown in equation (4).

At the last step of the numerical analysis, the boundary conditions are indicated. Since the cable is located in a closed environment, free convection is available on the surface of the cable. Equation (7) is used to calculate heat transfer coefficient, and the wind speed is assumed as zero. The temperature of the outer boundary of the solution region is defined as constant temperature. This value is an average ambient temperature measured during the experiment (297.78°K) and it was added to the model.

After all these definitions, the region is divided into elements and the numerical solution is performed. The entire region is divided into 7212 elements. As a result of numerical analysis performed by using finite element method, the temperature distribution in and around the cable, and equi-temperature lines are shown in Fig. 20 and Fig. 21, respectively.

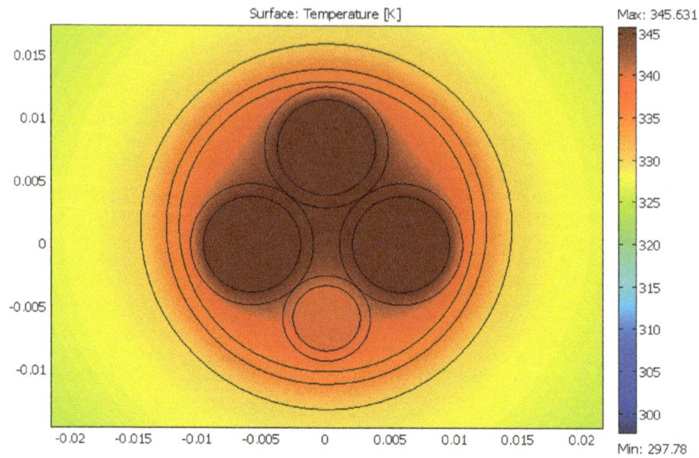

Fig. 20. Temperature distribution.

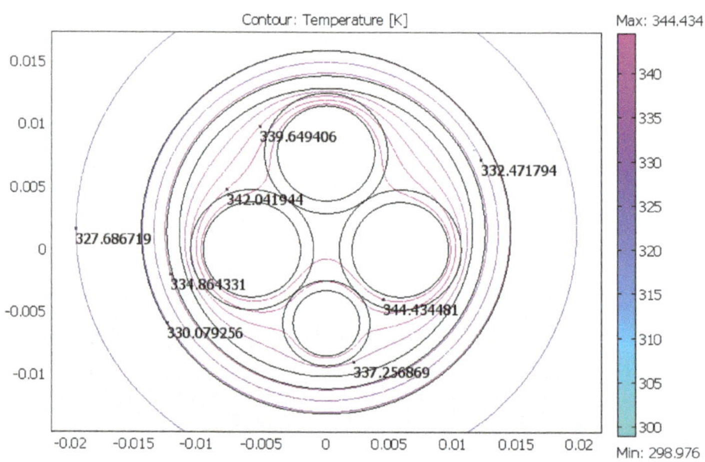

Fig. 21. Equi-temperature lines.

Temperature distribution during the balanced loading of the cable can be seen from the figures. In this case, there will be no current on the neutral conductor and the heat produced by currents passing through to three phase conductors will disperse to the surrounding environment. As seen in Fig. 20, the highest conductor temperature that can be reached was found to be 345.631°K (72.4°C). Steady state value of the average conductor temperature obtained from the experimental measurement is 70.1°C. Outer sheath temperature was

found as 329°K (55.8°C) by numerical analysis. The average sheath temperature obtained from the experimental measurements is approximately 52°C. The results obtained from the numerical analysis are very close to the experimental results.
At the second stage of the numerical model, the condition where the same cable is in the water has been taken into consideration. The model established in this case is the same with the model described above except the properties of the surrounding environment. However, in the numerical model the whole cable is assumed to be in the water. Thermal properties of the water are given in Table 8.

Material of the Cable	Density ρ (kg/m³)	Thermal Capacity c (J/kg·K)	Thermal Conductivity k (W/K·m)
Water	997.1	4181	k_water(θ)

Table 8. Thermal properties of the water.

As seen in Fig. 22, the thermal conductivity of the water depends on the temperature (Remsburg, 2001). This dependence has been included in the model as described for the power cable in air. In addition, temperature of the water is considered to be 24.1°C by calculating the average of measured values. After these definitions, the solution region is divided by finite elements, and then the numerical solution is carried out. As a result of performing the numerical analysis, the current-temperature curve for the power cable in the water environment is given in Fig. 23. As shown in the Fig. 23, the conductor temperature increases depending on the current passing through the cable. The ampacity of the cable was found to be 162.9 A considering the thermal strength of PVC material of 70°C.

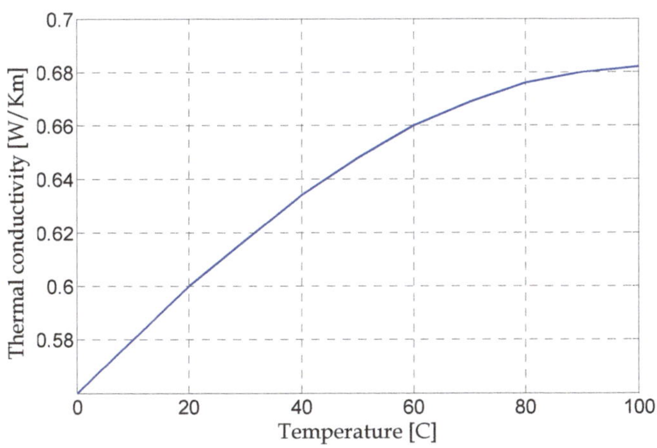

Fig. 22. Variation of thermal conductivity of the water with temperature.

This value is the value of the current carrying capacity where all of the cable is immersed in the water taking into account the water and environment temperature values in the laboratory conditions. In the experimental study, the current value to reach the value of the cable conductor temperature of 70°C is found as 135 A. In the experiment, 60% of the cable section is immersed in the water.

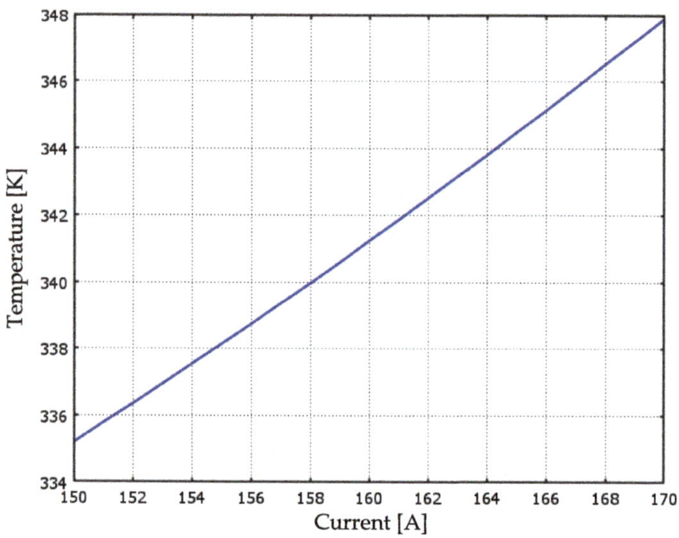

Fig. 23. Relation between current and temperature for the power cable immersed in water.

Therefore, the experimental study for the cable immersed in water can not be expected to give the actual behavior of the power cables. Beside this, the current value enabling the conductor to reach 70°C in the experimental study in water environment is higher than that of air environment. This indicates that power cables immersed in water has better cooling environment because of higher thermal conductivity of the water when compared with the air. Numerical analysis also confirms this result.

By using the numerical analysis common solution of electrical and thermal factors has been realized. Since electrical conductivity of the conductive material is temperature dependent, this increases the conductor temperature by 7°C as a result of numerical analysis. Similarly, thermal conductivity of surrounding environment is defined as a temperature dependent parameter in the numerical model. In the numerical analysis for the power cable immersed in water, 0.6°C decrease in the conductor temperature was seen when it was compared with the case, in which the thermal conductivity of water was taken as constant at 20°C. This is because of the fact that increases in cable temperature results in increase in thermal conductivity of the water and then heat disperses more effectively from the cable to surrounding environment.

4. Conclusion

The thermal analyze of power cable systems is very important especially in terms of determining the current carrying capacity of those cables. Cable temperature depends on many factors, such as current passing through the cable, cable structure and materials used in the manufacture of the cable, laying styles of other cables around that cable, thermal properties of the environment, and moisture of the surrounding soil.

In this study, in which the temperature distribution is studied by taking into account the electrical losses depending on current density and electric field in heat conduction equation, not only the usual temperature conditions but also electrical conditions are considered for the solution.

In 10 kV XLPE insulated cable taken into consideration as an example, the dielectric losses have been neglected due to being very low when compared to the current depended losses. Changes in the current carrying capacity of the cable were investigated by using the temperature distributions determined with the finite element method. Results indicate that current increases the temperature and increased temperature decreases the current carrying capacity of the cable. In this case, it was realized that because of the decreased current due to increased temperature, the temperature decreased, and thus leading to increase in the current again and at the end the stable values in terms of the temperature and current were achieved.

The current carrying capacity of a cable is closely linked with the thermal conductivity of the surrounding environment, such as soil which is the case for the mentioned cable example. Because this resistance has a role to transmit the heat generated in the cable to the environment. In the XLPE insulated cable model, in the range of thermal conductivity of soil encountered in practice, when the thermal conductivity is changed, as expected, the current carrying capacity is increased with the increased thermal conductivity; on the other hand the current carrying capacity is reduced with the decreased thermal conductivity. In the meantime, increase in thermal conductivity reduces the heat kept in the cable, therefore reduces the temperature of the cable.

Usually, there can be other laid cables next to or in the vicinity of the cables. The heat generated by a cable usually has a negative effect on heat exchange of the adjacent cables. As seen in XLPE insulated cable model, when three pieces of cables are laid side by side, the cable in the middle heats up more because of both not being able to transmit its heat easily and getting heat from the side cables. This also lowers the current carrying capacity of the center conductor. To reduce this effect it is necessary to increase the distance between cables. In the study that was conducted to see the effect of change in distance on the temperature distribution, it was seen a decrease in the cable temperature and an increase in the current carrying capacity when the distance was increased, as expected. At the end of our review, at least one cable diameter distance between the cables can be said to be appropriate in terms of the temperature and current conditions.

This study also reviewed the effect of cable burial depth on temperature distribution when a single XLPE insulated cable was considered. It was observed that the cables laid near to earth surface had increased current carrying capacity. This case is due to convection on earth. Reduction in the burial depth of cables provides better heat dissipation that is, better cooling for the cable. Again, in the case of a single cable example, analysis that was performed to study the effect of wind speed on the cable temperature indicated that the increase in wind speed slightly lowered the temperature of the cable. Considering the average wind speed for Istanbul for a power cable buried at a depth of 0.5 m, temperature value is 2°C less compared to cable buried at a depth of 1 m. The outcome of this case obtained from the numerical solution is that the cable can be loaded 11 A more. Based on this, in regions with strong wind, it can be seen that in order to operate the cable at the highest current carrying capacity, the wind speed is a parameter that can not be neglected.

Life of cables is closely linked with the operation conditions. Particularly temperature is one of the dominant factors affecting the life of a cable. For the three XLPE insulated cable model, change in cable life was examined with the temperature values found numerically using the expressions trying to establish a relationship between the temperature and cable life in our study. An increase in temperature shortens the life of the cable. Low temperatures increase both the life and the current carrying capacity of the cable.

Finally, experimental studies have been conducted to examine the relationship between the current and temperature in power cables. For this purpose, the conductor and sheath temperatures of 0.6/1 kV PVC insulated power cable in air and also in water have been studied. In the numerical model of the cable, the current value and environmental temperature obtained from the experiments were used as an input data and by adding temperature dependent electrical and thermal properties of both cable and surrounding environment to the model, the temperature distribution was determined for both the cable components and the surrounding environment. Temperature values obtained from the experimental measurements are in agreement with the results of the numerical solution.

As a result, running power cables in appropriate environmental and layout settings, and operating them in suitable working conditions, increase the cable life and its efficiency and make positive contribution to safety and economy of the connected power systems. This depends on, as in this study, effort put forward for modeling of cables closer to operating conditions, and further examining and evaluating.

5. Acknowledgment

The authors would like to thank to Prof. H. Selcuk Varol who is with Marmara University and Dr. Ozkan Altay who is with Istanbul Technical University, for their help and supports.

6. References

Hwang, C. C., Jiang, Y. H., (2003). "Extensions to the finite element method for thermal analysis of underground cable systems", *Elsevier Electric Power Systems Research*, Vol. 64, pp. 159-164.

Kocar, I., Ertas, A., (2004). "Thermal analysis for determination of current carrying capacity of PE and XLPE insulated power cables using finite element method", IEEE MELECON 2004, May 12-15, 2004, Dubrovnik, Croatia, pp. 905-908.

IEC TR 62095 (2003). Electric Cables - Calculations for current ratings - Finite element method, IEC Standard, Geneva, Switzerland.

Kovac, N., Sarajcev, I., Poljak, D., (2006). "Nonlinear-Coupled Electric-Thermal Modeling of Underground Cable Systems", *IEEE Transactions on Power Delivery*, Vol. 21, No. 1, pp. 4-14.

Lienhard, J. H. (2003). *A Heat Transfer Text Book*, 3rd Ed., Phlogiston Press, Cambridge, Massachusetts.

Dehning, C., Wolf, K. (2006). *Why do Multi-Physics Analysis?*, Nafems Ltd, London, UK.

Zimmerman, W. B. J. (2006). *Multiphysics Modelling with Finite Element Methods*, World Scientific, Singapore.

Malik, N. H., Al-Arainy, A. A., Qureshi, M. I. (1998). *Electrical Insulation in Power Systems*, Marcel Dekker Inc., New York.

Pacheco, C. R., Oliveira, J. C., Vilaca, A. L. A. (2000). "Power quality impact on thermal behaviour and life expectancy of insulated cables", *IEEE Ninth International Conference on Harmonics and Quality of Power, Proceedings*, Orlando, FL, Vol. 3, pp. 893-898.

Anders, G. J. (1997). *Rating of Electric Power Cables - Ampacity Calculations for Transmission, Distribution and Industrial Applications*, IEEE Press, New York.

Thue W. A. (2003). *Electrical Power Cable Engineering*, 2nd Ed., Marcel Dekker, New York.

Tedas (Turkish Electrical Power Distribution Inc.), (2005). Assembly (application) principles and guidelines for power cables in the electrical power distribution networks.

Internet, 04/23/2007. istanbul.meteor.gov.tr/marmaraiklimi.htm

Turkish Prysmian Cable and Systems Inc., Conductors and Power Cables, Company Catalog.

TS EN 50393, Turkish Standard, (2007). *Cables - Test methods and requirements for accessories for use on distribution cables of rated voltage 0.6/1.0 (1.2) kV.*

Remsburg, R., (2001). *Thermal Design of Electronic Equipment*, CRC Press LLC, New York.

Gouda, O. E., El Dein, A. Z., Amer, G. M. (2011). "Effect of the formation of the dry zone around underground power cables on their ratings", *IEEE Transaction on Power Delivery*, Vol. 26, No. 2, pp. 972-978.

Nguyen, N., Phan Tu Vu, and Tlusty, J., (2010). "New approach of thermal field and ampacity of underground cables using adaptive hp- FEM", 2010 IEEE PES Transmission and Distribution Conference and Exposition, New Orleans, pp. 1-5.

Jiankang, Z., Qingquan, L., Youbing, F., Xianbo, D. and Songhua, L. (2010). "Optimization of ampacity for the unequally loaded power cables in duct banks", 2010 Asia-Pacific Power and Energy Engineering Conference (APPEEC), Chengdu, pp. 1-4.

Karahan, M., Varol, H. S., Kalenderli, Ö., (2009). Thermal analysis of power cables using finite element method and current-carrying capacity evaluation, *IJEE* (*Int. J. Engng Ed.*), Vol. 25, No. 6, pp. 1158-1165.

10

Ultrahigh Strength Steel: Development of Mechanical Properties Through Controlled Cooling

S. K. Maity[1] and R. Kawalla[2]
[1]*National Metallurgical Laboratory,*
[2]*TU Bergademie,*
[1]*India*
[2]*Germany*

1. Introduction

Structural steels with very high strength are referred as ultrahigh strength steels. The designation of ultrahigh strength is arbitrary, because there is no universally accepted strength level for this class of steels. As structural steels with greater and greater strength were developed, the strength range has been gradually modified. Commercial structural steel possessing a minimum yield strength of 1380 MPa (200 ksi) are accepted as ultrahigh strength steel (Philip, 1990). It has many applications such as in pipelines, cars, pressure vessels, ships, offshore platforms, aircraft undercarriages, defence sector and rocket motor casings. The class ultrahigh strength structural steels are quite broad and include several distinctly different families of steels such as (a) medium carbon low alloy steels, (b) medium alloy air hardening steel, (c) high alloy hardenable steels, and (d) 18Ni maraging steel. In the recent past, developmental efforts have been aimed mostly at increasing the ductility and toughness by improving the melting and the processing techniques. Steels with fewer and smaller non-metallic inclusions are produced by use of selected advanced processing techniques such as vacuum deoxidation, vacuum degassing, vacuum induction melting, vacuum arc remelting (VAR) and electroslag remelting (ESR). These techniques yield (a) less variation of properties from heat to heat, (b) greater ductility and toughness especially in the transverse direction, and (c) greater reliability in service (Philip, 1978). The strength can be further increased by thermomechanical treatment with controlled cooling.

1.1 Medium carbon low alloy steel

The medium carbon low alloy family of ultra high strength steel includes AISI/SAE 4130, the high strength 4140, and the deeper hardening and high strength 4340. In AMS 6434, vanadium has been added as a grain refiner to improve the toughness and carbon is reduced slightly to improve weldability. D-6a contains vanadium as grain refiner, slightly higher carbon, chromium, molybdenum and slightly lower nickel than 4340. Other less widely used steels that may be included in this family are 6150 and 8640. Medium-carbon low alloy ultrahigh strength steels are hot forgeable, usually at 1060 to 1230°C. Prior to machining, the usual practice is to normalise at 870 to 925°C and temper at 650 to 675°C. These treatments yield moderately hard structures consisting of medium to fine pearlite. It is observed that maximum tensile strength and yield strength result when these steels are tempered at 200°C. With higher tempering temperature, the mechanical properties drop sharply. The mechanical properties obtained in oil-quenched and tempered conditions are shown in Table 1.

Designation	Tempering temperature (°C)	Tensile strength (MPa)	Yield strength (MPa)	Elongation (%)	Hardness (HB)	Izod impact (J)	Fracture toughness (MPa√m)
4130	205 425	1550 1230	1340 1030	11 16.5	450 360	- -	70
4140	205 425	1965 1450	1740 1340	11 15	578 429	15 28	49
4340	205 425	1980 1500	1860 1365	11 14	520 440	20 16	46-AM 60- VAR
300M	205 425	2140 1790	1650 1480	7.0 8.5	550 450	21.7 13.6	-
D - 6a	205 425	2000 1630	1620 1570	8.9 9.6	- -	15 16	99

Table 1. Mechanical properties of medium carbon alloy steel.

1.2 Medium alloy air hardening steel

The steels H11, Modified (H11 Mod) and H13 are included in this category. These steels are often processed through remelting techniques like VAR or ESR. VAR and ESR produced H13 have better cleanness and chemical homogeneity than air melted H13. This results in superior ductility, impact strength and fatigue resistance, especially in the transverse direction, and in large section size. Besides being extensively used in dies, these steels are also widely used for structural purposes. They have excellent fracture toughness coupled with other mechanical properties. H11 Mod and H13 can be hardened in large sections by air-cooling. The chemical compositions and the mechanical properties of these steels are given in Table 2.

Designation	C (%)	Mn (%)	Si (%)	Cr (%)	Mo (%)	V(%)
H11 Mod	0.37 – 0.43	0.20 – 0.40	0.80 – 1.00	4.74 – 5.25	1.20 – 1.40	0.40 – 0.60
H13	0.32 – 0.45	0.20 – 0.50	0.80 – 1.20	4.75 – 5.50	1.10 – 1.75	0.80 – 1.20

Designation	Tempering temperature (°C)	Tensile strength (MPa)	Yield strength (MPa)	Elongation (%)	Hardness (HRc)	Izod impact (J)
H11 Mod	565	1850	1565	11	52	26.4
H13	575	1730	1470	13.5	48	27

Table 2. Chemical compositions and mechanical properties of medium alloy air hardening ultra high strength steel.

1.3 High alloy hardenable steel

These steels were introduced by Republic Steel Corporation in the 1960's and have four weldable steel grades with high fracture toughness and yield strength in heat treated condition. These nominally contain 9% Ni and 4% Co and differ only in carbon content. The four steels designated as HP9-4-20, HP9-4-25, HP9-4-30 and HP9-4-45 nominally have 0.20, 0.25, 0.30 and 0.45%C respectively. Among these steels, HP9-4-20 and HP9-4-30 are produced in significant quantities and their chemical composition and mechanical properties are given in Table 3 (Philip, 1978). As the carbon content of these steels increases, attainable strength increases with corresponding decrease in both toughness and weldability. The high nickel content of 9% provides deep hardenability, toughness and some solid solution strengthening. If the steel contains only higher amount of nickel but no cobalt, there would be a strong tendency for retention of large amounts of austenite on quenching. This retained austenite would not decompose even by refrigeration and tempering. Cobalt increases the Ms temperature and counteracts austenite retention. Chromium and molybdenum content are kept low for improvement of toughness. Silicon and other elements are kept as low as practicable.

Designation	C (%)	Mn (%)	Si (%)	Cr (%)	Ni (%)	Mo (%)	V (%)	Others (%)
HP 9-4-20	0.16–0.23	0.20–0.40	0.20 max	0.65–0.85	8.50–9.50	0.90–1.10	0.06–0.12	4.25– 4.75 Co
HP 9-4-30	0.29–0.34	0.10–0.35	0.20 max	0.90–1.10	7.0 – 8.0	0.90–1.10	0.06–0.12	4.25– 4.75 Co

Designation	Tensile strength (MPa)	Yield strength (MPa)	Elongation (%)	Hardness (HRc)	Izod impact (J)
HP 9-4-20	1380	-	-	-	-
HP 9-4-30	1650	1350	14	49 - 53	39

Table 3. Chemical compositions and typical mechanical properties of high alloy hardenable ultra high strength steel.

1.4 18 Ni maraging steel

Steels belonging to this class of high strength steels differ from other conventional steels. These are not hardened by metallurgical reactions that involve carbon, but by the precipitation of intermetallic compounds at temperatures of about 480°C. The typical yield strengths are in the range 1030 MPa to 2420 MPa. They have very high nickel, cobalt and molybdenum and very low carbon content. The microstructure consists of highly alloyed low carbon martensites. On slow cooling from the austenite region, martensite is produced even in heavy sections, so there is no lack of hardenabilty. Cobalt increases the Ms transformation temperature so that complete martensite transformation can be achieved. The martensite is mainly body centred cubic (bcc), and has lath morphology. Maraging steel normally contains little or no austenite after heat treatment. The presence of titanium leads to precipitation of Ni_3Ti. It gives additional hardening. However, high titanium content favours formation of TiC at the austenite grain boundaries, which can severely embrittle the

age-hardened steel (Philip, 1978). The nominal chemical compositions of the commercial maraging steels are shown in Table 4. Typical tensile properties are shown in Table 5.
One of the distinguishing features of the maraging steels is their superior toughness compared to conventional steels. Maraging steels are normally solution annealed (austenitised) and cooled to room temperature before aging. Cooling rate after annealing has no effect on microstructure. Aging is normally done at 480°C for 3 to 6 hours. These steels can be hot worked by conventional steel mill techniques. Working above 1260°C should however be avoided (Floreen, 1978). Maraging steels have found varieties of applications including missile casing, aircraft forgings, special springs, transmission shafts, couplings, hydraulic hoses, bolts and punches and dies.

Grade	C (%)	Ni (%)	Mo (%)	Co (%)	Ti (%)	Al (%)	Other (%)
18Ni (200)	0.03 max	18	3.3	8.5	0.2	0.1	-
18Ni (250)	0.03 max	18	5.0	8.5	0.4	0.1	-
18Ni (300)	0.03 max	18	5.0	9.0	0.7	0.1	-
18Ni (350)	0.03 max	18	4.2	12.5	1.6	0.1	-
18Ni (cast)	0.03 max	17	4.6	10.0	0.3	0.1	-
18Ni (180)	0.03 max	12	3	-	0.2	0.3	5.0% Cr

Table 4. The nominal chemical compositions of maraging steel.

Grade	Heat treatment	Tensile strength (MPa)	Yield strength (MPa)	Elongation (%)
18Ni (200)	A	1500	1400	10
18Ni (250)	A	1800	1700	8
18Ni (300)	A	2050	2000	7
18Ni (350)	B	2450	2400	6
18Ni (cast)	C	1750	1650	8

A: solution treat 1h at 820°C, aging 3h at 480°C; B: solution treat 1h at 820°C, aging 12h at 480°C;
C: anneal 1h at 1150°C, aging 1h at 595°C, solution treat 1h at 820°C, aging 3h at 480°C.

Table 5. Mechanical properties of the heat treated maraging steel.

1.5 Issues and objective

In addition to high strength-to-weight ratio, ultra high strength steels should possess good ductility, toughness, fatigue resistance and weldability. Some of the currently employed steels, like maraging steels, are highly alloyed and are expensive. Search for less expensive steels with better properties, is therefore a continuing process. High strength in these alloys is obtained by exploiting all the strengthening mechanisms, by careful control of alloying and subsequent processing. Often when strength is raised by alloying and thermomechanical treatment, ductility and toughness suffer. Additionally one can have serious problems with fatigue properties. Many defects are introduced, and inferior properties are obtained during the solidification process. It is, therefore, advantageous to exercise great control during this process. Secondary refining processes like vacuum arc remelting (VAR) and electroslag refining (ESR) are often employed to obtain superior properties in these materials for critical applications. Electroslag refining is known to give low inclusion content, low macro-and micro-segregation, and low microporosity due to near-directional solidification from a small pool with application of controlled cooling. Many alloys for critical application now use this process to ensure reliability and good properties.

The material developed earlier at Indian Institute of Technology (IIT) Bombay and Vikram Saravai Space Center (VSSC), Trivandrum, India with a yield strength of 1450 MPa, is qualified as aerospace application (Suresh et al., 2003; Chatterjee et al., 1990). This was a medium-carbon low alloy steel used mostly in tempered condition. The chemical composition of the alloy is: 0.3% C, 1.0% Mn, 1.0% Mo, 1.5% Cr, 0.3% V and named as 0.3C-CrMoV (ESR) steel (Suresh et al, 2003). The microstructure of heat treated alloy primarily consists of tempered lath martensite. The primary objective of the present work is to develop an alloy with yield strength in excess of 1700 MPa with adequate ductility and impact toughness. It has been achieved through:

a. ESR processing of the alloys
b. Thermomechanical treatment with controlled cooling

1.6 Plan of investigation

UHSS is mostly developed by interplay of all strengthening mechanisms. Grain refinement is achieved either by fine precipitates which pin the austenite grain boundaries by micro alloys (Tanaka, 1981; Umemoto et al., 1987). Precipitation of carbides and carbonitrides both at high temperatures or during cooling and tempering helps to improve the mechanical properties for specific needs (Bleck et al., 1988). Ductility and toughness suffer in most methods of strengthening when one tries to increase strength. The approach in the present work, therefore, is to adjust the chemistry and optimise the production process to obtain clean steel with finer microstructures by special melting process. Therefore, it is advantageous to process these materials through a secondary refining process like electroslag refining (ESR), which ensures the cleanliness and chemical homogeneity (Shash, 1988; Choudhary & Szekely, 1981). Further improvement of mechanical properties is to be obtained by a control thermomechanical treatment (TMT). Melting and casting of alloys and subsequent processing like TMT are the two main aspects in this study.

In the first part of the study, the alloys were prepared with variation of chemical composition starting with a basic composition of 0.3%C, 4.2%Cr, 1%Mn, 1%Mo and 0.35%V. In the previous study, the effects addition of titanium and niobium, and increase of chromium and vanadium contents on the mechanical and microstructural properties were investigated (Maity et al., 2008a, 2008b). Most of these alloys in as cast tempered condition displayed minimum yield strengths of 1450 MPa with elongation of about 9-12% and impact toughness in many cases was in excess of 300 $kJ.m^{-2}$. For further improvement of mechanical properties especially to increase the toughness values, the basic steel is alloyed with 1-3% of nickel in this study. Nickel is generally added in many low alloy steels to improve low temperature toughness and hardenability (Maity et al., 2009). It also strengthens the steel by solid solution hardening, and is particularly effective when it is used in combination with chromium and molybdenum (Umemoto et al., 1987). Nickel is known to increase the resistance to cleavage fracture in steel and decreases ductile-to-brittle transition temperature. The medium-carbon low-alloy martensitic steel attains the best combination of properties in tempered condition owing to the formation of transition carbides

(Malakondaiah et al., 1997). It decreases the ductile-to-brittle transition temperature by promotion of a cross-slip of dislocations in ferritic as well as martensitic steels (Arsenault, 1967; Jolley, 1968; Norstr¨om & Vingsbo, 1979). This effect promotes deformation rather than cleavage fracture and therefore increases toughness. Also, nickel is known as the alloying element, which slightly increases the hardness of martensite and has a weak effect on hindering in decrease of hardness with the tempering temperature with tempered martensite and retained austenite (Grange, 1977).
In the second part of the investigation, it was attempted to further increase the strength and toughness by optimised schedule of thermomechanical treatment. It normally increases the tensile properties and toughness (Dhua, 2003) without reducing ductility or brittle fracture resistance (Akhlaghi, 2001; Jahazi & Egbali, 2000). With controlled rolling it is possible to refine the ferrite structures directly after finish rolling or by using additional accelerated cooling. The processes can be divided into the following stages (Kern et al., 1992): i) forming in the region in which the austenite matrix recrystallises, and/or ii) forming in a heterogeneous austenite-ferrite region after partial decomposition of austenite to ferrite followed by iii) a process of accelerated cooling after the controlled rolling (Umemoto et al., 1987; Kern et al., 1992). The essential hot rolling parameters of the thermomechanical process are: i) slab reheating temperature for dissolution of the precipitated carbonitrides, ii) roughing phase for producing a fine, polygonal austenite grain by means of recrystallisation, ii) final rolling temperature, and iv) degree of final deformation in the temperature range. For controlled cooling the additional parameters are: a) cooling rate, and b) cooling temperature. Controlled rolling and accelerated cooling play important role in the modification of final microstructures. The main result in the first phase of rolling is to increase of yield strength and toughness. This is attributed to the resultant fine grain microstructures. In the second phase (accelerated cooling), the contribution of the increase of the resultant mechanical properties is caused not only by refining of ferrite grains, but also by the change of the morphology of the various phases in the ferrite matrix (bainite or martensite). It is reported that if rolling is completed at a relatively high temperature (in the high temperature austenite range) and the sample is cooled in air, one gets a mixed microstructure of upper bainite and martensite (Tanaka, 1981). Accelerated cooling results in the formation of mostly martensite phase.
The combine influence of alloying elements and thermomechanical treatment allows to exploit of different mechanisms of strengthening, such as precipitation hardening, grain refinement, and transformation hardening by means of bainite and martensite transformations (Bleck et al., 1988). Although, application of thermomechanical treatment especially to high strength low alloy steel (HSLA) is known, little systematic work has been carried out with application of this technique to ultra high strength steels (UHSS) (Jahazi & Egbali, 2000). Present study, therefore is also aimed to produce ultrahigh strength steel through optimised schedule of the processes parameters of thermomechanical treatment so that such high strength materials can be rolled in the existing rolling mill and a minimum yield strength of 1700 MPa along with good impact toughness is achieved.

2. Experiment

2.1 Preparation of as-cast alloys

The alloys were produced by induction melting followed by electroslag refining (ESR). The electrodes, which were produced by induction melting, were remelted using the pre-fused flux of 70 CaF_2: 30 Al_2O_3. About 800g slag was used for each experiment and it was preheated in a muffle furnace at 800°C for 5-6 hours to eliminate free and combined moisture, before charging into the ESR furnace. Remelting was done in a water cooled steel mould of 80 mm diameter, with electrode connected to positive end of DC power. At equilibrium the current and voltage during this process were about 730 amps, and 25 ± 2V respectively, with mould water flow rate of 30 litre/minute and base-plate water flow rate of 20 litre/minute. After the ESR process, cooled ingots were taken out from the mould. ESR ingots were approximately 150 mm long and 75 mm in diameter. The ingots were annealed in a muffle furnace at 975°C for 8-9 h. After annealing, 20 mm and 10 mm lengths were discarded from the bottom and top of the ingot respectively. Samples for chemical analysis and mechanical tests specimens were taken from the ingot. The mechanical test specimens underwent for heat treatment. Heat treatment was organised in a tubular furnace of approximately 80 mm constant temperature (±2.5°C) zone. Argon atmosphere was provided to prevent any oxidation. The base alloy (ESR1) were austenitised at 975 °C, quenched in oil and tempered at 475°C (Chatterjee et al., 1990). For nickel containing steels, the specimens are austenitised at 930°C and tempered at 475°C (Maity et al., 2009). After heat treatment, the samples were prepared for mechanical tests by machining and grinding to final size.

2.2 Thermomechanical Treatment (TMT)

The size of the as-cast ESR ingots selected for rolling was about diameter of 75 mm and length of 60-65 mm. These ingots were soaked at 1200°C for about one hour pre rolled to □ 23.7 x 23.7 mm bars. During TMT, the bars were reheated again to 1200°C, soaked for 90 minutes, transported to the rolling mill and held there till it reached to 950°C, and then rolled to a size of approximately □ 16.5 x 26.5 mm. The second pass was applied in equal deformation as soon as the temperature reached to 850°C and finally it was rolled to □ 11 x 29 mm plates. Immediately thereafter, the samples were quenched in the different cooling mediums. Total reduction was approximately 30% in area and 50% in thickness. About 10 mm and 2 mm lengths were discarded from both the ends and the sides respectively and the samples were prepared for mechanical properties and microstructural studies correspond to the rolling direction. The sampling plan is shown in Figure 1.

2.3 Characterisation

Most of the chemical analysis was carried out by atomic absorption spectroscopy (AAS). Carbon was analysed in a Strohlien apparatus. Nitrogen and aluminium were analysed by a spectrometer. Sulphur and phosphorous were analysed in a SPECTROLAB analytical instrument. The heat treated specimens were analysed for various mechanical properties. For tensile test, round specimens of 4 mm diameter and 24 mm gauge length were prepared, as per DIN 50125-A 12 x 60, 1991 specification and tested at room temperature using the Servo Hydraulic UTM. Charpy U-notch impact toughness specimens were prepared as per DIN 50115-DVM, 1975 specification and also tested at room temperature. Hardness of quenched as well hardened and tempered materials was measured at on Rockwell C hardness tester with application of 15 kg load. Optical, SEM and TEM specimens were prepared by standard method. The TEM-carbon replica technique was employed to extract the precipitates from the specimens. The carbon replicas were examined using field emission electron microscope equipped with energy dispersive X-ray spectrometer (EDS).

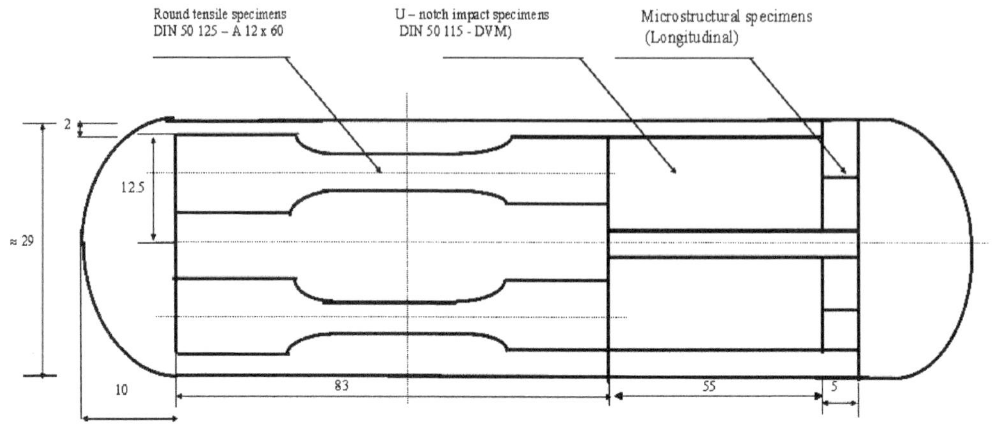

Fig. 1. Sample plan of the rolled plate in rolling direction.

3. Result and discussion

3.1 Properties of as-cast alloy

All ingots prepared by ESR process had smooth and bright surface with few blemishes. The loss of alloying elements was about 5% during ESR processes. Four alloys were prepared starting from a base alloy composition of 0.28%C, 1.0% Mn, 1.0% Mo, 0.35% V, 4.2% Cr. In other three alloys about 1% to 3% Ni was added with the basic composition. The chemical composition of the ESR ingots is illustrated in Table 6. The amount of nickel in ESR2 is about to 1%, 2% in ESR3 and 3.3% in ESR4, respectively. It can be also noticed that the amount of sulphur in all ESR ingots is substantially low. In our earlier work, it is reported that the inclusions in the electrodes were mainly of oxide and sulphide type, which were substantially removed during ESR process (Maity et al., 2006). Chemical homogeneity of the ESR steel is confirmed from glow discharge optical emission spectroscopy (GDOES) analysis and reported in our study (Maity et al., 2006). This study showed that the micro-segregation of chromium, carbon, silicon, manganese, and vanadium were minimal in ESR alloys.

Sample	Chemical composition of ESR ingot (in wt.%)										
	C	Mn	Cr	V	Mo	Si	Ni	Al	N	P	S
ESR 1	0.28	1.00	4.20	0.34	1.01	0.24	00	0.067	0.0161	0.031	0.011
ESR 2	0.28	0.91	4.50	0.35	0.97	0.19	1.07	0.057	0.0158	0.038	0.011
ESR 3	0.28	0.86	4.10	0.47	1.37	0.27	1.97	0.037	0.0128	0.034	0.010
ESR 4	0.30	1.20	4.60	0.47	0.94	0.19	3.29	0.110	0.0108	0.040	0.008

Table 6. Chemical composition of ESR ingot.

In this study, ESR1 is the basic steel. The mechanical properties of the as-cast and as-tempered alloy are shown in Table 7. The yield strength of this alloy is 1450MPa with good ductility and charpy impact toughness. The optical, SEM and TEM micrographs of base alloy (ESR1) are shown in Figure 2. The optical, SEM studies reveal that the microstructures of the tempered specimens mostly consist of lath martensites. The bright field TEM micrograph confirms that the inter lath martensite spacing is of the order of 550-700 nm. The carbon replica micrographs of this steel and the associated EDS analysis (also shown in Figure 2) show the precipitation of complex carbides. The precipitates are spherical in shape with rounded edges evenly distributed in the metal matrix. The EDS analysis of the precipitates shows that these are complex carbides/carbonitrides comprising vanadium, molybdenum and chromium. It can be noted that at the austenitising temperature of 975°C, the precipitates in ESR1 alloy are expected to contain very little amount of chromium and molybdenum because at this temperature most of these precipitates go into solution (Maity et al., 2006). Only vanadium carbonitrides remains partly undissolved in this temperature and chromium and molybdenum would have been precipitated into these pre-existing vanadium carbonitride precipitates during cooling and subsequent tempering.

Sample	Chemistry highlights	Room temperature mechanical properties					Grain Size (μm)
	Ni (wt%)	UTS (MPa)	YS (MPa)	Elonga-tion (%)	Impact strength ($kJ.m^{-2}$)	Hardness (HRc)	
ESR1	-	1660	1450	11.2	300	44	65
ESR2	1.07	1670	1500	9.5	400	45.5	51
ESR3	1.97	1712	1506	12.6	328	46.7	55
ESR4	3.29	1758	1542	9.6	274	46.5	56

Table 7. Mechanical properties of as-cast alloy in quenched-and-tempered condition.

It can be noticed from Table 7 that on addition of 1% nickel in ESR2 alloy, the grain size is marginally reduced to 51 μm. The yield strength has been increased, and there is a marginal increase in hardness and tensile strength. It may be observed that the impact toughness significantly increases with the increase of nickel content up to 1%. On further increasing nickel to 2% (ESR3) and 3.2% (ESR4), the tensile strength and yield strength progressively increases and the later reaches a value of 1542 MPa in 3.2 % nickel steel (ESR4). Impact toughness drops sharply. At the same time grain size remains unchanged. The trend of increase in impact toughness from base alloy to 1% nickel steel and the decrease of its values at higher nickel containing alloys are interesting. The optical micrographs of the lightly etched specimens of nickel steels are shown in Figure 3. It reveals that the microstructures of nickel containing steel differ significantly from 1% nickel steel (ESR2) to 3% nickel steel (ESR4). In ESR2 alloy, the microstructures consist of some amount of grain boundary ferrite (GBF) and acicular ferrite (AF) in the martensite matrix. When nickel content is further increased in ESR3 to ESR4 steel, the GBF and AF phases significantly decreases. It is interesting to note that ESR4 steel consists of predominantly lath martensite microstructures. The effect of the GBF phases and the AF phases on the toughness of the steel is discussed later. The SEM micrographs of these steel are shown in Figure 4. It can be noticed from this figure that all the specimens consist of tempered lath martensite. The lath is uniform and seems to be finer in higher nickel alloys. The thermodynamic stability of precipitates of these steel is estimated by CHEMSAGE software as shown in Figure 5. It can be noticed from the figure that at the austinitising temperature of 930°C all the precipitates except the vanadium carbides are dissolved. The fraction of undissolved vanadium is about 60% which has locked equivalent atomic percentage of carbon (considering the precipitate as VC) and forms corresponding vanadium carbides/carbonitrides. The calculated dissolved carbon contents at the austenitising temperature are: 0.15% in the alloy containing 2% nickel steel (ESR3) and similarly 0.19 % in the alloy containing 3.3% nickel steel (ESR4). The slight increase (0.04 wt.%) of dissolved carbon in 3.2% nickel steels might have some effect in strengthening the martensite. It might be one of the reasons for improvement of strength values apart from possible solid solution hardening effect of nickel.

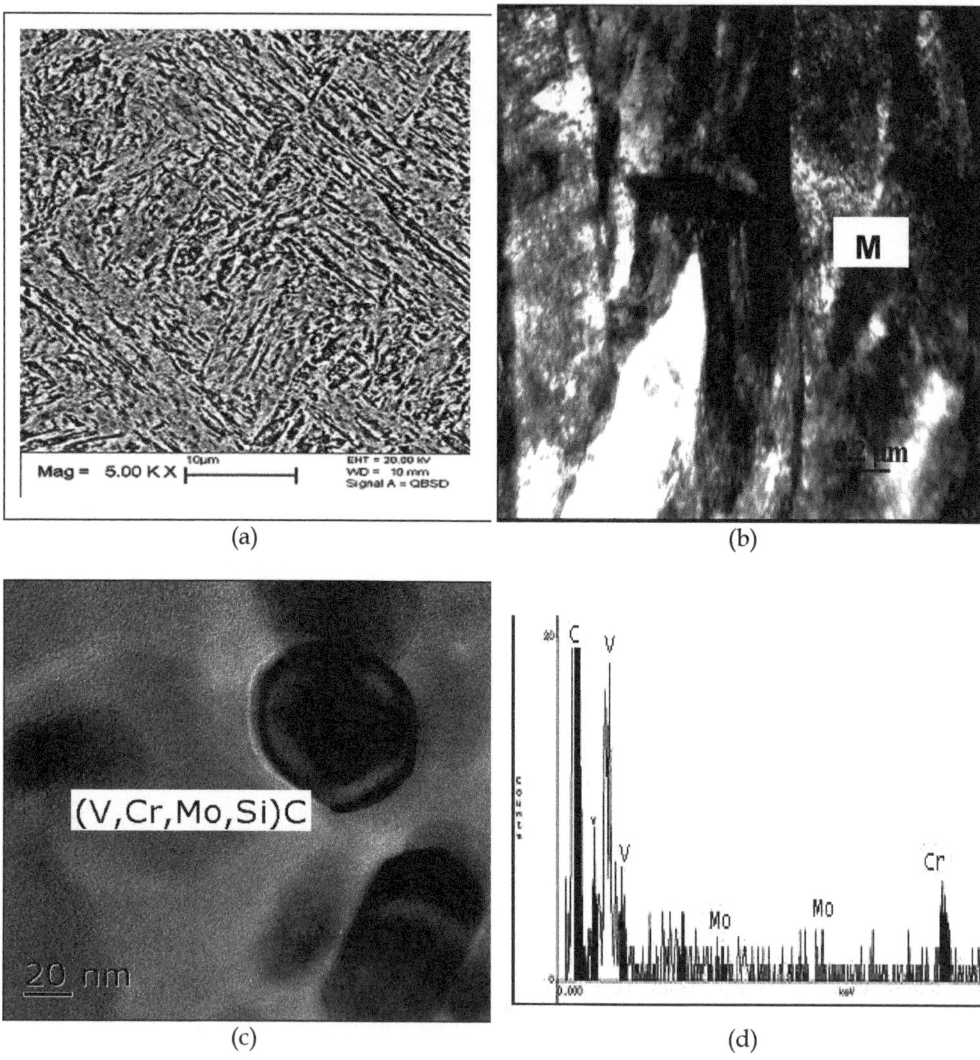

Fig. 2. (a) SEM image, (b) TEM bright field image, (c) TEM-carbon replica micrograph and (d) EDS analysis of the precipitates of as-cast base alloy (ESR1) sample quenched at 975°C in oil and tempered at 475°C.

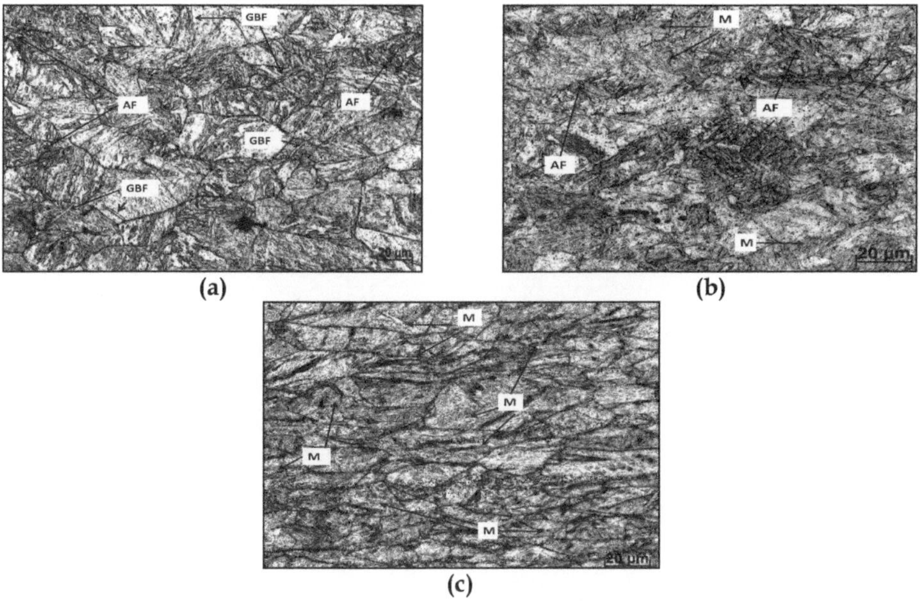

Fig. 3. Optical micrographs of nickel steels, showing the decreasing tendency of formation of acicular ferrite (AF) and grain boundary ferrite (GBF) in as-cast, quenched and tempered specimens of (a) ESR2, (b) ESR3, and (c) ESR4 alloy.

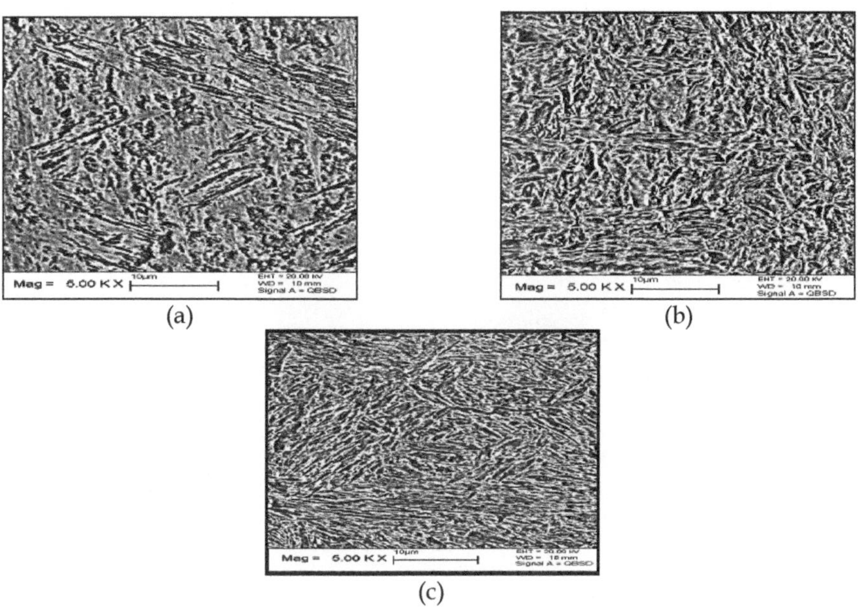

Fig. 4. SEM micrographs of steels, showing the effect of nickel on the fineness of martensite laths in (a) ESR2, (b) ESR3, and (c) ESR4 alloy in as-cast tempered condition.

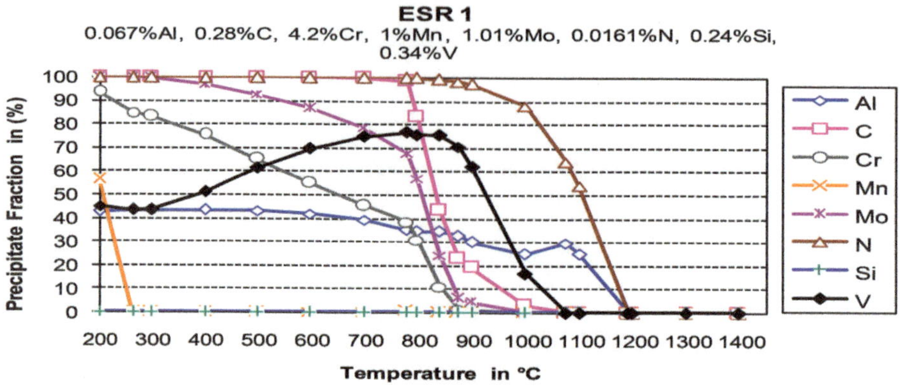

(a)

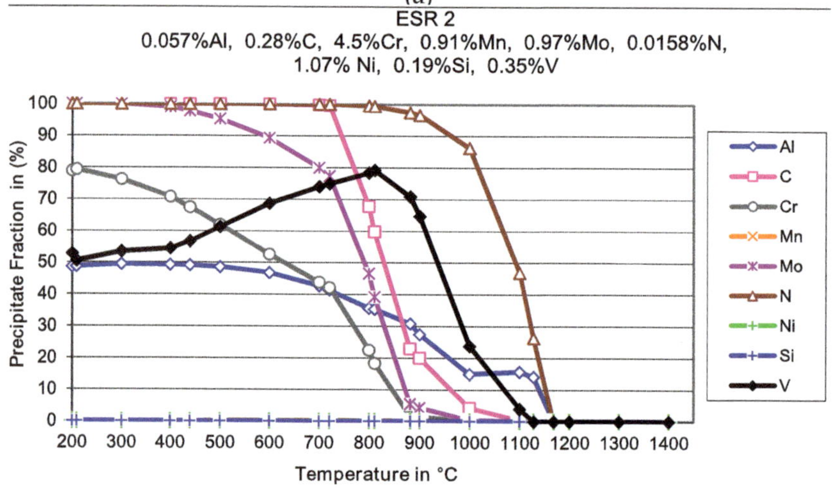

(b)

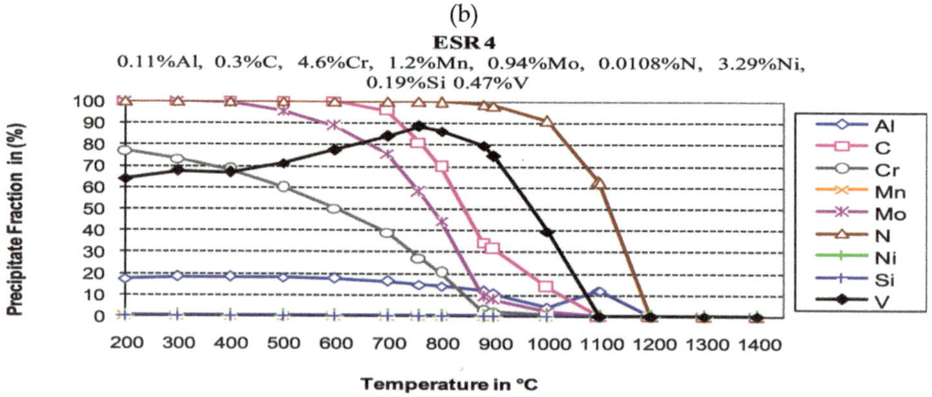

(c)

Fig. 5. Calculation of precipitate stability using CHEMSAGE software for ESR1, ESR2 and ESR4 alloy showing the volume fraction of precipitates.

3.2 Optimisation of processes parameters of TMT

It is possible to obtain optimum combination of strength and toughness by a control process parameters of thermomechanical treatment such as slab reheating temperature, deformation temperature, deformation per pass, cooling rate, etc (Kim et al., 1987). In the present study, it was attempted to optimise some of the process parameters like slab reheating temperature, deformation temperatures and the cooling rate of the cooling medium, etc which are discussed in the following section.

3.2.1 Soaking temperature

The initial stage of any hot rolling process usually consists of a selection of proper soaking temperature. At this temperature, attempt is normally made to dissolve all the carbides or carbonitrides present in the steel, so that these can be re-precipitated at smaller sizes in the later stage of the process. At the same time, too high soaking temperature leads to increase in austenite grain size, which controls the final microstructure. Therefore, it is necessary to select the appropriate soaking temperature at which the optimum results may be achieved. The microalloys form different carbides and carbonitrides, which go into solution at different temperatures, and therefore one needs to know these temperatures. Equilibrium stability of the carbides and carbonitrides in the alloys were calculated using CHEMSAGE software and the result are shown in Figure 5. The calculation is based on the chemical composition of the steel. Calculations were done for temperatures in the range of 200°C to 1400°C and in the intervals of 100°C. It may be noticed from these figures that the precipitates of carbides in ESR1, ESR2, ESR3, ESR4 are almost completely dissolved at around 900-1000°C and nitrides at 1200°C. The soaking temperature of these steels was therefore fixed at 1200°C.

3.2.2 Deformation and deformation temperature

Hot compression tests were performed to get an idea about the required load during hot rolling for a given amount of deformation. The specimen size was identical for all alloys. It was cylindrical in shape with 8 mm diameter and 14.4 mm height. The samples were reheated in a controlled atmosphere in a cast iron mould. The compression tests were performed at 1200°C with a strain rate of 1.0 s^{-1} with 50% total reduction. Result of hot compression test is represented by stress vs. degree of deformation (flow stress curve). The entire test was performed within 10 seconds. Visual observation showed that no major defect occurred in the compressed samples. Figure 6 shows the flow stress curves of ESR1 (base alloy), ESR2 (1% Ni), ESR3 (2 %Ni), and ESR4 (3.2% Ni). Except ESR3 alloy, the curves are similar for all the steels. The gradual increase of stress in all the alloys reflects the work hardening of the austenite. It can be inferred from Figure 6 that the required stresses for 50% hot deformation of the steels for all alloys are in the range 60 and 70 MPa, except in ESR3 (2% Ni) requiring the highest stress (80 MPa). TTT diagram of the base alloy (ESR1) has been predicted and reported using a model based on the chemistry of the metal (Maity et al., 2006). The calculated diagram for ESR1 steel is shown in Figure 7. This figure predicts that AC_1 temperature of this steel is about 825°C and martensite start transformation (Ms) temperature is above 300°C. Fast cooling below Ms temperature, could lead to transformation of martensite. Relatively slower cooling may result in a mixture of bainite and martensite. It was not possible to model the TTT diagram for the nickel containing alloys, as the γ-loop shifted extremely to the right. The diagram provides probable information regarding the beginning and end of transformation into stable and metastable phases. It was planed to roll the material in the two-phase α-γ region between AC_3 and AC_1 temperatures. As the α- phase in the two phase region being softer than the γ-phase in the stable γ-region (Yu et al., 2006), the high strength steels could then be rolled with the existing equipment. Additionally, if the first phase of rolling is done at a relatively high temperature in the two-phase region (above the recrystallisation temperature), one can get dynamic recrystallisation and finer austenite grains. The final pass can be made just above the AC_1 temperature so that recrystallisation can be limited and work hardening effect can be achieved (Kawalla & Lehnert, 2002). These arguments are based on equilibrium temperature. In reality, austenite to ferrite reaction may be sluggish enough throughout the rolling range. Small amount of ferrite may of course forms during rolling due to deformation induced transformation.

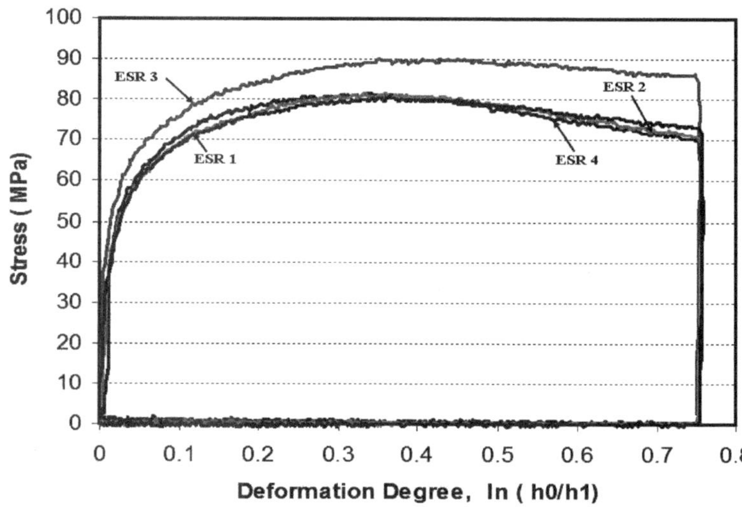

Fig. 6. Result of hot compression tests (50% reduction) on as-cast samples of ESR1, ESR2, ESR3 and ESR4 alloy.

3.2.3 Cooling rate of the medium

The cooling rate of the as-cast alloys was determined experimentally. The as-cast specimens were heated to 1200°C and after soaking at this temperature, the samples were held outside the furnace till it cooled to 850°C, and were then allowed to cool in different coolants. The selected coolants were air, oil, polymer-water mixture (1:1), polymer-water mixture (1:1.5) and the polymer-water mixture (1:2). The progress of cooling of the specimens in these coolants is shown in Figure 8. The figure shows that the rate of cooling is slowest in air, and polymer-water (1: 2) mixture results in the severest cooling. Cooling in oil is faster than the other two polymer-water mixtures down to a temperature of 250°C. The polymer-water (1:2) mixture was not selected for the final experiments, as it was considered too severe and therefore may lead to cracks. Use of the polymer-water (1:1) and (1:1.5) mixtures results in similar cooling profiles in the 300-700°C range. The polymer –water (1:1.5) mixture was used along with air and oil cooling in the final experiments. The average cooling rate for these coolants was estimated and it was 1.3°C.s^{-1} for air, 16°C.s^{-1} for polymer-water (1:1.5) mixture and 28°C.s^{-1} for oil, in the temperature range of 700°C-300°C. At temperatures below 300°C, oil cools slower than the polymer water solution.

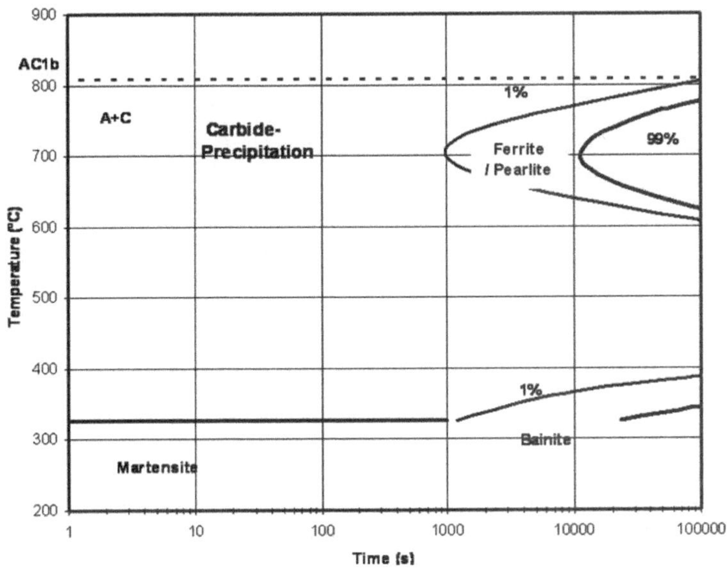

Fig. 7. Modelled TTT diagram of ESR1 (base alloy) showing AC_3 and M_S temperature.

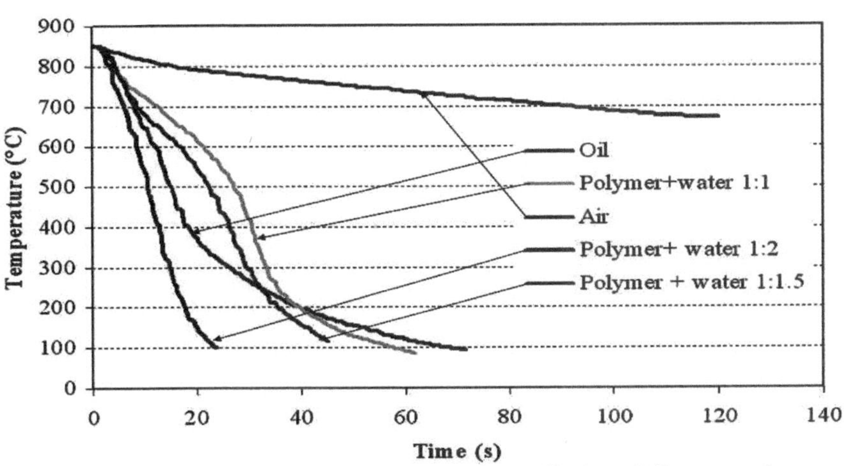

Fig. 8. Estimated average cooling rate of the ESR1 (base alloy) in different coolants.

3.2.4 Modelling of Continuous Cooling Transformation (CCT) diagram

Estimation of different phases was modelled to obtain a relationship of the phases to be appeared in different cooling conditions. The data predicts the transformation of various phases on application of continuous cooling conditions. The model used for this purpose was neural network based and claimed an error band of ± 14K for Ms temperature and ±10% for phase percentages (Ion, 1984; Doktorowski, 2002). Starting temperature for the model calculation has been considered as 900°C. The CCT diagram obtained by this model is shown in Figure 9. It predicts that at the slower cooling rate (less than 2-5K/s) the microstructures consist of a mixture of bainite, martensite and some amount of ferrite. Fast cooling (>10 K/s) on the other hand results in complete transformation to martensite. The results of these models are useful in analysing the results obtained after TMT.

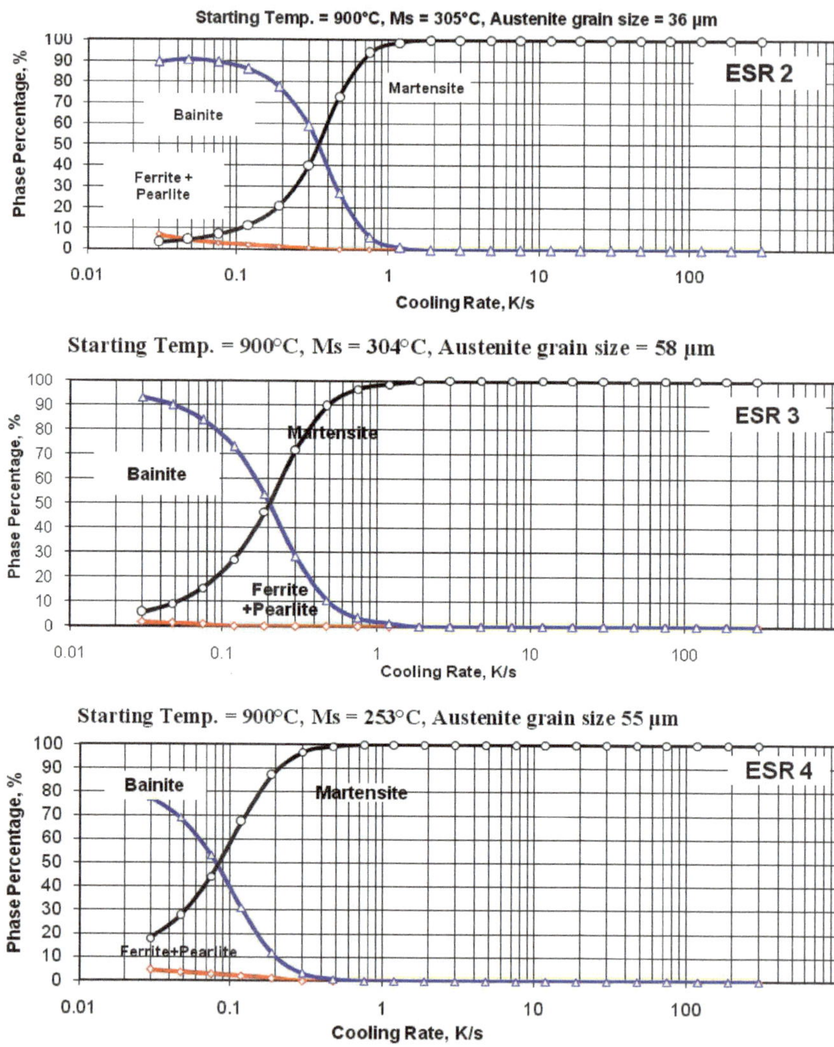

Fig. 9. Modelled CCT diagram predicts the microstructure constituents and Ms temperature for ESR 2, ESR3, and ESR4 alloys.

3.3 Properties of TMT plates

The summary of the observations during the hot rolling experiments is given in Table 8. The rolling stresses for each steel were calculated by the standard method (Zouhar, 1970). The calculated rolling stresses for the different alloys are illustrated in Figure 10. It can be noted that ESR1, base alloy, required the minimum stresses (113 MPa for 1st pass and 254 MPa for final pass). The three nickel containing steels, viz., ESR2, ESR3 and ESR4 required higher

steel	Initial		First pass				Final Pass				Cooling medium
	H_o (mm)	B_o (mm)	H_1 (mm)	B_1 (mm)	Fw_1 [kN]	Av σ_1 (MPa)	H_2 (mm)	B_2 (mm)	Fw_2 [kN]	Av σ_2 (MPa)	
ESR1	21.2	23.1	16.5	26.5	122	113	11.1	30.0	326	254	Air
	21.2	23.1	16.5	26.5	129		11.1	30.0	341		Oil
	21.2	23.1	16.5	26.5	120		11.1	30.1	340		Polymer
ESR2	21.0	22.6	16.5	26.5	131	126	11.1	29.9	327	254	Air
	21.0	22.6	16.5	26.5	135		11.2	30.0	328		Oil
	21.0	22.6	16.5	26.5	140		11.2	30.1	344		Polymer
ESR3	21.4	22.5	16.5	26.5	154	136	11.2	29.0	341	263	Air
	21.4	22.5	16.5	26.5	155		11.2	29.3	340		Oil
	21.4	22.5	16.5	26.5	148		11.2	29.3	324		Polymer
ESR4	21.0	22.3	16.5	26.5	143	141	11.2	29.8	345	267	Air
	21.0	22.3	16.5	26.5	156		11.2	29.9	337		Oil
	21.0	22.3	16.5	26.5	162		11.2	29.8	360		Polymer

Table 8. Experimental data of thermomechanical treatment.
Initial dimension of steel: 22.7 x 22.7 mm, final dimension of steels: plate 11 x 29 mm, temperature: 1st pass: 950°C, final pass: 850°C, ingot soaking temperature 1200C, soaking time: 90 minutes. Fw_1 is load, σ_1 is stress, H_o is initial height and B_o initial width.

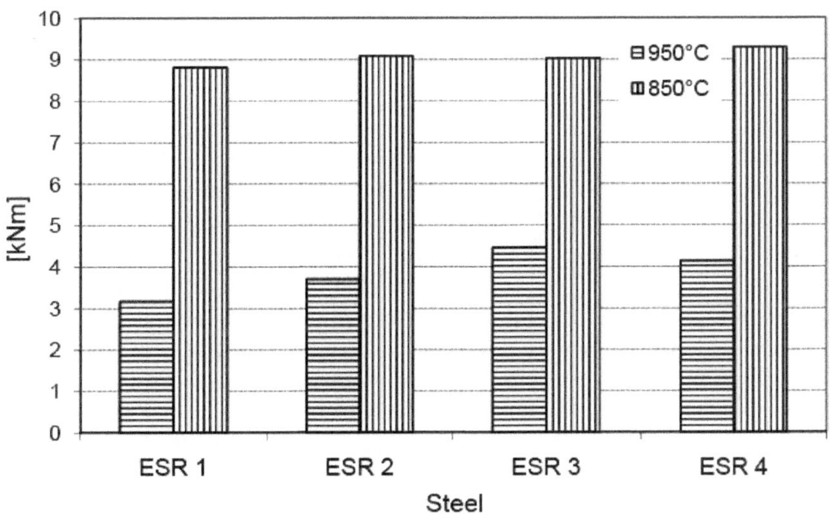

Fig. 10. Rolling stresses for first and final pass during hot rolling experiments.

stresses than that of ESR1. The result also shows that the stress for the final pass is much higher than that for the first pass in all samples. The rolling torque is also shown in Figure 11. The four selected grades of steels underwent hot rolling as mentioned in the experimental section, and were cooled in air, polymer-water mixture and oil after the final rolling. It produced total of 12 plate samples of 11 x 29 mm cross section. Preliminary investigation on the plates showed that no major surface defects like scaling, cracks, bends etc were present on the plates.

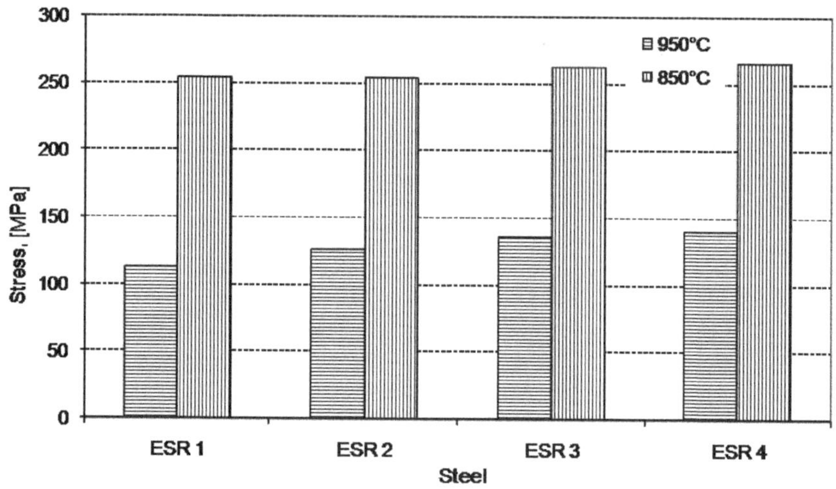

Fig. 11. Rolling torques for first and final pass during hot rolling experiment.

3.3.1 Effect of cooling rate

The tensile strengths, yield strengths and elongations of the hot rolled plates in the three cooling conditions are illustrated in Table 9. At the outset one can notice that in most of the cases the tensile strength and yield strength increase as the severity of cooling increases, best values being obtained with oil-cooled samples. It can also be seen that ductility is marginally improved in the oil-cooled samples. The hardness and impact toughness of the as rolled specimens in the three cooling conditions is shown in Table 10. It can be observed that for all steels, hardness increased as cooling became faster. Air-cooling resulted in the lowest hardness, and the highest hardness was observed in the oil cooled specimens. Among the samples, lowest and highest hardness were measured in ESR1 (base alloy) and ESR3 samples, respectively. Annealing of these samples resulted the decrease in hardness values compared to as rolled condition. It is also seen from table 10 that except of one or two cases, the impact toughness values also increase with increase of cooling rate. Highest impact toughness is observed in oil cooled specimens.

Sample	Air cooled			Polymer-water cooled			Oil cooled		
	UTS (MPa)	Y. S (MPa)	el (%)	UTS (MPa)	Y. S (MPa)	el (%)	UTS (MPa)	Y. S (MPa)	el (%)
ESR 1	1818	1525	8.8	1883	1550	8.1	2030	1615	10.7
ESR 2	1925	1600	9.8	1920	1703	9.5	2062	1721	10.4
ESR 3	1990	1667	8.9	2054	1705	9.6	2214	1750	9.9
ESR 4	1941	1635	9.8	2002	1684	9.3	2181	1715	10.1

UTS: ultimate tensile strength, Y.S: Yield strength, el: Elongation

Table 9. Tensile properties of TMT plates.

Sample	Hot-rolled, air-cooling		Hot-rolled, polymer cooling		Hot-rolled, oil-cooling	
	Hardness (HRc)	Impact toughness ($kJ.m^{-2}$)	Hardness (HRc)	Impact toughness ($kJ.m^{-2}$)	Hardness (HRc)	Impact toughness ($kJ.m^{-2}$)
ESR 1	44.3	391	45.7	421	48.0	516
ESR 2	48.6	629	48.1	655	49.3	742
ESR 3	48.3	496	51.2	467	52.5	564
ESR 4	48.4	439	50.9	546	51.7	516

Table 10. Impact strength and hardness of TMT plates.

It can be noticed that mechanical properties of the thermomechanically treated steels are greatly influenced by the quenching medium as in evident from Table 9 and Table 10. The mechanical properties are improved substantially with increase in cooling rate. After thermomechanical treatment the as-cooled plate displays significant increase in yield strength and toughness in compare to as-cast tempered alloys. The best combination of strength and toughness has been observed in oil cooled specimens of ESR2 steel. The optical metallography of one of the ESR2 alloy in three cooling conditions is given in Figure 12. It can be seen that the structure becomes progressively finer as cooling rate become faster. Figure 12 also reveals that in the slow cooling rate the microstructure consists of many more phases. There may be some lath martensites along with austenite and bainite in the matrix. Whereas, oil cooled plates consists of predominantly finer lath martensite structures. The SEM micrographs of ESR2 alloy are also shown in Figure 13. It can be seen that the microstructures of the specimens consist of lath martensites and more uniformity and homogeneity is observed in the specimens those are cooled in faster rate. Apparently it is also seen that the microstructures in oil cooled samples predominantly consist of finer lath martenisites. The TEM micrographs of ESR2 sample in air cooled and oil cooled samples are shown in Figure 14. The TEM micrograph reveals that air cooled sample consist of lath martensite, bainite and some retained austenites. In oil cooled sample the microstructure are mainly consist of lath martensites. The martesite interlath spacing in oil cooled is observed about 200-300 nm whereas, it is 300-400 nm in the air cooled sample. It can be noticed from Figure 15 that the specimens cooled at slower cooling rates showed segregation of carbon, which indicates the presence of retained austenite and bainite (Maity et al., 2008). It is also inline with the predicted phase transformation information as shown in Figure 9. According to CCT diagrams shown in Figure 9, all investigated alloys had enough hardenability to get full martensitic microstructure in cross-section of tested samples after oil quenching (cooling rate normally greater than 15K/s) and mixed microstructures in air cooing (cooling rate less than 1.5 K/s).

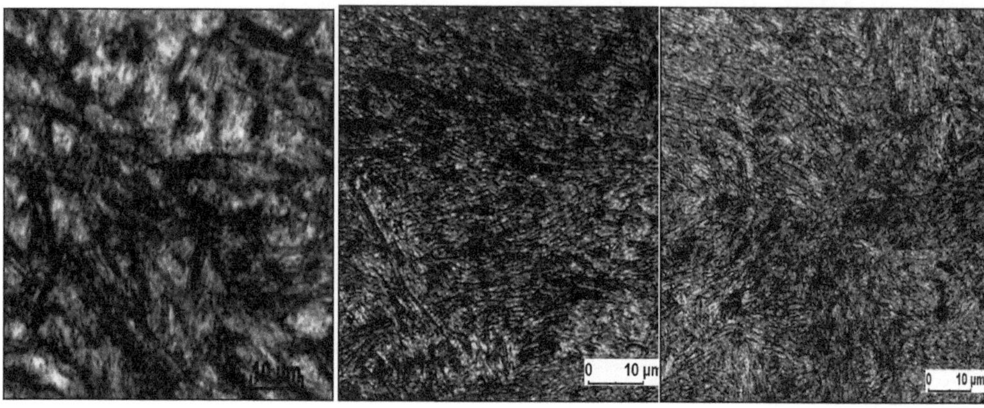

(a) Air cooling (b)Polymer+water cooling (c) Oil cooling

Fig. 12. Optical Micrographs of the TMT plates of ESR2 specimens cooled in different cooling medium.

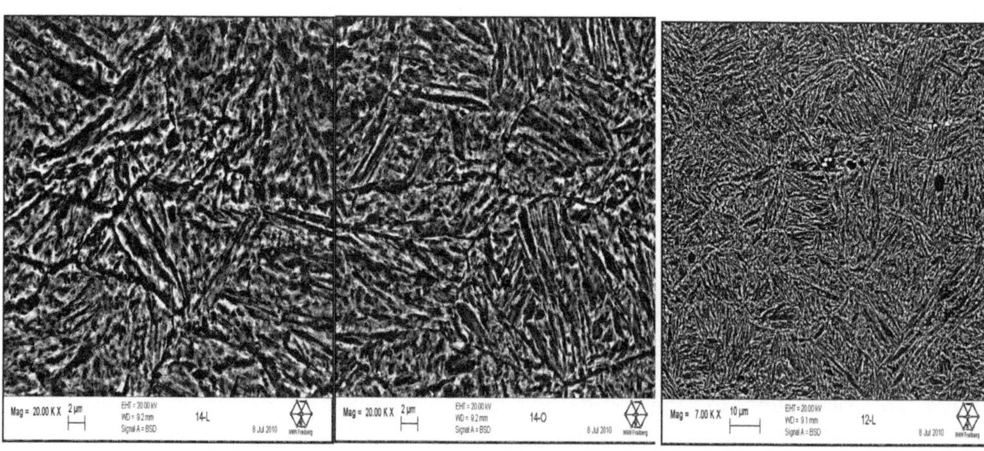

(a) Air cooled (b)Polymer - water cooled (c) Oil cooled

Fig. 13. SEM Micrographs of the TMT plates of ESR2 alloy cooled in different medium.

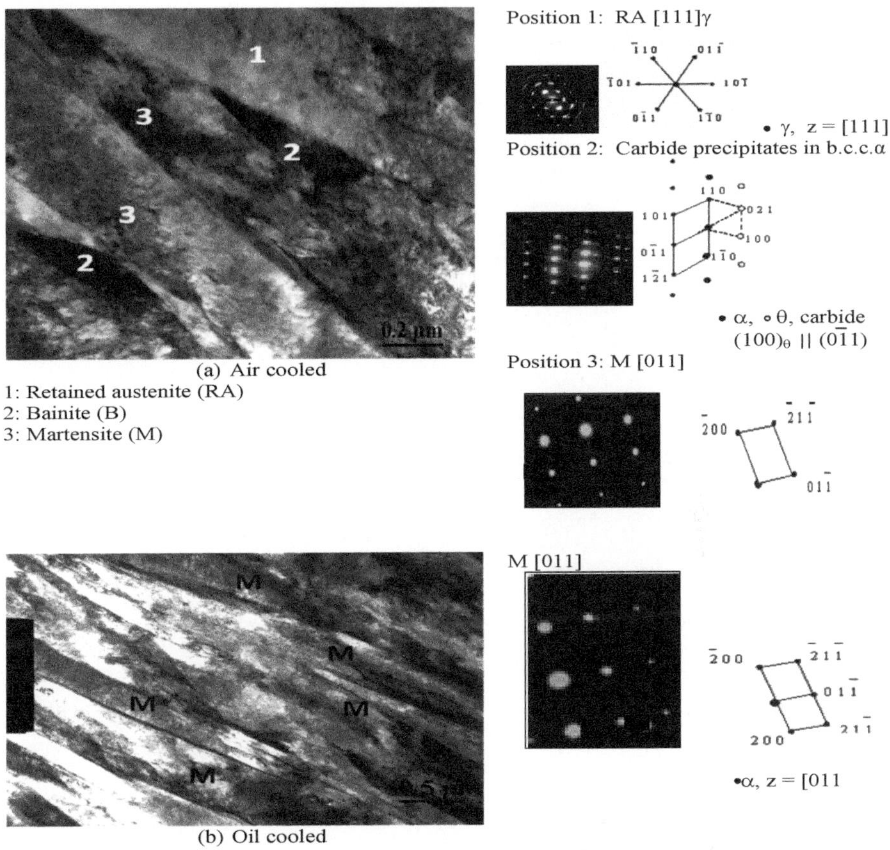

Fig. 14. TEM micrographs and diffraction pattern of TMT plates of ESR2 specimens cooled in air and oil showing:(a) the presence of martensite (M), retained austenite (RA) and bainite (B) in air cooled sample, and (b) predominantly martensite (M) in oil cooled specimens.

Evidences for phase identification are collected through EPMA and TEM studies. If during transformation, the temperature is high enough, carbon gets enough time to diffuse ahead of the transformation front. Higher carbon regions should be found at the boundaries of pockets of laths and retained austenite or in between upper bainite laths. Samples cooled in different quenching medium (air, oil and polymer) were subjected to EPMA analysis to reveal the segregation patterns, the results of which are presented in Figure 15 (Maity et al., 2008). One can clearly see that segregation of carbon decreases as the severity of quench increases. In the air-cooled sample, one can see peaks in carbon content nearly at regular intervals of about 15-25 µm. This may be due to retained austenite at the boundaries of packets of laths. The individual laths being less than a micron wide, inter lath segregation cannot be resolved in EPMA. In the specimen cooled at the intermediate quench rate (polymer-water 1:1.5 mixture), the extent of segregation is less indicating carbon had less time to diffuse. The interval between the peaks is also slightly less, indicating the size of packets of laths are smaller. This is in tune with the optical/SEM micrographs. The oil-cooled samples show very little long range segregation. Here the severity of quench has been high enough, and carbon could not diffuse out and austenite could not be retained. The improvement of mechanical properties in oil cooled specimens possibly due to the change of the morphology of the microstructural changes due to the change of cooling rate.

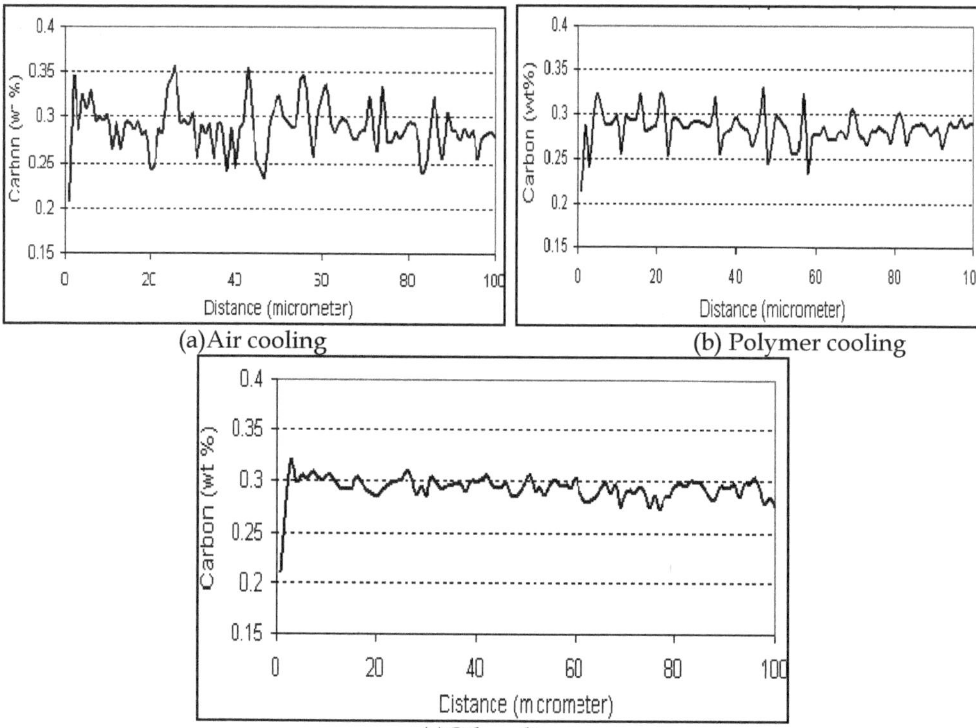

(a)Air cooling (b) Polymer cooling

(c)Oil cooling

Fig. 15. Electron probe microanalysis of the distribution of carbon in the central zone of the hot rolled steel under different cooling conditions.

3.3.2 Effect of nickel and other alloying elements

As discussed, in ESR2, ESR3 and ESR4 steel deliberately 1% to 3% nickel are added to the base composition of ESR1 alloy. It can be noticed from Table 9 and Table 10 that with increase of nickel content in TMT plates in three different cooling conditions, the tensile strength, and yield strength are progressively increased up to 0% to 2% with increase of nickel content. In 3% nickel steel the tensile properties are in reverse in trend. Highest tensile strength of 2214 MPa and yield strength 1750 MPa were obtained with 2% nickel in ESR3 steel. Other steels have also displayed tensile strength values of about 2000 MPa in oil cooled plates. As these steel has ductility values varies from 8-10%, so the change of elongation is not so prominent. The room temperature impact toughness of the rolled samples are shown in Table 10. It is interesting to see that the impact toughness in the most of the cases increases from 0% to 1% nickel steel and further increase of nickel content reduces the impact toughness. ESR1 (base alloy) displayed the lowest impact toughness and

ESR2 with 1% nickel gave the highest. All nickel containing steels showed higher impact toughness compared to the base alloy. This was the trend in the as-cast tempered steels too. Lower additions (up to 1% Ni) could give better toughness without sacrificing yield strength. In the alloys, all nickel containing as-cooled plate results better combination of tensile properties and toughness compare to base alloy. In the nickel alloys, one can also notice that the best combination of yield strength and toughness are obtained in the alloy containing 1% nickel (ESR2). Higher nickel contents had improved the yield strength but results comparatively lower impact toughness.

Generally nickel enlarges the γ phase region in Fe-C phase diagram, therefore it enables lower austenitizing temperature of steel, which can promote refinement of structure. Decrease in the martensite packet diameter, similar to the decrease of the grain size, improves the strength as well as the toughness of steel (Tomita & Okabayashi, 1986). Nickel can also influence increasing the stability of retained austenite (Rao & Thomas, 1980; Sarikaya et al., 1983) and the morphology of cementite precipitation at tempering (Peters, 1989). It is indeed happened in case of nickel steels. The SEM micrographs as shown Figure 4 reveal that the laths in martensite matrix are progressively finer with the increase of nickel content. Most of the cases, nickel increases toughness, but it is effective when its amount is controlled in the steel containing 1% Mn. Nickel increases the resistance to cleavage fracture of iron and decrease a ductile-to-brittle transition temperature (Bhole et al., 2006). It is also reported that increase of the nickel content, the grain boundary ferrite (GBF) and acicular ferrite (AF) decreases and as a result of the reduction of both AF and GBF, the impact toughness decreases (Bhole, 2006). It is also reported that when in C-Mn steel containing 1.4% Mn, the toughness drops if nickel content exceeds 2.25%. Kim et al. found that the combined presence of Ni and Mo decreases the volume fraction of GBF (Kim et al., 2000). This may be due to the improved wettability of the Ni as binder on the carbide phase due to the addition of Mo. Improved wettability results the decrease in micro-structural defects and an increase in the interphase bond strength and phase uniformity. The increase in nickel results in the reduction of impact toughness. It may be due to the significant reduction of the volume fraction of acicular ferrite or grain boundary ferrite. The optical micrograph (Figure 3) reveals the presence of substantial amount of acicular ferrite in ESR2 steel and trace amount in ESR3, but this phase could not be identified in ESR4 alloy. This may be one of the reason for the increase of impact toughness in ESR2 containing 1% nickel. It suggests that at the content of about 1% of nickel will have significant influence on notch toughness in these types of steels.

Nickel being an austenite stabilizer leads to retained austenite on one hand, and on the other hand it increases toughness, especially when the nickel content is low at about 1%. Nickel leads to grain refinement and improve toughness when it is used in optimum amount. As a result, all the alloys containing nickel showed high impact toughness after TMT and the one with 1% nickel shows a best combination of strength and toughness. On the other hand, hot rolling at temperatures just above AC_1, has been shown to be feasible and effective method to roll such high strength steel. It is also possible that ESR can be used effectively to reduce the major casting defects and can control the macro- and micro-segregation.

3.4 General discussion

The objective of the present work rose out of the requirement of developing an ultra high strength steel with a yield strength in excess of 1650 MPa, with a minimum elongation of 9-10%. This material is being developed primarily for application in the area of pressure vessels in aerospace vehicles. In such high strength alloys one needs to employ all modes of strengthening. There are heat treatable alloys where strength is obtained from finer martensites with additional precipitation hardening. The approach in the present work was to adjust the chemistry and the production process to obtain a optimum morphology in the microstructure in the as-cast steels. Further improvement was carried out by a optimised thermomechanical treatment with controlled cooling. These two aspects formed two parts of this work.

The primary alloying elements in this 0.3%C steel are chromium, molybdenum and vanadium, which are all carbide/carbonitride formers. At temperatures below about 500°C almost all carbon is in various precipitates at equilibrium. To obtain optimum properties one needs to balance the precipitation process between high and low temperatures. Precipitates at soaking temperatures are needed to limit austenite grain growth and modify the deformation processes. Management of precipitate size is extremely important here. Precipitation at lower temperatures, especially of carbides of chromium and molybdenum, can be coherent/semi-coherent and leads to large strength development during cooling and tempering. The alloys could only be developed because of ESR processing. Normally, most of the strengthening mechanisms lead to loss in ductility. The ability to ensure removal of all large and medium sized inclusions from near directional solidification under a high temperature gradient from a small liquid metal pool during the ESR process increases ductility, toughness and workability. Most of the defects like micro-and macro-segregations, micro porosities and looseness associated with solidification are nearly absent in ESR processed materials. Nickel containing alloys showed finer grain sizes compare to the basic steel. Addition of 1%Ni gave lower yield strength in combination with very high impact toughness. Some improvement in strength was indeed obtained at higher nickel contents. One reason for this behaviour may be the retention of austenite promoted by nickel. Softer austenite distributed in small amounts interferes the crack propagation and improves the impact toughness but decreases the strength at 1%Ni. Solid solution strengthening probably becomes important at higher percentages, more than compensating for loss due to larger proportion of retained austenite. These are the issues which need further exploration.

The thermomechanical treatment adopted, wherein the samples are rolled in the two phase region finishing the deformation just above AC_1, seems to have improved the properties enormously. This strategy permitted rolling to be done with the existing equipment, and to retain some work hardening effect to increase the strength. Controlled cooling allows one to optimise the final microstructure. It has been demonstrated that it is possible to obtain the optimum combination of strength and toughness by an appropriate control of processing parameters such as reheat temperature, deformation temperature, deformation per pass, cooling rate, etc. Cooling rate has large influence on the properties. Air-cooling generally gave lower strengths and oil cooling the highest. Interestingly oil-cooling also gave higher elongation, indicating the effect of auto-tempering. The microstructure in case of oil cooling seems to largely consist of finer lath martensite. At air cooling, there were clear evidences of retained austenite, bainite and martensite. It was also noticed that strength values increase with the increase in cooling rate and the highest yield strength were obtained in oil-cooled samples. Steels for aerospace and aircraft applications, need to possess ultrahigh strength coupled with high toughness to ensure high reliability. The ingots produced in this study are smaller size, however it should be brought to a practice of production of relevant level.

4. Conclusions

1. ESR processed ingots has low inclusion content and good microscopic homogeneity.
2. The base alloy consists of predominantly lath martensite microstructure, having lath sizes in the range of 550-700 nm. It contains complex carbonitrides precipitates of vanadium, chromium and molybdenum, of 25-70 nm size. The alloy displays a yield strength of about 1400 MPa, elongation of 11% and impact strength of 300 kJ/m^2.
3. The addition of 1 to 3 % nickel to the base alloy improves most of the mechanical properties. The yield strength of 1% nickel alloy is around 1500 MPa. The alloy containing 3% nickel results a yield strength value of 1542 MPa.
4. The process parameters for thermomechanical treatment were optimised based on model calculations and preliminary experiments. The treatment involved pre-rolling at 1200°C, followed by soaking at 1200°C and rolling in two passes starting from 950°C and 850°C respectively.
5. The thermomechanical treatment applied in the two phase region and finishing at just above AC_1, seems to improve the mechanical properties enormously. This strategy permits to roll this high strength steel with the existing equipment, and also helps to retain work hardening to obtain yield strength in excess of 1700 MPa in some alloys.
6. After thermomechnical treatment all the four alloys showed UTS values in the range of 1800-2200 MPa and yield strength in excess of 1700 MPa.
7. The increase of nickel content up to 1% results in increase of toughness in both as-cast tempered alloys and TMT plates. However, further increase of nickel did not beneficial in this composition of alloys. The best combination of tensile strength, yield strength, elongation and toughness are observed in 1% nickel alloy and may be the optimum composition in all alloys.
8. It can be noticed that cooling rate has large influence on the microstructure and thereby on the mechanical properties of the sample of thermomechanical treatment. It is found that the air cooled sample consists of martensite, bainite and retained austenite. The oil cooled sample consists of predominantly finer lath martensite. The air cooled sample results in low strengths compare to oil cooled plate.

5. Acknowledgement

The author wishes to thank the Director, CSIR-National Metallurgical Laboratory (NML), Jamshedpur, India. The authors are also thankful to DAAD and CSIR for facilitating the research work in TU Bergademie Freiberg, Germany. The authors are also thankful to the staffs of ferrous metallurgy of IIT Bombay and Dr. Klemn of Institute of Metal Forming of TU Freiberg for help during experimentation and for many useful discussions. The authors are also grateful to M. Chandra Shekhar, Manoj Gunjan, Dharambeer Singh and Anil Rajak.

6. References

Akhlaghi, S. & Yue, S. (2001). Effect of Thermomechanical Processing on the Hot Ductility of a Nb–Ti Microalloyed Steel. *The iron and Steel Institute of Japan International*, Vol.41, pp.1350-1356

Arsenault, R.J. (1967). The Double-Kink Model for Low-Temperature Deformation of B.C.C. Metals and Solid Solutions. *Acta Metllurgica, Vol.*15, pp.501-501

Bhole, S. D.; Nemade, J. B; Collins, L. & Liu, Cheng.(2006). Effect of Nickel and Molybdenum Additions on Weld Metal Toughness in a Submerged Arc Welded HSLA Line- Pipe Steel. *Journal of Material Processing Technology*, Vol.173, pp.92-100

Bleck, W.; Müschenborn,W. & Meyer, L. (1988). Recrystallisation and Mechanical Properties of Micro Alloyed Cold - Rolled Steel. *Steel Research*, Vol.59, pp.344-351

Chatterjee, M.; Balasubramanian, M. S.; Gupt, K. M. & Rao, P. K. (1990). Inoculation during Electroslag Remelting of 15CDV6 Steel. *Ironmaking Steelmaking*, Vol.17, pp.38-42

Choudhary, M. & Szekely, Z. (1981). Modelling of Fluid Flow and Heat Transfer in Industrial- Scale ESR System. *Ironmaking Steelmaking*, Vol.8, pp.225-232

Dhua, S. K.; Mukherjee, D. & Sarma, D. S. (2003). Influence of Thermomechanical Treatments on the Microstructure and Mechanical Properties of HSLA-100 Steel Plates. *Metallurgical and Material Transaction A*, Vol.34A, pp.241-253

Doktorowski. (2002). *Freiberger Forschungshefte*, Reihe B, Vol.319, pp.1-10

Floreen, S. (1978). Maraging steels, In: *Metal Handbook* (Vol.1, Ninth Edition), American Society for Metals,pp.445-452, ISBN 0-87170-377-7 (v.1), Ohio

Gladman, T.; Dulieu, D. & Mcivor, I. D. (1975). *Proceeding of Microalloying 75*, pp.25-25, Washington, 1975

Grange, R.A.; Hribal, C.R. & Porter, L.F. (1977). Hardness of Tempered Martensite in Carbon and Low Alloy Steels. *Metallurgical Transactions, Vol.8A*, pp.1775-1785

Ion, J. C.; Easterling, K. E. & Ashby, M. F.(1984). A Second Report on Diagrams of Microstructure and Hardness for Heat-Affected Zones in Welds. *Acta Metallurgica*, Vol.32, pp.1949-1955

Jahazi, M. & Egbali, B.(2000). The Influence of Hot Rolling Parameters on the Microstructure and Mechanical Properties of an Ultra-High Strength Steel. *Journal of Material Processing Technology*, Vol.103, pp.276-279

Jolley, W. (1968). Effect of Mn and Ni on Impact Properties of Fe and Fe-C Alloys. *Journal of Iron Steel Institute, Vol.*206, pp.170-173

Kawalla, R. & Lehnert, W. (2002). Hot Rolling in Ferrite Region. *Scandinavian Journal of Metallurgy*, Vol. 31, pp.281-287

Kern, A.; Degenkolbe, J.; Müsgen, B. & Schriever, U. (1992). Computer Modelling for the Prediction of Microstructure Development and Mechanical Properties of HSLA Steel Plates. *The iron and Steel Institute of Japan International*, Vol.32, pp.387-394

Kim, I.S.; Reichel, U. & Dahl, W. (1987). Effect of Bainite on the Mechanical Properties of Duel-Phase steels. *Steel Research*, Vol. 58, pp. 186-190

Kim, S.; Im, Y.R.; Lee, S.; Lee, H.C.; Oh, Y.J. & Hong, J.H. (2000). *Journal of the Korean Institute of Metals and Material*, Vol.38, pp.771-778

Maity, S. K.; Ballal, N. B. & Kawalla, R. (2006). Development of Ultrahigh Strength Steel by Electroslag Refining: Effect of Inoculation of Titanium on the Microstructures and Mechanical Properties. *The iron and Steel Institute of Japan International*, Vol.46, pp.1361-1370

Maity, S. K.; Ballal, N. B.; Goldhahn, G. & Kawalla, R. (2008a). Development of Low Alloy Titanium and Niobium Micro Alloyed Ultra High Strength Steel through Electroslag Refining. *Ironmaking Steelmaking*, Vol.35, pp.379-386

Maity, S. K.; Ballal, N. B.; Goldhahn, G. & Kawalla, R. (2008b). Development of Low Alloy Ultrahigh Strength Steel. *Ironmaking Steelmaking*, Vol.35, pp.228-240

Maity, S. K.; N. B. Ballal.; Goldhahn, G. & Kawalla, R. (2009). Development of Ultrahigh Strength Low Alloy Steel through Electroslag Refining Process. *The iron and Steel Institute of Japan International*, Vol.49, pp. 902-910

Malakondaiah, G.; Srinivas, M. & Rama-Rao, P. (1997). Ultrahigh-Strength Low-Alloy Steels with Enhanced Fracture Toughness. *Progress in Material Science, Vol.*42, pp. 209-242

Norstr¨om, L.-Å. & Vingsbo, O. (1979). Influence of Nickel on Toughness and Ductile-Brittle Transition in Low-Carbon Martensite Steels. *Metal Science, Vol.*13, pp.677-684

Peters, J. A.; Bee, J. V.; Kolk, B. & Garrett, G. G. (1989). On the Mechanisms of Tempered Martensite Embrittlement. *Acta Metallurgica,* Vol.37, pp.675-686

Phaniraj, M. P.; Behera, B. B. & Lahiri A. K. (2005). Thermo-Mechanical Modeling of two Phase Rolling and Microstructure Evolution in the Hot Strip Mill: Part I. Prediction of Rolling Loads and Finish Rolling Temperature. *Journal of Material Processing Technology*, Vol.170, pp.323-335

Philip,T.V.& McCaffy,T.J.(1990). Properties and Selection: Iron, Steels and High Performance Alloys, In: *Metals Handbook* (Vol.1, Tenth Edition), ASM International, pp. 431-448, ISBN 0-87170-377-7 (v.1), Ohio

Philip,T,V. (1978). Ultra High Strength Steel, Properties and Selection; Iron and Steels, In: *Metal Handbook* (Vol.1, Ninth Edition), American Society for Metals, pp. 421-443, ISBN 0-87170-007-7, Ohio

Philip, T. V. (1975). ESR: A means of Improving Transverse Mechanical Properties in Tool and Die Steel, In : *Metals Technology*, pp. 554-555

Rao, B. V. Narasimha & Thomas, G. (1980). Structure – Property Relations and Design of Fe-4Cr- C Base Structural Steel for High Strength and Toughness. *Metallurgical Transactions A*, Vol.11A, pp.441-457

Sarikaya, M.; Jhingan, A.K. & Thomas, G. (1983). Retained Austenite and Tempered Martensite Embrittlement in Medium Carbon Steel. *Metallurgical Transactions A,* Vol.14A, pp.1121-1131

Sellars, C.M.(1985). Proceeding on HSLA steels: Metallurgy and Applications, Ed. by J. M. Gray, ASM, Beijing

Shash, Y. M.; Gammal, T. E.; Salamoni, M. A. E. & Denkhaus, F. A.(1988). Improving Solidification Pattern of ESR Ingots Combined with Energy Savings, *Steel Research,* Vol.59, pp.269-274

Suresh, M. R.; Samajdar, I.; Ingle, A.; Ballal, N. B.; Rao, P. K. & Sinha, P. P. (2003). Structure-Property Changes during Hardening and Tempering of New Ultra High Strength Medium Carbon Low Alloy Steel. *Ironmaking Steelmaking,* Vol.30, pp.379-384

Tanaka, T. (1981). Controlled Rolling of Steel Plate and Strip. *International Metal Review,* Vol.26, pp.185-212

Tomita, Y. & Okabayashi, K. (1986). Effect of Micro Structure on Strength and Toughness of Heat-Treated Low Alloy Structural Steel. *Metallurgical Transactions,* Vol.17A, pp.1203-1209

Umemoto, M.; Guo, Z. H. & Tamura, I. (1987). Effect of Cooling Rate on Grain Size of Ferrite in a Carbon Steel. *Material Sciences and Technology, Vol.*3, pp.249-255

Yu, H.; Kang, Y.; Zhao, Z.; Wang, X. & Chen, L. (2006). Microstructural Characteristics and Texture of Hot Strip Low Carbon Steel Produced by Flexible Thin Slab Rolling with Warm Rolling Technology. *Material Characterisation*, Vol.56, pp.158-164

Zouhar, G. (1970). Grundlagen der Bildsamen Formunng, Lehrbrief No.2, TU Bergakademie Freiberg, Fernstudium, pp.1-7

11

Unsteady Heat Conduction Phenomena in Internal Combustion Engine Chamber and Exhaust Manifold Surfaces

G.C. Mavropoulos
Internal Combustion Engines Laboratory
Thermal Engineering Department, School of Mechanical Engineering
National Technical University of Athens (NTUA)
Greece

1. Introduction

Heat transfer to the combustion chamber walls of internal combustion engines is recognized as one of the most important factors having a great influence both in engine design and operation (Annand, 1963; Assanis & Heywood, 1986; Heywood, 1988; Rakopoulos et al., 2004). Research efforts concerning conduction heat transfer in reciprocating internal combustion engines are aiming, among other, to the investigation of thermal loading at critical combustion chamber components (Keribar & Morel, 1987; Rakopoulos & Mavropoulos, 1996) with the target to improve their structural integrity and increase their factor of safety against fatigue phenomena. The application of ceramic materials in low heat rejection (LHR) engines (Rakopoulos & Mavropoulos, 1999) is also among the large amount of examples where engine conduction heat transfer is a dominant factor. At the same time, special engine cases like the air-cooled (Perez-Blanco, 2004; Wu et al., 2008) or HCCI ones demand a special treatment for a successful description of the heat transfer phenomena involved.

Today, technology changes in the field of the internal combustion engines (mainly the diesel ones) are happening extremely fast. New demands are added towards the areas of controlled ignition of new and alternative fuels (Demuynck et al., 2009), reduction of tailpipe emissions (Rakopoulos & Hountalas, 1998) and improved engine construction that would ensure operation under extreme combustion chamber pressures (well above 200 bar). However, application of these revolutionary technologies creates several functional and construction problems and engine heat transfer is holding a significant share among them. Engine heat transfer phenomena are unsteady (transient), three-dimensional, and subject to rapid swings in cylinder gas pressure and temperatures (Mavropoulos et al., 2008), while the combustion chamber itself with its moving boundaries adds further to this complexity. In modern downsized diesel engines, the extreme combustion pressure and temperature values combined with increased speed values lead to increased amplitude of temperature oscillations and thus to enormous thermal loading of chamber surfaces (Rakopoulos et al., 1998). At the same time, transient engine operation (changes of speed and/or load) imposes a significant additional influence to the system heat transfer, which cannot (and should not) be neglected during the engine design stage (Mavropoulos, 2011). It is obvious that there is an urgent demand for simple and effective solutions that would allow the new technologies to enter marketplace as quick as possible. Likely, there are also available today several important tools (both theoretical and experimental) that help a lot the researchers towards solution of the above described problems.

Phenomena related to unsteady internal combustion engine heat transfer belong in two main categories:

- Short-term response ones, which are caused by the fluctuations of gas pressure and temperature during an engine cycle. As a result, temperature and heat flux oscillations are caused in the surface layers of combustion chamber in the frequency of the engine operating cycle (thus having a time period in the order of milliseconds). These are otherwise called cyclic engine heat transfer phenomena.
- Long-term response ones, resulting from the large time scale (in the order of seconds), non-periodic variations of engine speed and/or load. As a result, thermal phenomena of this category occur only during the transient engine operation. On the other hand, the short-term response phenomena are present under both engine operating modes.

Both the above categories of engine transient heat transfer phenomena have been investigated by the present author (Mavropoulos et al., 2009; Mavropoulos, 2011) and also other research groups. In (Keribar & Morel, 1987) the authors have studied the development of long-term temperature variations after a load or speed change. They used a convective heat transfer submodel based on in-cylinder flow accounting for swirl, squish, and turbulence, and a radiation heat transfer submodel based on soot formation.

Despite the large amount of existing studies with reference to the two categories of engine heat transfer phenomena, there is limited information (a few papers only) which examine both long- and short-term response categories together at the same time. In (Lin & Foster, 1989) the authors have reported experimental results concerning cycle resolved cylinder pressure, surface temperature and heat flux for a diesel engine during a step load change. They have also developed an analysis model to calculate heat flux during transient. In (Wang & Stone, 2008) the authors have studied the engine combustion, instantaneous heat transfer and exhaust emissions during the warm-up stage of a spark ignition engine. An one-dimensional model has been used to simulate the engine heat transfer during the warm-up stage. They have reported an increase in the measured peak heat flux as the combustion chamber wall temperature rises during warm-up.

However, several important issues still today remain under investigation. For example, and despite the significant progress made in this area during the last years (Mavropoulos et al., 2008, 2009; Mavropoulos, 2011) the interaction between long-term non-periodic variation of combustion chamber temperature caused during the transient engine operation and the short-term cyclic fluctuations of surface temperatures and heat fluxes needs to be further elucidated. It is of utmost importance to describe in detail, among other issues, the effect of this interaction on peak pressure, on the amplitude and phase change of temperature and heat flux oscillations etc. The answers to these questions would also reveal in what extend the transient heat transfer phenomena should be accounted for during the early design stage of an engine. In addition it needs to be clarified if the specific characteristics (time period, percentage of load and speed change) of any engine transient event influence the mechanism and characteristics of unsteady heat conduction in combustion chamber walls. An attempt to provide some insight to the above important phenomena would be performed by the author among other issues, in the present paper.

A special part of engine heat transfer studies concerns the gas exchange system. The phenomena of transient heat transfer in the intake and exhaust engine manifolds are of special difficulty due to the complex dynamic nature of gas flow inside both of them. Many interesting developments have been recently reported in (Sammut & Alkidas, 2007). An attempt to explore the combination of both short- and long-term unsteady conduction heat transfer effects in the exhaust manifold would also be among the subjects of the present paper.

The present author is participating as the main researcher in a general research program aiming to the investigation of heat transfer phenomena as they are developed in the Internal Combustion Engine chamber and exhaust manifold surfaces. This program was initiated more than fifteen years ago in the Internal Combustion Engine Laboratory (ICEL) of NTUA and is still today under progress. Within this framework, he has already reported detailed structural and thermodynamic analysis models (Rakopoulos & Mavropoulos, 1996), which are also capable to take into account the heat transfer behaviour of the insulated engine (Rakopoulos & Mavropoulos, 1998). He has also reported in detail the short-term variation of instantaneous diesel engine heat flux and temperature during an engine cycle (Rakopoulos & Mavropoulos, 2000, 2008, 2009). Among the most significant accomplishments of this investigation is the detailed description of the different phases of unsteady heat transfer and their accompanied phenomena as they are developed in the combustion chamber and exhaust manifold surfaces during an engine transient event. In addition, a prototype experimental measuring installation has been developed, specially configured for the investigation of the complex engine heat transfer phenomena. Using this installation they have been obtained experimental data during transient engine operation simultaneously for long- and short-term heat transfer variables' responses as they are developed in the internal surfaces of the combustion chamber. Similar experimental data have been obtained by the author and were presented for the first time in the relevant literature also from the exhaust manifold surfaces during transient engine operation (Mavropoulos et al., 2009; Mavropoulos, 2011).

In the present book chapter an overview is provided concerning several of the most important findings of engine heat transfer research as it was realized during a series of years in ICEL Laboratory of NTUA. It is especially examined the influence of transient engine operation (change of speed and/or load) on the short-term response cyclic oscillations as they are developed in the surface layers of combustion chamber and exhaust manifold. Among the factors influencing heat transfer, in the present investigation the effect of severity of the transient event as well as issues related with local heat transfer distribution would be considered. It is clearly displayed and quantified the significant influence of a transient event of speed and load change on engine cyclic temperatures and heat fluxes both for engine cylinder and exhaust manifold. Two phases (a thermodynamic and a structural one) are clearly distinguished in a thermal transient and the cases where such an event could endanger the engine structural integrity are emphasized.

2. Simulation model for unsteady wall heat conduction

2.1 Modelling cases

It should be emphasized that in the present work they are concerned only the phenomena related to unsteady engine heat transfer which present the highest degree of interest. Thus several heat transfer phenomena and corresponding modelling cases applicable to steady-state engine operation would not be mentioned in the following.

Under the above framework, the prediction of the temperature distribution in the metallic parts of combustion chamber involves the solution of the unsteady heat conduction equation with the appropriate boundary conditions. The following two different simulation cases were considered leading to the development of respective models:

a. Three-dimensional Finite Element (FEM) model developed and used for the overall description of thermal field as it is developed during a transient event in combustion chamber components. This is mainly applicable for the investigation of the long-term heat transfer phenomena which are extended throughout the whole volume of each component. However it can be as equally used for the investigation of the heat transfer phenomena developed in the short-term scale.
b. One-dimensional heat conduction model developed and used for the calculation of instantaneous heat flux through a certain location of the combustion chamber wall. This is only applicable for the investigation of heat transfer phenomena developed in the short-term scale which influence exclusively the surface layers of each component (the ones in contact with combustion gases) up to a distance of a few mm inside its metal volume.

Both previous models give satisfactory results with significant computer time economy. Boundary conditions are assumed to be of all three kinds, i.e. constant surface temperature, constant heat flux or constant heat-transfer coefficient and surrounding fluid temperature (convective conditions). Their successful application depends on the correct knowledge of the physical mechanisms that take place on the gas and cooling sides of the various combustion chamber parts. Details on all previous topics can be found in (Rakopoulos & Mavropoulos, 1996, 1998).

2.2 Three-dimensional FEM analysis of transient temperature fields

The heat conduction equation for a three-dimensional axisymmetric, time-dependent, problem takes the form (in the absence of heat sources and for constant thermal conductivity):

$$\frac{\partial T}{\partial t} = \frac{k}{\rho \cdot c}\left(\frac{\partial^2 T}{\partial r^2} + \frac{1}{r}\frac{\partial T}{\partial r} + \frac{\partial^2 T}{\partial z^2}\right) \tag{1}$$

The development of a Finite-Element formulation for the unsteady heat conduction equation with all three kinds of boundary conditions, was based on a "variational" approach i.e. minimization of an appropriate variational statement. This is a standard and well known procedure (Rakopoulos & Mavropoulos, 1996) leading for the unsteady heat conduction case to the following system of differential equations:

$$[C][\dot{T}] = -[[K] + [H_s]]\ [T] + [h_s] + [q_s] \tag{2}$$

with the additional assumption of linear temperature variation inside every finite element.

Following an "element by element" analysis and summing up over the whole region of interest, we obtain the final expression of the characteristic matrices (conduction, convection, heat flux etc.) as they are used in eq. (2). The procedure has been presented in detail in (Rakopoulos & Mavropoulos, 1996).

2.3 Boundary conditions at the combustion chamber components

A variety of thermal boundary conditions is necessary to complete the application of FEM models for the prediction of temperature and heat flux distributions on engine structure. Since the application of these conditions introduces a factor of uncertainty onto the final results, a detailed knowledge of the physical mechanisms becomes essential. To overcome these difficulties the author have tested with success and presented in the past (Rakopoulos & Mavropoulos, 1996) a detailed set of boundary conditions for all combustion chamber surfaces. For the gas-side of combustion chamber components, the analysis of experimental pressure measurements results to the calculation of the instantaneous values for heat transfer coefficient h_g and temperature T_g as a function of crank angle. From these, the time averaged equivalent values can be calculated for a four-stroke engine:

$$\bar{h}_g = \frac{1}{4\pi}\int_0^{4\pi} h_g d\phi \quad , \quad \bar{T}_g = \frac{1}{4\pi\bar{h}_g}\int_0^{4\pi} T_g h_g d\varphi \tag{3}$$

Special attention is needed when modelling the boundary conditions at the piston-ring-liner interfaces through which a large quantity of heat passes, under low thermal resistance conditions. Any attempt to evaluate a heat transfer coefficient between skirt and liner and especially between rings and ring-grooves requires a complete knowledge of dimensions and clearances among them. Following this information, two basic assumptions were made: (a) Flow through crevices is a fully developed laminar one (Couette), and (b) Clearances in the above areas are very small and as a result convection mechanisms can be neglected.

At the piston top land part and the lower part of skirt the heat transfer coefficient is mainly determined by the magnitude of the radial clearance (Δx) of the gas or oil film between piston and liner. For the various surfaces of the ring belt, however, a thermal network model was adopted and the corresponding thermal circuits were created for the upper, lower and side surface of each ring-groove, respectively.

For the piston undercrown surface, the heat transfer coefficient depends strongly on the piston design and cooling system used. For the majority of calculation cases a jet piston cooling is adopted, so that the heat transfer coefficient is calculated by the expression:

$$h_{oil} = 68.17\left[(r\omega)\frac{D_n}{\nu_b}\right]^{1/2} \quad (W/m^2K) \tag{4}$$

where h_{oil} is the heat transfer coefficient between oil and undercrown surface, r is the crank radius, D_n is the nozzle diameter of oil sprayer in m, ν_b is the oil kinematic viscosity at bulk oil temperature in m^2/sec and ω the engine angular speed.

In the case of air cooled engines the fins at the outside surface of cylinder head and liner form a number of parallel closed cooling passages via the cowling; here for the estimation of heat transfer coefficient Nusselt-type equations are considered, depending on the state of flow as follows:

a. For laminar flow with Reynolds numbers less than 2100 the Nusselt-type relation, based on the work by Sieder and Tate is (Annand, 1963)

$$Nu = 1.86\left[Re\,Pr\,\frac{D_1}{L}\right]^{1/3}\left(\frac{\mu_b}{\mu_s}\right)^{0.14} \tag{5}$$

where the air properties are evaluated at the bulk 'b' temperature which is the arithmetic mean of the inlet and outlet temperatures, whereas subscript 's' refers to the surface temperature. In addition, L is the length of the flow path and D_1 is the equivalent hydraulic diameter, $D_1 = 4A/f$, where A is the flow surface area in each cooling passage and f its internal (cooling) perimeter.

b. For the transition region with Reynolds numbers ranging from 2100 to 10000 the Nusselt-type relation, based on the work by Hausen is

$$Nu = 0.116\left[Re^{2/3} - 125\right] Pr^{1/3}\left(\frac{\mu_b}{\mu_s}\right)^{0.14}\left[1+\left(\frac{D_1}{L}\right)^{2/3}\right] \quad (6)$$

with the same symbol meanings as for the previous case, and

c. For turbulent flow with Reynolds numbers greater than 10000 the Nusselt-type relation, based on the work by Sieder and Tate is

$$Nu = 0.023\,(Re)^{0.8}\,Pr^{1/3}\left(\frac{\mu_b}{\mu_s}\right)^{0.14} \quad (7)$$

It is obvious from equations (5) to (7) that the air velocity through the engine fins is the most important factor in the engine cooling process. A detailed theoretical study in correlation with experimental results is necessary in order to determine the most accurate values of heat transfer coefficient for the particular engine type and operating conditions in hand.

2.4 One-dimensional unsteady wall heat conduction at surface layers

For the calculation of instantaneous heat flux through a certain location of the combustion chamber wall for a complete engine cycle during steady state operation, the time periodic (unsteady) heat conduction model is used. In this case, Fourier analysis is the standard calculation procedure as it is well-established by the present author (Mavropoulos et al., 2008, 2009; Mavropoulos, 2011) and other research groups (Demuynck et al., 2009). However, for the case of the transient engine operation the heat transfer phenomena that occur in combustion chamber (and also in exhaust manifold) surfaces are dynamic (that is time dependent) but non-periodic. So in that case, the basic assumption for the application of Fourier method, that is a time periodic problem, is not valid anymore.

To overcome this difficulty, the author has developed a modified version of Fourier analysis. Its basic principles have been described in detail in (Mavropoulos et al., 2009). It is based on the idea of separation of the continuous transient variation from its initial until its final steady state to a number of discrete steps N_c, each of them having the duration of the corresponding engine cycle. In other words, each engine cycle is decoupled from the rest of the transient event. Having adopted this approximation, each individual engine cycle is considered to be repeated for an infinite number of times, so that heat flow through the corresponding component becomes time periodic and thus Fourier analysis can be finally applied. It has been already validated from the results presented in (Mavropoulos et al., 2009) that such an approximation does not impose any significant errors.

Assuming that heat flow through the component is one-dimensional and that material properties remain constant, the corresponding expression of the unsteady heat conduction equation for the i-th cycle of the transient event is given by

$$\left(\frac{\partial T}{\partial t} = \alpha \frac{\partial^2 T}{\partial x^2}\right)_i \tag{8}$$

where i=1,...,Nc with Nc the total number of engine cycles during a transient event of engine speed and/or load change. Additionally, x is in this case the distance from the wall surface, $\alpha = k_w/\rho_w c_w$ is the wall thermal diffusivity, with ρ_w the density and c_w the specific heat capacity.

Following the steps used in the classic heat conduction Fourier analysis as presented in (Mavropoulos et al., 2008, 2009), the following expression is reached for the calculation of instantaneous heat flux on the combustion chamber surfaces during the transient engine operation

$$\begin{aligned} q_{w,i}(t) = \left(-k_w \left.\frac{\partial T}{\partial x}\right|_{x=0}\right)_i = \frac{k_w}{\delta}\left(T_{m,i} - \overline{T}_{\delta,i}\right) + \\ k_w \sum_{n=1}^{N} \phi_{n,i}\left[\left(A_{n,i} + B_{n,i}\right)\cos(n\omega_i t) + \left(B_{n,i} - A_{n,i}\right)\sin(n\omega_i t)\right] \end{aligned} \tag{9}$$

where δ is the distance from the wall surface of the in-depth thermocouple. Additionally, $T_{m,i}$ is the time averaged value of wall surface temperature $T_{w,i}$, $A_{n,i}$ and $B_{n,i}$ are the Fourier coefficients all of them for the i-th cycle, n is the harmonic number, N is the total number of harmonics, and ω_i (in rad/s) is the angular frequency of temperature variation in the i-th cycle, which for a four-stroke engine is half the engine angular speed. In the developed model, there is the possibility for the total number of harmonics N to be changed from cycle to cycle in case such a demand is raised by the form of temperature variation in any particular cycle.

3. Categories of unsteady heat conduction phenomena

Phenomena related to unsteady heat conduction in Internal Combustion Engines are often characterized in literature with the general term "thermal transients". In reality these phenomena belong to different categories considering their development in time. As a result and for systematic reasons a basic distribution is proposed for them as it appears in Fig. 1.

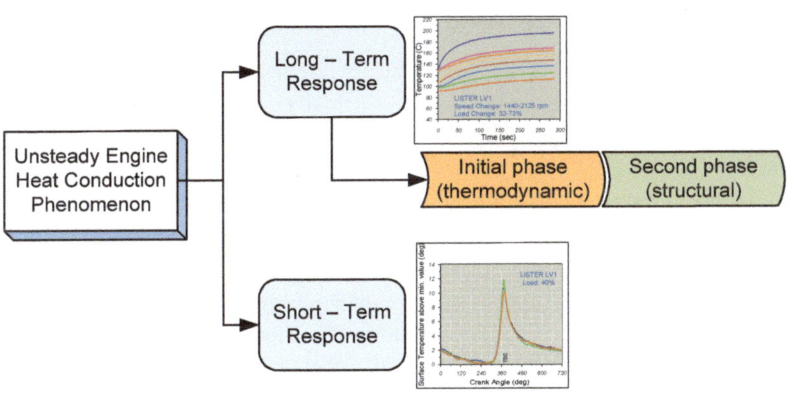

Fig. 1. Categories of engine unsteady heat conduction phenomena.

As observed any unsteady engine heat transfer phenomenon belongs in either of the following two basic categories:

- Short-term response ones, which are caused by the fluctuations of gas pressure and temperature during an engine cycle. These are otherwise called cyclic engine heat transfer phenomena and are developing during a time period in the order of milliseconds. Phenomena in this category are the result of the physical and chemical processes developing during the period of an engine cycle. They are finally leading to the development of temperature and heat flux oscillations in the surface layers of combustion chamber components. It is noted here that phenomena in this category should not normally mentioned as "transient" since they are mainly related with "steady state" engine operation. However their presence during transient engine operation is as equally important and this is considered in the present work. In addition the oscillating values of heat conduction variables around the surfaces of combustion chamber present a "transient" distribution in space since they are gradually faded out until a distance of a few mm below the surface of each component.
- Long-term response ones, resulting from the large time scale non-periodic variations of engine speed and/or load. As a result, thermal phenomena of this category have a time "period" in the order of several hundreds of seconds and are presented only during the transient engine operation.

Each case of long-term response thermal transient can be further separated in two different phases (Figs 1 and 2). The first of them involves the period from the start of variation until the instant in which all thermodynamic (combustion gas pressure and temperature, gas mixture composition etc.) and functional variables (engine torque, speed) reach their final state of equilibrium. This period lasts a few seconds (usually 3-20) depending on the type of engine and also on the kind of transient variation under consideration. This first phase of thermal transient is named as "thermodynamic".

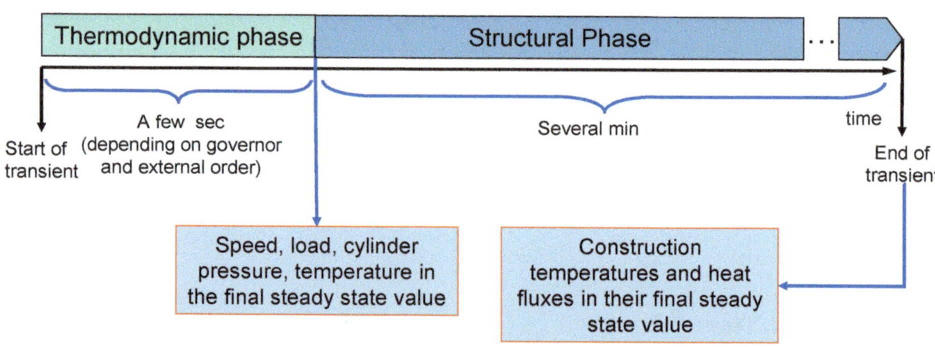

Fig. 2. Phases of long term response thermal transient event.

The upcoming second phase of the transient thermal variation is named as "structural" and its duration could in some cases overcome the 300 sec until all combustion chamber components have reached their temperatures corresponding to the final steady state. In the end of this second phase all variables related with heat conduction in the combustion chamber (temperatures, heat fluxes) and all heat transfer parameters of the fluids surrounding the combustion chamber (water, oil etc.) have reached their values corresponding to the final state of engine transient variation.

Specific examples from the above thermal transient variations are provided in the upcoming sections.

4. Test engine and experimental measuring installation

4.1 Description of the test engine

A series of experiments concerning unsteady engine heat transfer was conducted by the author on a single cylinder, Lister LV1, direct injection, diesel engine. The technical data of the engine are given in Table 1. This is a naturally aspirated, air-cooled, four-stroke engine, with a bowl-in-piston combustion chamber. All the combustion chamber components (head, piston, liner etc.) are made from aluminum. The normal speed range is 1000-3000 rpm. The engine is equipped with a PLN fuel injection system. A three-hole injector nozzle (each hole having a diameter of 0.25 mm) is located in the middle of the combustion chamber head. The engine is permanently coupled to a Heenan & Froude hydraulic dynamometer.

Engine type	Single cylinder, 4-stroke, air-cooled, DI
Bore/Stroke	85.73 mm/82.55 mm
Connecting rod length	148.59 mm
Compression ratio	18:1
Speed range	1000-3000 rpm
Cylinder dead volume	28.03 cm^3
Maximum power	6.7kW @ 3000 rpm
Maximum torque	25.0 Nm @ 2000 rpm
Inlet valve opening/ closing	15°CA before TDC /41°CA after BDC
Exhaust valve opening /closing	41°CA before BDC /15°CA after TDC
Inlet / Exhaust valve diameter	34.5mm / 31.5mm
Fuel pump	Bryce-Berger with variable-speed mechanical governor
Injector	Bryce- Berger
Injector nozzle opening pressure	190 bar
Static injection timing	28°CA before TDC
Specific fuel consumption	259 g/kWh (full load @ 2000 rpm)

Table 1. Engine basic design data of Lister LV1 diesel engine.

The engine experimental test bed was accompanied with the following general purpose equipment:

- Rotary displacement air-flow meter for engine air flow rate measurement
- Tank and flow-meter for diesel fuel consumption rate measurement
- Mechanical rpm indicator for approximate engine speed readings
- Hydraulic brake water pressure manometer, and
- Hydraulic brake water temperature thermometer.

4.2 Experimental measuring installation

4.2.1 General

A detailed description of the experimental installation that was used in the present investigation can be found in previous publications of the author (Mavropoulos et al., 2008,

2009; Mavropoulos, 2011). For that reason, only a brief description will be provided in the following.

The whole measuring installation was developed by the author in the ICEL Laboratory of NTUA and was specially designed for addressing internal combustion engine thermal transient variations (both short- and long-term ones). As a result, its configuration is based on the separation of the acquired engine signals into two main categories:

- Long-term response ones, where the signal presents a non-periodic variation (or remains essentially steady) over a large number of engine cycles, and
- Short-term response ones, where the corresponding signal period is one engine cycle.

To increase the accuracy of measurements, the two signal categories are recorded separately via two independent data acquisition systems, appropriately configured for each one of them. For the application in transient engine heat transfer measurements, the two systems are appropriately synchronized on a common time reference.

4.2.2 Long-term response installation

The long term response set-up comprises 'OMEGA' J- and K-type fine thermocouples (14 in total), installed at various positions in the cylinder head and liner in order to record the corresponding metal temperatures. Nine of those were installed on various positions and in different depths inside the metal volume on the cylinder head and they are denoted as "TH#j" (j=1,...9) in Fig. 3 (a and b). Thermocouples of the same type were also used for measuring the mean temperatures of the exhaust gas, cooling air inlet, and engine lubricating oil.

The extensions of all thermocouple wires were connected to an appropriate data acquisition system for recording. A software code was written in order to accomplish this task.

4.2.3 Short-term response installation

The short-term response installation is in general the most important as far as the periodic thermal phenomena inside the engine operating cycle are concerned. In general, it presents the greater difficulty during the set-up and also during the running stage of the experiments. It comprises the following components:

4.2.3.1 Transducers and heat flux probes

The following transducers were used to record the high-frequency signals during the engine cycle:

- "Tektronix" TDC marker (magnetic pick-up) and electronic 'rpm' counter and indicator.
- "Kistler" 6001 miniature piezoelectric transducer for measuring the cylinder pressure, flush mounted to the cylinder head. Its output signal is connected to a "Kistler" 5007 charge amplifier.
- Four heat flux probes installed in the engine cylinder head and the exhaust manifold, for measuring the heat flux losses at the respective positions. The exact locations of these probes (HT#1 to 4) and of the piezoelectric transducer (PR#1), are shown in the layout graph of Fig. 3a and also in the image of Fig. 3b.

The prototype heat flux sensors were designed and manufactured by the author at the Internal Combustion Engine Laboratory (ICEL) of (NTUA). Additional details and technical data about them can be found in (Mavropoulos et al., 2008, 2009). They are customized especially for this application as shown in the images of Fig. 4 where it is presented the whole instantaneous heat flux measurement system module created and used for the present investigation. They belong in two different types as described below:

- Heat flux sensors (HT#1-3 in Fig. 3a and 3b) installed on the cylinder head, consisting of a fast response, K-type, flat ribbon, "eroding" thermocouple, which was custom designed and manufactured for the needs of the present experimental installation, in combination with a common K-type, in-depth thermocouple. Each of the fast response thermocouples was afterwards fixed inside a corresponding compression fitting, together with the in-depth one that is placed at a distance of 6 mm apart, inside the metal volume. The final result is shown in Fig. 4.

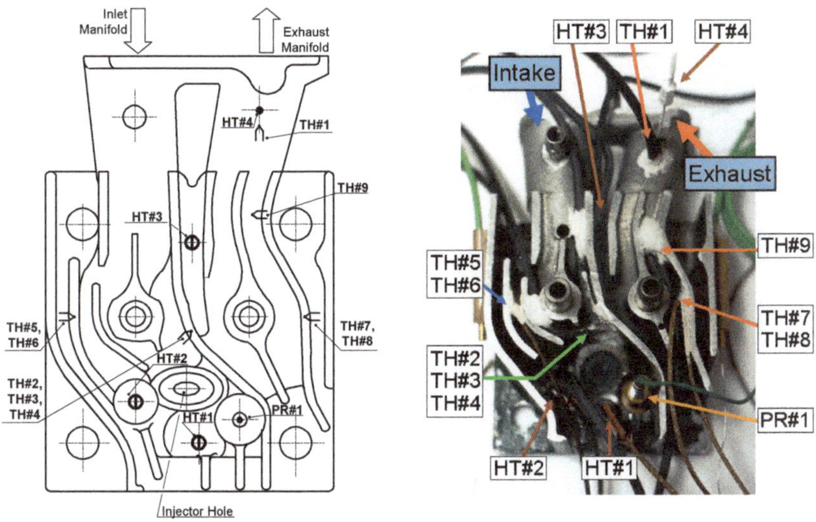

Thermoc. #	Distance from top head surface (mm)
TH#1	0.0
TH#2	53.4
TH#3	18.0
TH#4	4.0
TH#5	21.0
TH#6	4.0
TH#7	21.0
TH#8	5.0
TH#9	0.0

(a) (b)

Fig. 3. Graphical layout (a), and image (b), of the engine cylinder head instrumented with the surface heat flux sensors, the piezoelectric pressure transducer and the "long-term" response thermocouples at selected locations.

- The heat flux sensor installed in the exhaust manifold (HT#4 in Fig. 3a and 3b) has the same configuration, except that the fast response thermocouple used is a J-type, "coaxial" one. It is accompanied with a common J-type, in-depth thermocouple, located inside the compression fitting at a distance of 6 mm behind it. The sensor was flush-mounted on the exhaust manifold at a distance of 100 mm (when considered in a straight line) from the exhaust valve.

The heat flux sensors developed in this way displayed a satisfactory level of reliability and durability, necessary for this application. Also, special care was given to minimize distortion of thermal field in each position caused by the presence of the sensor. Before being placed to their final position in the cylinder head and exhaust manifold, all heat flux sensors were extensively tested and calibrated through a long series of experiments conducted in different engines, under motoring and firing operating conditions.

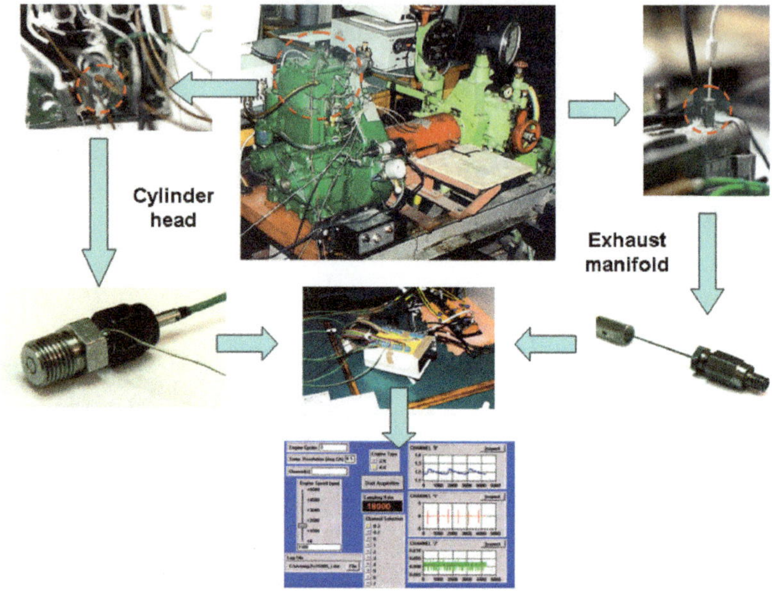

Fig. 4. Instantaneous heat flux measurement system module used in the cylinder head and exhaust manifold wall.

4.2.3.2 Signal pre-amplification and data acquisition system

In order to obtain a clear thermocouple signal when acquiring fast response temperature and heat flux data, the author had introduced the technique of an initial pre-amplification stage. This independent pre-amplification stage is applied on the sensor signal before the latter enters the data acquisition system. The need for such an operation emanates from the fact that this kind of measurements combines the low voltage level of a thermocouple signal output with an unusual high frequency. As a result, its direct acquisition using a common multi-channel data acquisition system creates a great percentage of uncertainty and in some cases it becomes even impossible. The introduction of pre-amplification stage solves the previous problems with only a small contribution to signal noise. For recording the fast response signals during the transient engine operation, the frequency used was in the range of 4500-6000 ksamples/sec/channel, which resulted in a corresponding signal resolution in the range of 1-2 deg CA dependent on the instantaneous engine speed.
The prototype preamplifier and signal display device (Fig. 4) was designed and constructed in the NTUA-ICEL laboratory, using commercially available independent thermocouple amplifier modules for the J- and K-type thermocouples, respectively. Ten of the above amplifiers were installed on a common chassis together with necessary selectors and

displays, forming a flexible device that can route the independent heat flux sensor signals either in the input of an oscilloscope for display and observation, or in the data acquisition system for recording and storage as it is displayed in Fig. 4. Additional details for the pre-amplifier can be found in (Mavropoulos et al., 2008, 2009, Mavropoulos, 2011). After the development of this device by the author, similar devices specialized in fast response heat flux signal amplification have also become commercially available.

The output signals from the thermocouple pre-amplifier unit, together with the magnetic TDC pick-up and piezoelectric transducer signals are connected to the input of a high-speed data acquisition system for recording. Additional details concerning the data acquisition system are provided in (Mavropoulos, 2011).

5. Presentation and discussion of the simulated and experimental results

5.1 Simulation process and experimental test cases considered

The theoretical investigation of phenomena related to the unsteady heat conduction in combustion chamber components was based on the application of the simulation model for engine performance and structural analysis developed by the author. The structural representation of each component is based on the 3-dimensional FEM analysis code developed especially for the simulation of thermal phenomena in engine combustion chamber. For the application of boundary conditions in the various surfaces of each component, a series of detailed physical models is used. As an example, for the boundary conditions in the gas side of combustion chamber a thermodynamic simulation model of engine cycle operation is used in the degree crank angle basis. A brief reference of the previous models was provided in subsections 2.2 and 2.3. Additional details are available in previous publications (Rakopoulos & Mavropoulos, 1996, 1999).

Like any other classic FEM code, the thermal analysis program developed consists of the following three main stages: (a) preprocessing calculations; (b) main thermal analysis; and (c) postprocessing of the results. An example of these phases of solution is provided in Fig. 5 (a-e) applied in an actual piston and liner geometry of a four stroke diesel engine. For each of the components a 3-dimensional representation (Fig. 5a) is first created in a relevant CAD system. In the next step the component is analysed in a series of appropriate 3d finite elements (Fig. 5b) and the necessary boundary conditions are applied in all surfaces. Then, during the main analysis the thermal field in each component is solved and this process could follow several solution cycles until an acceptable convergence in boundary conditions is achieved. It should be mentioned in this point that due to the complex nature of this application each combustion chamber component is not independent but it is in contact with others (for example the piston with its rings and liner etc.). This way the final solution is achieved when the heat balance equation between all components involved is satisfied. More details are provided in (Rakopoulos & Mavropoulos, 1998, 1999).

For the postprocessing step one option is a 3d representation of the thermal field variables (Fig. 5c and 5d). In alternative, a section view (Fig. 5e) is used to describe the thermal field in the internal areas of the structure in detail. This way the comparison with measured temperatures in specific points of the component (numbers in parentheses in Fig. 5e) is also available which is used for the validation of the simulated results.

For the needs of the present investigation several characteristic actual engine transient events were selected to demonstrate the results of the unsteady heat conduction simulation model both in the long-term and in the short-term time scale. All of them are performed in the test engine and the experimental installation described in section 4. For the long-term scale the following two variations are examined:

- A load increment (*"variation 1"*) from an initial steady state of 2130 rpm engine speed and 40% of full load to a final one of 2020 rpm speed and 65% of full load.

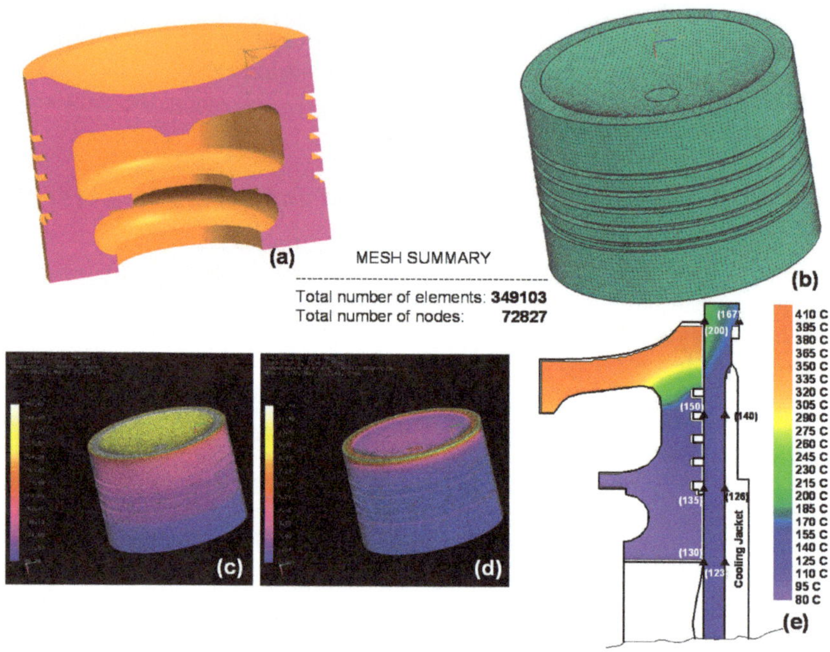

Fig. 5. Application of the simulation model for engine performance and structural analysis. A 3d engine piston geometry representation (a), its element mesh (b) and results of thermal field variables in three (c and d) and two dimensional representations (e).

- A speed increment (*"variation 2"*) from an initial steady state of 1080 rpm engine speed and 10% of full load to a final one of 2125 rpm speed and 40% of full load.

For the short-term scale the next two transient events are respectively considered:

- A change from 20-32% of full load (*"variation 3"*). During this change, engine speed remained essentially constant at 1440 rpm. Characteristic feature in this variation was the slow pace by which the load was imposed (in 10 sec, approximately). For this transient variation, a total of 357 consecutive engine cycles were acquired in a 30 sec period via the "short-term response" system signals. For the "long-term response" data acquisition system, the corresponding figures for this transient variation raised in 3417 consecutive engine cycles during a time period of 285 sec.
- Following the previous one, a change from 32-73% of full engine load (*"variation 4"*) with a simultaneous increase in engine speed from 1440 to 2125 rpm. In this variation, the load change was imposed rapidly in an approximate period of 2 sec. This was accomplished on purpose trying to imitate in the "real engine" the theoretical *ramp* variation of engine speed and load. For this transient variation and the "short-term response" system, 695 engine cycles were acquired in a period of 40 sec. The

corresponding figures for the "long-term response" signals raised in 5035 engine cycles in a time period of 285 sec.

For all the above transient variations, the initial and final steady state signals were additionally recorded from both the short- and long-term response installations. Selective results from the simulation performed and the experiments conducted concerning the previous four variation cases are presented in the upcoming sections.

5.2 Results concerning long-term heat transfer phenomena in combustion chamber

Before proceeding with the application of the model to transient engine operation cases, it was first necessary to calibrate the thermostructural submodel under steady state conditions, especially for the verification of the application of boundary conditions as described in 2.3. Several typical transient variations (events) of the engine in hand were then examined which involve increment or reduction of load and/or speed. Results concerning variation of engine performance variables under each transient event are not presented at the present work due to space limitations. They are available in existing publications of the author (Mavropoulos et al., 2009; Rakopoulos et al., 1998; Rakopoulos & Mavropoulos, 2009).

The Finite Element thermostructural model was then applied for the cylinder head of the Lister-LV1, air-cooled DI diesel engine for which relevant experimental data are available. For the needs of the present application a mesh of about 50000 tetrahedral elements was developed, allowing a satisfactory degree of resolution for the most sensitive points of the construction like the valve bridge area. For the early calculation stages it was found convenient to utilize a coarser mesh, which helps on the initial application of boundary conditions furnishing significant computer time economy. The final finer mesh can then be applied giving the maximum possible accuracy on the final result.

In Fig. 6a the experimental temperature values taken from three of the cylinder head thermocouples (TH#2-TH#4) during the load increment variation "1", are compared with the corresponding calculated ones at the same positions. The calculated curves follow satisfactorily the experimental ones throughout the progress of the transient event. The steepest slope between the different curves included in Fig. 6a is observed on the corresponding ones of thermocouple TH#2 (Fig. 3) placed at the valve bridge area, while the most moderate one is observed for thermocouple TH#4 placed at the outer surface of the cylinder head. As expected, the valve bridge is one of the most sensitive areas of the cylinder head suffering from thermal distortion caused by these sharp temperature gradients during a transient event (thermal shock). Many cases of damages in the above area have been reported in the literature, a fact which also confirms the results of the present calculations.

Similar observations can be made for the cylinder head temperatures in the case of the speed increment variation "2" presented in Fig. 6b. Again the coincidence between calculated and experimental temperature profiles is very good. Temperature levels for all positions present now smaller differences between the initial and final steady state; the steepest temperature gradient is again observed in the valve bridge area. The initial drop in the temperature value of thermocouple TH#4 is due to the increase in engine speed for the first few seconds of the variation which causes a corresponding increase in the air velocity through the fins and so in the heat transfer coefficient given by eqs (5) to (7) with a simultaneous decrease in air temperature. From the results presented in Fig. 6 it is concluded that the developed model manages to simulate satisfactorily the long-term response unsteady heat transfer phenomena as they are developed in the engine under consideration.

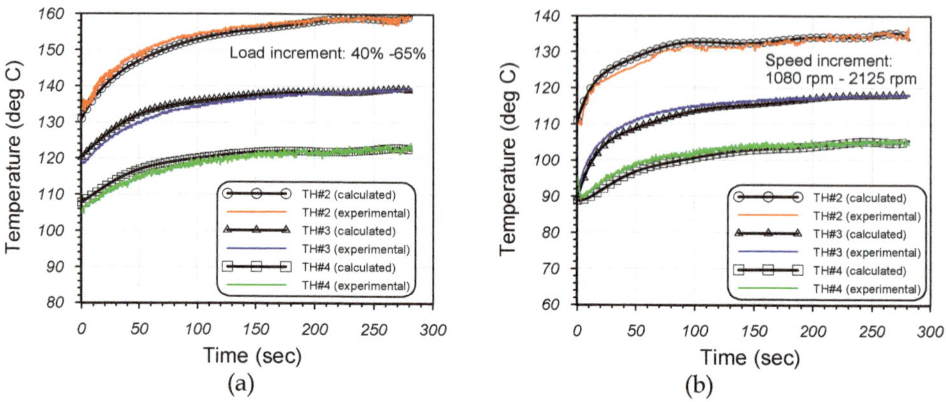

Fig. 6. Comparison between calculated and experimental temperature profiles vs. time for three of the cylinder head thermocouples, during the load increment variation "1" (a) and the speed increment variation "2" (b).

Figs 7 (a and b) present the results of temperature distributions at the whole cylinder head area in the form of isothermal charts, as they were calculated for the initial and final steady state of transient variation "1". Numbers inside squares denote experimental temperature values recorded from thermocouples. A significant degree of agreement is observed between the simulated temperature results and the corresponding measured values which confirms for the validity of the developed model. Similar charts could be drawn for any of the variations examined and at any specific moment of time during a transient event. They are presenting in a clear way the local temperature distinctions in the various parts of the construction, thus they are revealing the mechanism of heat dissipation through the structure. The observed temperature differences between the inlet and the exhaust valve side of the cylinder head (exceeding 150 ^{o}C for the full load case) are characteristic for air-cooled diesel engines, where construction leaves only small metallic common areas between the inlet and the exhaust side of head. Corresponding results reported in the literature confirm the above observation (Perez-Blanco, 2004; Wu et al., 2008).

5.3 Results concerning short-term heat transfer phenomena in combustion chamber

During the experiments conducted, the heat flux sensors HT#2 and HT#3 (installed on the cylinder head) were not able to operate adequately over most of the full spectrum of measurements taken. The reasons for this failure are described in detail in (Mavropoulos et al., 2008). Therefore, in this work the short-term results for the cylinder head will be presented only from sensor HT#1 together with the ones for the exhaust manifold from sensor HT#4.

In Figs 8 and 9 are presented the time histories for several of the most important engine performance and heat transfer variables during the first 2 sec from the beginning of the transient event for variations "3" and "4", respectively, which are examined in the present study. The number of cycles in the first 2 sec of each variation is different as it was expected.

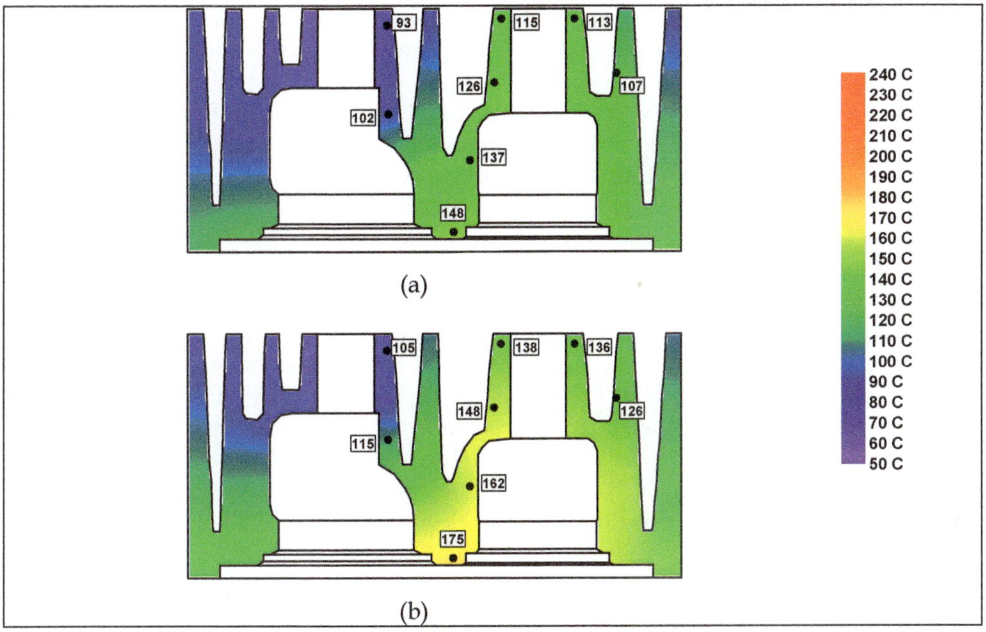

Fig. 7. Cylinder head temperature distributions, in deg. C, at the initial (a) and final (b) state of the load increment variation "1". Numbers in "squares" denote experimental temperature values taken from thermocouples.

The temporal response of cylinder pressure is presented for the two variations in Figs 8a and 9a, respectively. For variation "3", an increase of 1-1.5 bar is observed in the peak pressure during the first 3 cycles of the event. Variation in peak cylinder pressure becomes marginal after this moment, presents a slight fluctuation and reaches its final value almost 3 sec after initiation of the variation. For variation "4", the case is highly different from the previous one. Pressure changes rapidly and during the first four engine cycles after the beginning of the transient its peak value is increased linearly from 60 to 80 bar approximately. The 80 bar peak value is maintained afterwards almost constant for a period of slightly higher than 1 sec, when after approximately the 15th engine cycle it starts to decline in a slower pace to its final level of 70 bar which corresponds to the final steady state. The total time period the peak pressure demanded to settle in its final steady state value for this variation was evaluated to 5 sec. For both variations "3" and "4", the time instant after which peak pressure is settled to its final steady state value marks the end of the first phase of the thermal transient variation that was named as the "thermodynamic" one. As a result at the end of this phase, the combustion gas has reached its final steady state. The upcoming second phase of the transient thermal variation named as the "structural" one is expected to last much longer until all combustion chamber components have reached their temperatures corresponding to the final steady state. Additional details about these phases were provided by the author in (Rakopoulos and Mavropoulos, 1999, 2009). It is in general accepted that the duration of each period is primarily dependent on the respective duration and also on the magnitude

of speed and/or load change during each specific event. For the present case, the duration of "thermodynamic" phase is 3 sec for variation "3" and 5 sec for variation "4", respectively.
The time histories for the variation of measured wall surface temperature at the position of sensor HT#1 on cylinder head for the two transient events are presented in Figs 8b and 9b. In the same Figs they are observed the corresponding wall temperature variations for depths 1.0-3.0 mm below cylinder head surface inside the metal volume. The last variations were calculated using the modified one dimensional wall heat conduction model as described in 2.4. It is observed that wall surface temperature, as being a structural variable, continues to rise after 2 sec from the beginning of each transient event. However, this increase in surface temperature refers to its "long-term scale" variation and it is linear in the case of the moderate load increase of variation "3" (Fig. 8b), or exponential in the case of the ramp speed and load increase of variation "4" (Fig. 9b). By analysing the whole range of both experimental measurements it was concluded that the total duration of structural phase of the transient is estimated at 200 sec for variation "3", whereas it exceeds 300 sec in the case of variation "4". Similar values have been calculated theoretically by the author in the past using the simulation model for structural thermal field (Rakopoulos and Mavropoulos, 1999).
Of special importance are the results of measurements presented in Figs 8b and 9b related to the "short-term scale" that is with reference to the instantaneous cyclic surface temperatures. In the moderate load increase of variation "3", the amplitude of temperature oscillations remains essentially constant during the first 2 sec (and also during the rest of the event). On the contrary, in the case of the sudden ramp speed and load increase of variation "4", a gradual increase is observed in the amplitude of temperature oscillations during the first four cycles after the beginning of the transient following the corresponding increase of cylinder pressure in Fig. 9a. However, in the case of wall surface temperature (x=0.0), its peak values are presented rather unstable and amplitudes are far beyond the normal ones expected in the case of an aluminum combustion chamber surface. It is characteristic that the maximum amplitude of temperature oscillations as presented in Fig. 9b was 31 deg, which is inside the area of values observed in the case of ceramic materials in insulated engines (Rakopoulos and Mavropoulos, 1998). These extreme values of temperature oscillations is a clear indication of abnormal combustion, which occurs in the beginning of variation "4" and it likely lasts only for about 1.5 sec or the first 21 cycles after the beginning of the transient. After this period, surface temperature in the combustion chamber returns to its normal fluctuation and its amplitude is reduced to the value corresponding to the final steady state after approximately the 50th cycle from the beginning of the transient.
To obtain further insight into the mechanism of heat transfer during a transient operation, it is useful to examine the temporal development of temperature in the internal layers of cylinder wall up to a distance of a few mm below the surface. The results for the transient temperatures during variations "3" and "4" are presented in Figs 8b and 9b for values of depth x varying from 1.0-3.0 mm below the surface of the cylinder head. In Fig. 8b it is observed that for transient variation "3" there is no essential difference between the different engine cycles in each depth during the development of transient event. As expected the amplitude of temperature oscillations is highly reduced in the internal layers of

cylinder head volume and for x=3.0 mm below the combustion chamber surface practically there exists no temperature oscillation. On the other hand during transient variation "4" in Fig. 9b, the abnormal combustion indicated previously causes the development of a heat wave penetrating quickly in the internal layers of cylinder head. It is remarkable that during the first 20 cycles from the beginning of the event, temperature swings of 0.7 deg can be sensed even in a depth of x=3.0 mm below the surface of combustion chamber. The instant velocity of this penetration during the transient event "4" can also be estimated from the results presented in Fig. 9b. From the analysis of the results it was observed that the peak temperature in the depth of x=3.0 mm below the surface appears at an angle of 720 deg. As a consequence, during an approximate "time period" of 360 deg the thermal wave penetrates 3.0 mm inside the metallic volume of cylinder head. After the 20th cycle the temperature oscillations start to reduce and after a few more engine cycles are vanished in the depth of 3.0 mm below surface.

Following the above analysis for surface temperature, heat flux time histories for the point of measurement (HT#1) in the cylinder head and the two variations examined, are presented in Figs 8c and 9c. Heat flux histories are highly influenced by gas pressure and surface temperature variations, and their patterns are in general similar with them. In the case of variation "3", a mild increase in peak cylinder heat flux is observed during the first four cycles of the event and this is due to the similar increase observed in cylinder pressure during the same period. There is a marginal increase in peak values afterwards due to surface temperature increase and the final steady state peak value is reached after the 50th cycle, approximately. In variation "4", the heat flux is rather unstable following the pattern of surface temperatures. Due to the combustion instabilities described previously, measured peak heat flux values raised to almost three times higher than the ones observed during the normal engine operation, the highest of them reaching the value 9000 kW/m^2 corresponding to the same cycles in which the extreme surface temperature values have occurred. Peak heat flux is reduced afterwards at a slower pace to its final steady-state value, which is reached after the 200th cycle from the beginning of the event. A similar form of instantaneous heat flux variation during the first cycles of the warm-up period for a spark ignited engine was presented by the authors of (Wang & Stone, 2008).

5.4 Unsteady heat conduction phenomena in the engine gas exchange system

Phenomena related with the unsteady heat transfer in the inlet and exhaust engine manifolds are of special interest. In particular during the last years these phenomena have drawn special attention due to their importance in issues related with pollutant emissions during transient engine operation and especially the combustion instability which occurs in the case of an engine cold-starting event.

The variation of surface temperature and heat flux in the engine exhaust manifold follows in general the same trends as in the cylinder head. In this case, since the point of temperature and heat flux measurement was placed 100 mm downstream the exhaust valve (Figs 3 and 4), the corresponding phenomena are significantly faded out (Figs 10 and 11).

Increase of the amplitude of temperature oscillations is again obvious for variation "4" (Fig. 11a). However, there are no extreme amplitudes present in this case, as they have been absorbed due to the transfer of heat to the cylinder and manifold walls along the 100 mm distance from the exhaust valve to the point of measurement.

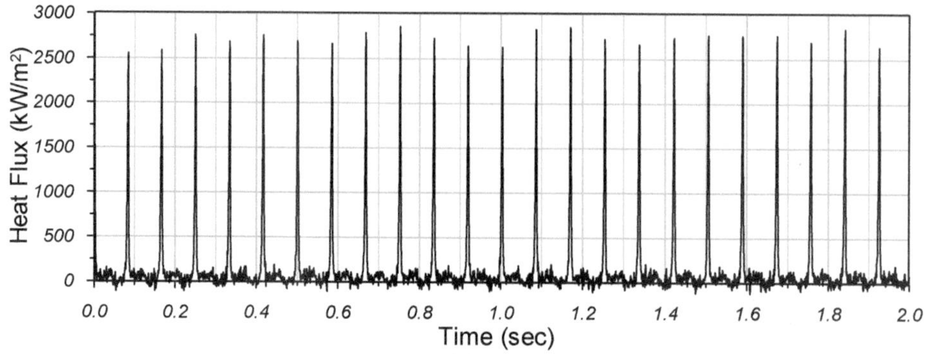

(c)

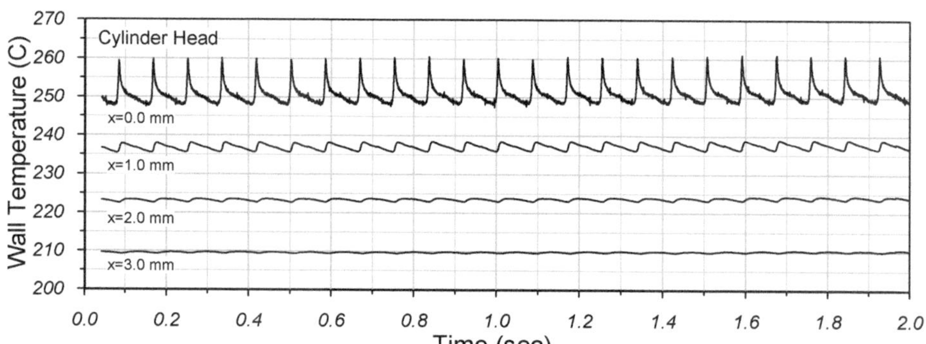

(b)

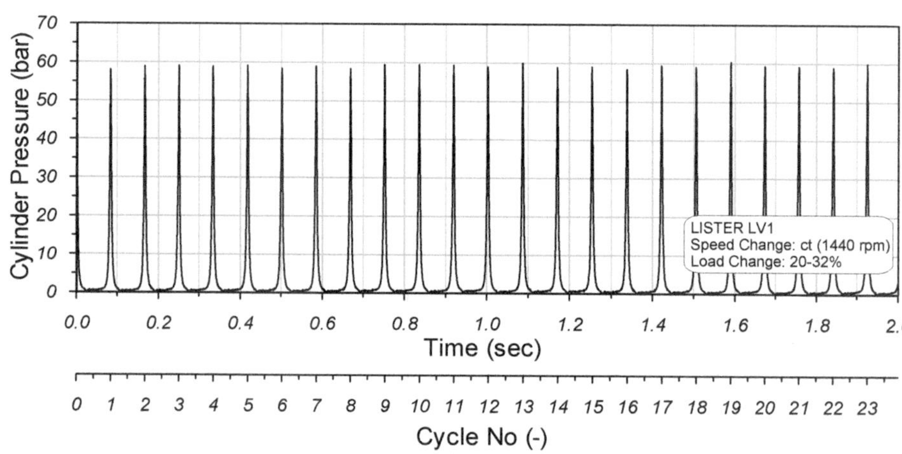

(a)

Fig. 8. Time histories of cylinder pressure (a), wall temperature for cylinder head on surface x=0.0 and three different depths inside the metal volume (b) and heat flux variation for cylinder head (c), for the first 2 sec of transient variation "3".

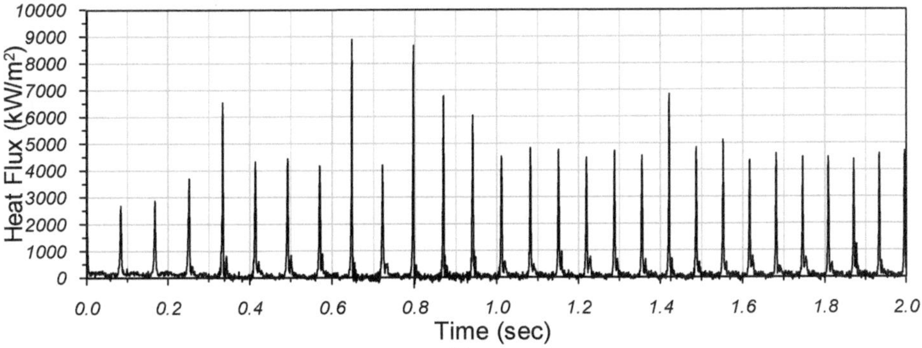

(c)

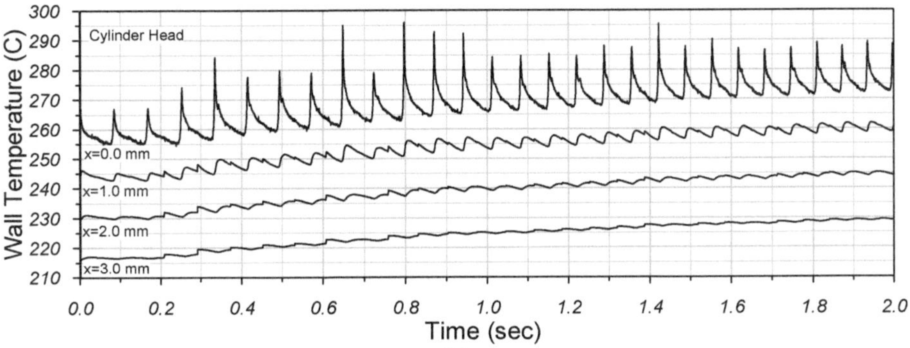

(b)

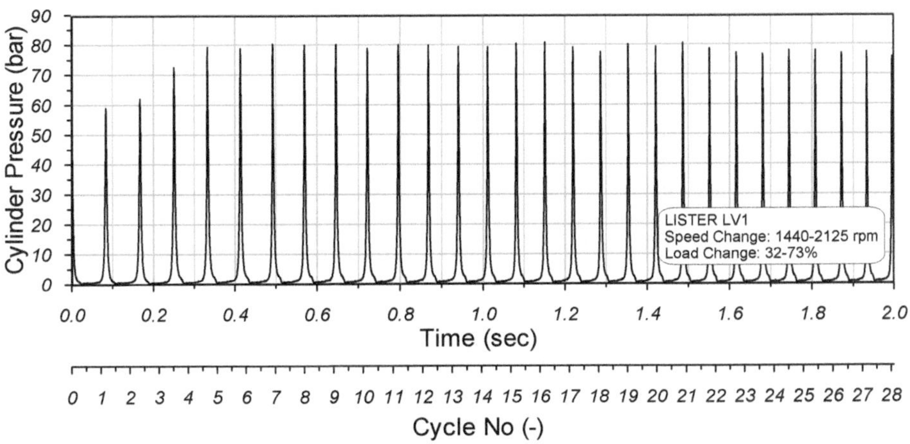

(a)

Fig. 9. Time histories of cylinder pressure (a), wall temperature for cylinder head on surface x=0.0 and three different depths inside the metal volume (b) and heat flux variation for cylinder head (c), for the first 2 sec of transient variation "4".

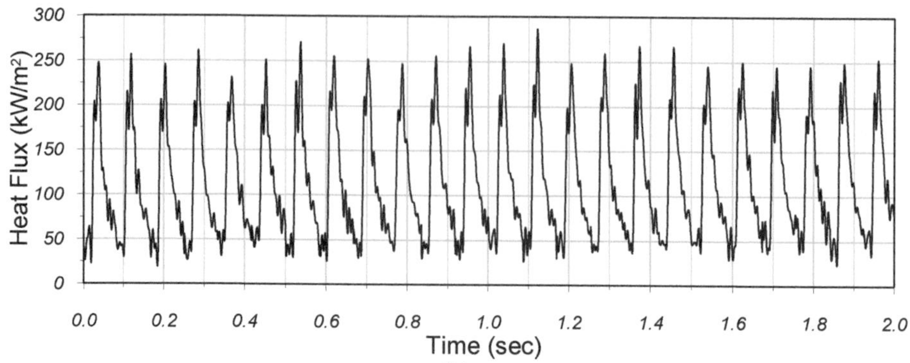

(b)

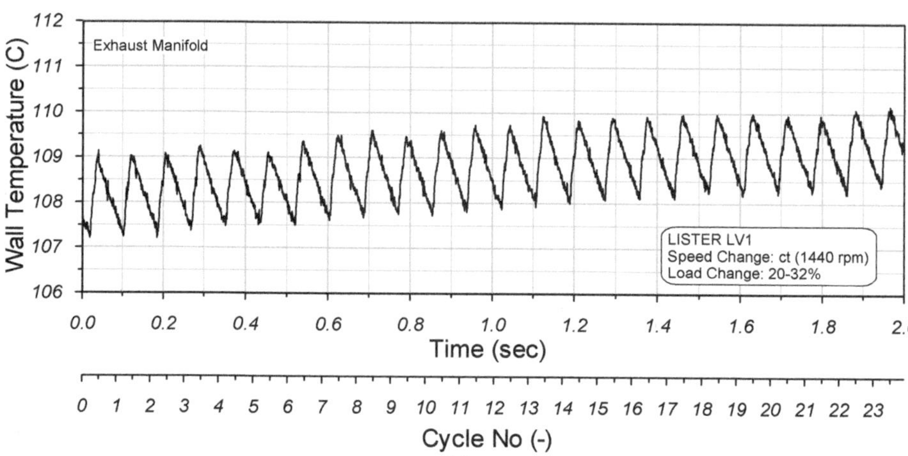

(a)

Fig. 10. Time histories of exhaust manifold wall surface temperature (a) and heat flux (b) at the position of sensor HT#4 for the first 2 sec of transient variation "3".

The corresponding results for heat flux time histories in the point of measurement on the exhaust manifold are presented in Figs 10b and 11b. In the case of variation "3", the moderate load increase is reflected as a marginal increase in exhaust manifold heat flux (a difference cannot be observed in time history of Fig. 10b). In the case of ramp variation "4" on the other hand, it is observed in Fig 11b a sudden increase in the amplitude of exhaust manifold heat flux, which starts 4 cycles after the beginning of the transient. In this case, there is no gradual increase of heat flux amplitude during the first four cycles, as it was the case for cylinder pressure and also cylinder head surface temperature and heat flux. Like the case of exhaust manifold surface temperature, this result is due to the heat transfer to combustion chamber and exhaust manifold walls until the point of measurement. It is observed that during the first 20 cycles of variation "4" the heat losses to exhaust manifold walls are increased beyond their normal level, due to increased engine speed and consequently gas velocity inside the exhaust manifold. The latter is the primary factor influencing heat losses in the exhaust manifold, as shown in (Mavropoulos et al., 2008). The increased level of heat losses during the gas exchange period of each cycle for the first 20 cycles is the reason for the appearance of negative heat fluxes in the results of Fig. 11b. Such a case is quite remarkable and could not appear in the position of measurement during steady state operation. Heat flux becomes negative (that is heat is transferred from manifold wall to the gas) for a short period of engine cycle after TDC. This coincides with the period during which combustion gas temperature at the distance of 100 mm downstream the exhaust valve inside the manifold reaches its minimum value. The combination of instantaneous exhaust gas temperature with gas velocity at the point of measurement is the reason for the final result concerning the time history of heat flux in the exhaust manifold.

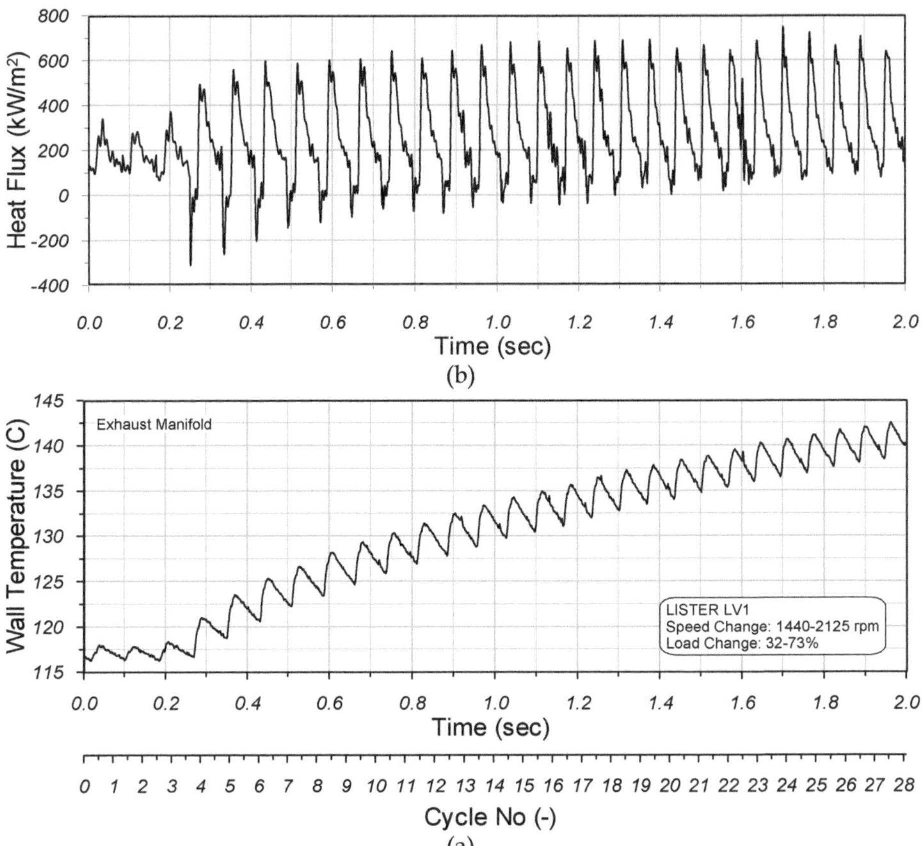

Fig. 11. Time histories of exhaust manifold wall surface temperature (a) and heat flux (b) at the position of sensor HT#4 for the first 2 sec of transient variation "4".

6. Conclusion

A theoretical simulation model accompanied with a comprehensive experimental procedure was developed for the analysis of unsteady heat transfer phenomena which occur in the combustion chamber and exhaust manifold surfaces of a DI diesel engine. The results of the study clearly reveal the influence of transient engine heat transfer phenomena both in the engine structural integrity as well as in its performance aspects. The main findings from the analysis results of the present investigation can be summarized as follows:

- Thermal phenomena related to unsteady heat transfer in internal combustion engines can be categorized as long- or short-term response ones in relation to the time period of their development. Each long-term response variation is further separated to a "thermodynamic" and a "structural" phase.
- Calculated temperature profiles from the Finite Element sub-model matched satisfactorily the corresponding experimental temperature profiles recorded by the thermocouples, revealing that the area between the two valves (valve bridge) is the most sensitive one towards the generation of sharp temperature gradients during each transient (thermal shock). The effect of air velocity in the cooling procedure of external surfaces is clearly revealed and analysed.
- A strong influence exists between the long-term non-periodic heat transfer variation resulting from engine transient operation and the instantaneous cyclic short-term responses of surface temperatures and heat fluxes. The results of this interaction influence primarily the combustion chamber and secondary the exhaust manifold surfaces.
- In the first cycles ("thermodynamic" phase) of a ramp engine transient, abnormal combustion occurred. The result is that the amplitude of surface temperature swings and the peak heat flux value for cylinder head surfaces were increased at extreme values, reaching almost 3 times the level of the corresponding ones that occur during steady state operation.
- The respective phenomena inside the exhaust manifold at a distance of 100 mm downstream the exhaust valve have a minor impact on the local surfaces. Temperature gradients are reduced in low levels due to heat losses. The gas velocity inside the exhaust manifold is the main factor influencing heat transfer and wall heat losses.

7. References

Annand, W.J.D. (1963). Heat transfer in the cylinders of reciprocating internal combustion engines. *Proceedings of the Institution of Mechanical Engineers*, Vol.177, pp. 973-990

Assanis, D. N. & Heywood, J. B. (1986). Development and use of a computer simulation of the turbocompounded diesel engine performance and component heat transfer studies. *Transactions of SAE, Journal of Engines*, Vol.95, SAE paper 860329

Demuynck, J., Raes, N., Zuliani, M., De Paepe, M., Sierens, R. & Verhelst, S. (2009). Local heat flux measurements in a hydrogen and methane spark ignition engine with a thermopile sensor. *Int. J Hydrogen Energy*, Vol.34, No.24, pp. 9857-9868

Heywood, J.B. (1998). *Internal Combustion Engine Fundamentals*, McGraw-Hill, New York

Keribar, R. & Morel, T. (1987). Thermal shock calculations in I.C. engines, SAE paper 870162

Lin, C.S. & Foster, D.E. (1989). An analysis of ignition delay, heat transfer and combustion during dynamic load changes in a diesel engine, SAE paper 892054

Mavropoulos, G.C., Rakopoulos, C.D. & Hountalas, D.T. (2008). Experimental assessment of instantaneous heat transfer in the combustion chamber and exhaust manifold walls of air-cooled direct injection diesel engine. *SAE International Journal of Engines,* Vol.1, No.1, (April 2009), pp. 888-912, SAE paper 2008-01-1326

Mavropoulos, G.C., Rakopoulos, C.D. & Hountalas, D.T. (2009). Experimental investigation of instantaneous cyclic heat transfer in the combustion chamber and exhaust manifold of a DI diesel engine under transient operating conditions, SAE paper 2009-01-1122

Mavropoulos, G.C. (2011). Experimental study of the interactions between long and short-term unsteady heat transfer responses on the in-cylinder and exhaust manifold diesel engine surfaces. *Applied Energy,* Vol.88, No.3, (March 2011), pp. 867-881

Perez-Blanco, H. (2004). Experimental characterization of mass, work and heat flows in an air cooled, single cylinder engine. *Energy Conv. Mgmt,* Vol.45, pp. 157-169

Rakopoulos, C.D. & Mavropoulos, G.C. (1996). Study of the steady and transient temperature field and heat flow in the combustion chamber components of a medium speed diesel engine using finite element analyses. *International Journal of Energy Research,* Vol.20, pp. 437-464

Rakopoulos, C.D. & Mavropoulos, G.C. (1998). Components heat transfer studies in a low heat rejection DI diesel engine using a hybrid thermostructural finite element model. *Applied Thermal Engineering,* Vol.18, pp. 301-316

Rakopoulos, C.D., Mavropoulos, G.C. & Hountalas, D.T. (1998). Modeling the structural thermal response of an air-cooled diesel engine under transient operation including a detailed thermodynamic description of boundary conditions, SAE paper 981024

Rakopoulos, C.D. & Hountalas, D.T. (1998). Development and validation of a 3-D multi-zone combustion model for the prediction of DI diesel engines performance and pollutants emissions. *Transactions of SAE, Journal of Engines,* Vol.107, pp. 1413-1429, SAE paper 981021

Rakopoulos, C.D. & Mavropoulos, G.C. (1999). Modelling the transient heat transfer in the ceramic combustion chamber walls of a low heat rejection diesel engine. *International Journal of Vehicle Design,* Vol.22, No.3/4, pp. 195-215

Rakopoulos, C.D. & Mavropoulos, G.C. (2000). Experimental instantaneous heat fluxes in the cylinder head and exhaust manifold of an air-cooled diesel engine. *Energy Conversion and Management,* Vol.41, pp. 1265-1281

Rakopoulos, C.D., Rakopoulos, D.C., Giakoumis, E.G. & Kyritsis, D.C. (2004). Validation and sensitivity analysis of a two-zone diesel engine model for combustion and emissions prediction. *Energy Conversion and Management,* Vol.45, pp. 1471-1495

Rakopoulos, C.D. & Mavropoulos, G.C. (2008). Experimental evaluation of local instantaneous heat transfer characteristics in the combustion chamber of air-cooled direct injection diesel engine. *Energy,* Vol.33, pp. 1084–1099

Rakopoulos, C.D. & Mavropoulos, G.C. (2009). Effects of transient diesel engine operation on its cyclic heat transfer: an experimental assessment. *Proc. IMechE, Part D: Journal of Automobile Engineering,* Vol.223, No.11, (November 2009), pp. 1373-1394

Sammut, G. & Alkidas, A.C. (2007). Relative contributions of intake and exhaust tuning on SI engine breathing-A computational study, SAE paper 2007-01-0492

Wang, X. and Stone, C.R. (2008). A study of combustion, instantaneous heat transfer, and emissions in a spark ignition engine during warm-up. *Proc. IMechE,* Vol.222, pp. 607-618

Wu, Y., Chen, B., Hsieh, F. & Ke, C. (2008). Heat transfer model for scooter engines, SAE paper 2008-01-0387

12

Heat Conduction for Helical and Periodical Contact in a Mine Hoist

Yu-xing Peng, Zhen-cai Zhu and Guo-an Chen
School of Mechanical and Electrical Engineering,
China University of Mining and Technology, Xuzhou,
China

1. Introduction

Mine hoist is the "throat" of mine production, which plays the role of conveying coal, underground equipments and miners. Fig. 1 shows the schematic of mining friction hoist. The friction lining is fixed outside the drum and the wire rope is hung on the drum. It is dependent on friction force between friction lining and wire rope to lift miner, coal and equipment during the process of mine hoisting. Accordingly, the reliability of mine hoist is up to the friction force between friction lining and wire rope. Therefore, the friction lining is one of the most important parts in mine hoisting system. In addition, the disc brake for mine hoist is shown in Fig. 1 and it is composed of brake disc and brake shoes. During the braking process, the brake shoes are pushed onto the disc with a certain pressure, and the friction force generated between them is applied to brake the drum of mine hoist. And the disc brake is the most significant device for insuring the safety of mine hoist. Therefore, several strict rules for disc brake and friction lining are listed in "Safety Regulations for Coal Mine" in China (Editorial Committee of Mine Safety Handbooks, 2004).

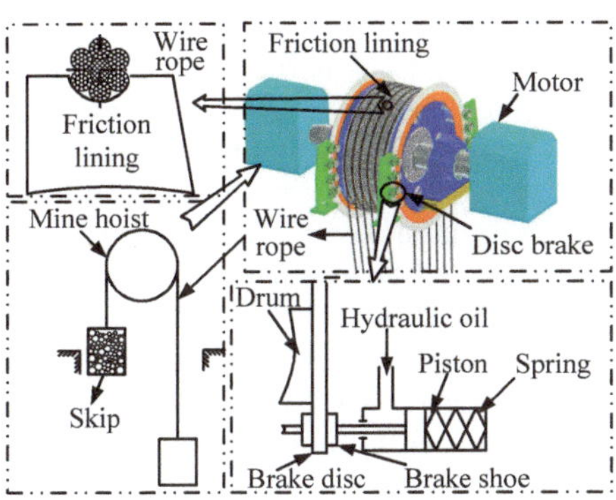

Fig. 1. Schematic of mine friction hoist

Under the condition of overload, overwinding or overfalling of a mine hoist, the high-speed slide occurs between friction lining and wire rope which will results in a serious accident. At this situation, the disc brake would be acted to brake the drum with large pressure, which is called a emergency brake. And a large amount of friction heat accumulates on the friction surface of friction lining and disc brake during the braking process. This leads to the decrease of mechanical property on the contact surface, which reduces the tribological properties and makes the hoist accident more serious. Therefore, it is necessary to study the heat conduction of friction lining and disc brake during the high-speed slide accident in a mine hoist.

The heat conduction of friction lining has been studied (Peng et al., 2008; Liu & Mei, 1997; Xia & Ge, 1990; Yang, 1990). However, the previous work neglected the non-complete helical contact between friction lining and wire rope. Besides, the previous results were based on the static thermophysical property (STP). But the friction lining is a kind of polymer and the thermophysical properties (specific heat capacity, thermal diffusivity and thermal conductivity) vary with the temperature (Singh et al., 2008; Isoda & Kawashima, 2007; He et al., 2005; Hegeman et al., 2005; Mazzone, 2005). Therefore, the temperature field calculated by STP is inconsistent with the actual temperature field. The methods solving the heat-conduction equation include the method of separation of variables (Golebiowski & Kwieckowski, 2002; Lukyanov, 2001), Laplace transformation method (Matysiak et al., 2002; Yevtushenko & Ivanyk, 1997), Green's function method (Naji & Al-Nimr, 2001), integral-transform method (Zhu et al., 2009), finite element method (Voldrich, 2007; Qi & Day, 2007; Thuresson, 2006; Choi & Lee, 2004) and finite difference method [Chang & Li, 2008; Liu et al., 2009], etc. The former three methods are analytic solution methods and it is difficult to solve the heat-conduction problem with the dynamic theromophysical property (DTP) and complicate boundary conditions. Though the integral-transform method is a numerical solution method and is suitable for solving the problem of non-homogeneous transient heat conduction, it is incapable of solving the nonlinear problem. Additionally, both the finite element method and finite difference method could solve nonlinear heat-conduction problem. However, the finite difference expression of the partial differential equation is simpler than finite element expression. Thereby, the finite difference method is adopted to solve the nonlinear heat-conduction problem with DTP and non-complete helical contact characteristics.

It is depend on the friction force between brake shoe and brake disc to brake the drum of mine hoist. So the safety and reliability of disc brake are mainly determined by the tribological properties of its friction pair. The tribological properties of brake shoe were studied (Zhu et al., 2008, 2006), and it was found that the temperature rise of disc brake affects its tribological properties seriously during the braking process, which in turn threatens the braking safety directly. Presently, most investigations on the temperature field of disc brake focused only on the operating conditions of automobile. The temperature field of brake disc and brake shoe was analyzed in an automobile under the emergency braking condition (Cao & Lin, 2002; Wang, 2001). The effects of parameters of operating condition on the temperature field of brake disc (Lin et al., 2006). Ma adopted the concept of whole and partial heat-flux, and considered that the temperature rise of contact surface was composed of partial and nominal temperature rise (Ma et al., 1999). And the theoretical model of heat-flux under the emergency braking condition was established by analyzing the motion of automobile (Ma & Zhu, 1998). However, the braking condition in mine hoist is worse than that in automobile, and the temperature field of its disc brake may show different behaviors. Nevertheless, there are a few studies on the temperature field of mine hoist's disc brake. Zhu investigated the temperature field of brake shoe during emergency braking in mine hoist (Zhu et al., 2009). Bao brought forward a new method of calculating the maximal surface temperature of brake shoe during mine hoist's emergency braking (Bao et al., 2009). And yet, the above studies were based on the invariable thermophysical properties of brake shoe, and the temperature field of brake disc hasn't been investigated.

In order to master the heat conduction of friction lining and improve the mine safety, the non-complete helical contact characteristics between friction lining and wire rope was analyzed, and the mechanism of dynamic distribution for heat-flow between friction lining and wire rope was studied. Then, the average and partial heat-flow density were analyzed. Consequently, the friction lining's helical temperature field was obtained by applying the finite difference method and the experiment was performed on the friction tester to validate the theoretical results. Furthermore, the heat conduction of disc brake was studied. The temperature field of brake shoe was analyzed with the consideration of its dynamic thermophysical properties. And the brake disc's temperature rise under the periodical heat-flux was also investigated. The research results will supply the theoretical basis with the anti-slip design of mine friction hoist, and our study also has general application to other helical and periodical contact operations.

2. Heat conduction for helical contact

2.1 Helical contact characteristics

In order to obtain the temperature rise of friction lining during sliding contact with wire rope, it is necessary to analyze the contact characteristic between friction lining and wire rope. The schematic of helical contact is shown in Fig. 2.

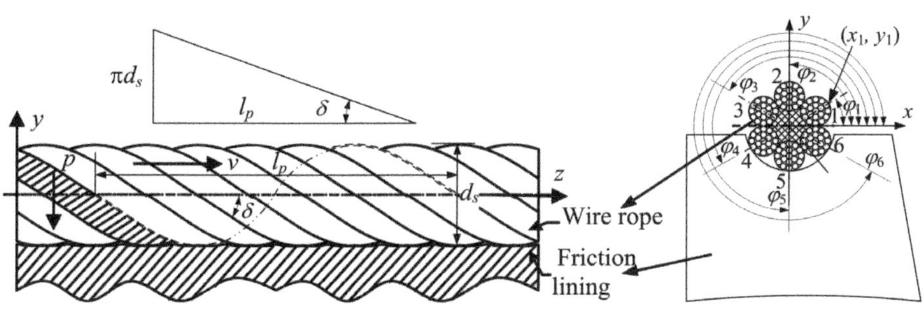

Fig. 2. Schematic of helical contact

For obtaining the exact heat-flow generated by the helical contact, the contact characteristics must be determined firstly. As is shown Fig. 2, the friction lining contacts with the outer strand of wire rope which is a helical structure and the helical equation is as follows

$$\begin{cases} x_i = \dfrac{d_s}{2} \cdot \cos \varphi_i = \dfrac{d_s}{2} \cdot \cos\left[\omega t + \dfrac{\Pi}{3} i - \dfrac{\Pi}{6}\right] \\ y_i = \dfrac{d_s}{2} \cdot \sin \varphi_i = \dfrac{d_s}{2} \cdot \sin\left[\omega t + \dfrac{\Pi}{3} i - \dfrac{\Pi}{6}\right] \\ z_i = v \cdot t \end{cases} \tag{1}$$

where j is the helix angle, i is the strand number in the wire rope (i=1, 2, 3,···,6), d_s is the diameter of the wire rope, and v is the relative speed between friction lining and wire rope. It is seen from Fig. 2(a) that, any point on the contact surface of friction lining contacts periodically with the outer surface of wire rope because of wire rope's helical structure, and the period for unit pitch is expressed as

$$T_P = \frac{l_p}{v} = \frac{\pi d_s}{\tan\delta \cdot v} \tag{2}$$

where l_p is the pitch of outer strand, d_s is the lay angle of strand ($\delta_s = 0.28$), and $\omega = 2\pi/T_P$ in Eq. (1).

The contact characteristics can be gained according to Eqs. (1) and (2). The variation of j_i corresponding to coordinates x_i and y_i is shown in Fig. 3.

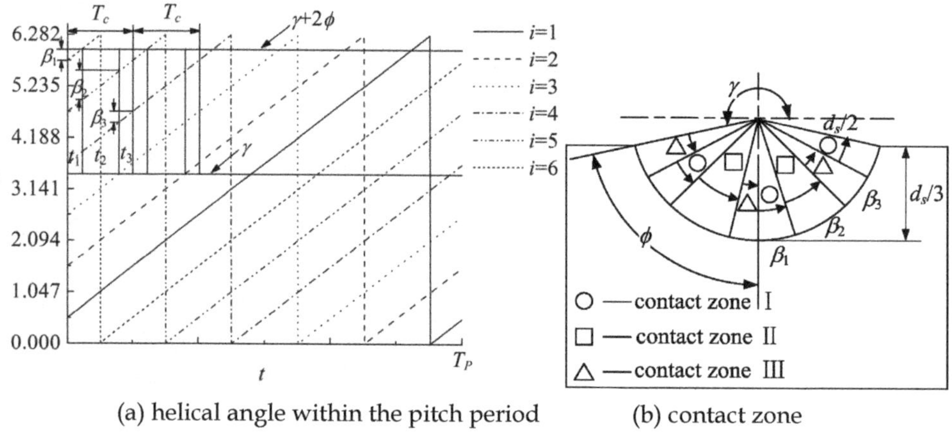

Fig. 3. Schematic of helical contact

From Figs. 3(a) and 3(b), it is observed that the contact period is T_c within the angle (g~g+2f) of the rope groove in the lining, and the contact zone is divided into three regions which is shown in Eq. (3).

$$\begin{aligned}
&\gamma_1 \in \left\{ \left(\frac{7}{6}\pi, \frac{7}{6}\pi + \beta_1\right), \left(\frac{3}{2}\pi, \frac{3}{2}\pi + \beta_1\right), \left(\frac{11}{6}\pi, \frac{11}{6}\pi + \beta_1\right) \right\}, \\
&\qquad t \in \left(mT_C,\ mT_C + t_1\right); \\
&\gamma_2 \in \left\{ \left(\frac{7}{6}\pi + \beta_1, \frac{7}{6}\pi + \beta_1 + \beta_2\right), \left(\frac{3}{2}\pi + \beta_1, \frac{3}{2}\pi + \beta_1 + \beta_2\right) \right\}, \\
&\qquad t \in \left(mT_C + t_1,\ mT_C + t_1 + t_2\right); \\
&\gamma_3 \in \left\{ \left(\gamma, \frac{7}{6}\pi\right), \left(\frac{7}{6}\pi + \beta_1 + \beta_2, \frac{3}{2}\pi\right), \left(\frac{3}{2}\pi + \beta_1 + \beta_2, \frac{11}{6}\pi\right) \right\}, \\
&\qquad t \in \left(mT_C + t_1 + t_2,\ mT_C + T_C\right),\ (m = 0,1,2,\cdots);
\end{aligned} \tag{3}$$

where b_s is the angle increment within t_s, t_s is the contact time, $t_s = b_s/w$ (s=1, 2, 3), $T_c = t_1+t_2+t_3$; where $\phi = 1.27$, $\beta_1 = \beta_3 = 0.22, \beta_2 = 0.6$. It is seen from Fig. 3(a) and Fig. 3(b), the lining groove contacts with the outside of wire rope and the number of contact point is two or three. And the contact arc length is unequal. At the certain speed, the contact arc length within t_2 is the longest and the contact arc length within t_2 and t_3 is equal.

2.2 Mechanism of dynamic distribution for heat-flow

2.2.1 Dynamic thermophysical properties of friction lining

At present, the linings G and K are widely used in most of mine friction hoists in China. The lining is kind of polymer whose thermophysical properties are temperature-dependent. In orer to master the friction heat, it is necessary to study their dynamic thermophysical properties. In this study, the selected sample G and K were analyzed, and its thermophysical properties were measured synchronistically on a light-flash heat conductivity apparatus (LFA 447). Given the friction lining's density *r*, the thermal conductivity is defined by

$$\lambda(T) = \rho \times C_p(T) \times \alpha(T) \tag{4}$$

where C_p is the specific heat capacity and *a* is the thermal diffusivity

It is seen from Fig. 4(a) that the C_p increases with the temperature and the lining G has higher value of C_p than lining K. In Fig. 4(b), the *a* decreases with the temperature nonlinearly whose value of lining G is obviously higher than that of K. As shown in Fig. 4(c), the *l* increases with the temperature below 90°C and keeps approximately stable above 90°C. And the *l* of lining G is about 0.45lw·m^{-1}k^{-1} within the temperature range (90°C~240°C), while that of lining K is only 0.3w·m^{-1}k^{-1}.

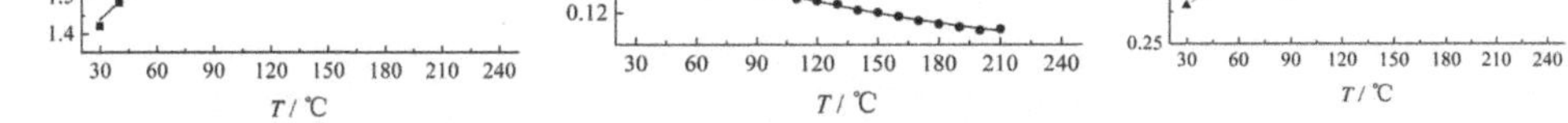

(a) Specific heat capacity (b) Thermal diffusivity (c) Thermal conductivity

Fig. 4. Dynamic thermophysical parameters of friction linings

According to the change rules of specific heat capacity and thermal diffusivity in Fig. 4, the polynomial fit and exponential fit are used to fit curves, and the fitting equations are as follows:

for lining G,

$$\begin{cases} C_p(T) = 1.344 + 8.48\times10^{-3}T - 4\times10^{-5}T^2 + 1.026\times10^{-7}T^3,\ r_0^{\,2} = 0.972 \\ \alpha(T) = 0.132 + 0.0832\times e^{-\frac{T-30.1}{148.749}}, r_0^{\,2} = 0.998 \end{cases} \tag{5}$$

for lining K,

$$\begin{cases} C_{\mathrm{p}}(T) = 1.272 + 6.31 \times 10^{-3} T - 2 \times 10^{-5} T^2 + 4.861 \times 10^{-8} T^3, \ r_0^{\ 2} = 0.961 \\ \alpha(T) = 0.104 + 0.0379 \times \mathrm{e}^{-\frac{T-29.6}{141.032}}, \ r_0^{\ 2} = 0.996 \end{cases} \tag{6}$$

where r_0^2 is the correlation coefficient whose value is close to 1, which indicates that the fitting curves agree well with the experiment results. Consequently, the fitting equation of thermal conductivity of Lining G is deduced by Eqs. (4) and (5).

2.1.2 Dynamic distribution coefficient of heat-flow

In order to master the real temperature field of the friction lining, the distribution coefficient of heat-flow must be determined with accuracy. Suppose the frictional heat is totally transferred to the friction lining and wire rope. According to the literature (Zhu et al., 2009), the dynamic distribution coefficient of heat-flow for the friction lining is obtained

$$k = \frac{q_{\mathrm{f}}}{q_{\mathrm{f}} + q_{\mathrm{W}}} = 1 - \frac{q_{\mathrm{W}}}{q_{\mathrm{f}} + q_{\mathrm{W}}} = 1 - \frac{1}{\frac{q_{\mathrm{f}}}{q_{\mathrm{W}}} + 1} = 1 - \frac{1}{\sqrt{\frac{\rho C_{\mathrm{p}} \lambda}{\rho_{\mathrm{w}} C_{\mathrm{pw}} \lambda_{\mathrm{w}}}} + 1} \tag{7}$$

where q_{f} and q_{w} are the heat-flow entering the friction lining and wire rope. r_{w}, C_{pw}, l_{w} and a_{w} are the density, special heat, thermal diffusivity and thermal conductivity of wire rope, respectively.

2.3 Heat-flow density

Determining the friction heat-flow accurately during the sliding process is the important precondition of calculating the temperature field of friction lining. In this study, according to the force analysis of friction lining under the experimental condition, the total heat-flow is studied. And the partial heat-flow on the groove surface of friction lining is gained with the consideration of mechanism of dynamic distribution for heat-flow and helical contact characteristic.

The sliding friction experiment is performed on the friction tester. As shown in Fig. 2, the average heat-flow entering the friction lining under the experiment condition is given as

$$q_{\mathrm{f}} = k \cdot q_{\mathrm{a}} = k \cdot f_1 \cdot p \cdot v \tag{8}$$

where f_1 is the coefficient of friction between friction lining and wire rope, p is the average pressure on the rope groove of friction lining, v is the sliding speed.

According to the helical contact characteristic, the contact period is divided into three time period. Therefore, the partial heat-flow at every time period is obtained on the basis of the contact time

$$q_{\mathrm{f1}} = q_{\mathrm{f}} \frac{t_1}{t_1 + t_2 + t_3}, \ q_{\mathrm{f2}} = q_{\mathrm{f}} \frac{t_2}{t_1 + t_2 + t_3}, \ q_{\mathrm{f3}} = q_{\mathrm{f}} \frac{t_3}{t_1 + t_2 + t_3} \tag{9}$$

2.4 Theoretical analysis on temperature field of friction lining

2.4.1 Theoretical model

On the basis of the above analysis of contact characteristics, it reveals that the temperature field is nonuniform due to the non-complete helical contact between friction lining and wire rope. Moreover, the heat conduction equation is nonlinear on account of DTP. Based on the heat transfer theory, the heat conduction equation, the boundary condition and the initial condition are obtained from Fig. 2:

$$\frac{\partial}{\partial r}\left(\lambda(T)\frac{\partial T}{\partial r}\right)+\frac{\lambda(T)}{r}\frac{\partial T}{\partial r}+\frac{1}{r^2}\frac{\partial}{\partial \theta}\left(\lambda(T)\frac{\partial T}{\partial \theta}\right)=\rho(T)C_p(T)\frac{\partial T}{\partial t} \tag{10}$$

$$\begin{aligned}
&-\lambda\frac{1}{r}\frac{\partial T}{\partial \theta}+h_1T=h_1T_0 && (\theta=-\phi,t\ge 0,r_1\le r\le r_2) && \text{(a)}\\
&\lambda\frac{1}{r}\frac{\partial T}{\partial \theta}+h_2T=h_2T_0 && (\theta=\phi,t\ge 0,r_1\le r\le r_2) && \text{(b)}\\
&-\lambda\frac{\partial T}{\partial r}+h_3T=q_f+h_3T_0 && (r=r_1,t\ge 0,-\phi\le\theta\le\phi) && \text{(c)}\\
&\lambda\frac{\partial T}{\partial r}+h_4T=h_4T_0 && (r=r_2,t\ge 0,-\phi\le\theta\le\phi) && \text{(d)}\\
&T(r,\theta,\alpha,t)=T_0 && (t=0,r_1\le r\le r_2,-\phi\le\theta\le\phi) && \text{(e)}
\end{aligned} \tag{11}$$

where h_m is the coefficient of convective heat transfer (m=1, 2, 3 , 4).

2.4.2 Solution

The finite difference method is adopted to solve Eqs. (10) and (11), because it is suitable to solve the problem of nonlinear transient heat conduction. Firstly, the solving region is divided into grid with mesh scale of Dr and Dq, and the time step is Dt. And then the friction lining's temperature can be expressed as

$$T(r,\theta,t)=T(r_1+i\Delta r,j\Delta\theta,n\Delta t)=T_{i,j}^{n} \tag{12}$$

The central difference is utilized to express the partial derivatives $\partial T/\partial r$ 、 $\partial(\lambda\partial T/\partial r)/\partial r$ and $\partial(\lambda\partial T/\partial\theta)/\partial\theta$, and their finite difference expressions are obtained

$$\begin{aligned}
&\frac{\partial T}{\partial r}=\frac{T_{i+1,j}-T_{i-1,j}}{2\Delta r}+O(\Delta r)^2 && \text{(a)}\\
&\frac{\partial}{\partial r}\left(\lambda\frac{\partial T}{\partial r}\right)=\frac{\lambda_{i-1/2,j}\left(T^{n+1}{}_{i-1,j}-T^{n+1}{}_{i,j}\right)-\lambda_{i+1/2,j}\left(T^{n+1}{}_{i,j}-T^{n+1}{}_{i+1,j}\right)}{(\Delta r)^2}+O(\Delta r)^2 && \text{(b)}\\
&\frac{\partial}{\partial \theta}\left(\lambda\frac{\partial T}{\partial \theta}\right)=\frac{\lambda_{i-1/2,j}\left(T^{n+1}{}_{i,j-1}-T^{n+1}{}_{i,j}\right)-\lambda_{i+1/2,j}\left(T^{n+1}{}_{i,j}-T^{n+1}{}_{i,j+1}\right)}{(\Delta \theta)^2}+O(\Delta\theta)^2 && \text{(c)}
\end{aligned} \tag{13}$$

Submit Eq. (13) into Eq. (10), the following equation is obtained

$$\frac{\lambda_{i-1/2,j}\left(T^{n+1}{}_{i-1,j}-T^{n+1}{}_{i,j}\right)-\lambda_{i+1/2,j}\left(T^{n+1}{}_{i,j}-T^{n+1}{}_{i+1,j}\right)}{(\Delta r)^2}+\lambda_{i,j}\frac{T^{n+1}_{i+1,j}-T^{n+1}_{i-1,j}}{2i\Delta r^2}$$
$$+\frac{\lambda_{i-1/2,j}\left(T^{n+1}{}_{i,j-1}-T^{n+1}{}_{i,j}\right)-\lambda_{i+1/2,j}\left(T^{n+1}{}_{i,j}-T^{n+1}{}_{i,j+1}\right)}{i^2(\Delta r\Delta\theta)^2}=\left(\rho C_{\mathrm{p}}\right)_{i,j}\frac{T^{n+1}_{i,j}-T^{n}_{i,j}}{\Delta t} \tag{14}$$

where the subscript (i-1/2) of l denotes the average thermal conductivity between note i and note i-1, and the subscript (i+1/2) is the average thermal conductivity between note i and note i+1. In the same way, the difference expressions of boundary condition can be gained by the forward difference and backward difference:

$$\begin{aligned}
&-\lambda_{i,-N}\frac{T^{n+1}_{i,-N+1}-T^{n+1}_{i,-N}}{i\Delta r\Delta\theta}+h_1T^{n+1}_{i,-N}=h_1T_0 \quad (j=-N) \quad \text{(a)}\\
&\lambda_{i,N}\frac{T^{n+1}_{i,N}-T^{n+1}_{i,N-1}}{i\Delta r\Delta\theta}+h_2T^{n+1}_{i,N}=h_2T_0 \quad (j=N) \quad \text{(b)}\\
&-\lambda_{0,j}\frac{T^{n+1}_{1,j}-T^{n+1}_{0,j}}{i\Delta r}+h_3T^{n+1}_{0,j}=q_{\mathrm{f}}+h_3T_0 \quad (i=0) \quad \text{(c)}\\
&-\lambda_{M,j}\frac{T^{n+1}_{M,j}-T^{n+1}_{M-1,j}}{i\Delta r}+h_4T^{n+1}_{M,j}=h_4T_0 \quad (i=M) \quad \text{(d)}\\
&T^{0}_{i,j}=T_0 \qquad (n=0) \qquad \text{(e)}
\end{aligned} \tag{15}$$

Combined with Eqs. (14) and (15), the friction lining's temperature field is obtained by the iterative computations.

2.5 Experimental study

At present, the non-contact thermal infrared imager is widely used to measure the exposed surface, while the friction surface contacts with each other and it is impossible to gain the

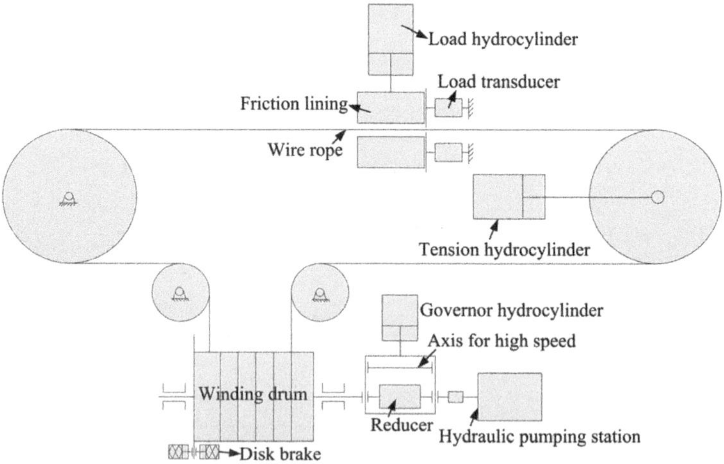

Fig. 5. Schematic of friction tester

surface temperature by the non-contact measurement. Presently, there is no better way to measure the temperature of friction contact surface. In this study, the thermocouple is used to measure to the surface layer temperature, which is embedded in the friction lining and closed to the friction surface. The experiment is performed on the friction tester to study the temperature of friction lining during the friction sliding process. In Fig. 5, the hydraulic pumping station drives the winding drum through the coupling device (axis for high speed or reducer for low speed), and the governor hydrocylinder controls which of the coupling devices would be connected. The wire rope is wrapped on the winding drum, and the motion of the drum leads to the cyclical motion of wire rope. Before the wire rope moves, the tension hydrocylinder makes the wire rope tense and the friction lining is pushed by the load hydrocylinder to clamp the wire rope. Consequently, as the wire rope moves, the friction force is measured by the load transducer and the normal force acted on the wire rope is deduced from the hydraulic pressure of the load hydrocyclinder.

2.5.1 Thermocouple layout

According to the helical contact characteristics, eight thermocouples with the diameter of 0.3mm were embedded in the friction lining which are close to the contact surface. The position of thermocouple is shown in Fig. 6. Firstly, the holes with the diameter of 1mm were drilled on the lateral side of friction lining. Then the oddment of the friction lining was filled in the hole after the thermocouple was embedded. In Fig. 8, points a, b and c were in the central line of rope groove, points d and h were in the middle of contact zone I, points e and f were in the middle of contact zone II and point g was in the middle of contact zone III. The distance from points c, d, f, g and h to the contact surface is about 2mm, and the distance between points e and f, a and b, b and c is about 2mm, too.

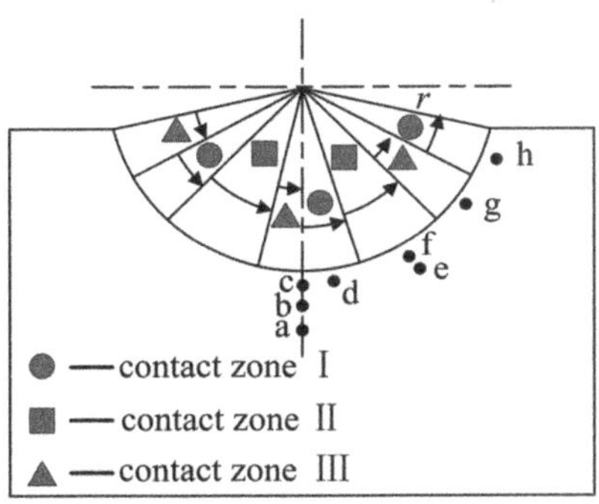

Fig. 6. Layout of thermocouple

2.5.2 Experimental parameters

The sliding speed and the equivalent pressure are the main factors affecting the temperature rise of friction lining during the sliding process. Therefore, the experiments were carried out with different sliding speeds and equivalent pressures. The sliding distance is about 20m. According to the friction experiment standard for friction lining (MT/T 248-91, 1991), the equivalent pressure is 1.5~3MPa. The parameters for the experiment are listed in Table 1.

	$v \leq 10$mm/s	$v > 10$mm/s
Equivalent pressure (MPa)	1.5, 2, 2.5, 3	1.5, 2.5
Speed (mm/s)	1, 3, 5, 7, 10	30, 100, 300, 500, 700, 1000

Table 1. Parameters for friction experiment

2.5.3 Experimental results

Fig. 7 shows the partial experiment results within the speed range of 1~10mm/s.

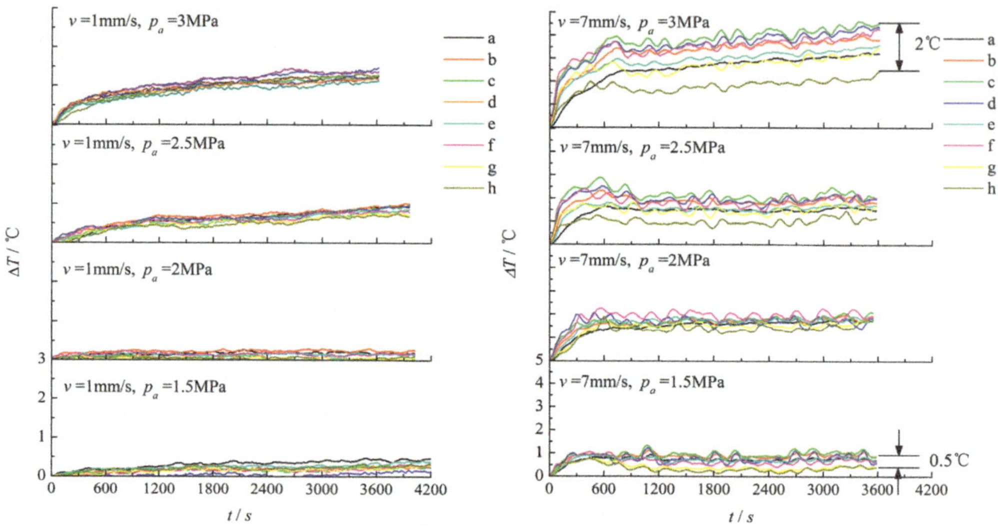

(a) 1mm/s (b) 7mm/s

Fig. 7. Variation of testing points' temperature

As shown in Fig. 7, the temperature rise is less than 5°C within 1 hour when the sliding speed is less than 10mm/s. Therefore, the temperature rise of friction lining can be neglected under the normal hoist condition. It is observed that the temperature increases wavily and the amplitude of waveform increases with the sliding speed. This is due to the periodical heat-flow resulting from the helical contact characteristics. In addition, the temperature difference among 8 points is small and it increases with the equivalent pressure: the temperature difference increases from 0.5°C to 2°C when the equivalent pressure increases to 3MPa. It is found that the temperature increases quickly at the beginning of the sliding process, and then it increases slowly.

In order to analyze the effect of the sliding speed and the equivalent pressure on the temperature, Fig. 8 shows the temperature rise of point c at different sliding speeds and equivalent pressures.

It is seen from Fig. 8 that the sliding speed has stronger effect than the equivalent pressure on the temperature. It is concluded that the sliding speed is more sensitive to the temperature. Therefore, only two equivalent pressures (1.5MPa and 2.5MPa) are selected at the high-speed experiment.

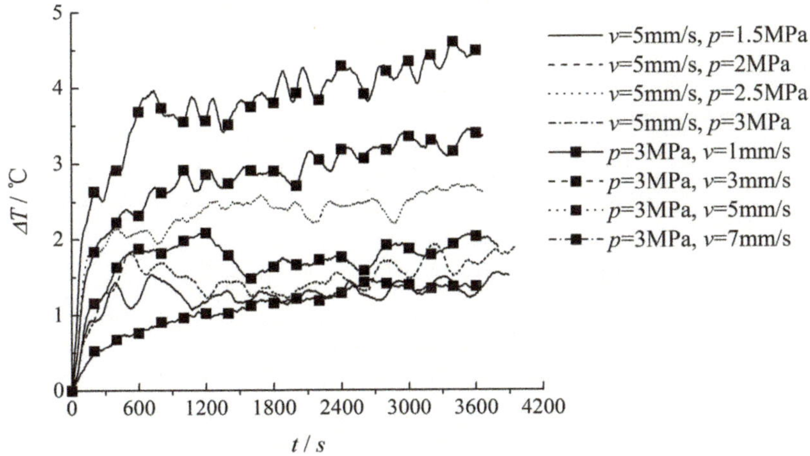

Fig. 8. Effect of speed and equivalent pressure on temperature within low speed

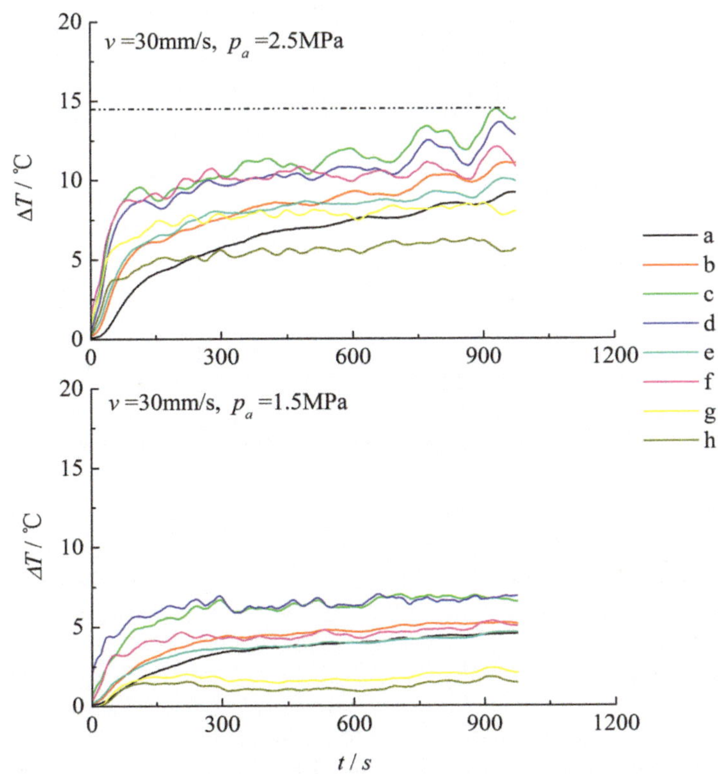

Fig. 9. Variation of testing points' temperature

As show in Fig. 9, the highest temperature rise increases to 15°C at the speed of 30mm/s. Additionally, it is observed that the temperature at points c, d and f is higher than that at other points. However, the temperature rise is not high enough to tell the highest among them.

	Distance between points and point to friction surface / mm (fs-friction surface)							
	a-b	b-c	c-fs	d-fs	e-f	f-fs	g-fs	h-fs
1.5Mpa	2	2	1.8	1.38	2	2.16	2.24	2.24
2.5MPa	1.98	2.06	1.7	2	1.9	1.7	1.48	1.48

Table 2. Position of testing points in friction lining

In order to master the surface temperature of friction lining, the friction experiment was performed with the increasing speed. Table 2 shows the distance of 8 points, and Fig. 10 shows the partial experiment results within the speed range 100~1000mm/s.
It is seen from Fig. 10 that the temperature rise of every point increases obviously. And the order of temperature rise at testing points is shown in Table 3.

Pressure / speed	Order of temperature rise Lining G（high→low）
1.5MPa / 100mm/s	d, c, f, b, e, a, g, h
1.5MPa / 400mm/s	d, c, f, b, e, h, g, a
1.5MPa / 600mm/s	d, f, c, g, h, e, b, a
1.5MPa / 800mm/s	d, c, f, g, e, h, b, a
1.5MPa / 1000mm/s	d, f, c, g, h, e, b, a
2.5MPa / 200mm/s	f, g, c, d, e, h, b, a
2.5MPa / 300mm/s	f, c, g, d, e, h, b, a
2.5MPa / 550mm/s	f, c, g, d, e, h, b, a
2.5MPa / 750mm/s	f, g, c, d, h, e, b, a
2.5MPa / 980mm/s	f, c, d, g, h, e, b, a

Table 3. Order of temperature rise at testing points

The highest temperature rise occurs at point d with p_a=1.5MPa while the highest temperature rise appears at point f with p_a=2.5MPa. This is because that point d is close to the friction surface with the minimize distance of 1.38mm while the distance from point f to friction surface is 2.16mm, which reveals the temperature gradient in the surface layer is high. Therefore, the temperature at point d is higher than that at point f. It is found from Table 3 that the temperature at points f, d and c is higher than that at other points, which is in accordance with the analytical results of the helical contact characteristics and partial heat-flow density. As shown in Fig. 3(b), the contact zone II is subject to the long-time heat-flow and the convection heat transfer of contact zone I at the bottom of the rope groove is worse than that of other zone. Consequently, the temperature at points c, d and f is higher.

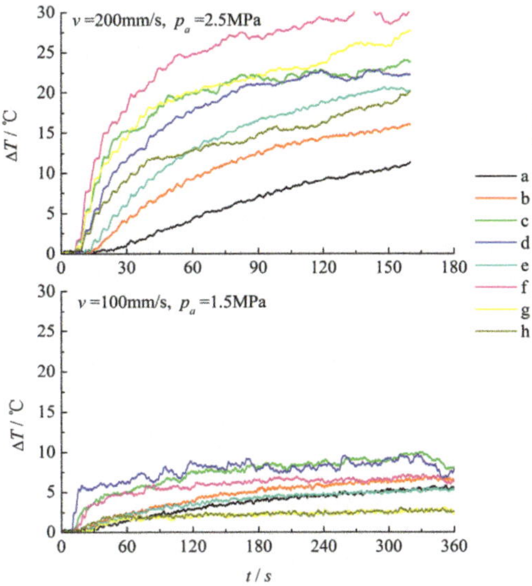

(a) 100-210mm/s

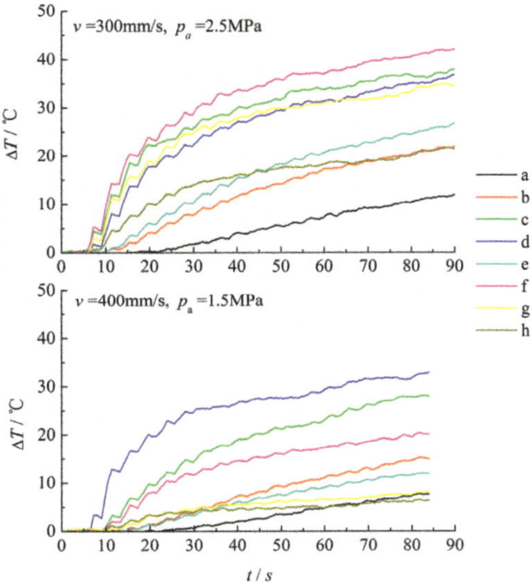

(b) 300-400mm/s

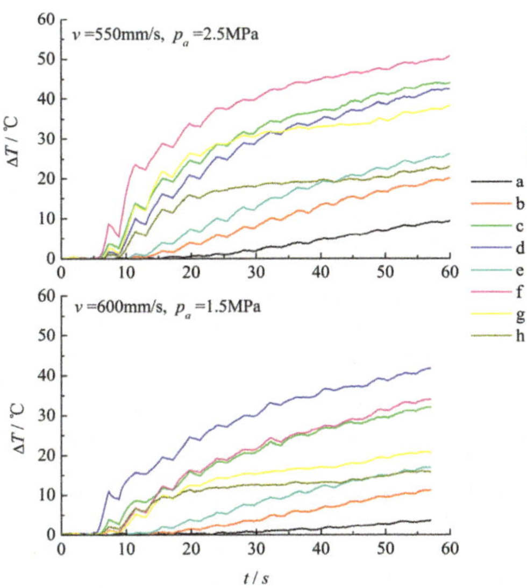

(c) 550-600mm/s

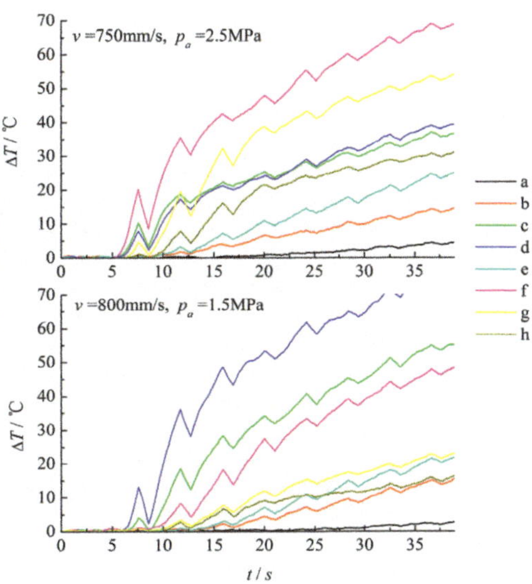

(d) 750-800mm/s

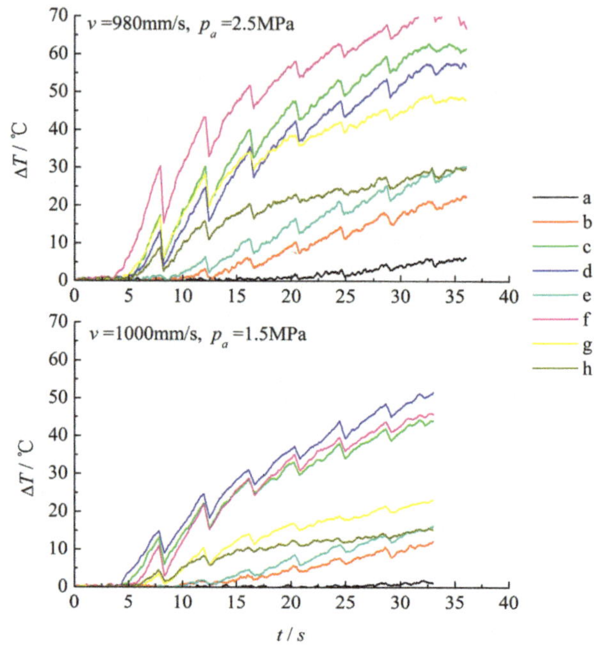

(e) 980-1000mm/s

Fig. 10. Variation of testing points' temperature

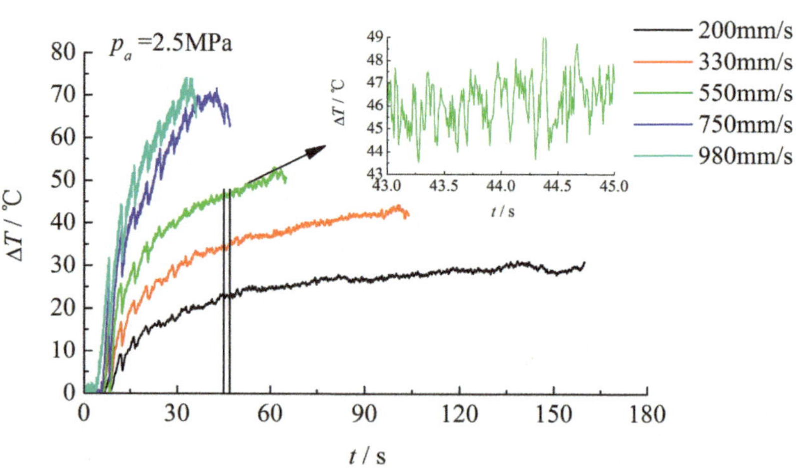

Fig. 11. Variation of temperature rise at point f with speed

The variation of the temperature at point f with the speed is shown in Fig. 11. It is found that the temperature at point f increases with the sliding speed. And the gradient of the temperature rise increases rapidly with the speed at the beginning of the sliding process. Additionally, from the drawing of partial enlargement 550mm/s, the temperature increases periodically which agrees with the helical contact characteristics.

Combining Fig. 10 and Fig. 11, it is found that the temperature rise increases wavily at the initial sliding stage, and the amplitude of wave increases with the sliding speed while it decreases with the time. The explanation for the reduced temperature in the wavy temperature is given as: (a) the rapid temperature rise results in decrease of the mechanical property, and the contact area is enlarged which accelerates the heat exchange between friction lining and wire rope, thus the temperature of friction lining reduces; (b) the increase of contact area reduces the equivalent pressure and the heat-flow decreases rapidly; (c) due to unstable and discontinuous speed at the initial sliding stage, the heat exchange between friction lining and wire rope increases and the temperature decreases in a short time. As the sliding distance increases, the heat exchange tends to balance which decrease the amplitude of the wave.

2.6 Analysis on numerical simulation and experimental results

In order to validate the theoretical model, the theoretical results are compared with the experiment results. The parameters for the experiment are: v=0.55m/s, p_a=2.5MPa. h_1=h_2=h_4=10W/m²K. Supposed that the contact surface of rope groove is only subjected to heat-flow, then h_3=0. r_1=0.014m, r_2=0.04m and T_0=30°C. And the thermophysical properties of wire rope are shown in Table 4.

	ρ_w/ (kg·m^{-3})	C_{pw}/ (J·kg^{-1}·K^{-1})	λ_w / (W·m^{-1}·K^{-1})
Wire rope	7866	473	53.2

Table 4. Thermophysical properties of wire rope

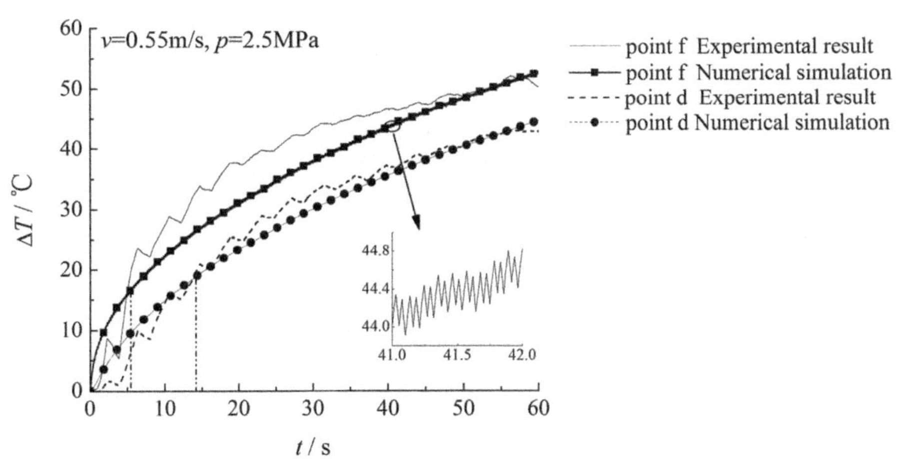

Fig. 12. Comparison between simulation and experimental results

As shown in Fig. 12, the temperature rise at point f increases to 52°C at 60s. From the drawing of the partial enlargement at point f, the simulation result shows that the temperature increases during the cycle of heat absorption and heat dissipation, which agrees with the experiment results in Fig. 11. At the beginning of the sliding process, the temperature of simulation result is higher than that of experiment results. And the experimental value fluctuates obviously. This is because that the thermocouple absorbs the heat, and the speed at the initial state of sliding process is unstable which leads to the heat conduction for longer time at the local zone. Thus the experimental data is low and the curve of temperature behaves serrasoidal. As the sliding process continues, the experimental value is higher than the simulation value and the both values tend to be equal in the end. This is because that the friction lining is subjected to temperature and stress and the temperature rise results in the increase of surface deformation in the rope groove, which makes the embedding thermocouple closer to the friction surface and leads to the rapid temperature rise of the measuring point. Therefore, the experimental value is higher than the theoretical value: the higher temperature rise at point f makes the deformation bigger and the thermocouple is closer to the surface, which makes the temperature difference between the experimental value and theoretical value at point f is higher than that at point d. When the heat conduction tends to achieve the balance, the temperature variation gets gently and the experimental value is consistent with the theoretical value. Compared the experimental value with the theoretical value in Fig. 12, both of them agree well with each other which validates the theoretical model of the temperature field.

The above analysis indicates that the theoretical model of temperature field is reasonable and correct. Due to difficultly obtaining the temperature of the friction surface by the way of the direct contact measurement, the temperature of the contact surface is simulated in Fig. 13.

As in Fig. 13, the temperature on the friction surface is much higher: though the distance between point f and friction surface is only 2mm, the temperature at friction surface is 18°C higher than that at point f. In addition, the radial temperature gradient at different time is simulated in Fig. 14. The temperature gradient at the surface layer is high and its maximum is about 35000°C/m. And the temperature gradient decreases with the radius. As the sliding process continues, the curve of temperature gradient at the surface layer tends to be flat. The

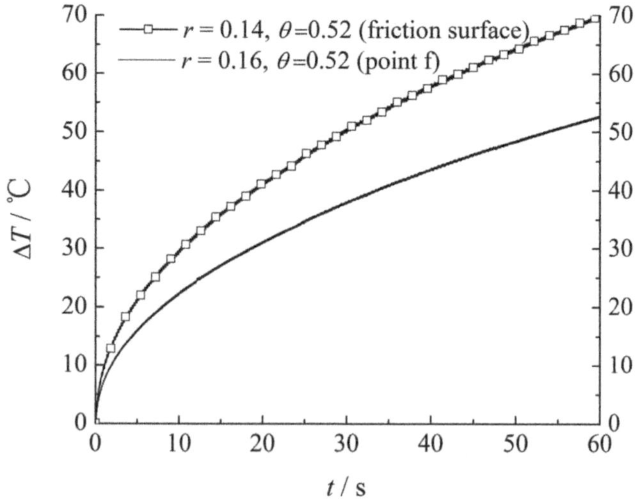

Fig. 13. Variation of temperature on friction surface

above analysis indicates that the heat-conducting property of friction lining is poor. Thus, in order to develop the new friction lining with good thermophysical properties, it is necessary to optimize the ratio of basic material and filler and select the component with good heat-conducting property.

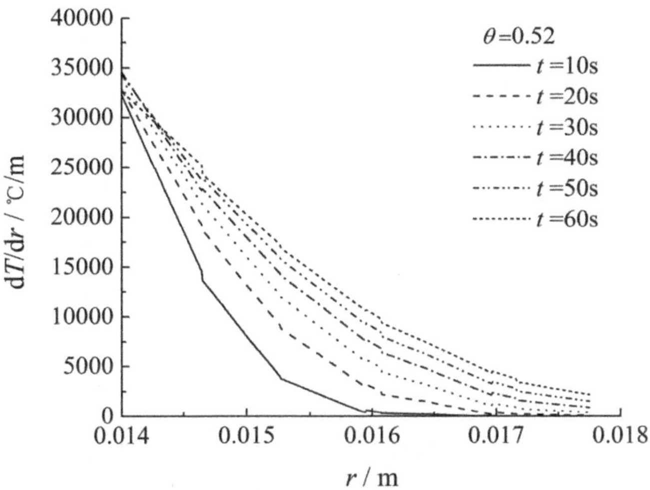

Fig. 14. Radial temperature gradient at different time

3. Heat conduction for periodical contact

3.1 Theoretical model

As shown in Fig. 1, the disc brake for mine hoist is composed of brake disc and brake shoes. During the braking process, the brake shoes are pushed onto the disc with a certain pressure, and the friction force generated between them is applied to brake the drum of mine hoist.

The heat energy caused by the friction between brake shoes and brake disc leads to their temperature rise. However, the two important parts have different behaviors of heat conduction: the brake shoe is subjected to continuous heat while the brake disc is heated periodically due to rotational motion. And the two types of temperature field will be discussed as follows.

3.1.1 Dynamic thermophysical properties of brake shoe

In order to obtain the temperature field of disc brake, it is necessary to obtain their thermophysical properties. As the brake shoe is a kind of composite material, its dynamic thermophysical properties (DTP) (specific heat capacity c_s, thermal diffusivity a_s and thermal conductivity l_s) vary with the temperature. And the testing results (Bao, 2009) are shown in Fig. 15.

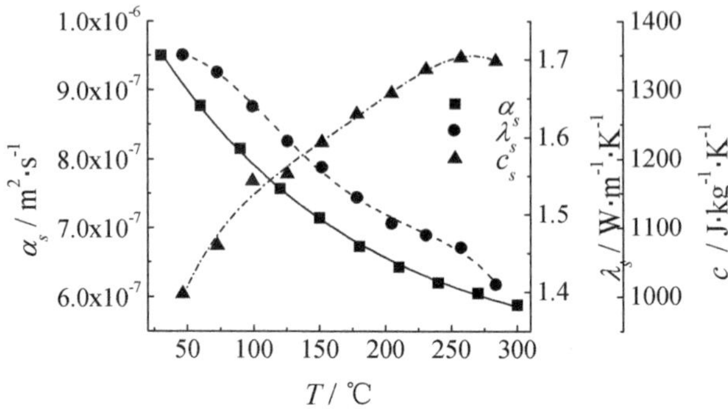

Fig. 15. Dynamic thermophysical properties of brake shoe

According to the data in Fig. 15, the fitting equations of DTP are gained by regression analysis

$$\begin{aligned}
&\alpha_s = 0.509 + e^{-\frac{T-30}{153.843}}, \\
&\lambda_s = 1.694+1.21\times 10^{-3}T - 3\times 10^{-5}T^2 + 1.436\times 10^{-7}T^3 - 2.144\times 10^{-10}T^4, \\
&c_s = 0.864+5.59\times 10^{-3}T - 4\times 10^{-5}T^2 + 1.56\times 10^{-7}T^3 - 2.3\times 10^{-10}T^4.
\end{aligned} \tag{16}$$

The brake disc is a kind of steel material, and it is generally assumed that its thermophysical properties are invariable during the braking process. And its static thermophysical parameters (STP) are shown in Table 5.

ρ_d [kg·m^{-3}]	c_d [J·kg^{-1}·K^{-1}]	λ_d [W·m^{-1}·K^{-1}]
7866	473	53.2

Table 5. Static thermophysical parameters of brake disc

Accordingly, the dynamic distribution coefficient of heat-flux is obtained (Zhu, 2009):

$$\begin{aligned}
&k_s = \frac{1}{1+\sqrt{\dfrac{\rho_d c_d \lambda_d}{\rho_s c_s \lambda_s}}}, \\
&k_d = 1 - k_s,
\end{aligned} \tag{17}$$

where k_s and k_d are the dynamic distribution coefficient of heat-flux for brake shoe and brake disc, respectively.

Combining Eq. (16) and Eq. (17), the dynamic distribution coefficient of heat-flux is plotted in Fig. 16. The curves show that the distribution coefficient of heat-flux for the brake disc always exceeds 0.88, and it absorbs most of heat energy. And k_d decreases with the temperature until 270°C, then it increases above 270°C. According to Eq. (17), k_s has the reverse variation.

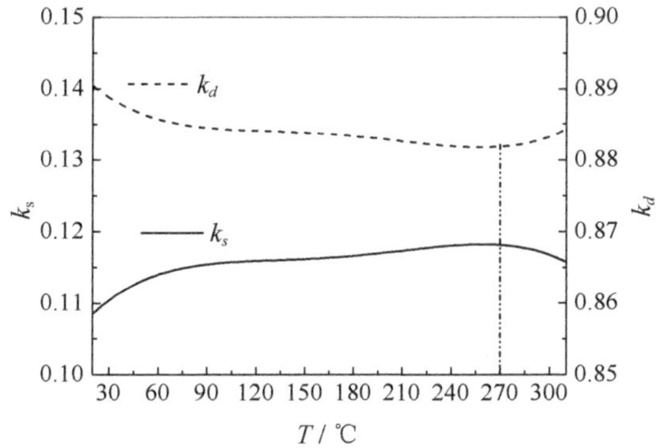

Fig. 16. Dynamic distribution coefficient of heat-flux

3.1.2 Partial differential equation of heat conduction

As shown in Fig. 17, the cylindrical coordinate is used to describe their geometry. Based on the theory of heat conduction, the transient models of disc brake's temperature field are as follows:

$$\rho_s c_s \frac{\partial T_s}{\partial t} = \frac{1}{r_s}\frac{\partial}{\partial t}\left(\lambda_s r_s \frac{\partial T_s}{\partial r_s}\right) + \frac{1}{r_s^2}\frac{\partial}{\partial \theta_s}\left(\lambda_s \frac{\partial T_s}{\partial \theta_s}\right) + \frac{\partial}{\partial z_s}\left(\lambda_s \frac{\partial T_s}{\partial z_s}\right), \tag{18}$$

$$\frac{\rho_d c_d}{\lambda_d}\frac{\partial T_d}{\partial t} = \frac{1}{r_d}\frac{\partial}{\partial t}\left(r\frac{\partial T_d}{\partial r_d}\right) + \frac{1}{r_d^2}\frac{\partial^2 T_d}{\partial \theta_d^2} + \frac{\partial^2 T_d}{\partial z_d^2}. \tag{19}$$

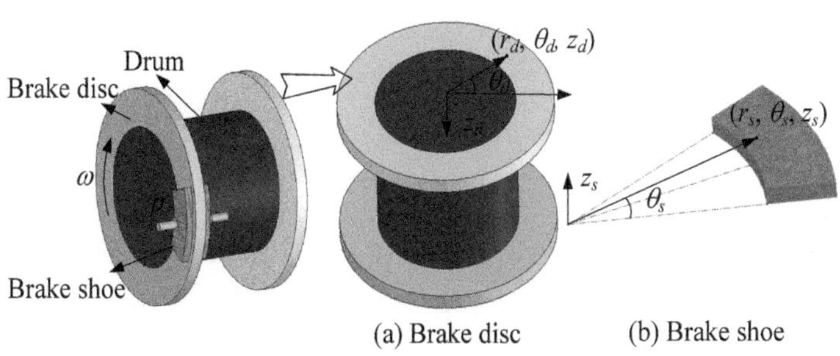

Fig. 17. Geometry Model of disc brake

3.1.3 Heat-flux

Suppose that the friction heat energy is absorbed completely by the brake shoe and brake disc

$$\begin{aligned} q &= q_d + q_s, \\ q_d &= k_d q, \; q_s = k_s q, \end{aligned} \tag{20}$$

where q is the whole heat-flux, q_d and q_s are the disc's and shoe's heat-flux, respectively. And the

$$q(r,t) = f \cdot p(t) \cdot v(t) = f \cdot p(t) \cdot \omega(t) \cdot r, \tag{21}$$

where f is the coefficient of friction between brake disc and brake shoe, p is the brake pressure, v and w are the linear and angular velocity. The brake pressure pushed onto the brake disc increases gradually, and the pressure is given as (Cao & Lin, 2002)

$$p(t) = p_0\left(1 - e^{-\frac{t}{t_z}}\right), \tag{22}$$

where p_0 is the initial brake pressure, and t_z is total braking time. In addition, the angular velocity is assumed to decrease linearly, then

$$\omega = \omega_0\left(1 - \frac{t}{t_z}\right), \tag{23}$$

where $\omega_0 = v_0/r_0$, v_0 is the initial linear velocity and r_0 is the average radius.

Combining Eq.(21), Eq.(22) and Eq.(23), the whole heat-flux is gained as following

$$q(r,t) = f \cdot p_0 \cdot \omega_0\left(1 - e^{-\frac{t}{t_z}}\right) \cdot \left(1 - \frac{t}{t_z}\right) \cdot r. \tag{24}$$

3.1.4 Boundary conditions

During the barking process, the brake disc rotates while the brake shoe keeps static. Therefore, the brake disc and brake shoe have different boundary conditions. For the brake shoe, its friction surface is subjected to the constant heat-flux.

$$q_{sf} = q_s, r_s \in (r_1, r_2), t \in (0, t_z), z_s = 0. \tag{25}$$

With regard to the fixed area in the brake disc, it is subject to the periodical heat-flux. In order to calculate the temperature rise of brake disc, the movable heat-flux is expressed by

$$\begin{aligned} &q_{df} = q_d, \ \ \theta_d \in (\theta_1 - \theta_0, \theta_1 + \theta_0), r_d \in (r_1, r_2), z_d = 0 \\ &\theta_1 = \int_0^t \omega(t)\mathrm{d}t = \omega_0\left(t - \frac{t^2}{2t_z}\right). \end{aligned} \tag{26}$$

where $\theta_1 = \theta_1 - 2n\pi$, $n = \theta_1 \bmod 2\pi$.

3.2 Results and discussion

According to the above theory analysis, the temperature rise of brake shoe and brake disc is simulated by finite element method. The dimension parameters are as follows: $r_1 = 2.35$,

$r_2 = 2.55$, $r_0 = (r_1 + r_2)/2$, $\theta_d \in (0, 2\pi)$, $\theta_0 = -0.0612$, $r_d \in (-0.03, 0)$, $r_s \in (0, 0.025)$. The braking parameters are shown in Table 6. And Figs. 18-23 show the simulation results.

f	p [MPa]	v_0 [m·s^{-1}]	t_z [s]
0.4	1.36	10	7.32

Table 6. Braking parameters for disc brake

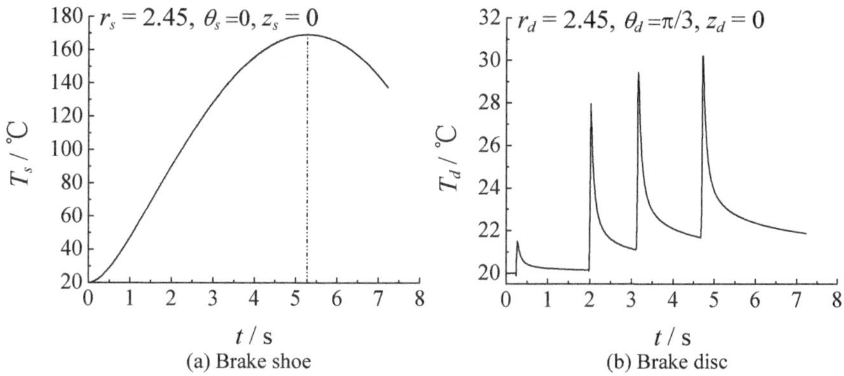

Fig. 18. Temperature rise of disc brake

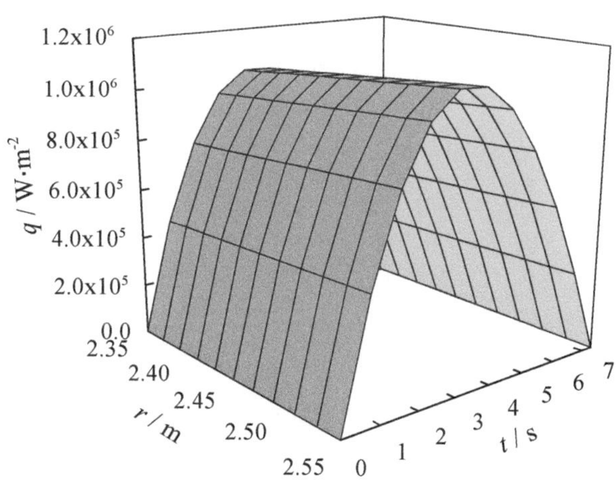

Fig. 19. Variation of heat-flux with t and r

It is seen from Fig. 18 that DTP has little effect on the temperature rise. The brake shoe's temperature calculated with DTP is a little higher than that with STP, while the temperature of brake disc with DTP is in accordance with that with STP. In Fig. 18, the temperature of brake shoe varies smoothly while the temperature of brake disc changes periodically. In Fig. 18(a), the temperature of brake shoe increases with the time and reaches the maximum 169°C at t = 5.3s, then it decreases. This result agrees with the variation of heat-flux in Fig. 19: there is a peak in the curve of heat-flux during the braking process. Though the brake disc absorbs most of friction heat, its maximal temperature, which is only 30°C, is much lower than brake shoe's. With regard to the fixed point in brake disc, it is subject to heat-flux within short time and convects with the air in most time of every circle. Thus, the temperature of brake disc is lower and varies periodically.
In addition, the temperature of fixed point at different q_d in a circle is simulated, and the simulation results are in Fig. 20. The peak of temperature occurs when the point on the disc contacts with the shoe. And the variation of temperature peak agrees well with the variation of heat-flux during the braking process. Furthermore, the variation of temperature with radius is shown in Fig. 21. The temperature of brake shoe increases with the radius slightly for minor difference between inner and outer radius. The temperature on both edges of the contact zone in brake disc is low, while it is high in the middle.

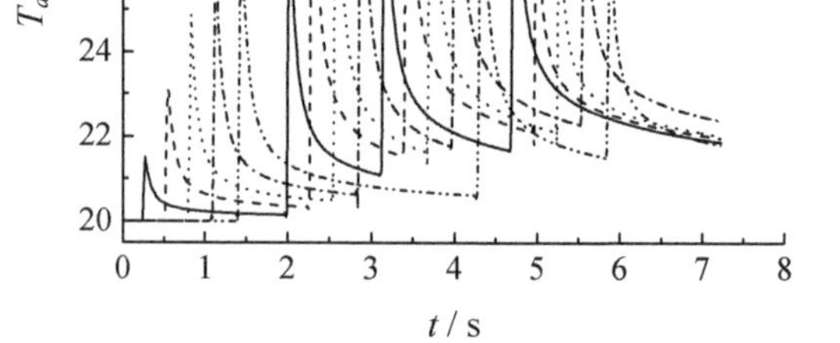

Fig. 20. Variation of T_d with q_d

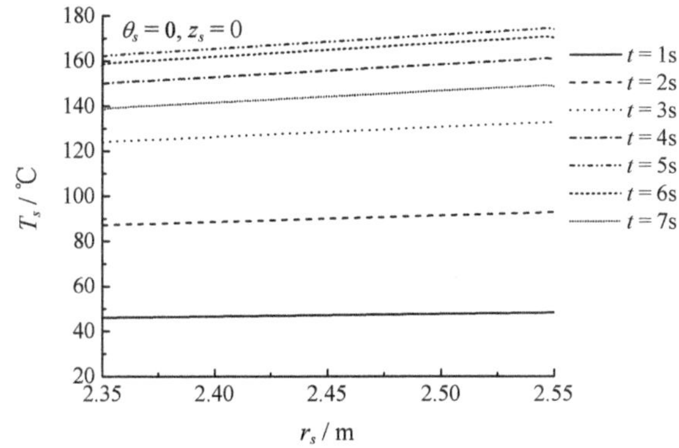

(a) brake shoe

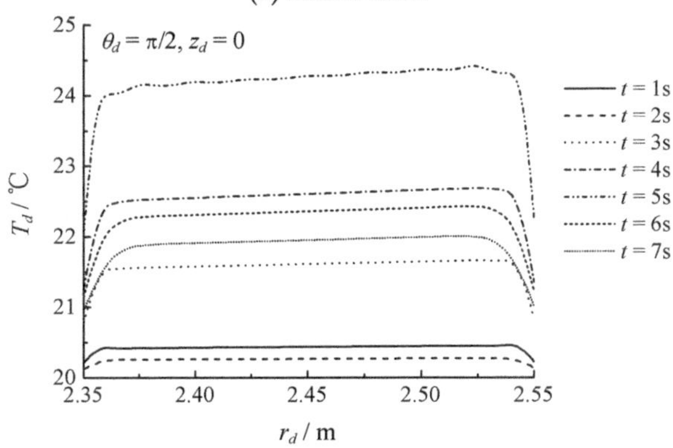

(b) brake disc

Fig. 21. Variation of temperature with radius

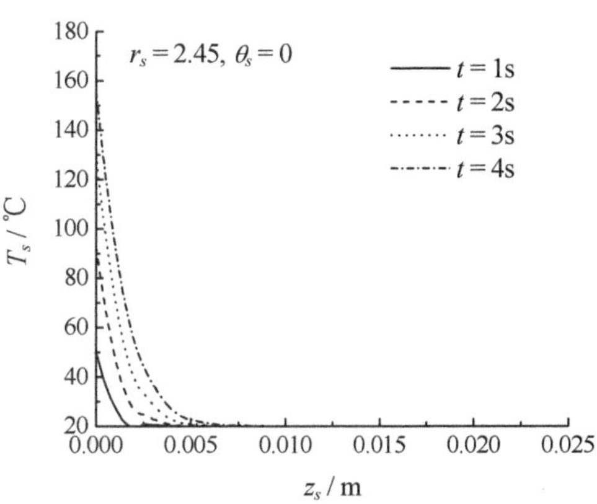

(a) brake shoe

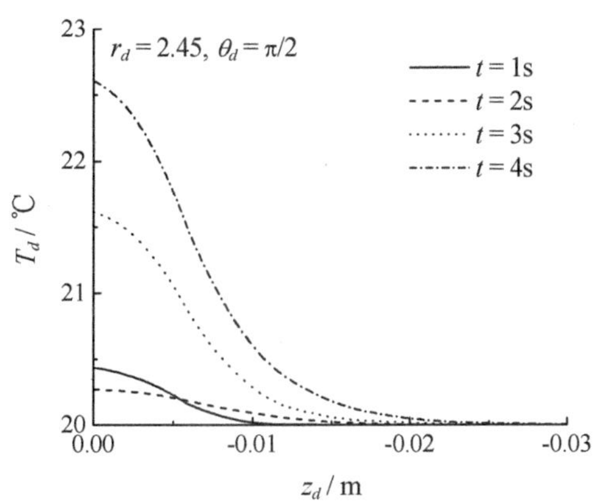

(b) brake disc

Fig. 22. Variation of temperature with depth

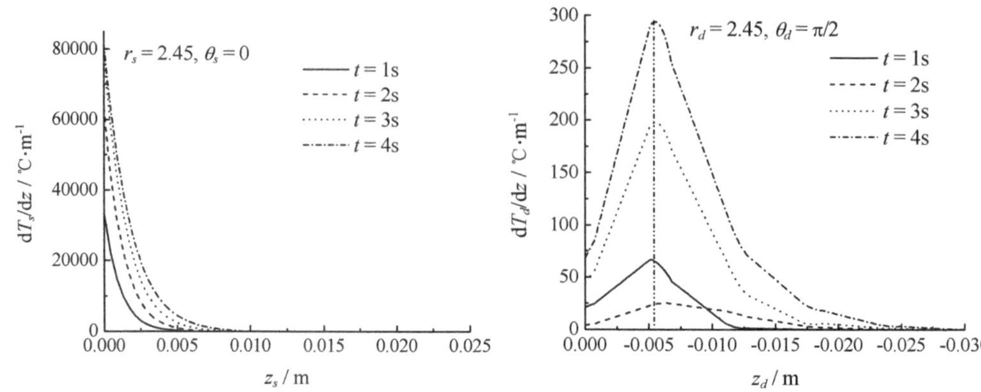

(a) brake shoe (b) brake disc

Fig. 23. Variation of temperature gradient with depth

From Figs. 22 and 23, the temperature of brake shoe decreases sharply with z_d: it reduced to zero at z_s=0.0075m. And the temperature gradient on the friction surface is the highest: the maximum of temperature gradient is up to 8×10^4°C m^{-1} at t = 4s. These results show that the friction heat energy concentrates on the surface layer and the brake shoe's heatconducting property is poor. However, the temperature rise of brake disc decreases to zero after z_d = -0.025m and its maximum is only 300°C m^{-1} at t = 4s. Additionally, it is found that the maximum of brake disc's temperature gradient is not on the friction surface and it appears at z_d = -0.006m. Due to the high thermal conductivity of the brake disc and surface heat convection, the speed of heat dissipation is fast on the surface. But the inner surface has the low speed of heat transfer. Therefore, though its surface temperature is the highest, but the maximum of temperature gradient occurs at z_d = -0.006m of the inner surface. These results reveal that the brake disc's heatconducting property is better than brake shoe's.

4. Conclusions

1. The friction lining contacts with the outside of rope strand periodically and the number of contact point is two or three. Additionally, the rope groove of friction lining is divided into three contact zones and the contact arcs are unequal: the contact arc in the contact zone II is the longest, and the contact arcs in the contact zones I and III are equal.
2. The thermophysical properties of lining K is lower than that of ling G. As the testing temperature increases, the specific heat capacity increases with the temperature, the thermal diffusivity decreases with the temperature nonlinearly and the thermal conductivity□increases with the temperature below 90°C and keeps approxistable above 90°C.
3. According to the force analysis of friction lining under the operating condition and experimental condition, the total heat-flow under the two situations is obtained. And

the partial heat-flow corresponding to every contact zone is obtained based on the helical contact characteristic.

4. With the consideration of the thermophysical properties and helical non-complete contact characteristics, the theoretical model of friction lining's transient temperature field is established, and the finite difference method is adopted to solve this problem.
5. The friction experiment indicates that the temperature of friction lining increases with the equivalent pressure and sliding speed and the sliding speed has stronger effect than the equivalent pressure on the temperature. During the low-speed sliding, the temperature rises gently and the temperature at the measuring points is approximately equal. The temperature rise is less than 5 °C within 1 hour when the sliding speed is less than 10mm/s. And the highest temperature rise increases to 15°C at 30mm/s. As the speed increases to 1000mm/s, the temperature rise at every point increases obviously. Additionally, the temperature rise increases wavily at the initial sliding stage and the amplitude of the wave increases with the sliding speed while it decreases with the time. Furthermore, the temperature increases periodically which agrees with the helical contact characteristics.
6. The simulation result agrees with the experiment result, which validate the theoretical model of the temperature field. The temperature on the contact surface and the temperature gradient were simulated under the experimental condition (v=0.55m/s, p_a=2.5MPa). The simulation result indicates that the friction heat focuses on the contact surface layer and the temperature gradient on the surface layer is the highest. In addition, the heat-conducting property of friction lining is poor. In order to develop the new friction lining with good thermophysical properties, it is necessary to optimize the ratio of basic material and filler and selected the component with good heat-conducting property.
7. Combining dynamic thermophysical properties of brake shoe and dynamic distribution coefficient of heat-flux, the theoretical models of brake shoe and brake disc's temperature field were established. And the static and periodical heat-flux on the friction surface of brake shoe and brake disc were obtained.
8. DTP of brake shoe has little effect on the temperature. The temperature of brake shoe varies smoothly while that of brake disc changes periodically. The maximal temperature of the brake disc is much lower than that of the brake shoe during the braking process. And the variation of temperature rise's peak at different $\square_d$ agrees well with the variation of heat-flux during the braking process. Additionally, the temperature of disc brake increases with the radius slightly.
9. The temperature of brake shoe and brake disc decreases with the depth, and the temperature gradient of brake shoe is much higher than that of brake disc. In addition, the maximum of brake disc's temperature gradient is not on the friction surface and it appears at z_d = -0.006m. The friction heat energy concentrates on the surface layer of brake shoe and the heatconducting property of brake disc is better than that of brake shoe.

5. Acknowledgements

This project is supported by the National Natural Science Foundation of China (Grant No. 51105361), the China Postdoctoral Science Foundation funded project (Grant No. 20100481179), Fundamental Research Funds for the Central Universities (Grant No.

2010QNA25), National High Technology Research and Development Program of China (863 Program) (Grant No. 2009AA04Z415), National Natural Science Foundation of China (Grant No. 50875253) and Natural Science Foundation of Jiangsu Province (Grant No. BK2008127).

6. References

Editorial Committee of Mine Safety Handbooks, Editor, (2004). *Safety Regulations for Coal Mine*, China Coal Industry Publishing House, Beijing.

Peng,Y.X., Zhu, Z.C., Chen, G.A. (2008). Numerical simulation of lining's transient temperature field during friction hoist's sliding. *J. Chin. Univ. Min. Technol.*, Vol. 37, No. 4, (2008), pp. 526-531. (in Chinese)

Liu, D.P., Mei, S.H. (1997). Approximate method of calculating friction temperature in friction winder lining. *J. Chin. Univ. Min. Technol.*, Vol. 26, No. 1, (1997), pp. 70-72. (in Chinese)

Xia, R.H. Ge S.R. (1990). Calculation of temperature rise of lining of friction winder. *J. Chin. Coal Soc.*, Vol. 15, No. 2, (1990), pp. 1-9. (in Chinese)

Yang, Z.J. (1990). Theoretical calculation of the lining's temperature field of multi-rope friction winder. *J. Shanxi Min. Inst.*, Vol. 8, No. 4, (1990), pp. 304-314. (in Chinese)

Singh, K., Singh, A.K. Saxena, N.S. (2008). Temperature dependence of effective thermal conductivity and effective thermal diffusivity of $Se_{90}In_{10}$ bulk chalcogenide glass. *Curr. Appl. Phys.*, Vol. 8, No. 2, (2008), pp. 159-162.

Isoda, H., Kawashima, R. (2007). Temperature dependence of thermal property for lead nitrate crystal. *J. Phys. Chem. Solids.*, Vol. 68, No. 4, (2007), pp. 561-563.

He, W., Liao, G.X., Liu, C. (2005). Thermal and dynamic mechanical properties of PPESK/PTFE blends. *Chin. J. Mater. Res.*, Vol. 19, No. 5, (2005), pp. 464-470. (in Chinese)

Hegeman, J.B.J., van der Laan, J.G., van Kranenburg, M., Jong, M., d'Hulst, D., ten Pierick, P. (2005). Mechanical and thermal properties of SiC_f/SiC composites irradiated with neutrons at high temperatures. *Fusion Eng. Des.*, Vol. 75-79, (2005), pp. 789-793.

Mazzone, A.M. (2005). Thermal properties of clustered systems of mixed composition: the temperature response of Si-Al clusters studied quantum mechanically. *Comput. Mater. Sci.*, Vol. 34, No. 1, (2005), pp. 64-69.

Golebiowski, J., Kwieckowski, S. (2002). Dynamics of three-dimensional temperature field in electrical system of floor heating. *Int. J. Heat Mass Transfer*, Vol. 45, No. 12, (2002), pp. 2611-2622.

Lukyanov, S. (2001). Finite temperature expectation values of local fields in the sinh-Gordon model. *Nucl. Phys. B.*, Vol. 612, No. 3, (2001), pp. 391-412.

Matysiak, S.J., Yevtushenko, A.A., Ivanyk, E.G. (2002). Contact temperature and wear of composite friction elements during braking. *Int. J. Heat Mass Transfer*, Vol. 45, No. 1, (2002), pp. 193-199.

Yevtushenko, A.A., Ivanyk, E.G. (1997). Determination of temperatures for sliding contact with applications for braking systems. *Wear*. Vol. 206, No. 1-2, (1997), pp. 53-59.

Naji, M., Al-Nimr, M. (2001). Dynamic thermal behavior of a brake system. *Int. Commun. Heat Mass Transfer.*, Vol. 28, No. 6, (2001), pp. 835-845.

Zhu, Z.C., Peng, Y.X., Shi, Z.Y., Chen, G.A. (2009). Three-dimensional transient temperature field of brake shoe during hoist's emergency braking. *Appl. Therm. Eng.*, Vol. 29, No. 5-6, (2009), pp. 932-937.

Voldrich, J. (2007). Frictionally excited thermoelastic instability in disc brakes-transient problem in the full contact regime. *Int. J. Mech. Sci.*, Vol. 49, No. 2, (2007), pp. 129-137.

Qi, H.S. Day, A.J. (2007). Investigation of disc/pad interface temperatures in friction braking. *Wear*, Vol. 262. No. 5-6, (2007), pp. 505-513.

Thuresson, D. (2006). Stability of sliding contact-comparison of a pin and a finite element model. *Wear*, Vol. 261, No. 7-8, (2006), pp. 896-904.

Choi, J.H., Lee, I. (2004). Finite element analysis of transient thermoelastic behaviors in disk brakes. *Wear*, Vol. 257, No. 1-2, (2004), pp. 47-58.

Chang, L.Z., Li, B.Z. (2008). Numerical simulation of temperature fields in electroslag remelting slab ingots. *Acta Metall. Sinica.*, Vol. 21, No. 4, (2008), pp. 253-259.

Liu, X., Yao, J., Wang, X., Zou, Z., Qu, S. (2009). Finite difference modeling on the temperature field of consumable-rod in friction surfacing. *J. Mater. Process. Technol.*, Vol. 209, No. 3, (2009), pp. 1392-1399.

Zhu, Z.C., Shi, Z.Y., Chen, G.A. (2008). Experimental Study on Friction Behaviors of Brake Shoes Materials for Hoist Winder Disc Brakes. *J. Harbin Inst. Techol.*, Vol. 40, No. 3, (2008), pp. 462-465. (In Chinese)

Zhu, Z.C., Shi, Z.Y., Chen, G.A. (2006). Tribological Behaviors of Asbestos-free Brake Shoes for Hoist Winder Disc Brakes. *Lubr. Eng.*, No. 12, (2006), pp. 99-101. (In Chinese)

Cao, C.H., Lin, X.Z. (2002). Transient Temperature Field Analysis of a Brake in a Non-axisymmetric Three-dimensional Model. *J. Mater. P. T.*, Vol. 129, No. 1-3, (2002), pp. 513-517.

Wang, Y., Cao, X.K., Yao, A.Y., Li, L.Z. (2001). Study on the Temperature Field of Disc Brake Friction Flake. *J. Wuhan U. T.*, Vol. 23, No. 7, (2001), pp. 22-24. (In Chinese)

Lin, X.Z., Gao, C.H., Huang, J.M. (2006). Effects of Operating Condition Parameters on Distribution of Friction Temperature Field on Brake Disc. *J. Eng. Des.*, Vol. 13, No. 1, (2006), pp. 45-48. (In Chinese)

Ma, B.J., Zhu, J. (1999). Contact Surface Temperature Model for Disc Brake in Braking. *J. Xian Inst. Tech.*, Vol. 19, No. 1, (1999), pp. 35-39. (In Chinese)

Ma, B.J., Zhu, J. (1998). The Dynamic Heat Flux Model for Emergency Braking. *Mech. Sci. Tech.*, Vol. 17, No. 5, (1998), pp. 698-700. (In Chinese)

Bao, J.S., Zhu, Z.C., Yin, Y., Peng, Y.X. (2009). A Simple Method for Calculating Maximal Surface Temperature of Mine Hoister's Brake Shoe During Emergency Braking. *J. Comput. Theor. Nanosci.*, Vol. 6, No. 7, (2009), pp. 1566-1570.

MT/T 248-91, (1991). *Testing method for coefficient of friction of lining in friction hoist* [China Coal Industry Standards] (in Chinese)

Bao, J.S. (2009). *Tribological Performance and Its Catastrophe Behaviors of Mine Hoister's Brake Shoe During Emergency Braking*. [Ph.D. Dissertation], China University of Mining and Technology, Xuzhou (2009). (in Chinese)

13

Mathematical Modelling of Dynamics of Boiler Surfaces Heated Convectively

Wiesław Zima
Cracow University of Technology
Poland

1. Introduction

In order to increase the efficiency of electrical power production, steam parameters, namely pressure and temperature, are increased. Changes in the superheated steam and feed water temperatures in boiler operation are also caused by changes in the heat transfer conditions on the combustion gases side. When the waterwalls of the furnace chamber undergo slagging up, the combustion gases temperature at the furnace chamber outlet increases, and the superheaters and economizers take more heat. In order to maintain the same temperature of the superheated steam at the outlet, the flow of injected water must be increased. Upon cleaning the superheater using ash blowers, the heat flux taken by the superheater also increases, which in turn changes the coolant mass flow. Changes of the superheated steam and feed water temperatures caused by switching off some burners or coal pulverizers or by varying the net calorific value of the supplied coal may also be significant. Precise modelling of superheater dynamics to improve the quality of control of the superheated steam temperature is therefore essential. Designing the mathematical model describing superheater dynamics is also very important from the point of view of digital control of the superheated steam temperature. A crucial condition for its proper control is setting up a precise numerical model of the superheater which, based on the measured inlet and outlet steam temperature at the given stage, would provide fast and accurate determination of the water mass flow to the injection attemperator. Such a mathematical model fulfils the role of a process "observer", significantly improving the quality of process control (Zima, 2003, 2006). The transient processes of heat and flow occurring in superheaters and economizers are complex and highly nonlinear. That complexity is caused by the high values of temperature and pressure, the cross-parallel or cross-counter-flow of the fluids, the large heat transfer surfaces (ranging from several hundred to several thousand square metres), the necessity of taking into account the increasing fouling of these surfaces on the combustion gases side, and the resulting change in heat transfer conditions. The task is even more difficult when several heated surfaces are located in parallel in one combustion gas duct, an arrangement which is applied quite often. Nonlinearity results mainly from the dependency of the thermo-physical properties of the working fluids and the separating walls on the pressure and temperature or on the temperature only. Assumption of constancy of these properties reduces the problem to steady state analysis. Diagnosis of heat flow processes in power engineering is generally based on stabilized temperature conditions. This is due to the absence of mathematical models that apply to big power units under transient thermal conditions (Krzyżanowski & Głuch, 2004). The existing attempts to model steam superheaters and economizers are based on greatly simplified one-dimensional models or models with lumped parameters (Chakraborty & Chakraborty, 2002; Enns, 1962; Lu, 1999; Mohan et al., 2003). Shirakawa presents a dynamic simulation tool that facilitates plant and control system design of thermal power plants (Shirakawa, 2006). Object-oriented modelling techniques are used to model individual plant components. Power plant components can also be modelled using a modified neural network structure (Mohammadzaheri et al., 2009). In the paper by Bojić and Dragićević a linear programming model has been developed to optimize the performance and to find the optimal size of heating surfaces of a steam boiler (Bojić & Dragićević, 2006). In this chapter a new mathematical method for modelling transient processes in convectively heated surfaces of boilers is proposed. It considers the superheater or economizer model as one with distributed parameters. The method makes it possible to model transient heat transfer processes even in the case of fluids differing considerably in their thermal inertias.

2. Description of the proposed model

Real superheaters and economizers are three-dimensional objects. The basic assumptions of the proposed model refer to the parameters of the working fluids. It was assumed that there are no changes in combustion gases flow and temperature in the arbitrary cross-section of the given superheater or economizer stage (Dechamps, 1995). The same applies to steam and feed water. When the real heat exchanger is operating in cross-counter-flow or cross-parallel-flow and has more than four tube rows, its one-dimensional model (double pipe heat exchanger), represented by Fig. 1, can be based on counter-flow or parallel-flow only (Hausen, 1976). In the proposed model, which has distributed parameters, the computations are carried out in the direction of the heated fluid flow in one tube. The tube is equal in size to those installed in the existing object and is placed, in the calculation model, centrally in a larger externally insulated tube of assumed zero wall thickness (Fig. 1). The cross-section A_{cg} of the combustion gases flow results, in the computation model, from dividing the total free cross-section of combustion gases flow by the number of tubes. The mass flows of the working fluids are also related to a single tube.

A precise mathematical model of a superheater, based on solving equations describing the laws of mass, momentum, and energy conservation, is presented in (Zima, 2001, 2003, 2004, 2006). The model makes it possible to determine the spatio-temporal distributions of the mass flow, pressure, and enthalpy of steam in the on-line mode. This chapter presents a model based solely on the energy equation, omitting the mass and momentum conservation equations. Such a model results in fewer final equations and a simpler form. Their solution is thereby reached faster. The short time taken by the computations (within a few seconds) is very important from the perspective of digital temperature control of superheated steam. In the papers by Zima that control method was presented for the first time (Zima, 2003, 2004, 2006). In this case the mathematical model fulfils the role of a process "observer", significantly improving the quality of process control. The omission of the mass and momentum balance equations does not generate errors in the computations and does not constitute a limitation of the method. The history of superheated steam mass flow is not a

rapidly changing one. Also taking into consideration the low density of the steam, it is possible to neglect the variation of steam mass existing in the superheater. Feed water mass flow also does not change rapidly. Moreover the water is an incompressible medium. The results of the proposed method are very similar to results obtained using equations describing the laws of mass, momentum, and energy conservation (Zima, 2001, 2004).
The suggested in this chapter 1D model is proposed for modelling the operation of superheaters and economizers considering time-dependent boundary conditions. It is based on the implicit finite-difference method in an iterative scheme (Zima, 2007).

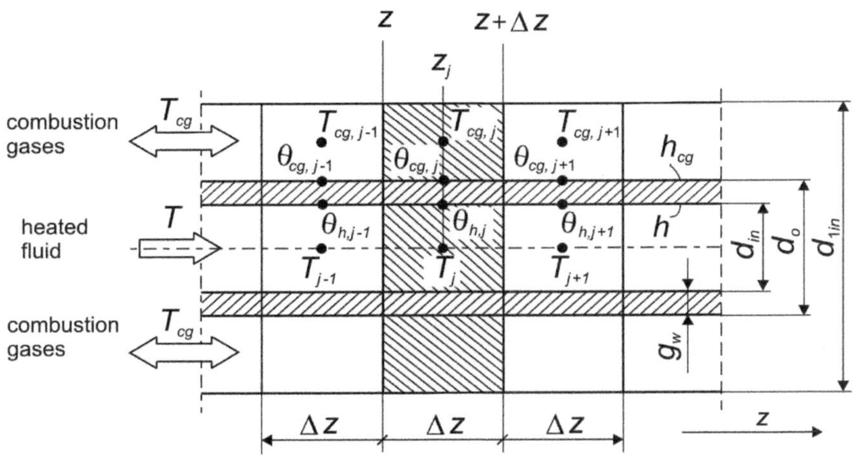

Fig. 1. Analysed control volume of double-pipe heat exchanger

Every equation presented in this section is based on the geometry shown in Fig. 1 and refers to one tube of the heated fluid. The Cartesian coordinate system is used.
The proposed model shows the same transient behaviour as the existing superheater or economizer if:

a. the steam or feed water tube has the same inside and outside diameter, the same length, and the same mass as the real one
b. all the thermo-physical properties of the fluids and the material of the separating walls are computed in real time
c. the time-spatial distributions of heat transfer coefficients are computed in the on-line mode, considering the actual tube pitches and cross-flow of the combustion gases
d. the appropriate free cross-sectional area for the combustion gases flow is assumed in the model:

$$A_{cg} = \frac{A_{cg,t}}{n} = \frac{\pi\left(d_{1in}^2 - d_o^2\right)}{4} \tag{1}$$

e. mass flow of the heated fluid is given by:

$$\dot{m} = \frac{\dot{m}_t}{n} \tag{2}$$

f. mass flow of the combustion gases is given by:

$$\dot{m}_{cg} = \frac{\dot{m}_{cg,t}}{n}. \tag{3}$$

In the above equations:

$A_{cg,t}$ - total free cross-section of combustion gases flow, m^2,

$\dot{m}_{cg,t}$ - total combustion gases mass flow, kg/s,

$\dot{m}_t$ - total heated fluid mass flow, kg/s,

n - number of tubes.

The temperature θ of the separating wall is determined from the equation of transient heat conduction:

$$c_w(\theta)\rho_w(\theta)\frac{\partial \theta}{\partial t} = \frac{1}{r}\frac{\partial}{\partial r}\left[r k_w(\theta)\frac{\partial \theta}{\partial r}\right], \tag{4}$$

where:

c_w - specific heat of the tube wall material, J/(kg K),

k_w - thermal conductivity of the tube wall material, W/(mK),

ρ_w - density of the tube wall material, kg/m^3.

In order to obtain greater accuracy of the results, the wall is divided into two control volumes. This division makes it possible to determine the temperature on both surfaces of the separating wall, namely θ_{cg} at the combustion gases side and θ_h at the heated medium side (Fig. 2).

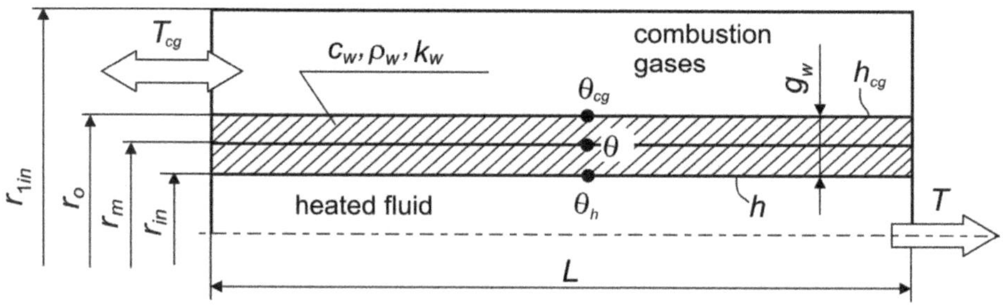

Fig. 2. Tube wall divided into two control volumes

After some transformations, the following formulae are obtained from Equation 4:

$$c_w(\theta_h)\rho_w(\theta_h)\frac{\left(r_m^2 - r_{in}^2\right)}{2}\frac{\partial \theta_h}{\partial t} = \left[r k_w(\theta)\frac{\partial \theta}{\partial r}\right]\Bigg|_{r=r_m} - \left[r k_w(\theta)\frac{\partial \theta}{\partial r}\right]\Bigg|_{r=r_{in}}, \tag{5}$$

$$c_w(\theta_{cg})\rho_w(\theta_{cg})\frac{\left(r_o^2 - r_m^2\right)}{2}\frac{\partial \theta_{cg}}{\partial t} = \left[r k_w(\theta)\frac{\partial \theta}{\partial r}\right]\Bigg|_{r=r_o} - \left[r k_w(\theta)\frac{\partial \theta}{\partial r}\right]\Bigg|_{r=r_m}. \tag{6}$$

Taking into consideration the boundary conditions:

$$k_w(\theta)\frac{\partial\theta}{\partial r}\bigg|_{r=r_{in}} = h\left(\theta\big|_{r=r_{in}} - T\right) = h(\theta_h - T), \tag{7}$$

$$k_w(\theta)\frac{\partial\theta}{\partial r}\bigg|_{r=r_m} = k_w(\theta_m)\frac{\theta_{cg}-\theta_h}{r_o - r_{in}}, \tag{8}$$

$$k_w(\theta)\frac{\partial\theta}{\partial r}\bigg|_{r=r_o} = h_{cg}\left(T_{cg} - \theta\big|_{r=r_o}\right) = h_{cg}\left(T_{cg} - \theta_{cg}\right), \tag{9}$$

where:
h and h_{cg} - heat transfer coefficients at the sides of heated fluid and combustion gases, respectively, W/(m^{2}K),
the following ordinary differential equations are obtained:

$$\frac{\mathrm{d}\theta_h}{\mathrm{d}t} = B\left(\theta_{cg} - \theta_h\right) + C(T - \theta_h), \tag{10}$$

$$\frac{\mathrm{d}\theta_{cg}}{\mathrm{d}t} = D\left(T_{cg} - \theta_{cg}\right) + E\left(\theta_h - \theta_{cg}\right). \tag{11}$$

In the above equations:

$$B = \frac{\pi d_m k_w(\theta_m)}{A_h c_w(\theta_h)\rho_w(\theta_h) g_w},\quad d_m = \frac{d_{in}+d_o}{2},\quad \theta_m = \frac{\theta_{cg}+\theta_h}{2},\quad C = \frac{h\pi d_{in}}{A_h c_w(\theta_h)\rho_w(\theta_h)},$$

$$D = \frac{h_{cg}\pi d_o}{A_{cg} c_w\left(\theta_{cg}\right)\rho_w\left(\theta_{cg}\right)},\ E = \frac{\pi d_m k_w(\theta_m)}{A_{cg} c_w\left(\theta_{cg}\right)\rho_w\left(\theta_{cg}\right) g_w},\ A_h = \frac{\pi\left(d_m^2 - d_{in}^2\right)}{4},\ \text{and } A_{cg} = \frac{\pi\left(d_o^2 - d_m^2\right)}{4}.$$

The transient temperatures of the combustion gases and heated fluid are evaluated iteratively, using relations derived from the equations of energy balance. In these equations, the change in time of the total energy in the control volume, the flux of energy entering and exiting the control volume, and the heat flux transferred to it through its surface are taken into consideration.
The energy balance equations take the following forms (Fig. 1):
- combustion gases

$$\Delta z A_{cg} c_{cg}\left(T_{cg}\right)\rho_{cg}\left(T_{cg}\right)\frac{\Delta T_{cg}}{\Delta t} = \pm\left(\dot{m}_{cg} i_{cg}\Big|_{z+\Delta z} - \dot{m}_{cg} i_{cg}\Big|_{z}\right) + h_{cg}\pi d_o \Delta z\left(\theta_{cg} - T_{cg}\right), \tag{12}$$

- feed water or steam

$$\Delta z A c(T,p)\rho(T,p)\frac{\Delta T}{\Delta t} = \dot{m}\, i\Big|_{z} - \dot{m}\, i\Big|_{z+\Delta z} + h\pi d_{in}\Delta z(\theta_h - T), \tag{13}$$

where:
i - specific enthalpy, J/kg,

p – pressure, Pa,

$$A_{cg} = \frac{\pi d_{1in}^2 - \pi d_o^2}{4}, \text{ and } A = \frac{\pi d_{in}^2}{4}.$$

After rearranging and assuming that $\Delta t \to 0$ and $\Delta z \to 0$, the following equations are obtained from (12) and (13), respectively:

$$\frac{\partial T_{cg}}{\partial t} = \pm F \frac{\partial T_{cg}}{\partial z} + G\left(\theta_{cg} - T_{cg}\right), \tag{14}$$

$$\frac{\partial T}{\partial t} = -H \frac{\partial T}{\partial z} + J(\theta_h - T). \tag{15}$$

In the above equations:

$$F = \frac{\dot{m}_{cg}}{A_{cg}\rho_{cg}\left(T_{cg}\right)}, \quad G = \frac{h_{cg}\pi d_o}{A_{cg}c_{cg}\left(T_{cg}\right)\rho_{cg}\left(T_{cg}\right)}, \quad H = \frac{\dot{m}}{A\rho(T,p)} \text{ and } J = \frac{h\pi d_{in}}{Ac(T,p)\rho(T,p)}.$$

The sign "+" in Equations (12) and (14) refers to counter-flow, and the sign " – " to parallel-flow. The implicit finite-difference method is proposed to solve the system of Equations (10) to (11) and (14) to (15). The time derivatives are replaced by a forward difference scheme, whereas the dimensional derivatives are replaced by the backward difference scheme in the case of parallel-flow and the forward difference scheme in the case of counter-flow.
After some transformations the following formulae are obtained:

$$\theta_{h,j}^{t+\Delta t} = \frac{1}{K\Delta t}\theta_{h,j}^{t} + \frac{C}{K}T_j^{t+\Delta t} + \frac{B}{K}\theta_{cg,j}^{t+\Delta t}, \qquad j = 1, \ldots, M; \tag{16}$$

$$\theta_{cg,j}^{t+\Delta t} = \frac{1}{L\Delta t}\theta_{cg,j}^{t} + \frac{D}{L}T_{cg,j}^{t+\Delta t} + \frac{E}{L}\theta_{h,j}^{t+\Delta t}, \qquad j = 1, \ldots, M; \tag{17}$$

$$T_{cg,j}^{t+\Delta t} = \frac{1}{P\Delta t}T_{cg,j}^{t} + \frac{F}{P\Delta z}T_{cg,j\pm 1}^{t+\Delta t} + \frac{G}{P}\theta_{cg,j}^{t+\Delta t}, \tag{18}$$

$$T_j^{t+\Delta t} = \frac{1}{Q\Delta t}T_j^{t} + \frac{H}{Q\Delta z}T_{j-1}^{t+\Delta t} + \frac{J}{Q}\theta_{h,j}^{t+\Delta t}, \qquad j = 2, \ldots, M; \tag{19}$$

where:
M – number of cross-sections,

$$K = \frac{1}{\Delta t} + B + C, \quad L = \frac{1}{\Delta t} + D + E, \quad P = \frac{1}{\Delta t} + \frac{F}{\Delta z} + G, \text{ and } Q = \frac{1}{\Delta t} + \frac{H}{\Delta z} + J.$$

In Equation (18), $j = 2, \ldots, M$ for parallel-flow (sign "−") and $j = 1, \ldots, M-1$ for counter-flow (sign "+").
Considering the small temperature drop on the thickness of the wall (≈ 3–4 K), Equation (4) can also be solved assuming only one control volume. The result will be a formula determining only the mean temperature θ of a wall (Fig. 2).

In this case, after some transformations, Equation (4) takes the following form:

$$c_w(\theta)\rho_w(\theta)\frac{\left(r_o^2 - r_{in}^2\right)}{2}\frac{\partial\theta}{\partial t} = \left[rk_w(\theta)\frac{\partial\theta}{\partial r}\right]\Bigg|_{r=r_o} - \left[rk_w(\theta)\frac{\partial\theta}{\partial r}\right]\Bigg|_{r=r_{in}} . \quad (20)$$

Taking into consideration the boundary conditions described by Equations (7) and (9), the following ordinary differential equation is obtained:

$$\frac{\mathrm{d}\theta}{\mathrm{d}t} = U\left(T_{cg} - \theta\right) + V\left(T - \theta\right). \quad (21)$$

Replacing the time derivative by the forward difference scheme, after rearranging we obtain:

$$\theta_j^{t+\Delta t} = \frac{1}{W\Delta t}\theta_j^t + \frac{U}{W}T_{cg,j}^{t+\Delta t} + \frac{V}{W}T_j^{t+\Delta t}, \quad (22)$$

where:

$$U = \frac{h_{cg}d_o}{c_w(\theta)\rho_w(\theta)g_w d_m}, \quad V = \frac{hd_{in}}{c_w(\theta)\rho_w(\theta)g_w d_m}, \quad d_m = \frac{d_o + d_{in}}{2} \text{ and } W = \frac{1}{\Delta t} + U + V .$$

The suggested method is also suitable for modelling the dynamics of several surfaces heated convectively, often placed in parallel in a single gas pass of the boiler.
As an example of these surfaces it was assumed that the feed water heater and superheater are located in parallel in such a gas pass (Fig. 3). Additionally, the flow of combustion gases is in parallel-flow with feed water and simultaneously in counter-flow to steam.
The equation of transient heat conduction (Equation 4) takes the following forms (the walls of steam and feed water pipes are divided into two control volumes):

- wall of steam pipe

$$c_w(\theta_{1s})\rho_w(\theta_{1s})\frac{\left(r_m^2 - r_{in}^2\right)}{2}\frac{\partial\theta_{1s}}{\partial t} = \left[rk_w(\theta_1)\frac{\partial\theta_1}{\partial r}\right]\Bigg|_{r=r_m} - \left[rk_w(\theta_1)\frac{\partial\theta_1}{\partial r}\right]\Bigg|_{r=r_{in}}, \quad (23)$$

$$c_w\left(\theta_{1cg}\right)\rho_w\left(\theta_{1cg}\right)\frac{\left(r_o^2 - r_m^2\right)}{2}\frac{\partial\theta_{1cg}}{\partial t} = \left[rk_w(\theta_1)\frac{\partial\theta_1}{\partial r}\right]\Bigg|_{r=r_o} - \left[rk_w(\theta_1)\frac{\partial\theta_1}{\partial r}\right]\Bigg|_{r=r_m}, \quad (24)$$

- wall of economizer pipe

$$c_w\left(\theta_{2fw}\right)\rho_w\left(\theta_{2fw}\right)\frac{\left(r_{2m}^2 - r_{2in}^2\right)}{2}\frac{\partial\theta_{2fw}}{\partial t} = \left[rk_w(\theta_2)\frac{\partial\theta_2}{\partial r}\right]\Bigg|_{r=r_{2m}} - \left[rk_w(\theta_2)\frac{\partial\theta_2}{\partial r}\right]\Bigg|_{r=r_{2in}}, \quad (25)$$

$$c_w\left(\theta_{2cg}\right)\rho_w\left(\theta_{2cg}\right)\frac{\left(r_{2o}^2 - r_{2m}^2\right)}{2}\frac{\partial\theta_{2cg}}{\partial t} = \left[rk_w(\theta_2)\frac{\partial\theta_2}{\partial r}\right]\Bigg|_{r=r_{2o}} - \left[rk_w(\theta_2)\frac{\partial\theta_2}{\partial r}\right]\Bigg|_{r=r_{2m}} . \quad (26)$$

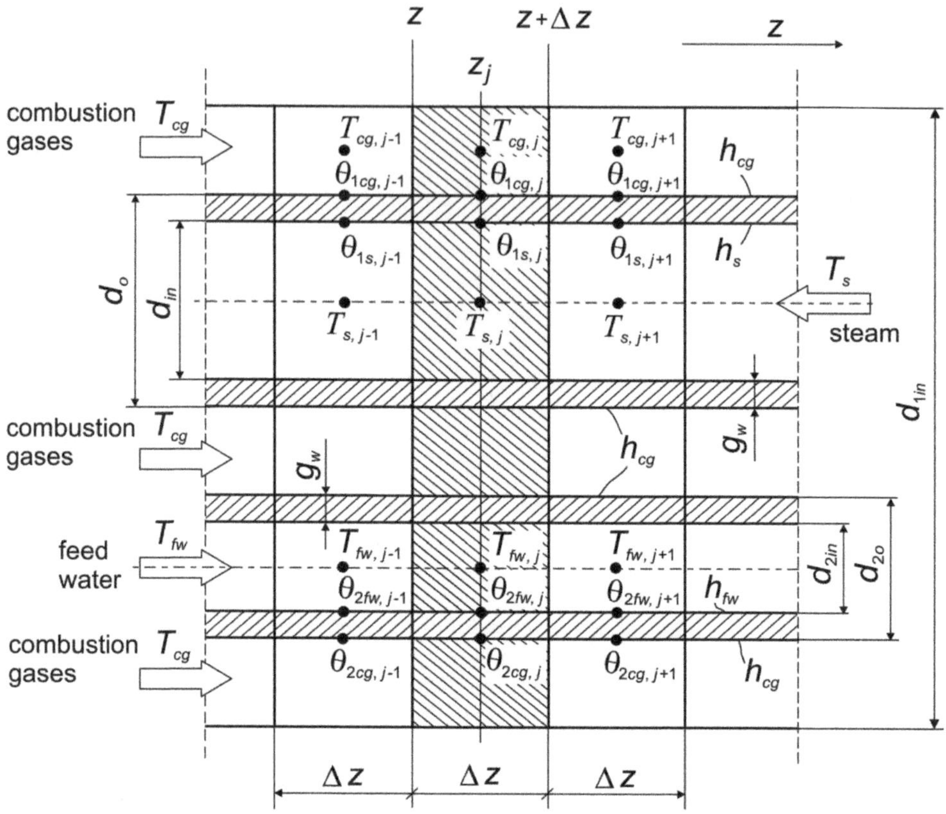

Fig. 3. Analysed control volume of several surfaces heated convectively, placed in parallel in a single gas pass

Substituting the appropriate boundary conditions, the following differential equations are obtained after some transformations:

$$\frac{d\theta_{1s}}{dt} = B_1\left(\theta_{1cg} - \theta_{1s}\right) + C_1\left(T_s - \theta_{1s}\right), \tag{27}$$

$$\frac{d\theta_{1cg}}{dt} = D_1\left(T_{cg} - \theta_{1cg}\right) + E_1\left(\theta_{1s} - \theta_{1cg}\right), \tag{28}$$

$$\frac{d\theta_{2fw}}{dt} = F_1\left(\theta_{2cg} - \theta_{2fw}\right) + G_1\left(T_{fw} - \theta_{2fw}\right), \tag{29}$$

$$\frac{d\theta_{2cg}}{dt} = H_1\left(T_{cg} - \theta_{2cg}\right) + J_1\left(\theta_{2fw} - \theta_{2cg}\right). \tag{30}$$

In the above equations:

$$B_1 = \frac{k_w(\theta_{1m})\pi d_m}{A_{1s}c_w(\theta_{1s})\rho_w(\theta_{1s})g_w}, \quad C_1 = \frac{h_s\pi d_{in}}{A_{1s}c_w(\theta_{1s})\rho_w(\theta_{1s})}, \quad D_1 = \frac{h_{cg}\pi d_o}{A_{1cg}c_w(\theta_{1cg})\rho_w(\theta_{1cg})},$$

$$E_1 = \frac{k_w(\theta_{1m})\pi d_m}{A_{1cg}c_w(\theta_{1cg})\rho_w(\theta_{1cg})g_w}, \quad F_1 = \frac{k_w(\theta_{2m})\pi d_{2m}}{A_{2fw}c_w(\theta_{2fw})\rho_w(\theta_{2fw})g_w}, \quad G_1 = \frac{h_{fw}\pi d_{2in}}{A_{2fw}c_w(\theta_{2fw})\rho_w(\theta_{2fw})},$$

$$H_1 = \frac{h_{cg}\pi d_{2o}}{A_{2cg}c_w(\theta_{2cg})\rho_w(\theta_{2cg})}, \quad J_1 = \frac{k_w(\theta_{2m})\pi d_{2m}}{A_{2cg}c_w(\theta_{2cg})\rho_w(\theta_{2cg})g_w}, \quad \theta_{1m} = \frac{\theta_{1s}+\theta_{1cg}}{2},$$

$$\theta_{2m} = \frac{\theta_{2fw}+\theta_{2cg}}{2}, \quad d_m \frac{d_{in}+d_o}{2}, \quad d_{2m} \frac{d_{2in}+d_{2o}}{2}, \quad A_{1s} = \frac{\pi\left(d_m^2-d_{in}^2\right)}{4}, \quad A_{1cg} = \frac{\pi\left(d_o^2-d_m^2\right)}{4},$$

$$A_{2fw} = \frac{\pi\left(d_{2m}^2-d_{2in}^2\right)}{4}, \text{ and } A_{2cg} = \frac{\pi\left(d_{2o}^2-d_{2m}^2\right)}{4}.$$

The energy balance equations take the following forms (Fig. 3):

- combustion gases

$$\Delta z A_{cg}c_{cg}(T_{cg})\rho_{cg}(T_{cg})\frac{\Delta T_{cg}}{\Delta t} = \dot{m}_{cg}i_{cg}\Big|_z - \dot{m}_{cg}i_{cg}\Big|_{z+\Delta z} + h_{cg}\pi d_o\Delta z\left(\theta_{1cg}-T_{cg}\right) + h_{cg}\pi d_{2o}\Delta z\left(\theta_{2cg}-T_{cg}\right) \tag{31}$$

- steam

$$\Delta z A_s c_s(T_s,p_s)\rho_s(T_s,p_s)\frac{\Delta T_s}{\Delta t} = \dot{m}_s i_s\Big|_{z+\Delta z} - \dot{m}_s i_s\Big|_z + h_s\pi d_{in}\Delta z\left(\theta_{1s}-T_s\right), \tag{32}$$

- feed water

$$\Delta z A_{fw}c_{fw}(T_{fw},p_{fw})\rho_{fw}(T_{fw},p_{fw})\frac{\Delta T_{fw}}{\Delta t} = \dot{m}_{fw}i_{fw}\Big|_z - \dot{m}_{fw}i_{fw}\Big|_{z+\Delta z} + h_{fw}\pi d_{2in}\Delta z\left(\theta_{2fw}-T_{fw}\right), \tag{33}$$

where:

$$A_{cg} = \frac{\pi d_{1in}^2}{4} - \left(\frac{\pi d_o^2}{4} + \frac{\pi d_{2o}^2}{4}\right), \quad A_s = \frac{\pi d_{in}^2}{4}, \text{ and } A_{fw} = \frac{\pi d_{2in}^2}{4}.$$

After rearranging and assuming that $\Delta t \to 0$ and $\Delta z \to 0$, the following formulae were obtained (from Equations (31)–(33), respectively):

$$\frac{\partial T_{cg}}{\partial t} = K_1\left(\theta_{1cg}-T_{cg}\right) + L_1\left(\theta_{2cg}-T_{cg}\right) - P_1\frac{\partial T_{cg}}{\partial z}, \tag{34}$$

$$\frac{\partial T_s}{\partial t} = Q_1(\theta_{1s} - T_s) + R_1 \frac{\partial T_s}{\partial z}, \quad (35)$$

$$\frac{\partial T_{fw}}{\partial t} = S_1\left(\theta_{2fw} - T_{fw}\right) - U_1 \frac{\partial T_{fw}}{\partial z}, \quad (36)$$

where:

$$K_1 = \frac{h_{cg}\pi d_o}{A_{cg}c_{cg}\left(T_{cg}\right)\rho_{cg}\left(T_{cg}\right)},\ L_1 = \frac{h_{cg}\pi d_{2o}}{A_{cg}c_{cg}\left(T_{cg}\right)\rho_{cg}\left(T_{cg}\right)},\ P_1 = \frac{\dot{m}_{cg}}{A_{cg}\rho_{cg}\left(T_{cg}\right)},\ R_1 = \frac{\dot{m}_s}{A_s\rho_s(T_s, p_s)},$$

$$Q_1 = \frac{h_s\pi d_{in}}{A_s c_s(T_s, p_s)\rho_s(T_s, p_s)},\ S_1 = \frac{h_{fw}\pi d_{2in}}{A_{fw}c_{fw}\left(T_{fw}, p_{fw}\right)\rho_{fw}\left(T_{fw}, p_{fw}\right)},\ \text{and}\ U_1 = \frac{\dot{m}_{fw}}{A_{fw}\rho_{fw}\left(T_{fw}, p_{fw}\right)}.$$

To solve the system of Equations (27) to (30) and (34) to (36) the implicit finite-difference method was used. After some transformations the following dependencies were obtained:

$$\theta_{1s,j}^{t+\Delta t} = \frac{1}{V\Delta t}\theta_{1s,j}^{t} + \frac{B_1}{V}\theta_{1cg,j}^{t+\Delta t} + \frac{C_1}{V}T_{s,j}^{t+\Delta t}, \quad j = 1, ..., M; \quad (37)$$

$$\theta_{1cg,j}^{t+\Delta t} = \frac{1}{V_1\Delta t}\theta_{1cg,j}^{t} + \frac{D_1}{V_1}T_{cg,j}^{t+\Delta t} + \frac{E_1}{V_1}\theta_{1s,j}^{t+\Delta t}, \quad j = 1, ..., M; \quad (38)$$

$$\theta_{2fw,j}^{t+\Delta t} = \frac{1}{W\Delta t}\theta_{2fw,j}^{t} + \frac{F_1}{W}\theta_{2cg,j}^{t+\Delta t} + \frac{G_1}{W}T_{fw,j}^{t+\Delta t}, \quad j = 1, ..., M; \quad (39)$$

$$\theta_{2cg,j}^{t+\Delta t} = \frac{1}{W_1\Delta t}\theta_{2cg,j}^{t} + \frac{H_1}{W_1}T_{cg,j}^{t+\Delta t} + \frac{J_1}{W_1}\theta_{2fw,j}^{t+\Delta t}, \quad j = 1, ..., M; \quad (40)$$

$$T_{cg,j}^{t+\Delta t} = \frac{1}{X_1\Delta t}T_{cg,j}^{t} + \frac{K_1}{X_1}\theta_{1cg,j}^{t+\Delta t} + \frac{L_1}{X_1}\theta_{2cg,j}^{t+\Delta t} + \frac{P_1}{X_1\Delta z}T_{cg,j-1}^{t+\Delta t}, \quad j = 2, ..., M; \quad (41)$$

$$T_{s,j}^{t+\Delta t} = \frac{1}{Y_1\Delta t}T_{s,j}^{t} + \frac{Q_1}{Y_1}\theta_{1s,j}^{t+\Delta t} + \frac{R_1}{Y_1\Delta z}T_{s,j+1}^{t+\Delta t}, \quad j = 1, ..., M\text{-}1; \quad (42)$$

$$T_{fw,j}^{t+\Delta t} = \frac{1}{Z_1\Delta t}T_{fw,j}^{t} + \frac{S_1}{Z_1}\theta_{2fw,j}^{t+\Delta t} + \frac{U_1}{Z_1\Delta z}T_{fw,j-1}^{t+\Delta t}, \quad j = 2, ..., M. \quad (43)$$

In the above equations:

$$V = \frac{1}{\Delta t} + B_1 + C_1,\ V_1 = \frac{1}{\Delta t} + D_1 + E_1,\ W = \frac{1}{\Delta t} + F_1 + G_1,\ W_1 = \frac{1}{\Delta t} + H_1 + J_1,$$

$$X_1 = \frac{1}{\Delta t} + K_1 + L_1 + \frac{P_1}{\Delta z},\ Y_1 = \frac{1}{\Delta t} + Q_1 + \frac{R_1}{\Delta z},\ \text{and}\ Z_1 = \frac{1}{\Delta t} + S_1 + \frac{U_1}{\Delta z}.$$

In view of the iterative character of the suggested method, the computations should satisfy the following condition:

$$\frac{\left|Y_{j,(k+1)}^{t+\Delta t} - Y_{j,(k)}^{t+\Delta t}\right|}{Y_{j,(k+1)}^{t+\Delta t}} \le \vartheta \tag{44}$$

where Y is the currently evaluated temperature in node j; ϑ is the assumed tolerance of iteration; and $k = 1, 2, \ldots$ is the next iteration counter after a single time step.
Additionally, the following condition - the Courant-Friedrichs-Lewy stability condition over the time step - should be satisfied (Gerald, 1994):

$$|\beta| \le 1, \quad \Delta t \le \frac{\Delta z}{w}, \tag{45}$$

where: $\beta = \frac{w\Delta t}{\Delta z}$ is the Courant number.
When satisfying this condition, the numerical solution is reached with a speed $\Delta z/\Delta t$, which is greater than the physical speed w.

3. Computational verification

The efficiency of the proposed method is verified in this section by the comparison of the results obtained using the method and from the corresponding analytical solutions. Exact solutions available in the literature for transient states are developed only for the simplest cases. In this section a step function change of the fluid temperature at the tube inlet and a step function heating on the outer surface of the tube are analysed.

3.1 Analytical solutions for transient states

The available analytical dependencies allow the following to be determined (Serov & Korolkov, 1981):

- the time-spatial temperature distribution of the tube wall, insulated on the outer surface, as the tube's response to the temperature step function of the fluid at the tube inlet,
- the time-spatial temperature distribution of the fluid in the case of a heat flux step function on the outer surface of the tube.

3.1.1 Temperature step function of the fluid at the tube inlet

The analysed step function is assumed as follows (Fig. 4):

$$\Delta T(t) = \begin{cases} 0 & \text{for } t < 0, \\ 1 & \text{for } t \ge 0. \end{cases} \tag{46}$$

For this step function, the dimensionless dependency determining the increase of the tube wall temperature takes the following form:

$$\frac{\Delta\theta}{\Delta T} = V_1 - V_0, \tag{47}$$

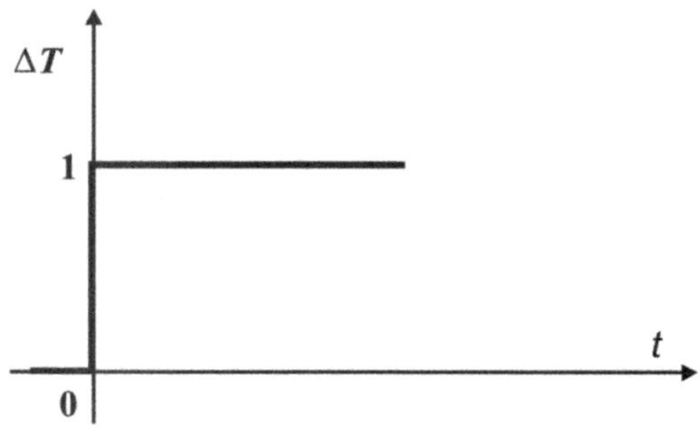

Fig. 4. Temperature step function of the fluid at the tube inlet

where:

$$V_1 = e^{-(\zeta+\eta)} U(\zeta,\eta), \tag{48}$$

$$V_0 = e^{-(\zeta+\eta)} I_0\left(2\sqrt{\zeta\eta}\right). \tag{49}$$

The $U(\zeta,\eta)$ function is described by the following dependency:

$$U(\zeta,\eta) = \sum_{n=0}^{\infty}\sum_{k=0}^{n} \frac{\eta^{n}\zeta^{k}}{n!k!}, \tag{50}$$

and the Bessel function:

$$I_0\left(2\sqrt{\zeta\eta}\right) = \sum_{k=0}^{\infty} \frac{(\zeta\eta)^k}{(k!)^2}. \tag{51}$$

Values ζ and η present in Formulae (48)–(51) are the dimensionless variables of length and time respectively, expressed by the following dependencies:

$$\zeta = \frac{z}{F_2}; \qquad \eta = \frac{t - t_{TP}(z)}{D_2}, \tag{52}$$

where:

$$t_{TP}(z) = B_2\zeta = \frac{z}{w}. \tag{53}$$

Coefficients B_2, D_2, and F_2 are described in Section 3.2.

3.1.2 Heat flux step function on the outer surface of the tube

A dimensionless time-spatial function describing the increase of the fluid temperature ΔT, caused by the heat flux step function Δq on the outer surface of the tube, is expressed as:

$$\varphi_1 = \frac{\Delta T}{-\frac{c}{1-c}E_2 \Delta q} = \frac{t}{D_2} - \frac{1}{1-c}\varphi_0 - V_2 \,. \tag{54}$$

In the above formula:
$c = - D_2/B_2$; Δq and coefficient E_2 are described in Section 3.2.
Functions φ_0 and V_2 are described by the following dependencies:

$$\varphi_0 = 1 - e^{-(1-c)\frac{t}{D_2}} - V_1 + V_{00} \,, \tag{55}$$

$$V_2 = e^{-(\zeta+\eta)}\left[(\eta-\zeta)\ U(\zeta,\eta) + \zeta\, I_0\left(2\sqrt{\zeta\eta}\right) + \sqrt{\zeta\eta} I_1\left(2\sqrt{\zeta\eta}\right)\right], \tag{56}$$

where:

$$I_1\left(2\sqrt{\zeta\eta}\right) = \sum_{k=0}^{\infty} \frac{(\zeta\eta)^{\frac{2k+1}{2}}}{(k!)\,(k+1)!} \,. \tag{57}$$

Function V_{00} present in Formula (55) is expressed as:

$$V_{00} = e^{-(\zeta+\eta)} U\left(\frac{\zeta}{c}, c\eta\right). \tag{58}$$

The analytical dependencies (47) and (54) presented above allow the time-spatial temperature increases, $\Delta\theta$ for the tube wall and ΔT for the fluid, to be determined for any selected cross-section. The results are obtained beginning from time $t_{TP}(z) = z/w$, that is, from the moment this cross-section is reached by the fluid flowing with velocity w. For example, if the flow velocity equals 1m/s, then the analytical solutions allow the temperature changes for the cross-section located 10 m away from the inlet of the tube to be determined only after 10 s.

3.2 Application of the proposed method for the purpose of verification

In order to compare the results obtained using the suggested method with the results of analytical solutions for transient states, the appropriate dependencies are derived for the control volume shown in Fig. 5.
Assuming one control volume of the tube wall, Equation (4) takes the form of Equation (20).
Taking into consideration the boundary conditions:

$$k_w(\theta)\left.\frac{\partial\theta}{\partial r}\right|_{r=r_o} = q \,, \tag{59}$$

and

$$k_w(\theta)\left.\frac{\partial\theta}{\partial r}\right|_{r=r_{in}} = h\left(\left.\theta\right|_{r=r_{in}} - T\right) = h(\theta - T), \tag{60}$$

the following differential equation is obtained:

$$D_2 \frac{d\theta}{dt} = T - \theta + E_2 \Delta q \,. \tag{61}$$

In the above equation:

$$D_2 = \frac{c_w(\theta)\rho_w(\theta)\, d_m\, g_w}{h d_{in}}, \quad E_2 = \frac{1}{h\pi d_{in}}, \text{ and } \quad d_m = \frac{d_o + d_{in}}{2}.$$

Moreover, the heat flux step function is described as:

$$\Delta q = q \cdot s \,, \tag{62}$$

where:
q - heat flux, W/m²,
s - actual tube pitch, m.

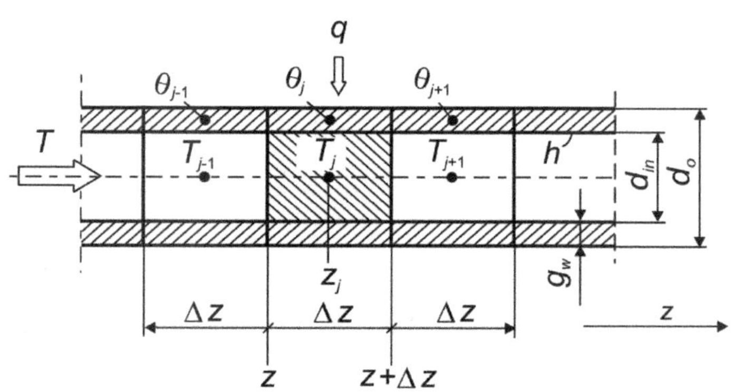

Fig. 5. Analysed control volume

On the side of the working fluid the energy balance equation takes the form of Equation (13), in which the mean wall temperature θ is used instead of θ_{in}:

$$\Delta z A c(T,p)\rho(T,p)\frac{\Delta T}{\Delta t} = \dot{m}\, i\Big|_z - \dot{m}\, i\Big|_{z+\Delta z} + h\pi d_{in}\Delta z(\theta - T). \tag{63}$$

Assuming that $\Delta t \to 0$ and $\Delta z \to 0$, the following equation is obtained:

$$B_2 \frac{\partial T}{\partial t} = \theta - T - F_2 \frac{\partial T}{\partial z}, \tag{64}$$

where:

$$B_2 = \frac{A c(T,p)\rho(T,p)}{h\pi d_{in}}, \quad F_2 = \frac{\dot{m} c(T,p)}{h\pi d_{in}} \quad \text{and} \quad A = \frac{\pi d_{in}^{\;2}}{4}.$$

To solve the system of Equations (61) and (64), the implicit finite difference method was used, and after transformations we obtain:

$$\theta_j^{t+\Delta t} = \left(\frac{D_2}{D_2+\Delta t}\right)\theta_j^t + \left(\frac{\Delta t}{\Delta t + D_2}\right)\left(T_j^{t+\Delta t} + E_2 \Delta q_j^{t+\Delta t}\right), \quad j = 1, ..., M \tag{65}$$

$$T_j^{t+\Delta t} = \frac{\theta_j^{t+\Delta t} + \frac{B_2}{\Delta t} T_j^t + \frac{F_2}{\Delta z} T_{j-1}^{t+\Delta t}}{\frac{B_2}{\Delta t} + \frac{F_2}{\Delta z} + 1}, \quad j = 2, ..., M. \tag{66}$$

3.3 Results and discussion

As an illustration of the accuracy and effectiveness of the suggested method the following numerical analyses are carried out:

- for the tube with the temperature step function of the fluid at the tube inlet,
- for the tube with the heat flux step function on the outer surface.

The results obtained are compared afterwards with the results of analytical solutions. In both cases the working fluid is assumed to be water. The heat transfer coefficient is taken as constant and equals $h = 1000$ W/(m²K). Because the exact solutions do not allow the temperature dependent thermo-physical properties to be considered, the following constant water properties were assumed for the computations: $\rho = 988$ kg/m³ and $c = 4199$ J/(kgK). For both cases it was also assumed that the tube is $L = 131$ m long, its external diameter equals $d_o =$ 0.038 m, the wall thickness is $g_w = 0.0032$ m, and the tube is made of K10 steel of the following properties: $\rho_w = 7850$ kg/m³ and $c_w = 470$ J/(kgK). Satisfying the Courant condition (45), the following were taken for the computations: $\Delta z = 0.5$ m, $\Delta t = 0.1$ s and $w = 1$ m/s ($\dot{m} = 0.775$ kg/s).

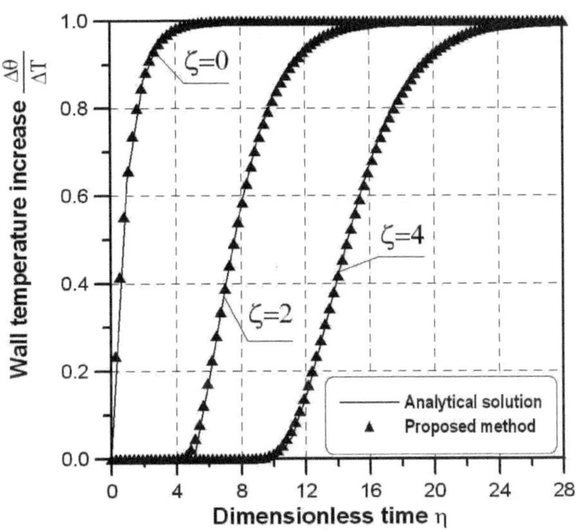

Fig. 6. Dimensionless histories of tube wall temperature increase

In the first numerical analysis it was assumed that water of initial temperature $T = 20$ °C flows through the tube. Also, the tube wall for the initial time $t = 0$ has the same initial temperature. Beginning from the next time step, the fluid of temperature $T = 100$ °C appears at the inlet. The temperature step function is thus $\Delta T = 80$ K. The results of the computations are presented in Fig. 6. The presented dimensionless coordinates $\zeta = 0$, 2, and 4 correspond with the dimensional coordinates $z = 0$, 65.5 m, and 131 m respectively. An analysis of the comparison shows satisfactory convergence of the exact solution results with the results obtained using the presented method.

In the second case it was assumed that the working fluid and the tube at time $t = 0$ take the initial temperature $T = \theta = 70$ °C. Starting from the next time step, the heat flux step function ($\Delta q = q{\cdot}s$) appears on the outer surface of the tube. The assumed heat load is the heat flux $q = 10^5$ W/m^2 and the tube pitch $s = 0.041$ m. The selected results of the numerical analysis, comprising a comparison of the dimensionless histories of the fluid temperature increase for the same cross-sections as in the first case, are shown in Fig. 7.

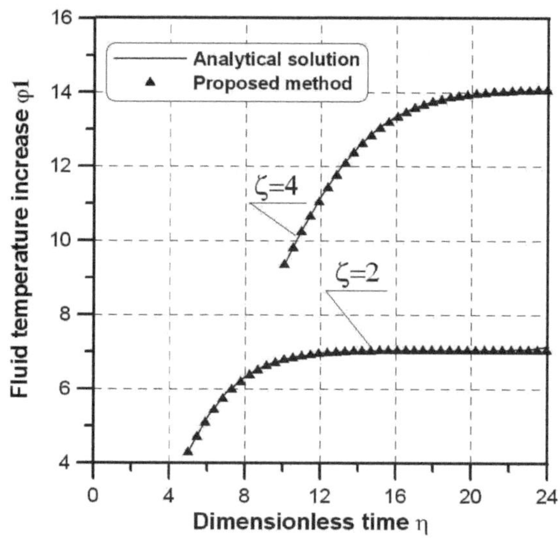

Fig. 7. Histories of dimensionless fluid temperature increase

These histories begin from the time instants $\eta = 5.04$ ($t = 65.5$ s), and $\eta = 10.08$ ($t = 131$ s), respectively, that is, from the moment the analysed cross-sections were reached by the fluid flowing with the velocity $w = 1$ m/s. A satisfactory convergence of the results of the analytical solution with the results obtained using the suggested method was achieved.

4. Experimental verification

This section describes the experimental verification of the proposed method for modelling transient processes which occur in power boilers surfaces heated convectively. Transient state operation of the platen superheater during the start-up of an OP-210 boiler was analysed. The boiler capacity is $210{\cdot}10^3$ kg/h of live steam with 9.8 MPa pressure and 540^{+5}_{-10} °C temperature. The platen superheater (Figs. 8 and 13) consists of 14 vertical screens installed with 520 mm transversal pitch. Each screen consists of 13 tubes (ϕ 32 × 5 mm) placed with 36 mm longitudinal pitch. The heated surface of the superheater is 406 m^2 (n = 182 tubes) and the total free cross-section of the combustion gases flow is $A_{cg,t}$ = 64.5 m^2. The tubes, each L = 26.3 m long, are made of 12H1MF steel and placed in 52 rows. As the analysed platen superheater is operating in cross-parallel-flow, a parallel-flow arrangement was assumed for numerical modelling.

The time-spatial heat transfer coefficients for steam and combustion gases were computed in the on-line mode using dependencies published in (Kuznetsov et al., 1973). Moreover, based on the data given by (Meyer et al., 1993; Kuznetsov et al., 1973; Wegst, 2000) appropriate functions were created. These functions allow the thermo-physical properties of the steam, combustion gases and the material of the tube wall to be computed in real time.

The platen superheater tube was divided into M = 16 cross-sections (Δz = 1.75 m). The time step of computations was taken at Δt = 0.1 s.

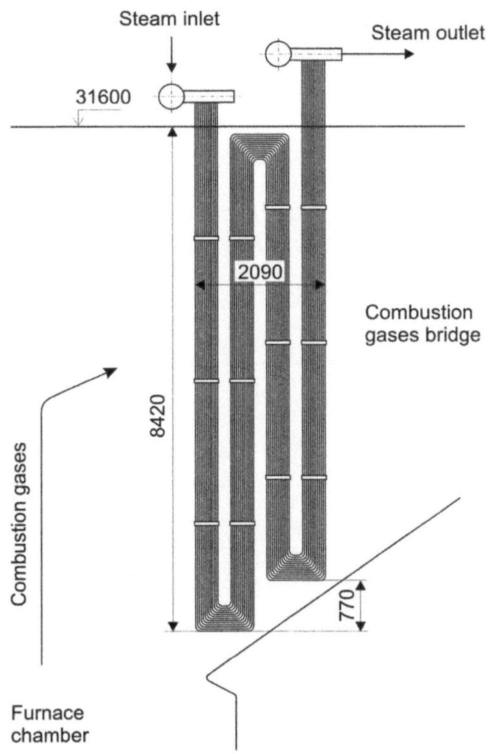

Fig. 8. Location of platen superheater

In order to model the dynamics of the platen superheater it is necessary to know the transient values of temperature, pressure, and total mass flow of steam and combustion gases at the superheater inlet. On the steam side, these values were known from measurements and are shown in Figs. 9 and 11 (curve b), whereas at the combustion gases side they were computed (Fig. 10). To calculate the pressure drop of the steam (in the direction of the steam flow), the Darcy-Weisbach equation was used.

The selection of a platen superheater for verification was not accidental. It is, namely, located in the combustion gas bridge, just behind the furnace chamber (Fig. 8). The computed values of combustion gases temperature and mass flow at the furnace chamber outlet therefore constituted the input data for modelling the platen superheater operation. In order to compute these transient values, the fuel mass flow should be determined first. To find it, a method based on the known characteristics of the coal dust feeder in function of its number of revolutions was used (Cwynar, 1981). The total mass flow of combustion gases at the furnace chamber outlet was computed using stoichiometric combustion equations and the known mass flow of combustion coal. The combustion gases temperature at the furnace chamber outlet was determined by solving the equations of energy and heat transfer for the boiler furnace chamber using the CKTI method (Kuznetsov et al., 1973). The computed values of combustion gases temperature and mass flow are shown in Fig. 10.

The measurements carried out on the real object were disturbed by errors resulting from the degree of inaccuracy of the measuring sensors and converters.

These errors, related to the maximum measuring ranges, were as follows:

a. ± 3.3 °C in the superheated steam temperature readings (measuring range: 0–600 °C; level of sensor inaccuracy: 0.25; level of converter inaccuracy: 0.3),

b. ± $96 \cdot 10^3$ Pa in the superheated steam pressure readings (measuring range: 0–16 MPa; level of sensor inaccuracy: 0.6),

c. ± 0.799 kg/s in the mass flow of superheated steam (measuring range: 0–69.44 kg/s; level of measuring orifice inaccuracy: 1; level of converter inaccuracy: 0.15).

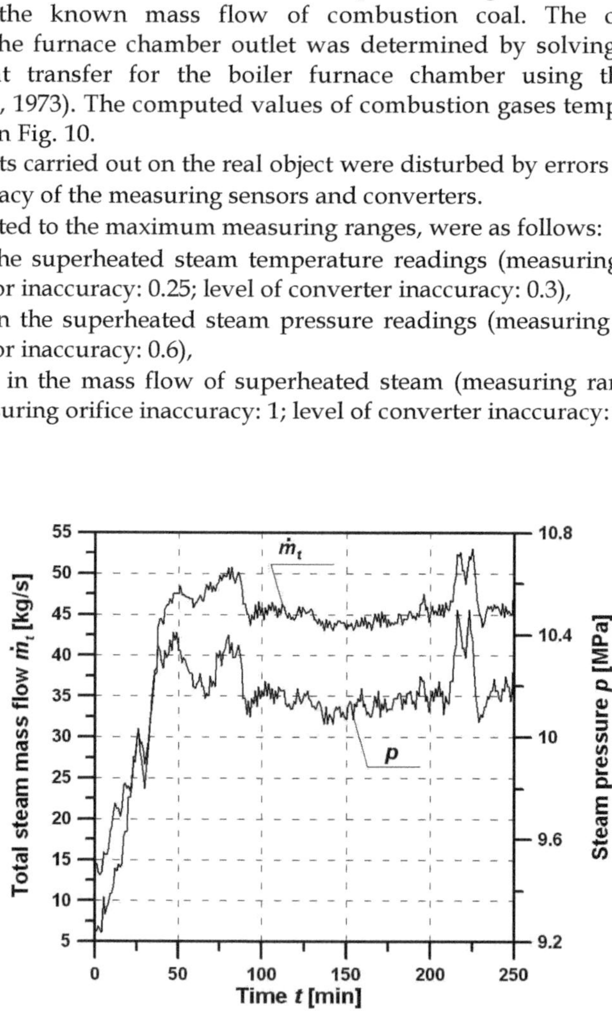

Fig. 9. Histories of the measured steam pressure and total mass flow at the platen superheater inlet

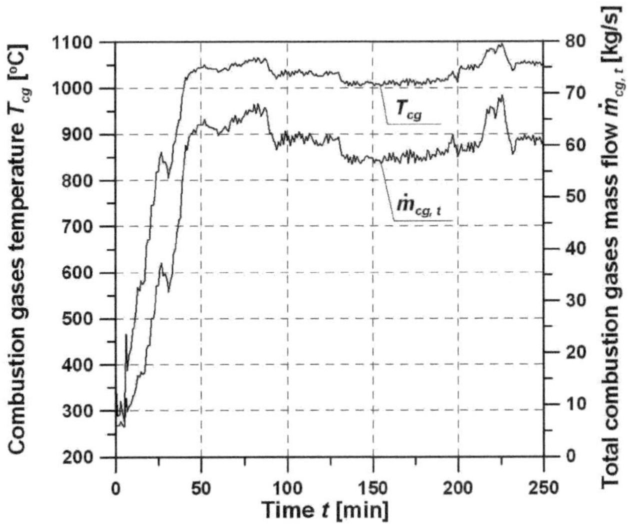

Fig. 10. Histories of the computed combustion gases temperature and total mass flow at the platen superheater inlet (at the furnace chamber outlet)

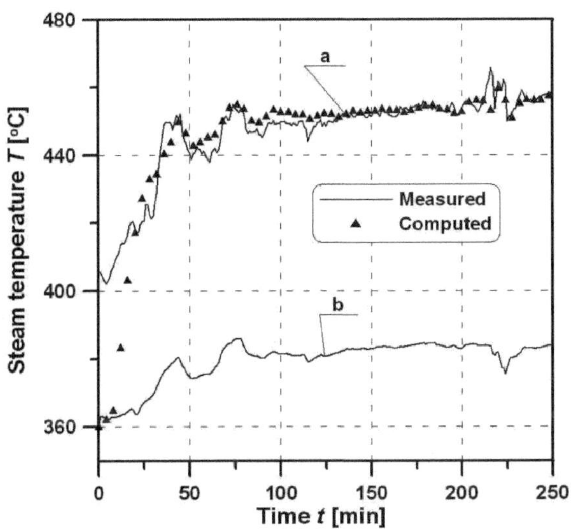

Fig. 11. Comparison of the measured and computed steam temperatures at the superheater outlet (a) and history of the measured steam temperature at the superheater inlet (b)

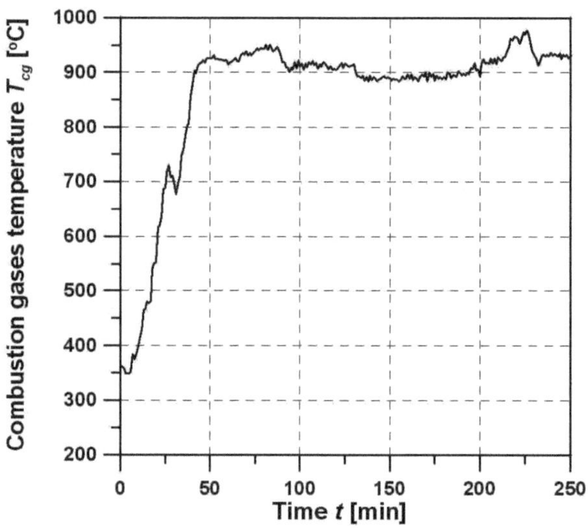

Fig. 12. History of the computed combustion gases temperature at the superheater outlet

When comparing the results of steam temperature measurement at the platen superheater outlet with the results of numerical computation, fully satisfactory convergence is found (Fig. 11 – curve a). The divergences visible in Fig. 11 (curve a), in the range of 0 to about 30 min, result from the assumption in the calculation model that the initial temperature of the analysed steam superheating system at time $t = 0$ is equal to the measured steam temperature at the superheater inlet, that is, $T = T_{cg} = \theta_h = \theta_{cg} = 359$ °C.

The computed combustion gases temperature at the platen superheater outlet (Fig. 12) can be used for modelling the dynamics of steam superheaters located after it (Fig. 13). The two stages, KPP-2 and KPP-3, of the superheater are installed parallel to each other in one gas pass. The superheater KPP-2 operates in counter-flow, and KPP-3 operates in parallel-flow to combustion gases. A comparison of the measured and computed steam temperature histories at the KPP-3 outlet is presented in the paper (Zima, 2003).

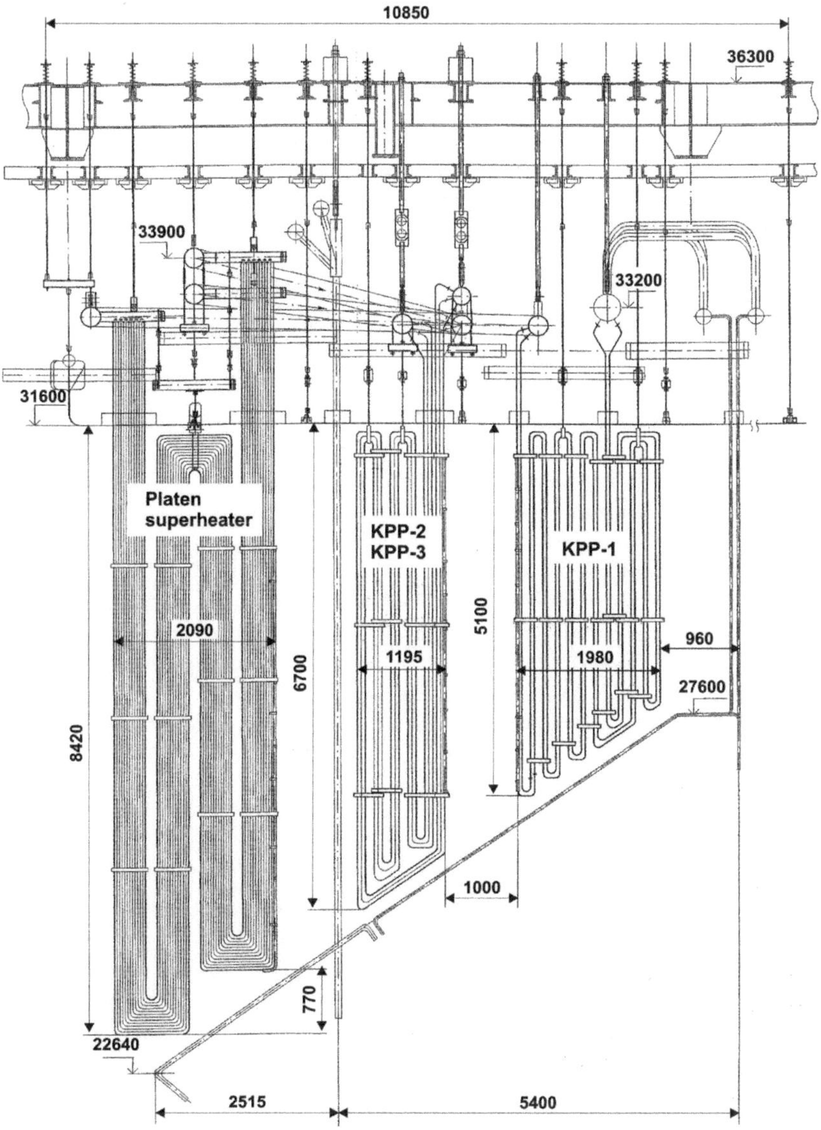

Fig. 13. Location of the analysed platen superheater and three stages of convective steam superheater (KPP-1, KPP-2, and KPP-3)

The selected results of modelling the dynamics of the economizer installed in the convective duct of the OP-210 boiler are presented in the paper (Zima, 2007). In the computations the fins were considered on the combustion gases side, and the heat transfer coefficient was calculated according to (Taler & Duda, 2006). The measured history of feed water temperature at the economizer outlet was compared with the computational results and satisfactory agreement was achieved.

5. Conclusions

The chapter presents a method for modelling the dynamics of boiler surfaces heated convectively, namely steam superheaters and economizers. The proposed method comprises solving the energy equations and considers the superheater or economizer model as one with distributed parameters. The proposed model is one-dimensional and is suitable for pendant superheaters and economizers. In this model, the boundary conditions can be time-dependent. The computations are carried out in the direction of the heated fluid flow in one tube. The time-spatial temperature history of the separating wall is determined from the equation of transient heat conduction. As the time-spatial heat transfer coefficients at the working fluids sides are computed in the on-line mode considering the actual tube pitches and cross-flow of combustion gases, the physics of the phenomena occurring in the superheaters and economizers does not change. All the thermo-physical properties of the fluids and the material of the separating walls are also computed in real time. In order to prove the accuracy and effectiveness of the proposed method, computational and experimental verifications were carried out. The analysis of the presented comparisons demonstrates fully satisfactory convergence of the results obtained using the suggested method with the results of analytical solutions and with measured temperature history. When analysing the presented comparisons it should be considered that many parameters affect the final result of the operation of the surfaces heated convectively (e.g. ones resulting from gradual fouling of these surfaces). Not all these parameters can be fully taken into consideration in the calculation algorithm.

6. References

Bojić, M. & Dragićević, S. (2006). Optimization of steam boiler design. *Proceedings of the Institution of Mechanical Engineers, Part A: Journal of Power and Energy*, Vol. 220, No. 6 (September 2006), pp. 629–634, ISSN 0957-6509

Chakraborty, N. & Chakraborty, S. (2002). A generalized object-oriented computational method for simulation of power and process cycles. *Proceedings of the Institution of Mechanical Engineers, Part A: Journal of Power and Energy*, Vol. 216, No. 2 (April 2002), pp. 155–159, ISSN 0957-6509

Cwynar, L. (1981). *Start-up of Power Boilers* (in Polish), Scientific and Technical Publishing Company, ISBN 83-204-0416-9, Warsaw

Dechamps, P.J. (1995). Modelling the transient behaviour of heat recovery steam generators. *Proceedings of the Institution of Mechanical Engineers, Part A: Journal of Power and Energy*, Vol. 209, No. A4 (January 1995), pp. 265–273, ISSN 0957-6509

Enns, M. (1962). Comparison of dynamic models of a superheater. *ASME Transactions – Journal of Heat Transfer*, Vol. 84, No. 4, pp. 375–385

Gerald, C.F. & Wheatley, P.O. (1994). *Applied numerical analysis*, Addison-Wesley Publishing Company, ISBN 0-201-56553-6, New York

Hausen, H. (1976). *Wärmeübertragung im Gegenstrom, Gleichstrom und Kreuzstrom* (2nd ed.), Springer Verlag, ISBN 3540075526, Berlin

Krzyżanowski, J. & Głuch, J. (2004). *Heat-Flow Diagnostics of Energetic Objects* (in Polish), Polish Academy of Sciences, ISBN 83-88237-65-9, Gdansk

Kuznetsov, N.V.; Mitor, V.V.; Dubovskij, I.E. & Karasina, E.S. (1973). *Standard Methods of Thermal Design for Power Boilers* (in Russian), Central Boiler and Turbine Institute, Energija, UDK 621.181.001.24:536.7, Moscow

Lu, S. (1999). Dynamic modelling and simulation of power plant systems. *Proceedings of the Institution of Mechanical Engineers, Part A: Journal of Power and Energy*, Vol. 213, No. 1 (February 1999), pp. 7–22, ISSN 0957-6509

Meyer, C. A. et al. (1993). *ASME Steam Tables*, American Society of Mechanical Engineers, ISBN 0791806324, New York

Mohammadzaheri, M.; Chen, L.; Ghaffari, A. & Willison, J. (2009). A combination of linear and nonlinear activation functions in neural networks for modeling a de-superheater. *Simulation Modelling Practice & Theory*, Vol. 17, No. 2 (February 2009), pp. 398–407, ISSN 1569190X

Mohan, M.; Gandhi, O.P. & Agrawal, V.P. (2003). Systems modelling of a coal-based steam power plant. *Proceedings of the Institution of Mechanical Engineers, Part A: Journal of Power and Energy*, Vol. 217, No. 3 (June 2003), pp. 259–277, ISSN 0957-6509

Serov, E.P. & Korolkov, B.P. (1981). *Dynamics of Steam Generators* (in Russian), Energoizdat, UDK 621.181.016.7, Moscow

Shirakawa, M. (2006). Development of a thermal power plant simulation tool based on object orientation. *Proceedings of the Institution of Mechanical Engineers, Part A: Journal of Power and Energy*, Vol. 220, No. 6 (September 2006), pp. 569–579, ISSN 0957-6509

Taler, J. & Duda, P. (2006). *Solving Direct and Inverse Heat Conduction Problems*, Springer, ISBN 978-3-540-33470-5, Berlin

Wegst, C.W. (2000). *Key to Steel*, Verlag Stahlschlüssel Wegst GmbH, ISBN 3922599176, Marbach

Zima, W. (2001). Numerical modeling of dynamics of steam superheaters. *Energy*, Vol. 26, No. 12, (December 2001), pp. 1175–1184, ISSN 0360-5442

Zima, W. (2003). Mathematical model of transient processes in steam superheaters. *Forschung im Ingenieurwesen*, Vol. 68, No. 1 (July 2003), pp. 51–59, ISSN 0015-7899

Zima, W. (2004). *Simulation of transient processes in boiler steam superheaters* (in Polish), Monograph 311, Publishing House of Cracow University of Technology, ISSN 0860-097X, Cracow

Zima, W. (2006). Simulation of dynamics of a boiler steam superheater with an attemperator. *Proceedings of the Institution of Mechanical Engineers, Part A: Journal of Power and Energy*, Vol. 220, No. 7 (November 2006), pp. 793–801, ISSN 0957-6509

Zima, W. (2007). Mathematical modelling of transient processes in convective heated surfaces of boilers. *Forschung im Ingenieurwesen*, Vol. 71, No. 2 (June 2007), pp. 113–123, ISSN 0015-7899

Part 3

Air Cooling of Electronic Devices

14

Air Cooling Module Applications to Consumer-Electronic Products

Jung-Chang Wang[1] and Sih-Li Chen[2]
[1]*National Taiwan Ocean University*
[2]*National Taiwan University*
Taiwan, R.O.C.

1. Introduction

The purpose of this chapter is to describe how a air-cooling thermal module is comprised with single heat sink, two-phase flow heat transfer modules with high heat transfer efficiency, to effectively reduce the temperature of consumer-electronic products as Personal Computer (PC), Note Book (NB), Server including central processing unit (CPU) and graphic processing unit (GPU), and LED lighting lamp of smaller area and higher power. The research design concentrates on several air-cooling thermal modules. For air cooling, the extended surface, such as fin is usually added to increase the rate of heat removal. The heat capacity from heat source conducted and transferred through heat sink to the surroundings by air convection. Thus, the aim of adding fin is to help dissipate heat flow from heat source. The air convection heat transfer mechanism was shown in the figure 1, which can be separated into forced and free/nature convection through dynamic fluid device as fan. The chapter is divided into three parts; first part discusses optimum, performance analysis and verification of a practical convention parallel plate-fin heat sink. Second part employs two-phase flow heat transfer devices, such as heat pipe, thermosyphon and vapor chamber comprised with heat sink to consumer-electronic products. The last part utilizes air-cooling thermal module in other industrial areas including injection mold and large motor.

A conventional plate-fin heat sink is composed of a plate-fin heat sink and a fan. Thermal resistance network is often employed to analyze the thermal model and system in the industry. The overall thermal resistance includes interface resistance, base-conduction resistance, and convective resistance. It is worth developing a model for a conventional air-cooling device that takes heat sink configuration and airflow conditions into account in order to predict the device's thermal performance when developing laminar-, transition-, and turbulent-flow regimes. Although, solving the high heat capacity of electronic components has been to install a heat sink with a fan directly on the heat source, removing the heat through forced convection. Increasing the fin surface and fan speed are two direct heat removal heat sink in order to solve the ever increasing high heat flux generated by heat source from consumer-electronic products. They can reduce the total thermal resistance from 0.6 °C/W to 0.3 °C/W. Lin & Chen (2003) and Wu et al. (2011) has been developed an analytical all-in-one asymptotic model to predict the hydraulic and thermal performance of a practical heat sink including a rectangular base plate and parallel fins with a non-uniform heat source, which proposed for a wide range of Reynolds and Nusselt numbers, including laminar, transition, and turbulent flows. However, increasing the surface area results in an increase in cost and boosting the fan speed results in noise, vibration and more power consumption, which increases the probability of failure to consumer-electronic components. Its total thermal resistance is usually over 0.3 °C/W not adjust high heat capacity; A two-phase flow heat transfer module with high heat transfer efficiency, to effectively reduce the temperature of heat sources of smaller area and higher power.

Fig. 1. Air convection mechanism

In recent years, technical development related with the application of two-phase flow heat transfer assembly to thermal modules has become mature and heat pipe-based two-phase flow heat transfer module is one of the best choices (Wang, 2008). A heat sink with embedded heat pipes transfers the total heat capacity from the heat source to the base plate with embedded heat pipes and fins sequentially, and then dissipates the heat flow into the surrounding air. Wang et al. (2007) have experimentally investigated the thermal resistance of a aluminium heat sink with horizontal embedded two and four U-shape heat pipes of 6 mm diameter; they showed that two heat pipes embedded in the base plate carry 36 percent of the total dissipated heat capacity from Central Process Unit (CPU), while 64% of heat was delivered from the base plate to the fins. Furthermore, when the CPU power was 140 W, the total thermal resistance was at its minimum of 0.27 °C/W. And using four embedded heat pipes carry 48 percent of the total dissipated heat capacity from CPU; the total thermal resistance is under 0.24 °C/W. The total thermal resistance of the heat sink with embedded heat pipes is only affected by changes in the base to heat pipes thermal resistance and heat pipes thermal resistance over the heat flow path; that is, the total thermal resistance varies according to the functionality of the heat pipes. If the temperature of the heat source is not allowed to exceed 70 °C, the total heating powers of heat sink with two and four embedded heat pipes will not exceed 131 Wand 164 W respectively. The superposition principal analytical method for the thermal performance of the heat sink with embedded heat pipes is completely established (Wang, 2009). The thermal performance of a heat sink with embedded heat pipes has been developed a Windows program for rapidly calculating through Visual Basic commercial software (Wang, 2010a). The computing core of this Windows program employs the theoretical thermal resistance analytical approach with iterative convergence to obtain a numerical solution. The estimation error between the numerical and experimental solutions is less than ±5%. The optimum inserting heights with total fin height are also obtained through the fitting curves generated in the program. From this Windows program, the optimum height of the embedded heat pipes inserted through fins is 21mm and 15mm for one pair and two pairs of embedded heat pipes, respectively. If the heat sink is considered in different orientations with respect to gravity, the results may be different. Finally, this Windows program has the advantage of rapidly calculating the thermal performance of a heat sink with embedded heat pipes installed horizontally with a processor by inputting simple parameters.

Moreover, one set of risers of the L-shape heat pipes were functioning as the evaporating section while the other set acted as condensing section. Six L-type heat pipes are arranged vertically in such a way that the bottom acts as the evaporating section and the risers act as the condensing section (Wang, 2011b). It describes the design, modeling, and test of a heat sink with embedded L-shaped heat pipes and plate fins. This type of heat sink is particularly well suited for cooling electronic components such as microprocessors using forced convection. The mathematical model includes all major components from the thermal interface through the heat pipes and fins. It is augmented with measured values for the heat pipe thermal resistance. A Windows-based computer program also uses an iterative superposition method to predict the thermal performance. The sum of the bypass heating power ratios is 14.4% for Q1 and Q2, 20.8% for Q3 and Q4, and 52% for Q5 and Q6, obtained using both the experimental results and the software program based on VB6.0. Thermal performance testing shows that a representative heat sink with six heat pipes will carry 160W and has reached a minimum thermal resistance of 0.22 °C/W. The computer software predicted a thermal resistance of 0.21 °C/W, which was within 5% of the measured value. Moreover, the total thermal resistance of the heat sink with six embedded L-type heat pipes is only affected by changes in the base to heat pipes thermal resistance and heat pipes thermal resistance over the heat flow path. That is, the total thermal resistance varies according to the functionality of the L-type heat pipes. The index of the thermal performance of a heat pipe for a thermal module manufacturer is the temperature difference between the evaporation and condensation sections of a single heat pipe and maximum heat capacity. The maximum heat capacity reaches the highest point, as the amount of the non-condensation gas of a heat pipe is the lowest value and the temperature difference between evaporation and condensation sections is the smallest one. The temperature difference is under 1°C while the percentage of the non-condensation gas is less than 8×10^{-5}%, and the single heat pipe has the maximum heat capacity (Wang, 2011a). To establish a practical quick methodology that can effectively and efficiently determine the thermal performances of heat pipes so as to substitute the use of the conventional steady-state test. A novel dynamic test method is originated and developed (Tsai et al. 2010a). With a view toward shortening the necessary time to examine the thermal performances of heat pipes, a novel dynamic test method is originated and compared to the conventional steady-states test. The dynamic test can be adopted as a serviceable method to determine thermal performances of heat pipes. Only 10-15 min is necessary to examine a heat pipe using the dynamic test. This is much more efficient than the steady-state test and would be greatly beneficial to the notebook PC industry or other heat dissipation technologies that use heat pipes.

Liquid cooling technology employs the excellent thermal performance of liquid to quickly take away the heat capacity from a heat source. The method by which liquid contacts the heat source can be divided into two types, including immediacy and mediacy. And thermoelectric cooler (TEC) has been applied to electronic cooling with its advantages of sensitive temperature control, quietness, reliability, and small size. Thermoelectric cooler is regarded as a potential solution for improving the thermal performances of cooling devices on the package. Huang et al. (2010) have combined TEC and water-cooling device to investigate the thermal performance. An analytical model of the thermal analogy network is provided to predict the cooling capability of the thermoelectric device. The prediction by the theoretical model agrees with the experimental results. Increasing the electric current not only enhances the Peltier effect, but also increases Joule heat generation of the TEC. Therefore, an optimum electric current of 7 A is determined to achieve the lowest overall thermal indicator at a specific heat load. A water-cooling device with a TEC is helpful to enhance the thermal performance when the heat load is below 57 W. Comparing with thermoelectric air-cooling module for electronic devices (Chang et al. 2009), the optimum input currents are from 6 A to 7 A at the heat loads from 20W to 100 W. The result also demonstrates that the thermoelectric air-cooling module performs better performance at a lower heat load. The lowest total temperature difference-heat load ratio is experimentally estimated as 0.54 °C/W at the low heat load of 20 W, while it is 0.664 °C/W at the high heat load of 100 W. In some conditions, the thermoelectric air-cooling module performs worse than the air cooling heat sink only.

In indirect liquid cooling technology, the outer surface of the chamber containing the working fluid makes contact for the required cooling of the electronic components. The heat capacity transfers to the working fluid through the chamber for heat dissipation. The driving force can be divided into active and passive by the way of the working fluid. The main objective of a passive liquid cooling system is not to use components, such as a pump, to drive the working fluid cycle. At present, the development of passive and indirect liquid cooling technology includes heat pipes, and vapor chambers composed of thermosyphon thermal modules, which have been applied in a variety of high heat-flux electronic components. Due to the demand for different heat transfer components, a two-phase thermosyphon can be divided into closed-loop and closed types. Two-phase closed-loop thermosyphon thermal modules are all two-phase change heat transfer components, their operating principle is to transfer heat capacity for cooling purposes by boiling and condensation of the phase change of the working fluid. Thus, finding how to enhance the boiling mechanism and reduce the thickness of condensation film will determine the operating thermal performance of the thermal module. This module offers the same vapor and liquid flow direction without the limitations of traditional heat pipes. Dissipation of the heat capacity of the heat source is conducted by forced convection to the atmosphere around the condenser section. This is because the vapor pressure in the evaporator section through the connecting pipe to condensation caused by the pressure drop. Therefore, the two-phase closed-loop thermosyphon thermal module has a water level difference within the evaporator and condenser. Furthermore, the different cooling fin groups in the thermal module and the condensing capacity of the evaporator section and condenser section are in contact with the working fluid of the different cross-sectional areas. Therefore, the water level is significantly different on the left and right sides of the evaporation section and the condensation section of the internal working fluid of the two-phase closed-loop thermosyphon thermal module. Therefore, it is important to note the vapor pressure difference caused by the water level in the design of this type of thermal module.

The two-phase closed thermosyphon cooling system is combined with a vapor-chamber formed evaporator to gain the advantages of vapor chamber (Chang et al. 2008; Tsai et al. 2010b). The facility allows different structured surfaces to be applied, and the effects of heating powers, fill ratios of working fluid, and types of evaporation-enhanced surfaces on the performance of the two-phase closed thermosyphon vapor-chamber system are investigated and discussed. A thermal resistance net work is developed in order to study the effects of heating power, fill ratio of working fluid, and evaporator surface structure on the thermal performance of the system. Other words, the experimental parameters are different evaporation surfaces, fill ratios of working fluid and input heating powers. The results indicate that either a growing heating power or a decreasing fill ratio decreases the total thermal resistance, and the surface structure also influences the evaporator function prominently. An optimum overall performance exists at 140W heating power and 20% fill ratio with sintered surface, and the corresponding total thermal resistance is 0.495 °C/W. A growing fill ratio significantly enlarges the saturation pressure and temperature of the system, and results in worse performance of the condenser. The heat transfer mechanism of the three surfaces all can be ranked as boiling dominated. The result shows that the evaporation resistance and the condensation resistance both grow with increasing heating power and decreasing fill ratio. Flooding is found at the fill ratio of 20% with the evaporation surface noted Etched Surface 2 when heating power is above 120 W. Flooding phenomenon is caused by the opposite flow direction of vapor and liquid in a closed two-phase system. According to the result, the lowest total thermal resistance is 0.65 °C/W by the evaporation surface noted Etched Surface 2 at 30% fill ratio. Flooding phenomenon occurs as the system operated at low fill ratio and high input heating power. The flooding operation point for this system has been predicted by correlation, and the prediction is closed to the experimental results.

Vapour has advantages of fast, large amount and safety. Another two-phase heat transfer device is the Vapour Chamber (V.C.) inside vapour-liquid working, which has better thermal performance than metallic material in a large footprint heat sink. The overall operating principle of V.C. is defined as follows: at the very beginning, the interior of the vapor chamber is in the vacuum, after the wall face of the cavity absorbing the heat from its source, the working fluid in the interior will be rapidly transformed into vapour under the evaporating or boiling mechanism and fill up the whole interior of the cavity, and the resultant vapor will be condensed into liquid by the cooling action resulted from the convection between the fins and fan on the outer wall of the cavity, and reflow to the place of the heat source along the capillary structure. The effectiveness and better thermal performance of vapour chamber has been already confirmed according up-to-date researches and mass production application in server system and VGA thermal module. Moreover, vapour chamber-based thermal module has existed in the thermal-module industry for a year or so especially in server application (Wang & Chen 2009; Wang 2010; Wang et al. 2011a). A novel formula for effective thermal conductivity of vapor chamber has been developed by use of dimensional analysis in combination with thermal-performance experimental method (Wang & Wang 2011b). It respectively discussed these values of one, two and three-dimensional effective thermal conductivity and compared them with that of metallic heat spreader. For metallic materials as the heat spreaders, their thermal conductivities have constant values when the operating temperature varies not large. The thermal conductivities of pure cooper and aluminum as heat spreaders are 401 W/m°C and 237 W/m°C at operating temperature of 27 °C, respectively. When the operating temperature is 127 °C, they are 393 W/m°C and 240 W/m°C, respectively. Results show that the two and three-dimensional effective thermal conductivities of vapor chamber are above two times higher than that of the copper and aluminum heat spreaders, proving that it can effectively reduce the temperature of heat sources. The maximum heat flux of the vapor chamber is over 800,000 W/m^2, and its effective thermal conductivity will increase with input power increasing. Thermal performance of V.C. is closely relate to its dimensions and heat-source flux, in the case of small area vapour chamber and small heat-source flux, the thermal performance will be less than that of pure copper material. It is deduced from the novel formula that the maximum effective thermal conductivity is above 800 W/m°C, and comparing it with the experimental value, the calculating error is no more than ±5%.

A vapour chamber is a two-phase heat transfer components with a function of spreading and transferring uniformly heat capacity so that it is ideal for use in non-uniform heating conditions especially in LEDs (Wang 2011c). The solid-state light emitting diode (SSLED) has attracted attention on outdoor and indoor lighting lamp in recent years. LEDs will be a great benefit to the saving-energy and environmental protection in the lighting lamps region. A few years ago, the marketing packaged products of single die conducts light efficiency of 80 Lm/W and reduces the light cost from 5 NTD/Lm to 0.5 NTD/Lm resulting in the good market competitiveness. These types of LED lamps require combining optical, electronic and mechanical technologies. Wang & Wang (2011a) introduce a thermal-performance experiment with the illumination-analysis method to discuss the green illumination techniques requesting on LEDs as solid-state luminescence source application in relative light lamps. The temperatures of LED dies are lower the lifetime of lighting lamps to be longer until many decades. The thermal performance of the LED vapour chamber-based plate is many times than that of LED copper- and aluminum- based plate (Wang & Huang 2010). The results are shown that the experimental thermal resistance values of LED copper- and vapour chamber-based plate respectively are 0.41 °C/W and 0.38 °C/W at 6 Watt. And the illumination of 6 Watt LED vapour chamber-based plate is larger 5 % than the 6 Watt. Thus, the LED vapour chamber-based plate has the best thermal performance above 5 Watt. The thermal performance of the LED vapor chamber-based plate is worse than that of the LED copper-based plate of less than 4 Watts. In additional to having the best thermal performance above 5 Watts, the luminance of the LED vapor chamber-based plate is the highest. The temperature of the LED rises about 12 °C per Watt (Wang 2011d). Wang et al. (2010b) utilizes experimental analysis with window program VCTM V1.0 to investigate the thermal performance of the vapor chamber and apply to 30 Watt high-power LEDs. Results show that the maximum effective thermal conductivity is 870 W/m°C, and comparing it with the experimental value, the calculating error is no more than ±5%. And the LED vapor chamber-based plate works out hot-spot problem of 30 Watt high-power LEDs, successfully. Thermal performance of the thermal module with the vapour chamber can be determined within several seconds by using the window program VCTM V1.0, exactly. The maximum heat flux of the vapor chamber is over 100 W/cm^2, and the thermal performance of the LED vapor chamber-based plate is better than that of the LED aluminum based-plate above 10 °C and has the highest effective thermal conductivity of 965 W/m°C at 187.5 W/cm^2.

In last, air-cooling thermal module in large-scale industrial enclosed air-to-air cooled motor with a capacity of 2350 kW is experimentally and numerically investigated (Chen et al. 2009; Chang et al. 2010). The models of the fan and motor have been implemented in a Fluent/Flow-3D software packages to predict the flow and temperature fields inside the motor. The modified design can decrease the temperature rise by 6 °C in both the stator and rotor. Wang et al. (2011b) uses the local heating mechanism, along with the excellent thermal performance of vapour chamber, to analyze and enhance the strength of products formed after insert molding process. These results indicate that, the product formed by the local heating mechanism of vapour chamber can reduce the weld line efficiency and achieve high strength, which passed the standard of 15.82 N-m torque tests, with a yield rate up to 100%. In this study, a vapor chamber-based rapid heating and cooling system for injection molding to reduce the welding lines of the transparent plastic products is proposed. Tensile test parts and multi-holed plates were test-molded with this heating and cooling system. The results indicate that the new heating and cooling system can reduce the depth of the V-notch as much as 24 times (Tsai et al. 2011; Wang & Tsai 2011). The key results show that the proper air-cooling modules are important. There are several theoretical models of air-cooling modules developed to predict their thermal performance respectively. Finally, these results show that the prediction by the model agrees with the experimental data. The theoretical models with empirical formula have coded by Virtual Basic version 6.0 to develop window programs and convenience for industrials in this chapter.

2. Optimum and performance analysis of a parallel plate-fin heat sink

A conventional air-cooling device combining a plate-fin heat sink and forced convection with cost-competitive advantages and simple and reliable manufacturing processes, has been widely used in electronic cooling for the past several years. This practical heat sink is composed of a plate-fin heat sink and a fan, which attached to the heat source with proper thermal interface material and is cooled by airflow caused by the fan. The heat is conducted through the thermal interface, spreads into the base plate and the fins of the heat sink, and then transfers into the environment by airflow. A thermal resistance network includes interface resistance, base-conduction resistance, and convective resistance. It is worth developing a model for a conventional air-cooling device that takes heat sink configuration and airflow conditions into account in order to predict the device's thermal performance when developing laminar-, transition-, and turbulent-flow regimes. A plate-fin heat sink has been developed to predict the hydraulic and thermal performance in the following paragraphs. This all-in-one asymptotic model was proposed for a wide range of Reynolds and Nusselt numbers, including laminar, transition, and turbulent flows. It can predict pressure drops with accuracy within 6% and clarify the heat transfer coefficients within 15% of error range. Furthermore, optimization in geometry with the present model is achieved. The optimal values contain fin height, spacing and thickness, and base thickness, width and length. Using constrictive ratio and apparent interface ratio to interpret equivalent heat source area and maximal heat flux on source-to-sink contacting surface, the non-uniform heat source problem can be simplified as an equivalent uniform heat source problem. Then by using the existed correlations the present model can calculate the overall thermal resistance as a function of heat sink geometry, properties, interface conditions, and airflow velocity. An experimental investigation is performed to verify the theoretical model. Prediction results show good agreement with experimental measurements over a number of testing units.

2.1 Practical pressure drops model for a conventional plate-fin heat sink

The analysis for the extended fins array is conducted first to derive a working fluid-pressure drop across the heat sink and the effective convection coefficient. Analysis of the friction factor and heat-transfer coefficient in the channel flow evaluates whether the coolant flow is laminar or turbulent. The critical Reynolds number, Re_c, is used to determine the flow regime. It has been found that the laminar to turbulent transition is not a sudden phenomenon, but occurs over a range of Reynolds numbers, $Re_c < Re < 4000$, for rectangular-duct flow. For numerical calculation purposes, a curve-fitting correlation by abrupt-entrance data is presented below as Eq. (1). The fitting error is within 8% to 1.6%, in a range of $0< \alpha < 1$:

$$\mathrm{Re}_c = 1.6(\ln\alpha)^3 + 6.6(\ln\alpha)^2 - 161.6(\ln\alpha) + 2195 \tag{1}$$

Where α is the aspect ratio, and b and a are the fin spacing and the fin height, respectively. The pressure drop in the plate heat sink can be divided into two parts: the friction term, and the term due to the change of flow section. The heat sink pressure drop is considered as Eq. (2), where f_{app} is the fanning-friction factor, K_c and K_e are contraction and expansion pressure loss coefficients, u_m is the coolant velocity in the channel, and L and D_h are the channel length and hydraulic diameter, respectively.

$$\Delta P = \left(4f_{app}\frac{L}{D_h} + K_c + K_e\right)\cdot\frac{1}{2}\rho u_m^2 \tag{2}$$

In order to obtain a general friction factor correlation for the rectangular-duct flow over developing laminar-, transition-, and turbulent-flow regimes, an asymptotic solution is given as

$$f_{app} = \left(f_{lam}^n + f_{turb}^n\right)^{1/n} =$$

$$\left\{\left[\frac{24}{\mathrm{Re}}\left(1 - 1.4\alpha + 1.9\alpha^2 - 1.7\alpha^3 + \alpha^4 - 0.3\alpha^5\right)\right]^3 + \left[0.1\mathrm{Re}^{-0.2}\left(\frac{x}{D_h}\right)^{-0.175}\right]^3\right\}^{1/3} \tag{3}$$

Where Re is the Reynolds number according to the hydraulic diameter. This correlation is predictive to within 14% to 6% accuracy.

Pressure loss due to an abrupt cross section change in the heat sink has been studied. Correlations of K_e and K_c for laminar-parallel plate ductsare used in this paper to evaluate pressure loss:

$$K_c = 0.8 - 0.4\left(\frac{b}{p}\right)^2 \tag{4}$$

$$K_e = \left(1 - \left(\frac{b}{p}\right)^2\right) - 0.4\left(\frac{b}{p}\right) \tag{5}$$

Where p is the fin pitch.

2.2 Heat transfer model for a conventional plate-fin heat sink

Using the same idea, an asymptotic approach is also conducted to obtain a general Nusselt number correlation for the laminar, transition, and turbulent flows. The asymptotic solution is given as

$$Nu = \left(Nu_{lam,dev}^{n} + Nu_{turb,dev}^{n} \right)^{1/n} \tag{6}$$

In order to obtain $Nu_{lam,dev}$ with three heating walls in a channel in Eq.(6), the present work first determined $Nu'_{lam,dev}$ with four heating walls. A general model evaluating Nu in the laminar-thermally developing regime of a noncircular duct with four heating walls is employed. The solution is given as the above equation predicts the heat-transfer capacity when four walls are being heated. However, in the present case, only three walls are heated, and the top wall is insulated. The Nusselt number for low Reynolds numbers with three walls heated is valid for $0 < \alpha < 1$ and predicts accuracy within 5%. $Nu_{turb,dev}$ employs the correlation provided and substitutes an equivalent diameter for the hydraulic diameter: In contrast to the laminar flow, no compensation for heat-transfer coefficient for three wall heating is required, since the turbulent mixing among the channel sections does quite well and significantly reduces the influence of the asymmetric boundary condition. After the prediction of Nu is obtained, h_m can also be determined. The bulk convection heat-transfer coefficient, h_m, is based on the log-mean temperature difference. According to the same power dissipation, the definitions of these two are exhibited below: convective heat-transfer coefficient (h_i) refers to the inlet temperature for predicting convection thermal resistance. The coefficient is based on the temperature difference between the heat sink and the inlet-fluid temperatures. Moreover, energy balance between the heat-transfer rate from the heat sink wall to the fluid and the increment rate of fluid enthalpy is considered, and then obtains the result in a relationship between h_m and h_i: The above-mentioned correlations of the Nusselt number and the friction factor employ fluid properties corresponding to the mean bulk temperature.

The above model combines two correlations of laminar and turbulent flows, and provides the friction factor and the Nusselt number for developing laminar, transition, and turbulent flows. Using the friction factor and the loss coefficient formula of parallel plates, this model can predict a pressure drop in a heat sink within errors from 13.71% to 8.47% against experimental data with an aspect ratio from 0.05 to 0.2. The Nusselt number has been proposed for convective heat-transfer prediction. This asymptotic solution predicts a Nusselt number within acceptable errors for transition and turbulent flows, namely 15% to 12%, but it predicts a higher Nusselt number of 7%. This study considers the convection heat transfer of fins, pressure drop, fin efficiency, effective heat-transfer coefficient, and conduction problems of the heat sink base, then derives a practical model predicting the thermal performance of the plate-fin heat sink, and develops a numerical program for calculating. Within acceptable accuracy, this model is useful to obtain a set of parameters for designing a plate-fin heat sink with expected performance.

3. Two-phase flow heat transfer devices

Liquid cooling technology employs the excellent thermal performance of liquid to quickly take away the heat capacity from a heat source. The method by which liquid contacts the heat source can be divided into two types, including immediacy and mediacy. Direct liquid cooling technology was first applied in the large area of a super computer in the 1960s. Which this type of technology, it is necessary to pay attention to the thermal shock effect. In indirect liquid cooling technology, the outer surface of the chamber containing the working fluid makes contact for the required cooling of the electronic components. The heat capacity transfers to the working fluid through the chamber for heat dissipation. The driving force can be divided into active and passive by the way of the working fluid. The active type is like a pump in the liquid cooling system that drives the circulation loop of the working fluid, as used in laptop computer cooling systems. The disadvantages for the use of a pump are cost, lifespan, vibration, noise and other issues. Moreover, additional water-cooled assembly between the components increases the cost of thermal modules and reduces the reliability of electronic components. The main objective of a passive liquid cooling system is not to use components, such as a pump, to drive the working fluid cycle.

At present, the development of passive and indirect liquid cooling technology includes heat pipes, and vapor chambers composed of thermosyphon thermal modules, which have been applied in a variety of high heat-flux electronic components. These thermal modules also name two-phase flow heat transfer devices. This type of technology offers the following five benefits: 1. The module can transfer a lot of heat capacity with very small temperature gradient by the latent heat between the two-phase changes of liquid and gas. It has excellent thermal performance. 2. The flow in the system is driven by buoyancy and gravity/capillary force. These two driving forces are self-induced, so a flow-driven pump component is not necessary in this system. The thermal module itself has high reliability. 3. The two main components, including the evaporator and condenser, are separated and connected with pipes. Therefore, the arrangement of each component is much more flexible. Furthermore, fans can take advantage of existing systems. 4. The thermal module is composed of three parts, including the evaporator, connected piping and condenser. In addition, each of the three sections can be individually manufactured, providing high potential for extension. 5. It is highly feasible to combine more evaporators and condensers in a series of connections or parallel connections in the cooling system should extension be necessary. Therefore, this technology provides low cost manufacturing and installation of the modular.

3.1 Heat pipe

The heat pipe may mainly be differentiated for the evaporation section, the adiabatic section, and the condensation section and be regarded as the passive component of a self-sufficient vacuum closed system. The system which includes the capillarity structure and the filling of working fluid usually needs to soak through the entire capillarity structure. This research adopted a conventional steady-state test similar to previous works to investigate the steady-state thermal performances of heat pipes, and a detailed description of the experimental procedures and set-ups is introduced as follows. When the heat pipe works, the evaporation section of the heat absorption occurs in the capillarity structure. The working fluid gratifies as the vapor. The high temperature and pressure vapour is produced at the condensation section after the adiabatic section. The vapor releases heat and then condenses as the liquid. The working fluid in the condensation section is brought back to the evaporation section because of the capillary force in the capillarity structure. The capillary force has a relation with the capillarity structure, the viscosity, the surface tension, and the wetting ability. However, the capillary force and the high vapour pressure are the main driving force to make the circulation in the heat pipe. The working fluid in the heat pipe exists in the two-phase model of the vapor and liquid. The interface of the vapor and liquid's coexistence is regarded as the saturated state. It is usually assumed that the temperatures of the vapor and liquid are equal.

The heat pipe uses the working fluid with much latent heat and transfers the massive heat from the heat source under minimum temperature difference. Because the heat pipe has certain characteristics, it has more potential than the heat conduction device of a single-solid-phase. Firstly, due to the latent heat of the working fluid, it has a higher heat capacity and uniform temperature inside. Secondly, the evaporation section and the condensation section belong to the independent individual component. Thirdly, the thermal response time of the two-phase-flow current system is faster than the heat transfer of the solid. Fourthly, it does not have any moving components, so it is a quiet, reliable and long-lasting operating device. Finally, it has characteristics of smaller volume, lighter weight, and higher usability. Although the heat pipe has good thermal performance for lowering the temperature of the heat source, its operating limitation is the key design issue called the critical heat flux or the greatest heat capacity quantity. Generally speaking, we should use the heat pipe under this limit of the heat capacity curve.

There are four operating limits which are described as following. Firstly, the capillary limit, which is also called the water power limit, is used in the heat pipe of the low temperature operation. Specific wick structure which provides for working fluid in circulation is limiting. It can provide the greatest capillary pressure. Secondly, the sonic limit is that the speed of the vapour flow increases when the heat source quantity of heat becomes larger. At the same time, the flow achieves the maximum steam speed at the interface of the evaporation and adiabatic sections. This phenomenon is similar to the flux of the constant mass flow rate at conditions of shrinking and expanding in the nozzle neck. Therefore, the speed of flow in this area is unable to arrive above the speed of sound. This area is known for flow choking phenomena to occur. If the heat pipe operates at the limited speed of sound, it will cause the remarkable axial temperature to drop, decreasing the thermal performance of the heat pipe. Thirdly, the boiling limit often exists for the traditional metal, wick structured heat pipe. If the flow rate increases in the evaporation section, the working fluid between the wick and the wall contact surface will achieve the saturated temperature of the vapor to produce boiling bubbles. This kind of wick structure will hinder the vapour bubbles to leave and have the vapor layer of the film encapsulated. It causes large, thermal resistance resulting in the high temperatures of the heat pipe. Fourthly, the entrainment limit is that when the heat is increased and the vapor's speed of flow is higher than the threshold value, forcing it to bear the shearing stress in the liquid; vapor interface being larger than the surface tension of the liquid in the wick structure. This phenomenon will lead to the entrainment of the liquid, affecting the flow back to the evaporation section. Besides the above four limits, the choice of heat pipe is also an important consideration. Usually the work environment can have high temperature or low temperature conditions which will require a high temperature heat pipe or a low temperature heat pipe, accordingly. After deciding the operating environment, the material, internal sintered body, and type of working fluid for the heat pipe are determined. In order to prevent the heat pipe`s expiration, the consideration of the selection is very important.

3.2 Thermosyphon

This paragraph experimentally investigates a two-phase closed-loop thermosyphon vapor-chamber system for electronic cooling. A thermal resistance net work is developed in order to study the effects of heating power, fill ratio of working fluid, and evaporator surface structure on the thermal performance of the system. This study explored the relationship between the vapor pressure and water level inside a two-phase closed-loop thermosyphon thermal module to acquire a theoretical model of the water level height difference of the thermal module through the analysis of basic condensing and boiling theory. Figure 2 shows the internal vapor pressure and water level through the heat source with the heating power Q, based on the entire experimental system. The internal vapour pressure and water level through the heat source with the heating power Q based on the entire experimental system. The entire physical system can be divided into four control volumes to resolve the vapour pressure and the friction loss of steam from the first control volume (C.V.1) to the third control volume (C.V.3), as revealed by formula (7). Furthermore, the liquid static pressure balance of the fourth control volume (C.V.4) is exhibited by formula (8). The range of C.V.1 is from the vapor chamber, including the area from the connecting pipe to the entrance of the condenser region, which encompasses the loss of steam pressure through the connecting pipe of the insulation materials. The range of C.V.2 is from the entrance to the outlet of the condenser, which involves a loss of steam pressure after the condenser. The scope of C.V.3 is from the outlet of the condenser to the connection surface of the vapor chamber, which entails a loss of steam pressure through the connecting pipe. The scope of C.V.4 is from the connection surface of the vapor chamber to the same high level in the connecting pipe of the vapor chamber.

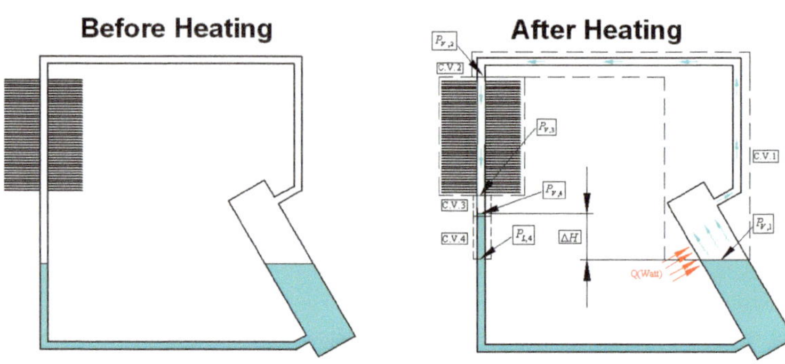

(a) Initial Condition (b) Steady State

Fig. 2. Relationship between vapour pressure and water level

$$P_{V,i} = P_{V,i+1} + \Delta P_{f,i} \tag{7}$$

Where $P_{V,i}$ is the vapor pressure of the ith control volume in this system, $P_{V,i+1}$ is the vapor pressure for the steam into (i+1)th control volume through ith control volume of the connecting pipe and $\Delta P_{f,i}$ is the friction loss of the pressure of steam flow.

$$P_{l,4} = P_{V,4} + \gamma_w \Delta H \tag{8}$$

where $P_{l,4}$ is the hydrostatic pressure of the C.V.4 of liquid, γ_w is the specific weight of liquid, ΔH is the height difference of the water level between the internal water level of the vapor chamber and the connecting pipe connected to the condenser.

The equations represented by C.V.1 to C.V.3 are all added up, and $P_{l,4}$ is equal to $P_{V,4}$ and substituting it into equation (8), ΔH can be obtained as shown in equation (9).

$$\Delta H = \left(\frac{1}{\gamma_w} \right) \cdot \sum_{i=1}^{3} \Delta P_{f,i} \tag{9}$$

From the equation (9), if there is no pressure drop loss for $\Delta P_{f,1}$ and $\Delta P_{f,3}$ of the pipeline and $\Delta P_{f,2}$ of the condenser, then the water level inside the vapour chamber and that connected to the condensation inside condenser will be the same. That is, ΔH is equal to zero.

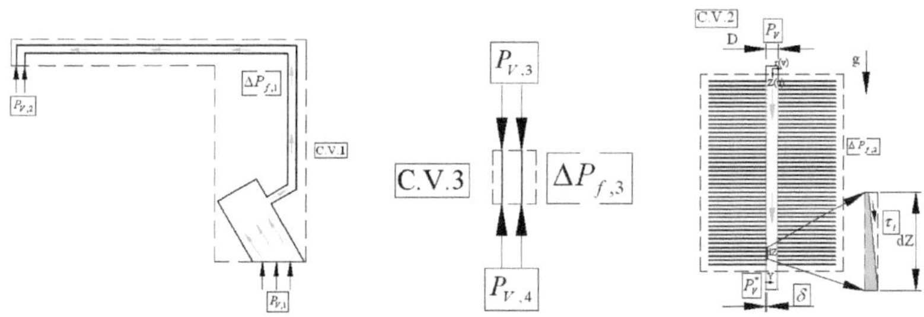

Fig. 3. Schematic diagram of the calculation of pressure drop loss (a) Pressure drop loss of the connecting pipe of C.V.1 (b) Pressure drop loss of the connecting pipe of C.V.3 (c) Pressure drop loss of the condenser

Figure 3(a) and 3(b) show the estimated method for $\Delta P_{f,1}$ and $\Delta P_{f,3}$ of the connecting pipe. According to a previous study, this can be calculated by formula (10).

$$\Delta P_{f,i} = f_i \cdot \frac{L_i}{D_i} \cdot \frac{1}{2} \cdot \rho_{v,i} \cdot V_{v,i}^2 \tag{10}$$

Where f_i is the friction coefficient generated by the steam flow through the pipes, L_i represents the equivalent length of the connecting pipe, D_i is the diameter of the connecting pipe and $\rho_{v,i}$ and $V_{v,i}$ represent the vapour density and speed respectively.
According to figure 3(c) and previous studies, the method for calculating $\Delta P_{f,2}$ considers the shear stress or the friction force at the gas-liquid interface with small control volume. Formula (11) can be attained based on momentum conservation.

$$(\delta - y) \cdot dZ \left(\rho_w g - \frac{dP}{dz} \right) + \tau_i \cdot dZ = \mu_w \left(\frac{dV}{dy} \right) \cdot dZ \tag{11}$$

Where δ is the film thickness of the liquid inside the condenser tube, ρ_w is the liquid density, μ_w is the dynamic viscosity of the liquid, τ_i is the shear stress at the gas-liquid interface, (dP/dz) is the pressure drop loss generated by the steam flow through the gas-liquid interface at the condenser, which can be expressed as equation (12).

$$-\left(\frac{dP}{dz} \right) = -(\rho_v g) + \left(\frac{4\tau_i}{D - 2\delta} \right) + \left(\frac{d(\dot{m}_v V_v + \dot{m}_w V_w)}{dz} \right) \tag{12}$$

In which, $\dot{m}_v$ and $\dot{m}_w$ represent the mass flow rate of the steam and liquid, respectively. V_v and V_w denote the speed of vapour and liquid. τ_i is the shear stress of the gas-liquid interface, as shown in equation (13) below.

$$\tau_i = 0.005\left(1+\frac{300\cdot\delta}{D}\right)\cdot\left(\frac{G^2\cdot x^2}{2\cdot\rho_v\cdot(1-4\frac{\delta}{D})}\right) \tag{13}$$

$\left(\frac{d(\dot{m}_v V_v + \dot{m}_w V_w)}{dz}\right)$ is the pressure drop produced by the mass flow rate of the gas-liquid interface, which can be expressed as in equation (14).

$$\frac{d(\dot{m}_v V_v + \dot{m}_w V_w)}{dz} = G^2\cdot\frac{d}{dz}\left[\left(\frac{x^2}{\rho_v\alpha}\right)+\left(\frac{(1-x)^2}{\rho_w(1-\alpha)}\right)\right] \tag{14}$$

Where G is the mass flow rate flux, x is the mass flow rate fraction and α is the ratio of the gas channel. Substituting equation (14) into equation (12), the integral of the range from zero to Z can be obtained by formula (15) as follows.

$$(P_V^* - P_V) = (\rho_v g Z) - \int_0^Z\left(\frac{4\tau_i}{D-2\delta}\right)dz - G^2\left(\int_0^Z \frac{d}{dz}\left[\left(\frac{x^2}{\rho_v\alpha}\right)+\left(\frac{(1-x)^2}{\rho_w(1-\alpha)}\right)\right]\right)dz \tag{15}$$

Substituting $\alpha = \frac{A_v}{A} = \left(\frac{D-2\delta}{D}\right)$ into the above equation, we can obtain the formula (16) after integration as follows.

$$\Delta P_{f,2} = (P_v - P_v^*) = G^2 D\left[\left(\frac{x^2}{\rho_v(D-2\delta)}\right)+\left(\frac{(1-x)^2}{2\rho_w\delta}\right)\right] + \int_0^Z\left(\frac{4\tau_i}{D-2\delta}\right)dz - (\rho_v g Z) \tag{16}$$

To calculate the right side of the integral term $\int_0^Z\left(\frac{4\tau_i}{D-2\delta}\right)dz$ of the above formula (16), first, assume that the internal film growth equation of the liquid is linear. Therefore, the assumed slope of SP can attain formula (17) as follows.

$$\delta = SP * Z \tag{17}$$

And let

$$\xi = \frac{\delta}{D} \tag{18}$$

Substituting equations (17) and (18) into equation (14), we can obtain formula (19) as follows.

$$\tau_i = 0.0001\cdot\left(\frac{G^2\cdot x^2}{2\cdot\rho_v}\right)\cdot\left(\frac{1+300\xi}{1-4\xi}\right) \tag{19}$$

Substituting equation (19) into the right side of the integral term $\int_0^Z \left(\frac{4\tau_i}{D-2\delta}\right)dz$ of equation (16), we can obtain formula (20) as follows.

$$\int_0^Z \left(\frac{4\tau_i}{D-2\delta}\right)dz = \left(\frac{0.005\cdot G^2\cdot x^2}{Sp\cdot\rho_v}\right)\left[151\cdot\ln(1-\frac{2\cdot Sp\cdot Z}{D})-76\cdot\ln(1-\frac{4\cdot Sp\cdot Z}{D})\right] \quad (20)$$

Finally, by substituting equation (20) back into formula (16), we can obtain $\Delta P_{f,2}$ with formula (21) as shown below.

$$\Delta P_{f,cv2} = G^2 D\left[\left(\frac{x^2}{\rho_v(D-2\delta)}\right)+\left(\frac{(1-x)^2}{2\rho_w\delta}\right)\right]-(\rho_v gZ)+ \left(\frac{0.005\cdot G^2\cdot x^2}{Sp\cdot\rho_v}\right)\left[151\cdot\ln(1-\frac{2\cdot Sp\cdot Z}{D})-76\cdot\ln(1-\frac{4\cdot Sp\cdot Z}{D})\right] \quad (21)$$

The film thickness δ can be calculated by the formula (22) as follows.

$$\frac{4\cdot Z\cdot\mu_w\cdot q''}{\rho_w\cdot(\rho_w-\rho_v)\cdot g\cdot h'_{fg}} = \delta^3 + \frac{4\cdot\tau_i\cdot\delta^2}{3\cdot(\rho_w-\rho_v)\cdot g} \quad (22)$$

In which

$$h'_{fg} = h_{fg}\cdot\left[1+\left(\frac{3}{8}\right)\cdot\frac{C_{pw}\cdot\delta\cdot q''}{h_{fg}\cdot k_w}\right] \quad (23)$$

Where h_{fg} is the latent heat of the working fluid, C_{pw} is the constant pressure of the specific volume of the liquid; q'' is the input heat flux of the heat source and k_w is the thermal conductivity of the liquid.

We use Microsoft® Visual Basic™ 6.0 to write the computing interface resulting from the above empirical formula and calculated the thermal performance and the water level deficit inside the thermal module of the two-phase closed-loop thermosyphon. The programming flow chart is shown in Figure 4(a) and the final operation interface is shown in Figure 4(b). This study discusses the thermal performance of the two-phase closed-loop thermosyphon thermal module, and indirectly confirms that the working fluid reflows into the condenser by measuring the wall temperatures of the condenser, which results in the water level difference phenomenon within the system. Figure 5 shows the theoretical curve of the water level height difference for the entire closed thermal module system. The solid black line in the figure is the theoretical water level height difference based on the heat transfer theory of pool nucleate boiling and film condensation in this study. Comparing the two curves, we can accurately predict the same level with the height difference between the experimental curve before the heating power is less than 60W; however, beyond 60W, the water level height difference obtained in the experimental curve has tended to be horizontal, while the theoretical curve will increases with the heating power, the water level height difference increases only slightly.

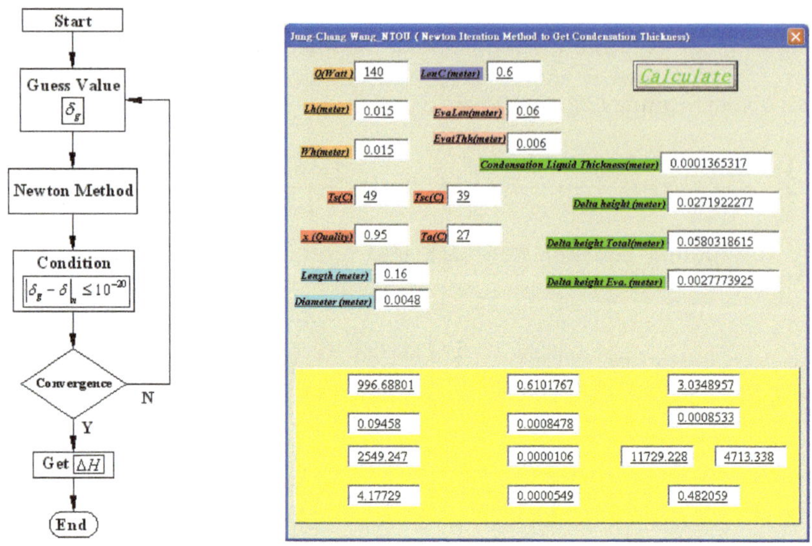

(a) Programming flow chart (b) Operator interface

Fig. 4. Programming and the operator interface

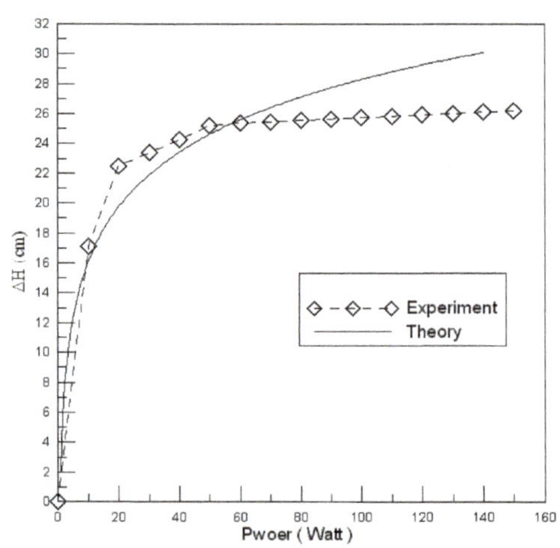

Fig. 5. The theoretical value of water level difference of vertical type

For the two-phase closed-loop thermosyphon cooling system, the micro-scale water level difference phenomenon resulting from the condensing and boiling vapor pressure difference between the evaporator and condenser sections based on the theories of pool nucleate boiling and film condensation and the validation of experimental method to measure the wall temperature of condenser. The height of the condenser of the two-phase closed-loop thermosyphon system can be shortened by 3.14cm by using the theoretical water level difference model. The working fluid within the two-phase closed-loop thermosyphon system has different heights resulting from the vapor pressure difference between the evaporator and the condenser sections. This should be noted in the design of such two-phase heat transfer components. Finally, this study has established a theoretical height difference model for two-phase closed-loop cooling modules. This can serve as a reference for future researchers.

3.3 Vapor chamber

This study derives a novel formula for effective thermal conductivity of a vapor chamber using dimensional analysis in combination with a thermal-performance experimental method. The experiment selected water as the working fluid filling up in the interior of vapour chamber. The advantages of water are embodied in its thermal-physics properties such as extremely high latent heat and thermal conductivity and low viscosity, as well as its non-toxicity and incombustibility. The overall operating principle of the experiment is defined as follows: at the very beginning, the interior of the vapour chamber is in vacuum, after the wall face of the cavity absorbs the heat from its source, the working fluid in the interior will be rapidly transformed into vapour under the evaporating or boiling mechanism and fill up the whole interior of the cavity. The resultant vapour will be condensed into liquid by the cooling action resulted from the convection between the fins and fan on the outer wall of the cavity, and condensate will reflow to the wall at the heat source along the capillary structure as shown in figure 6.

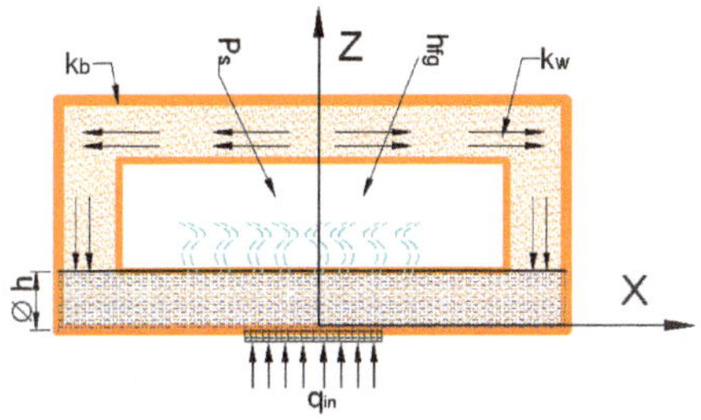

Fig. 6. Drawing of the vapor chamber

It discusses these values of one, two and three-dimensional effective thermal conductivity and compares them with that of metallic heat spreaders. Equation (24) indicates the effective thermal conductivity k_{index} of the vapor chamber, which is the result of the input heat flux q''_{in} multiplied thickness (t) of the vapour chamber divided by the temperature difference ΔT_{index}. The one-dimensional thermal conductivity (k_z) is when the index is equal to z and the temperature difference ΔT_z equals the central temperature (T_{dc}) on the lower surface minus that (T_{uc}) on the upper surface. The two-dimensional thermal conductivity (k_{xyd}) is when the index is equal to xyd and the temperature difference ΔT_{xyd} equals the central temperature (T_{dc}) on the lower surface minus mean surface temperature (T_{da}). The two-dimensional thermal conductivity (k_{xyu}) is when the index is equal to xyu and the temperature difference ΔT_{xyu} equals the central temperature (T_{uc}) on the upper surface minus mean surface temperature (T_{ua}). The three-dimensional thermal conductivity (k_{xyz}) is when the index is equal to xyz and the temperature difference ΔT_{xyz} equals mean surface temperature (T_{da}) on the lower surface minus that (T_{ua}) on the upper surface.

$$k_{index} = q''_{in} \cdot t \Big/ \Delta T_{index} \tag{24}$$

One of major purposes of this study is to deduce the thermal performance empirical formula of the vapour chamber, and find out several dimensionless groups for multiple correlated variables based on the systematic dimensional analysis of the [F.L.T.θ.] in Buckingham Π Theorem, as well as the relationship between dimensionless groups and the effective thermal conductivity. Figure 6 is the abbreviated drawing of related variables of the vapour chamber to be confirmed in this article, and the equation (25) is the functional expression deduced based on related variables in Figure 6. The symbol k_{eff} in the equation is the value of effective thermal conductivity of the vapour chamber, the k_b is the thermal conductivity of the material made of the vapour chamber, the symbol k_w is the value of effective thermal conductivity of the wick structure of the vapour chamber, the unit of these thermal conductivities are W/m°C. The symbol h_{fg} is latent heat of working fluid which has unit of J/K. The P_{sat} is saturated vapour pressure of working fluid with unit of N/m^2. The t is the thickness of vapour chamber. Their unit is m. The symbol A is the area of vapour chamber and its unit is m^2.

$$K_{eff} = Function\left\{k_b, k_w, q''_{in}, h_{fg}, P_{sat}, t, A, \phi h\right\} \tag{25}$$

It can be inferred from equation (25) that there are nine related variables (symbol m equalling to 9), and the following equation (26) can be inferred by making use of [F.L.T.θ.] system (symbol r equalling to 4) to do a dimensional analysis of various parameters in the above-mentioned equation and combining the analysis result with the equation (25).

$$\left(\frac{k_{eff}}{k_b}\right) = \alpha \cdot \left(\frac{k_w}{k_b}\right)^{\beta} \cdot \left(\frac{q''_{in}}{P_{sat} \cdot h_{fg}^{\ 0.5}}\right)^{\gamma} \cdot \left(\frac{A}{t^2}\right)^{\lambda} \cdot \left(\frac{\phi h}{t}\right)^{\tau} \tag{26}$$

The α,β,γ,λ,τ in the equation (26) indicate the constants determined based on the experimental parameters. We can know from the said equation (26) that effective thermal conductivity of the vapour chamber is related to controlling parameters of the experiment, fill-up number of the working fluid influencing Φh, volume of the cavity influencing t, input power and area of the heat source influencing q_{in}, area of the vapour chamber influencing A. Thus, this study is designed to firstly use thermal-performance experiment to determine the thermal performance and related experimental controlling parameters of the vapour chamber-based thermal, and sort them into the database of these experimental data, then combine with equation (26) to obtain the constants of the symbols α,β,γ,λ,τ. Let the constant α be 1. And these constants β,γ,λ,τ are equivalent to 0.13, 0.28, 0.15, and -0.54 based on some specified conditions in this research, respectively. This window program VCTM V1.0 was coded with Microsoft Visual Basic™ 6.0 according to the empirical formula and calculated the thermal performance of a vapor chamber-based thermal module in this study. These parameters affect its thermal performance including the dimensions, thermal performance and position of the vapor chamber. Thus it is very important for the optimum parameters to be selected to receive the best thermal performance of the vapor chamber-based thermal module. The program contains two main windows. The first is the selection window adjusted in the program as the main menu as shown in Fig. 7. In this window, the type of the air direction can be chosen separately. The second window has five main sub-windows. There are four sub-windows of the input parameters for the thermal module as shown in Fig. 7. The first sub-window is the simple parameters of the vapor chamber including dimensions and thermal performance. Fig. 7 shows the second sub-window involving detail dimensions of a heat sink. The third and fourth sub-windows are the simple parameters containing input power of heat source, soldering material, and materials of thermal grease and performance curve of fan. All the input parameters required for this study of the window program were given and the window program starts. Later, the program examines the situation by pressing calculated icon. The fifth sub-window is the window showing the simulation results. In this sub-window, when it is pressed at calculate icon for making analysis of the thermal performance of a vapor chamber-based thermal module, we can see a figure as it is shown in Fig. 7.

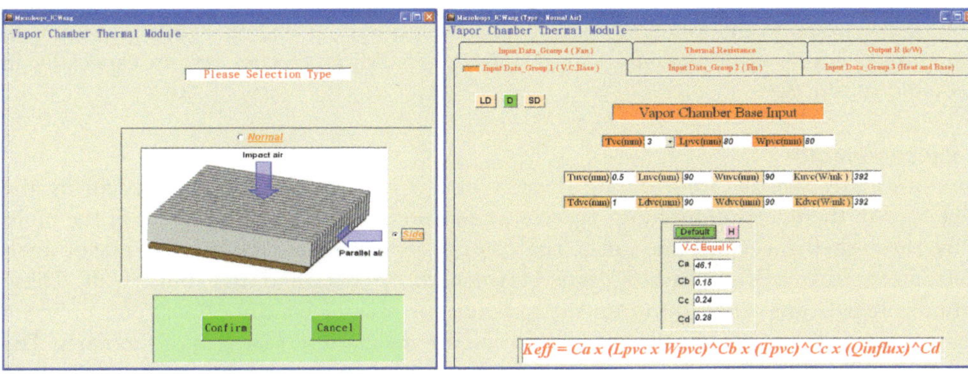

Fig. 7. Window program VCTM V1.0

Results show that the two and three-dimensional effective thermal conductivities of vapor chamber are more than two times higher than that of the copper and aluminum heat spreaders, proving that it can effectively reduce the temperature of heat sources. The maximum heat flux of the vapor chamber is over 800,000 W/m^2, and its effective thermal conductivity will increase with input power increases. It is deduced from the novel formula that the maximum effective thermal conductivity is above 800W/m°C. Certain error necessarily exist between the data measured during experiment, value deriving from experimental data and actual values due to artificial operation and limitation of accuracy of experimental apparatus. For this reason, it is necessary take account of experimental error to create confidence of experiments before analyzing experimental results. The concept of propagation of error is introduced to calculate experimental error and fundamental functional relations for propagation of error. During the experiment, various items of thermal resistances and thermal conductivities are utilized to analyze the heat transfer characteristics of various parts of thermal modules. The thermal resistance and thermal conductivity belong to derived variable and includes temperature and heating power, which are measured with experimental instruments. The error of experimental instruments is propagated to the result value during deduction and thus become the error of thermal resistance and thermal conductivity values. An experimental error is represented with a relative error and the maximum relative errors of thermal conductivities defined are within ±5% of k_{index}. This study answered how to evaluate the thermal-performance of the vapor chamber-based thermal module, which has existed in the thermal-module industry for a year or so. Thermal-performance of the thermal module with the vapor chamber can be determined within several seconds by using the final formula deduced in this study. One- and two-dimensional thermal conductivities of the vapor chamber are about 100 W/m°C, less than that of most single solid-phase metals. Three-dimensional thermal conductivity of the vapor chamber is up to 910 W/m°C, many times than that of pure copper base plate. The effective thermal conductivities of the vapor chamber are closely relate to its dimensions and heat-source flux, in the case of small-area vapor chamber and small heat-source flux, the effective thermal conductivity are less than that of pure copper material.

4. Air-cooling thermal module in other industrial areas

Air-cooling thermal module in other industrial areas as large-scale motor and LEDs lighting lamp are discussed in the following paragraphs. And a vapour chamber for rapid-uniform heating and cooling cycle was used in an injection molding process system especially in inset mold products.

4.1 Injection mold

There are many reasons for welding lines in plastic injection molded parts. During the filling step of the injection molding process, the plastic melt drives the air out of the mold cavity through the vent. If the air is not completely exhausted before the plastic melt fronts meet, then a V-notch will form between the plastic and the mold wall. These common defects are often found on the exterior surfaces of welding lines. Not only are they appearance defects, but they also decrease the mechanical strength of the parts. The locations of the welding lines are usually determined by the part shapes and the gate locations. In this paragraph, a heating and cooling system using a vapour chamber was developed. The vapor chamber was installed between the mold cavity and the heating block as shown in Fig. 8. Two electrical heating tubes are provided. A P20 mold steel block and a thermocouple are embedded to measure the temperature of the heat insert device. The mold temperature was raised above the glass transition temperature of the plastic prior to the filling stage. Cooling of the mold was then initiated at the beginning of the packing stage. The entire heating and cooling device was incorporated within the mold. The capacity and size of the heating and cooling system can be changed to accommodate a variety of mold shapes.

According to the experimental results, after the completion of molding, 10% of Type1 samples did not pass torque test, while all Type2 and Type3 samples passed the test. After thermal cycling test, the residual stress of the plastics began to be released due to temperature change, so the strength of product at the position of weld line was reduced substantially. Only 30% of Type1 products passed the 15.82 N-m torque tests after thermal cycling test, followed by 50% of Type2 products and 100% of Type3 products. This study proved that, among existing insert molding process, the temperature of inserts has impact on the final assembly strength of product. In this study, the local heating mechanism of vapor chamber can control the molding temperature of inserts; and the assembly strength can be improved significantly if the temperature of inserts prior to filling can be increased over the mold temperature, thus allowing the local heating mechanism to improve the weld line in the insert molding process. In this study, a vapour chamber based rapid heating and cooling system for injection molding to reduce the welding lines of the transparent plastic products is proposed. Tensile test parts and multi-holed plates were test-molded with this heating and cooling system. The results indicate that the new heating and cooling system can reduce the depth of the V-notch as much as 24 times.

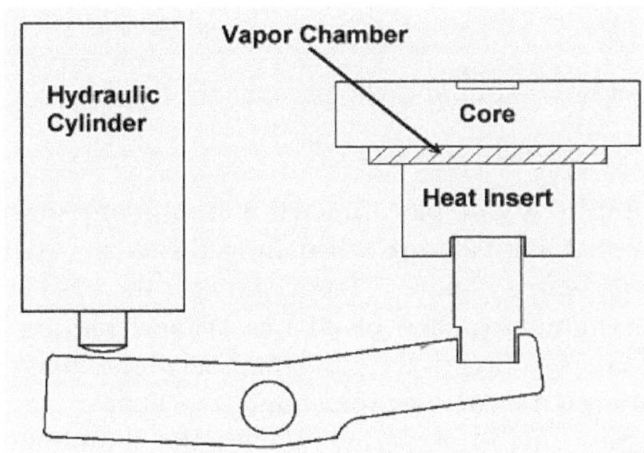

Fig. 8. Mechanics of heating and cooling cycle system with vapor chamber

4.2 Large motor

In this study, the 2350-kW completely enclosed air-to air cooled motor with dimensions 2435mm × 1321mm × 2177mm, as shown in Figure 9, is investigated. The motor includes a centrifugal fan, two axial fans, a shaft, a stator, a rotor, and a heat exchanger with 637 cooling tubes. There are two flow paths in the heat exchanger: the internal and external flows. As shown with the blue arrows, the external flow is driven by the rotation of the centrifugal fan, which is mounted externally to the frame on the motor shaft. The external air flows through the 637 tubes of the staggered heat exchanger mounted on top of the motor. The red arrows in Figure 8 show the internal air circulated by two axial fans on each side of the shaft and cooled by the heat exchanger. This study experimentally and numerically investigates the thermal performance of a 2350-kW enclosed air-to air cooled motor. The fan performances and temperatures of the heat exchanger, rotor, and the stator are numerically determined, which are in good agreement with the experimental data. Due to the non-uniform behaviours of the external air and air leakage of the internal air, the original motor design cannot operate at the best conditions. The designs with modified guide vanes and optimum clearance between the rotor and the axial fan demonstrate that the temperatures of the rotor and stator can decrease 5°C. The new design of the guide vanes makes the flow distributions uniform. Two axial fans with optimal distance operate at the maximum flow rate into the shaft, stator, and rotor, which increases the cooling ability. The present results provide useful information to designers regarding the complex flow and thermal interactions in large-scale motors.

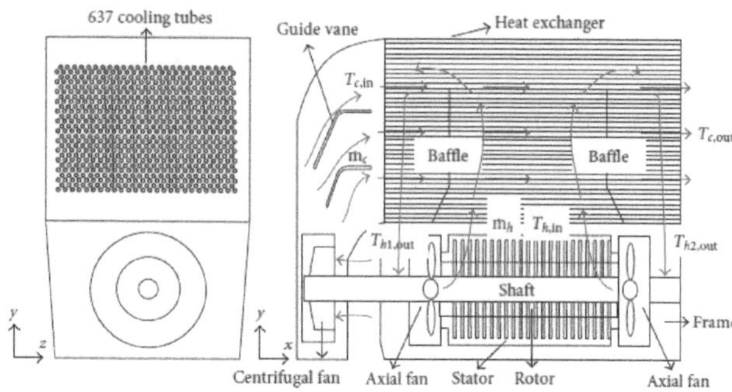

Fig. 9. Schematic view of flow paths and components for the motor.

4.3 LED lighting

The solid-state light emitting diode has attracted attention on outdoor and indoor lighting lamp in recent years. LEDs will be a great benefit to the saving-energy and environmental protection in the lighting lamps region. A few years ago, the marketing packaged products of single die conducts light efficiency of 80 Lm/W and reduces the light cost from 5 NTD/Lm to 0.5 NTD/Lm resulting in the good market competitiveness. These types of LED lamps require combining optical, electronic and mechanical technologies. This article introduces a thermal-performance experiment with the illumination-analysis method to discuss the green illumination techniques requesting on LEDs as solid-state luminescence source application in relative light lamps. The temperatures of LED dies are lower the lifetime of lighting lamps to be longer until many decades. We have successfully applied on LED outdoor lighting lamp as street lamp and tunnel lamp. In the impending future, we do believe that the family will install the LED indoor light lamps and lanterns certainly to be more popular generally.

LED light-emitting principle is put forward by the external bias on the P-Type and N-Type semiconductor, prompting both electron and electricity hole can be located through the depletion region near the P-N junction, and then were into the acceptor P-type and donor N-type semiconductor; and combine with another carrier, resulting in electron jumping and energy level gap in the form of energy to light and heat release, which the carrier concentration and to increase the luminous intensity of one of the factors. Therefore, LED can be a component of converting electrical energy into light energy, including the wavelength of light emitted by the infrared light, visible light and UV. The chemical family group IIIA in the periodic table (B, Al, Ga, IN, Th) and the VA family group (N, P, As, Sb, Bi) or IIA family group (Zn, Cd, Hg) and family group VIA (O, S, Se, Ti, Po) elements composed of compound semiconductor, and connected at the ends to the metal electrode (ohmic contact point), is the basic LED P-N junction structure.

The wavelength of light emitted can be obtained from the formula by Albert Einstein, who used Planck description of photoelectric effect of the quantum theory in 1921. Because the composition of materials for each energy level of semiconductor energy gap is different, its light wavelength generated by them is not the same as shown in Equation (27).

$$\lambda = {(h \times c)}\big/{E_\lambda} \cong \frac{1.988 \times 10^{-9}}{E'_\lambda}(nm) \cong \frac{1240}{E_\lambda}(nm) \tag{27}$$

Where λ is the light wavelength of LED (nm), h is the Planck constant 6.63 x 10-34 J · s , c is the vacuum velocity of light 2.998 x 10^8 m · s^{-1} and E_λ is the photon energy (eV).

Currently, one of the most serious problems is the thermal management for use of high-power LED lighting lamp, so the overall design and analysis of the thermal performance of LED lighting lamps is important. The following paragraphs will research in the thermal management for some commonly used methods applied to different kinds of LED lighting lamps. The heat-sink numerical analysis is a subject belonging to the computational fluid dynamics (CFD), in which fluid mechanics, discrete mathematics, numerical method and computer technology are integrated. Conventional numerical methods for the flow field are the Finite Element Method (F.E.M.), Finite Volume Method (F.V.M.) and Finite Difference Method (F.D.M.). A vapour chamber has uniquely high thermal performance and an isothermal feature; it has been developed and fabricated at a low-cost due to the mature manufacturing process. Fig. 10 shows a vapor chamber with above 800W/m°C, which is size of 80 x 80 x 3 mm^3 with light weight and antigravity characteristics to substitute for the present fine metal or the embedded heat pipe metal based plate, thus creating a new generation LED based plate. The device reduces the temperature of LEDs and enhances their lifetime. From the Fig.10, the spreading thermal performance of a vapor chamber is obviously better than a Copper plate after 60 seconds at the same operating conditions through thermograph. Its experimental results are shown in Table 1. T_a, T_{vc} and T_{AL} are the temperatures of surroundings, vapour chamber and aluminium based-plate, respectively. R_t is the thermal resistance of vapour chamber-based plate.

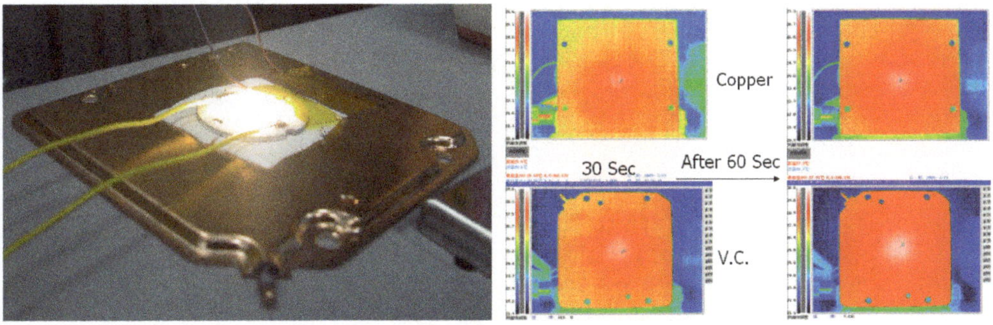

Fig. 10. LED vapour chamber-based plate and temperature distribution

Power (Watt)	Temperature(°C) / Thermal Resistance (°C/W)			
	T_a	T_{vc}	T_{AL}	R_t
5.236	24.5	51	54	5.63
7.100	24.8	68.9	70.9	6.49
8.614	24	75.6	79.2	6.41

Table 1. Experimental result for LED vapour chamber-based plate

Fig. 11 shows the temperature distributions of 12 pieces of LED up to 30Watt AL die-casting heat sink with asymmetry radial fins. A LEDs vapor chamber-based plate is placed on the heat sink and its size is a diameter of 9cm and a thickness of 3mm with thermal conductivity above 1500W/m°C according to the window program VCTM V1.0. To get the numerical results, we supposed that the coefficient of natural convection h is equal to 5W/m^2°C and 10W/m^2°C and ambient temperature is 25°C. The input power per die is 1.5Watt, 2Watt and 2.5Watt, respectively. Table 2 is the final simulation results.

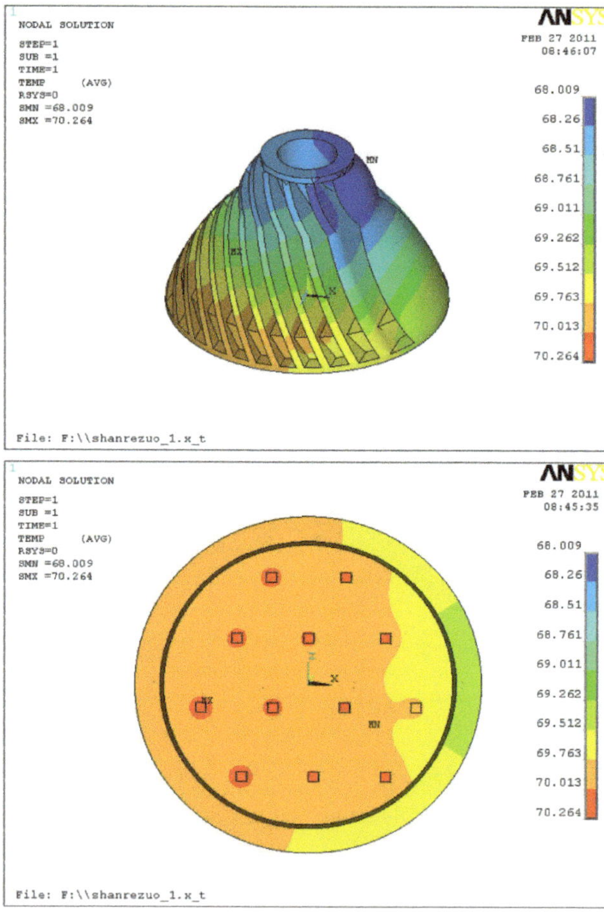

Fig. 11. Temperature distribution of 30 Watt LEDs at h=10

The light bar can be used as indoor living room lighting or outdoor architectural lighting. They are reduced the temperature T_j employed an extruded aluminum strip heat sink. Figure 12 shows a LED table lamp prototype, after a long test, the temperature of internal heat sink at 56°C or less. This table lamp prototype is divided into six parts including lamp body, LEDs, LEDs driver, aluminum based plates, heat sinks and spreading-brightness enhancement film. The illumination of the prototype is 600 lumens (Lm) and the input power is 12Watt. The luminosity is 1600 Lux measured by a photometer at a distance of 30cm from table lamp. Lastly, according to design and analyze the table lamp prototype, we draw four types of future LED table lamps utilizing above 15Watt or more as shown in Fig. 12. For centuries, all mankind have applied light generated by thermal radiation on many lighting things; now through progress rapidly of semiconductor and solid-state cold light technologies in recent decades, make mankind forward to green environmental protection and energy-saving lighting world in the 21st century. This article describes many indoor and outdoor lighting in features, analysis and design using lot types of heat sinks to address the high-brightness or high-power LEDs combined with optical, mechanical, and electric areas of lighting lamp. The authors are looking for contributing to the LED industry, government and academia for the green energy-saving lamps.

Total Power (Watt)	h=5(W/m^2°C)		h=10(W/m^2°C)	
	Ave. Temp. (°C)	Max. Temp. (°C)	Ave. Temp. (°C)	Max. Temp. (°C)
18	68.86	69.66	51.48	52.16
24	83.38	84.45	60.25	61.15
30	97.30	98.62	69.14	70.26

Table 2. Simulation situations for AL die-casting heat sink

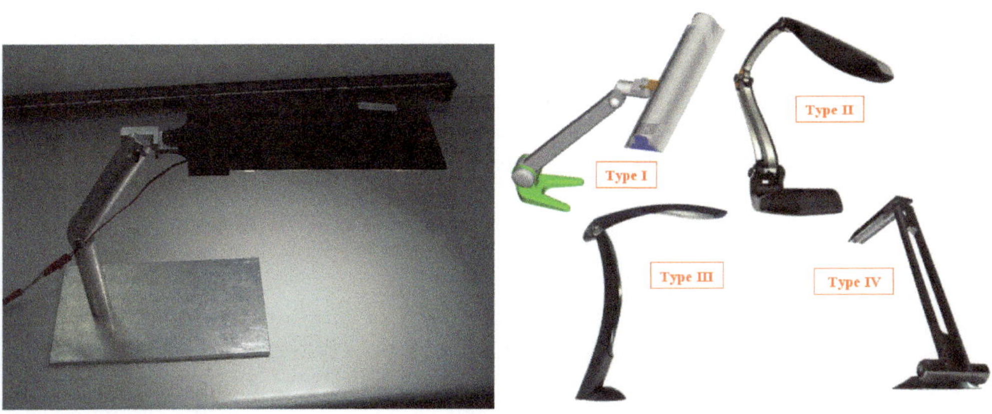

Fig. 12. 3D 12Watt table lamps

5. Conclusion

The air cooling module applies to consumer-electronic products involving automobiles, communication devices, etc. Recently, consumer-electronic products are becoming more complicated and intelligent, and the change occurs faster than ever. To recall the author's early experience in various consumer-electronic products, the heat/thermal problems play an important role in two decades. This chapter investigates all methodologies of Personal Computer (PC), Note Book (NB), Server including central processing unit (CPU) and graphic processing unit (GPU), and LED lighting lamp of smaller area and higher power in the consumer electronics industry. This approach is expected to help them make decisions related to the lifetime and reliability of their products in a right, reasonable and systematic way. The authors are looking for contributing to the LED industry, government and academia for the green energy-saving lamps. The author's future efforts could be dedicated to developing a LED green energy-saving lamps system. It is also desired that the evaluation method for thermal module be extended to other categories of consumer LED products such as home appliances, office-automation, personal communication devices, automobile interior design, and so on. Finally, the authors would like to mention a few points as the contribution of this study. This can serve as a reference for future researchers.

6. Acknowledgment

This chapter originally appeared in these References and is a major revised version. Some of the materials presented in this chapter were first published in these References. The authors gratefully acknowledge Prof. S.-L. Chen and his Energy Lab., Prof. J.-C. Wang and his Thermo-Illuminanace Lab for guidance their writings to publish and permission to reprint the materials here. The work and finance were supported by National Science Council (NSC), National Taiwan University (NTU) and National Taiwan Ocean University (NTOU). Finally, the authors would like to thank all colleagues and students who contributed to this study in the Chapter.

7. References

Chang, C.-C. ; Kuo, Y.u-F. ; Wang, J.-C. & Chen, S.-L. (2010). Air Cooling for a Large Scale Motor. *Applied Thermal Engineering,* Vol. 30, Issue 11-12, pp.1360-1368.

Chang, Y.-W. ; Cheng, C.-H. ; Wang, J.-C. & Chen, S.-L. (2008). Heat Pipe for Cooling of Electronic Equipment. *Energy Conversion and Management,* Vol. 49, pp.3398-3404.

Chang, Y.-W.; Chang, C.-C.; Ke, M.-T. & Chen, S.-L. (2009). Thermoelectric air-cooling module for electronic devices. *Applied Thermal Engineering,* Vol. 29, No. 13, pp.2731-2737.

Chen, S.-L.; Chang, C.-C.; Cheng, C.-H. & Ke, M.-T. (2009). Experimental and numerical investigations of air cooling for a large-scale motor. *International Journal of Rotating Machinery,* Vol. 2009, Article ID 612723, 7 pages.

Huang, H.-S.; Weng, Y.-C.; Chang, Y.-W.; Chen, S.-L. & Ke, M.-T. (2010). Thermoelectric water-cooling device applied to electronic equipment. *International Communications in Heat and Mass Transfer,* Vol. 37, No. 2, pp.140-146.

Lin, V. & Chen, S.-L. (2003). Performance analysis, optimum and verification for parallel plate heat sink associated with single non-uniform heat source, *ASME 2003 International Electronic Packaging Technical Conference and Exhibition,* InterPACK2003, Vol. 2, pp.229-236.

Tsai, T.-E.; Wu, G.-W.; Chang, C.-C; Shih, W.-P. & Chen, S.-L. (2010a). Dynamic test method for determining the thermal performances of heat pipes. *International Journal of Heat and Mass Transfer,* Vol. 53, No. 21-22, pp.4567-4578.

Tsai, T.-E.; Wu, H.-H.; Chang, C.-C. & Chen, S.-L. (2010b). Two-phase closed thermosyphon vapor-chamber system for electronic cooling. *International Communications in Heat and Mass Transfer,* Vol. 37, No. 5, pp.484-489.

Tsai, Y.-P. ; Wang, J.-C. & Hsu, R.-Q. (2011). The Effect of Vapor Chamber in an Injection Molding Process on Part Tensile Strength. *EXPERIMENTAL TECHNIQUES,* Vol. 35, Issue 1, pp.60-64.

Wang R.-T. & Wang, J.-C. (2011a). Green Illumination Techniques applying LEDs Lighting, *Proceedings of GETM 2011 May 28,* pp.1-7, Changhua, Taiwan.

Wang, J.-C. & Chen, T.-C. (2009). Vapor chamber in high performance server. *Microsystems IEEE 2010 Packaging Assembly and Circuits Technology Conference (IMPACT), 2009 4th International,* pp.364-367.

Wang, J.-C. & Huang, C.-L. (2010). Vapor chamber in high power LEDs. *IEEE 2011 Microsystems Packaging Assembly and Circuits Technology Conference (IMPACT), 2010 5th International,* pp.1-4.

Wang, J.-C. & Tsai,Y.-P. (2011). Analysis for Diving Regulator of Manufacturing Process. *Advanced Materials Research,* Vol. 213, pp.68-72.

Wang, J.-C. & Wang R.-T. (2011b). A Novel Formula for Effective Thermal Conductivity of Vapor Chamber, *EXPERIMENTAL TECHNIQUES,* DOI: 10.1111/j.1747-1567.2010.00652.x, early view.

Wang, J.-C. (2009). Superposition Method to Investigate the Thermal Performance of Heat Sink with Embedded Heat Pipes. *International Communication in Heat and Mass Transfer,* Vol. 36, Issue 7, pp.686-692.

Wang, J.-C. (2008). Novel Thermal Resistance Network Analysis of Heat Sink with Embedded Heat Pipes. *Jordan Journal of Mechanical and Industrial Engineering,* Vol. 2, No. 1, , pp. 23-30.

Wang, J.-C. (2010). Development of Vapour Chamber-based VGA Thermal Module. *International Journal of Numerical Methods for Heat & Fluid Flow,* Vol. 20, Issue 4, pp.416-428.

Wang, J.-C. (2011a). Investigations on Non-Condensation Gas of a Heat Pipe. *Engineering,* Vol. 3, pp.376-383.

Wang, J.-C. (2011b). L-type Heat Pipes Application in Electronic Cooling System. *International Journal of Thermal Sciences,* Vol. 50, Issue 1, pp.97-105.

Wang, J.-C. (2011c). Applied Vapor Chambers on Non-uniform Thermo Physical Conditions. *Applied Physics,* Vol. 1, pp.20-26.

Wang, J.-C. (2011d). Thermal Investigations on LEDs Vapor Chamber-Based Plates. *International Communication in Heat and Mass Transfer,* DOI: 10.1016/j.icheatmasstransfer.2011.07.002, Article in Press, Corrected Proof.

Wang, J.-C. ; Huang, H.-S. & Chen, S.-L. (2007). Experimental Investigations of Thermal Resistance of a Heat Sink with Horizontal Embedded Heat Pipes, *International Communications in Heat and Mass Transfer,* Vol. 34, Issue 8, pp.958-970.

Wang, J.-C. ; Wang, R.-T. ; Chang, C.-C. & Huang, C.-L. (2010a). Program for Rapid Computation of the Thermal Performance of a Heat Sink with Embedded Heat Pipes. *Journal of the Chinese Society of Mechanical Engineers,* Vol. 31, Issue 1, pp.21-28.

Wang, J.-C. ; Wang, R.-T.; Chang, T.-L. & Hwang, D.-S. (2010b). Development of 30 Watt High-Power LEDs Vapor Chamber-Based Plate. *International Journal of Heat and Mass Transfer,* Vol. 53, Issue 19/20, pp.3900-4001.

Wang, J.-C.; Chang T.-L. ; Tsai Y.-P. ; & Hsu R.-Q. (2011a). Experimental Analysis for Thermal Performance of a Vapor Chamber Applied to High-Performance Servers, *Journal of Marine Science and Technology-Taiwan,* Article in Press, Corrected Proof.

Wang, J.-C.; Li, A.-T.; Tsai,Y.-P. & Hsu, R.-Q. (2011b). Analysis for Diving Regulator Applying Local Heating Mechanism of Vapor Chamber in Insert Molding Process. *International Communication in Heat and Mass Transfer,* Vol.38, Issue 2, pp.179-183.

Wu, H.-H.; Hsiao, Y.-Y.; Huang, H.-S.; Tang, P.-H. & Chen, S.-L. (2011). A practical plate-fin heat sink model. *Applied Thermal Engineering,* Vol.31, Issue 5, pp.984-992.

15

Multi-Core CPU Air Cooling

M. A. Elsawaf, A. L. Elshafei and H. A. H. Fahmy
Faculty of Engineering, Cairo University
Egypt

1. Introduction

High speed electronic devices generate more heat than other devices. This chapter is addressing the portable electronic device air cooling problem. The air cooling limitations is affecting the portable electronic devices. The Multi-Core CPUs will dominate the Mobile handsets Platforms in the coming few years. Advanced control techniques offer solutions for the central processing unit (CPU) dynamic thermal management (DTM). This chapter objective is to minimize air cooling limitation effect and ensure stable CPU utilization using fuzzy logic control. The proposed solution of the air cooling limitation focuses on the design of a DTM controller based on fuzzy logic control. This approach reduces the problem design time as it is independent of the CPU chip and its cooling system transfer functions. On-chip thermal analysis calculates and reports thermal gradients or variations in operating temperature across a design. This analysis is increasingly important for the advanced digital integrated circuits (ICs). At today's 65nm and 45nm technologies, adding cores to CPU chip increases its power density and leads to thermal throttling. Advanced control techniques give a solution to the CPU thermal throttling problem. Towards this objective, a thermal model similar to a real IBM CPU chip containing 8 cores is built. This thermal model is integrated to a semiconductor thermal simulator. The open loop response of the CPU chip is extracted. This CPU chip thermal profile illustrates the CPU thermal throttling. The proposed DTM controller design is based on 3D fuzzy logic. There are many cores within CPU chip, each of them is a heat source. The correlation between these cores temperatures and their operating frequencies improves the DTM response and reduce the air cooling limitation effect. The 3D fuzzy controller takes into consideration these correlations. This chapter presents a new DTM technique called "Thermal Spare Core" algorithm (TSC). Thermal Spare Core (TSC) is a completely new DTM algorithm. The thermal spare cores (TSC) is based on the reservation of cores during low CPU utilization and activate them during thermal crises. The reservation of some cores as (TSC) doesn't impact CPU over all utilization. These cores are not activated simultaneously due to the air cooling limitations. The semiconductor technology permits more cores to be added to CPU chip. That means there is no chip area wasting in case of TSC. The TSC is a solution of the Multi-Core CPU air cooling limitations.

2. The CPU air cooling limitations

We live in a computer controlled epoch. We do not even realize how often our lives depend on machines and their programming. For example, mobile handsets, portable electronic devices, laptops, medical instruments, and many other devices all depend on digital processors in our everyday lives. There is no doubt that the size and the weight of these portable equipments is affecting their utilization. Unfortunately, there are many factors affecting the portability of electronic systems. The power consumption is affecting battery. Efficient cooling of portable electronic devices is becoming a problem due to air cooling limitations.

On-chip temperature gradient is a design challenge. Many technology factors affect the chip temperature gradients. In terms of the technology factors, power density (power per unit area) is increasing with each new technology node (ITRS , 2006). After all, smaller geometries enable more functionality to be fit within the same area of a chip which can result in high thermal gradients (Huangy et al., 2006). As shown in Fig.1A, adding more cores to the CPU chip increase the total power consumption. Fig.1B illustrates the maximum number of cores per chip and their maximum operating frequencies (D.D.Kim et al., 2008).

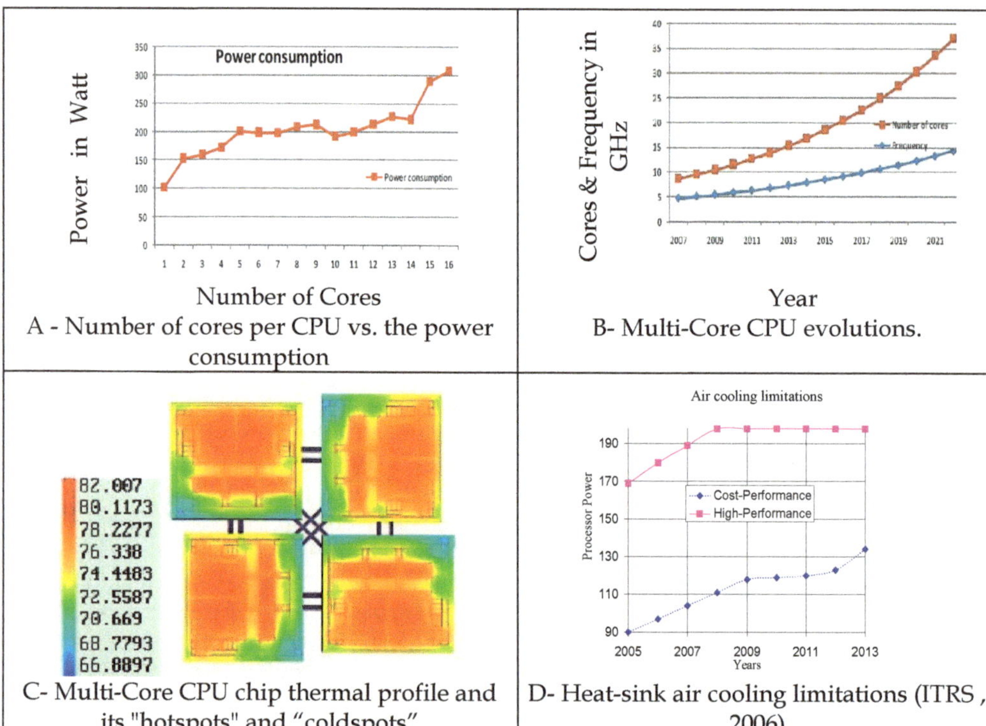

A - Number of cores per CPU vs. the power consumption

B- Multi-Core CPU evolutions.

C- Multi-Core CPU chip thermal profile and its "hotspots" and "coldspots"

D- Heat-sink air cooling limitations (ITRS , 2006)

Fig. 1. Multi-Core CPU evolutions

The CPU cores run relatively hot while on-chip memory tends to run relatively cold. The result is an ever-varying mish-mash of "hot" and "cold" spots that depend on the mode of operation. A cell phone is a good example of this type of design. The act of creating a text message will exercise certain functionality, which creates a specific thermal profile. But the act of transmitting this message will exercise different functionality, which results in a different profile. The same can be said for using the cell phone to make a voice call, play an

mp3 file, take a picture, and so forth. The resulting temperature variation across a chip is typically around 10° to 15°C. If this temperature distribution is not managed; then temperature variation will be as high as 30° to 40°C (Mccrorie, 2008).
The CPU power dissipation comes from a combination of dynamic power and leakage power (S.Kim et al., 2007). Dynamic power is a function of logic toggle rates, buffer strengths, and parasitic loading. The leakage power is function of the technology and device characteristics. Thermal-analysis solutions must account for both causes of power. In Fig.1C the thermal profile of a CPU chip is showing the temperature variation across the chip surface. This phenomenon is due to the variation of the power density according to each function block design. This power density distribution generates "hotspots" and "coldspots" areas across the CPU chip surface (Huangy et al., 2006). The high CPU operating temperature increases leakage current degrades transistor performance, decreases electro migration limits, and increases interconnect resistively (Mccrorie, 2008). In addition, leakage current increases the power consumption.

3. The CPU thermal throttling problem

The fabrication technology permits the addition of more cores to the CPU chip having higher speed and smaller size devices. But adding more cores to a CPU chip increases the power density and generates additional dynamic power management challenges. Since the invention of the integrated circuit (IC), the number of transistors that can be placed on an integrated circuit has increased exponentially, doubling approximately every two years (Moore, 1965). The trend was first observed by Intel co-founder Gordon E. Moore in a 1965 paper. Moore's law has continued for almost half a century! It is not a coincidence that Moore was discussing the heat problem in 1965: "will it be possible to remove the heat generated by tens of thousands of components in a single silicon chip?" (Moore, 1965). The static power consumption in the IC was neglected compared to the dynamic power for CMOS technology. The static power is now a design problem. The millions of transistors in the CPU chip exhaust more heat than before. The CPU cooling system capacity limits the number of cores within the CPU chip (ITRS , 2008).
The International Technology Roadmap for Semiconductors (ITRS) is a set of documents produced by a group of semiconductor industry experts. ITRS specifies the high-performance heat-sink air cooling maximum limits; which is 198 Watt (ITRS, 2006). The chip power consumption design is limited by cooling system level capacity. We already reached the air cooling limitation in 2008 as shown in Fig.1D.
As shown in Fig.2A; the CPU reaches the maximum operational temperature after certain time due to maximum CPU utilization. Thus the CPU utilization is reduced to the safe utilization in order not to exceed. This phenomenon is called CPU thermal throttling. Fig.2B shows the comparison between the ideal case "no thermal constrains", "low power consumption with thermal constraints" case and "high power consumption with thermal constraints" case. The addition of more cores to the CPU chip doesn't increase the CPU utilization. The curve drifts to lower CPU utilization due to the CPU thermal limitation in case of low power consumption. In case of high power consumption; the CPU utilization decreases by adding more cores to the CPU chip. Thus the CPU utilization improvement is not proportional to its number of cores.

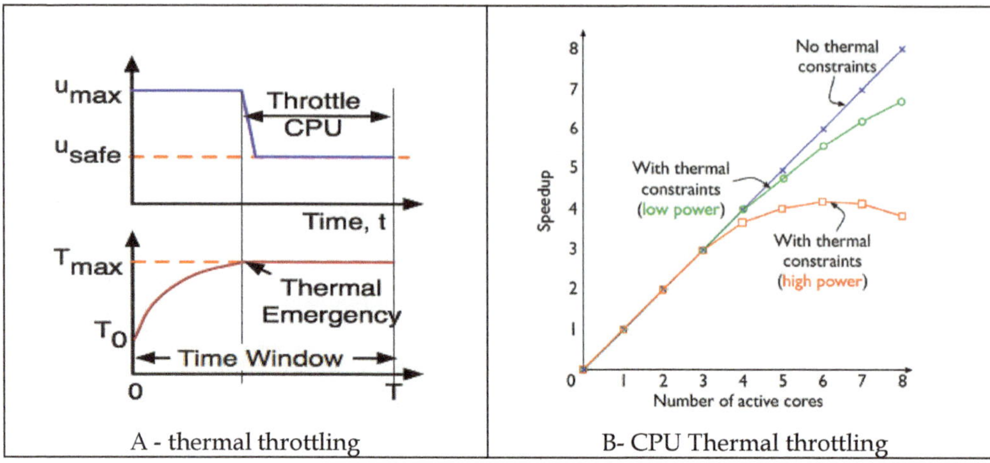

Fig. 2. CPU thermal throttling (Passino & Yurkovich, 1998)

4. The advance DTM controller design

The advanced dynamic thermal management techniques are mandatory to avoid the CPU thermal throttling. The fuzzy control provides a convenient method for constructing nonlinear controllers via the use of heuristic information. Such heuristic information may come from an operator who has acted as a "human-in-the-loop" controller for a process. The fuzzy control design methodology is to write down a set of rules on how to control the process. Then incorporate these rules into a fuzzy controller that emulates the decision-making. Regardless of where the control knowledge comes from, the fuzzy control provides a user-friendly and high-performance control (Patyra et al., 1996).

The DTM techniques are required in order to have maximum CPU resources utilization. Also for portable devices the DTM doesn't only avoid thermal throttling but also preserves the battery consumption. The DTM controller measure the CPU cores temperatures and according selects the speed "operating frequency" of each core. The power consumed is a function of operating frequency and temperature. The change in temperature is a function of temperature and the dissipated power.

The dynamic voltage and frequency scaling (DVFS) is a DTM technique that changes the operating frequency of a core at run time (Wu et al., 2004). Clock Gating (CG)or stop-go technique involves freezing all dynamic operations(Donald & Martonosi, 2006). CG turns off the clock signals to freeze progress until the thermal emergency is over. When dynamic operations are frozen, processor state including registers, branch predictor tables, and local caches are maintained (Chaparro et al., 2007). So less dynamic power consumed during the wait period. GC is more like suspend or sleep switch rather than an off-switch. Thread migration (TM) also known as core hopping is a real time OS based DTM technique. TM reduces the CPU temperature by migrating core tasks "threads" from an overheated core to another core with lower temperature. The current traditional DTM controller uses proportional (P controller) or proportional-integral (PI controller) or proportional-integral-derivative (PID controller) to perform DVFS (Donald & Martonosi, 2006; Ogras et al., 2008).

The fuzzy logic is introduced by Lotfi A. Zadeh in 1965 (Trabelsi et al., 2004). The traditional fuzzy set is two-dimensional (2D) with one dimension for the universe of discourse of the variable and the other for its membership degree. This 2D fuzzy logic controller (FC) is able to handle a non linear system without identification of the system transfer function. But this 2D fuzzy set is not able to handle a system with a spatially distributed parameter. While a three-dimensional (3D) fuzzy set consists of a traditional fuzzy set and an extra dimension for spatial information. Different to the traditional 2D FC, the 3D FC uses multiple sensors to provide 3D fuzzy inputs. The 3D FC possesses the 3D information and fuses these inputs into "spatial membership function". The 3D rules are the same as 2D Fuzzy rules. The number of rules is independent on the number of spatial sensors. The computation of this 3D FC is suitable for real world applications.

5. DTM evaluation index

An evaluation index for the DTM controller outputs is required. As per the thermal throttling definition, "the operating frequency is reduced in order not to exceed the maximum temperature". Both frequency and temperature changes are monitored as there is a non linear relation between the CPU frequency and temperature. One of the DTM objectives is to minimize the frequency changes. The core theoretically should work at open loop frequency for higher utilization. But due to the CPU thermal constrains the core frequency is decreased depending on core hotspot temperature.

The second DTM objective is to decrease the CPU temperature as much as possible without affecting the CPU utilization. A multi-parameters evaluation index ζ_t is proposed. It consists of the summation of each parameter evaluation during normalized time period. This index is based on the weighted sum method. The objective of multi-parameters evaluation index shows the different parameters effect on the CPU response. Thus the designer selects the suitable DTM controller that fulfils his requirements. The multi-parameters evaluation index permits the selection of DTM design that provides the best frequency parameter value without leading to the worst temperature parameter value.

The DTM evaluation index ζ_t calculation consists of 5 phases:

1. Identify the required parameters
2. Identify the design parameters ranges
3. Identify the desired parameters values of each range $\sigma_{ij}^{\text{Desired}}$
4. Identify the actual parameters values of each range $\sigma_{\text{ij}}^{\text{Actual}}$
5. Evaluate each parameter and the over all multi- parameter evaluation index

$$\zeta_t = \sum_{i=1}^{l} \lambda_i \tag{1}$$

The parameter λ_i value during the evaluation time period is the summation of the evaluation ranges divided by the number of ranges m_i.

$$\lambda_i = \frac{1}{m_i} \sum_{j=1}^{m_i} \sigma_{ij} \tag{2}$$

Each evaluation range σ_{ij} is evaluated over a normalized time period

$$\sigma_{ij} = \frac{\sigma_{\mathrm{ij}}^{\mathrm{Actual}}}{\sigma_{ij}^{\mathrm{Desired}}} \qquad (3)$$

$\sigma_{\mathrm{ij}}^{\mathrm{Actual}}$ is the actual percentage of time the CPU runs at that range

$\sigma_{ij}^{\mathrm{Desired}}$ is the desired percentage of time the CPU runs at that range

The λ_i value should be 1 or near 1. If $\lambda_i <$ 1 then the CPU runs less time than the desired within this range. If $\lambda_i >$ 1 then the CPU runs more time than the desired within this range. Thus the multi-parameters evaluation index equation is:

$$\zeta_t = \sum_{i=1}^{l} \frac{1}{m_i} \sum_{j=1}^{m_i} \left(\frac{\sigma_{\mathrm{ij}}^{\mathrm{Actual}}}{\sigma_{ij}^{\mathrm{Desired}}} \right) \qquad (4)$$

The DTM controller evaluation index desired value should be $\zeta_t = l$ or near l, where l is the number of parameters. The Multi-parameters evaluation index permit the designer to evaluate each rang independent on the other ranges and also evaluate the over all DTM controller response.

The multi-parameters evaluation index is flexible and accepts to add more evaluation parameters. This permits the DTM controller designer to add or remover any parameter without changing the evaluations algorithm. Fig.3 shows an example of the parameter λ_i calculation. In this example the parameter λ_i is the temperature. The temperature curve is divided into 3 ranges: High (H) - Medium (m) - Low (L), these ranges are selected as follow: High "greater than78 °C", Medium "between 74 °C and 78 °C", and Low "lower than 72 °C". The actual parameters values of each range $\sigma_{\mathrm{ij}}^{\mathrm{Actual}}$ is calculated as follow: $\sigma_{\mathrm{i\,High}}^{\mathrm{Actual}}$ = 20.5%, $\sigma_{\mathrm{i\,Medium}}^{\mathrm{Actual}}$ = 76%, and $\sigma_{\mathrm{i\,Low}}^{\mathrm{Actual}}$ =3.5%

6. Thermal spare core

As a CPU is not 100% utilized all time, thus some of the CPU cores could be reserved for thermal crises. Consider Fig.4A, when a core reaches the steady state temperature T_1, the cooling system is able to dissipate the exhausted heat outside the chip. However, if this core is overheated, the cooling system is not able to exhaust the heat outside the chip. Thus the core temperature increases until it reaches the thermal throttling temperature T_3 (Rao & Vrudhula, 2007).

The same thermal phenomena, as shown in Fig.4A, occur due to faults in the cooling system (Ferreira et al., 2007). The semiconductor technology permits more cores to be added to CPU chip. While the total chip area overhead is up to 27.9 % as per ITRS (ITRS , 2009). That means there is no chip area wasting in case of TSC. So reserving cores as thermal spare core (TSC) doesn't impact CPU over all utilization. These cores are not activated simultaneously due to thermal limitations. According to Amdahl's law: "parallel speedups limited by serial portions" (Gustafson , 1988). So adding more cores to CPU chip doesn't speedup due to the serial portion limits. Thus not all cores are fully loaded or even some of them are not even

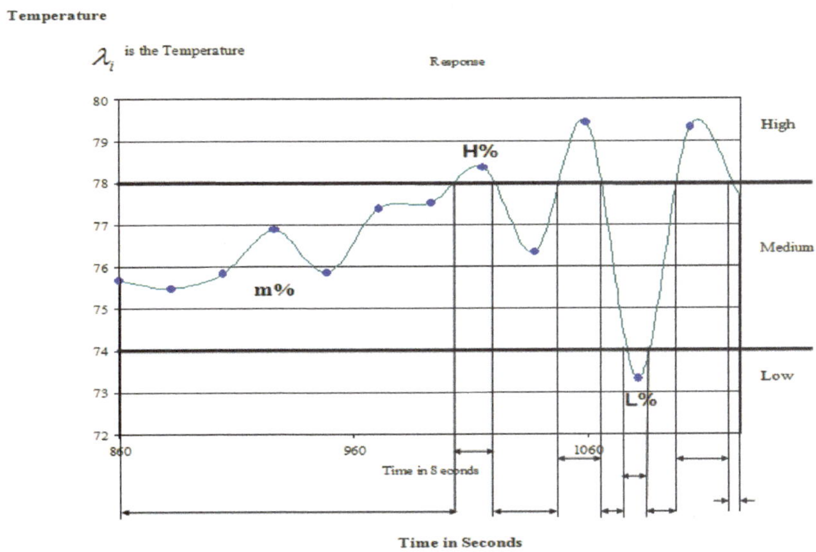

Fig. 3. Example of actual parameter value calculation

utilized if parallelism doesn't exist. The TSC concept uses the already existing chip space due to semiconductor technology. From the thermal point of view; the horizontal heat transfer path has for up to 30% of CPU chip heat transfer (Stan et al., 2006). The TSC is a big coldspot within the CPU area that handles the horizontal heat transfer path. The cold TSC reduces the static power as the TSC core is turned off. Also the TSC is used simultaneous with other DTM technique. The equation (5) calculates number of TSCs cores. The selection of TSC cores number is dependant on the number of cores per chip and maximum power consumed per core as follow:

$$N_{TSC} = | \{ (P_{mx} N_C - 198) / 198 \} | \tag{5}$$

where N_{TSC} : minimum number of TSCs, P_{mx} : maximum power consumed per core, N_C : total number of cores, 198 Watts is the thermal limitation of the air cooling system. Fig.4A shows core profile where lower curve is normal thermal behavior. The upper curve is the overheated core, T_1 is the steady state temperature, T_1 = 80 C corresponds to the temperature at t_1 . t_2 is required time for a thermal spare core to takeover threads from the overheated core, T_2 = 100 C corresponds to the temperature at t_2 . T_3 is the throttling temperature, and T_3 = 120 C corresponds to the temperature at t_3 .

TSC technique uses the already existing cores within CPU chip to avoid CPU thermal throttling as follow: Hot TSC: is a core within the CPU powered on but its clock is stopped. It only consumes static power. It is a fast replacement core. However, it is still a heat source. Cold TSC: is a core within the CPU chip powered off (no dynamic or static power consumed). It is not a heat source, but it is a slow replacement core. Its activation needs more time than hot TSC. But the cold TSC reduces the static power dissipation. Also cold TSC generates cold spot with relative big area that helps exhausting the horizontal heat transfer path out of the chip.

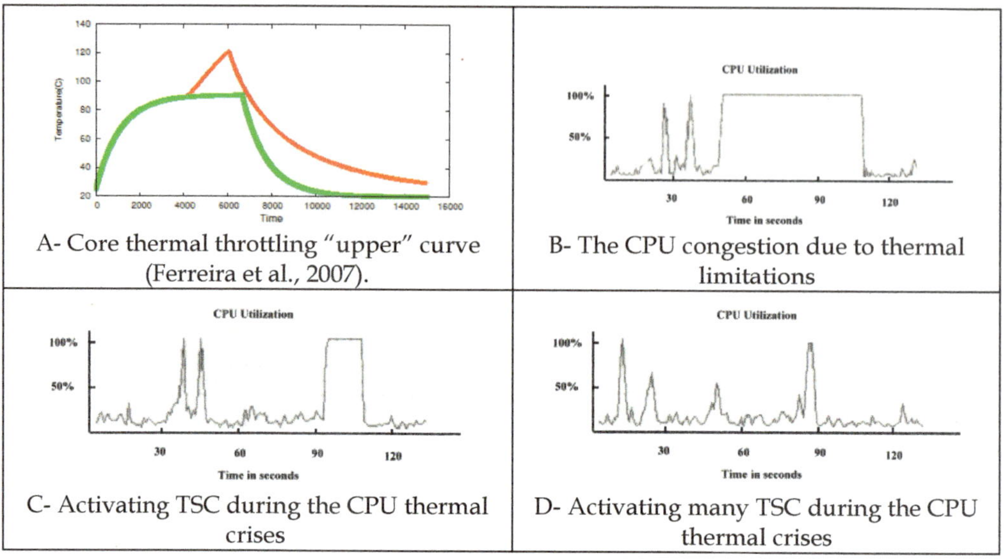

A- Core thermal throttling "upper" curve (Ferreira et al., 2007).

B- The CPU congestion due to thermal limitations

C- Activating TSC during the CPU thermal crises

D- Activating many TSC during the CPU thermal crises

Fig. 4. TSC Illustration

Defining T_{tsc} as the TSC activation temperature as follow:

$$T_{ss} \leq T_{tsc} \leq T_{th} \tag{6}$$

$$t_{tsc} = \min \{ (t_{th} - t_{CT}) , (t_{th} - t_{TM}) \} \tag{7}$$

Where: T_{ss} : core steady state temperature. T_{tsc} : The temperature that triggers TSC process. T_{th} : CPU throttling temperature. t_{tsc} : The time of activating TSC. t_{th} : The time required to reach thermal throttling. t_{CT} : The estimated time required for completing the current tasks within the over heated core. This information is not always accurate at run time. t_{TM} : Time required migrating threads from over heated core to TSC. If any core reaches T_{tsc} then the DTM controller will inform the OS to stop assigning new tasks to this overheated core. Thus the OS doesn't assign any new task to the overheated core. Therefore, T_{tsc} is not predefined constant temperature but variable temperature between T_{ss} and T_{th} . The DTM selects T_{tsc} depending on the minimum time required to evacuate the over heated core.

6.1 TSC illustration

This section illustrates the thermal spare cores (TSC) technique

As shown in Fig.4B, the CPU is 100% utilized for duration about 50 seconds. The OS realizes that the CPU congestion. The CPU executes its tasks slowly. In fact the CPU suffers from thermal throttling. This CPU utilization curve shows CPU congestion from OS point of view due to thermal limitations.

As shown in Fig.4C, The DTM controller detected the CPU high temperature. Thus the DTM controller executes the TSC algorithm. At 40 seconds time line, a TSC core replaces a hot core. The handover between the hot core the TSC core lead to a CPU peak. But The CPU improves its speed after that peak; as the TSC is still cold relatively and operates at higher

frequency. At 86 seconds, the CPU reaches thermal throttling again. Thus the CPU reaches congestion again. So the activation of a TSC core during the CPU thermal crises decreases the duration of the CPU degradation from 50 seconds to 15 seconds duration.
As shown in Fig.4D, the activation of 3 TSC cores during the thermal crises at 25 seconds, 45 seconds and 85 seconds time lines respectively increases the CPU utilization. The CPU executes its tasks normally without congestion rather than some CPU peaks. AS this CPU chip has many spare cores; the DTM controller activates the required TSC during the CPU thermal crises. So the CPU avoids the thermal throttling theoretically.

6.2 3D Fuzzy DTM controller

The 3D fuzzy control is able to handle the correlation between the different variable parameters of a distributed parameter system (Li & Li, 2007). Thus the 3D fuzzy logic is able to process the Multi-Core CPU correlation information. The 3D fuzzy control demonstrates its potential to a wide range of engineering applications. The 3D fuzzy control is feasible for real-time world applications (Li & Li, 2007). The thermal management process is a distributed parameter systems. The thermal management process is represented by the nonlinear partial differential equations (Doumanidis & Fourligkas, 2001).

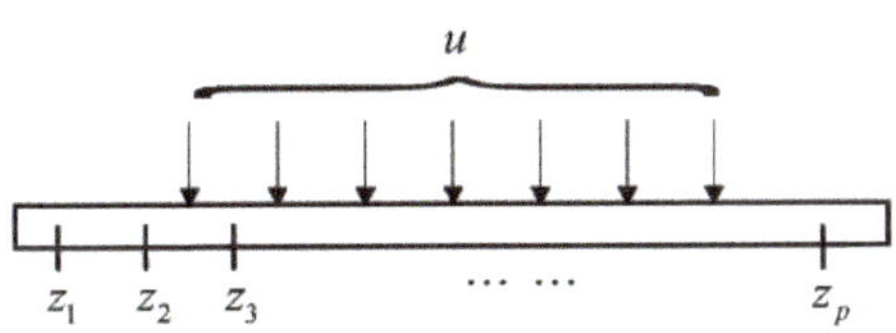

Fig. 5. Actuator u and the measurement sensors at p point.

Fig.5. presents a nonlinear distributed parameter system with one actuator ($\gamma = 1$). Where p point measurement sensors are located at $z_1, z_2, ..., z_p$ in the one-dimensional space domain respectively and an actuator u with some distribution acts on the distributed process. Inputs are measurement information from sensors at different spatial locations. i.e., deviations $e_1, e_2, ..., e_p$ and deviations change $\Delta e_1, \Delta e_2, ..., \Delta e_p$ where $e_1 = y_d(z_i) - y(z_i, n)$, $\Delta e_i = e_i(n) - e_i(n-1)$ $y_d(z_i)$ denotes the measurement value from location z_i, $n, n-1$ denote the n and $n-1$ sample time input. The output relationship is described by fuzzy rules extracted from knowledge. Since p sensors are used to provide $2p$ inputs.

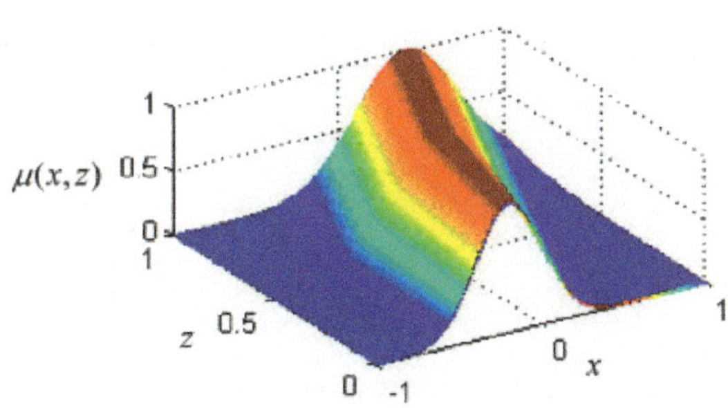

Fig. 6. 3D fuzzy set (Li & Li, 2007)

The 3D fuzzy control system is able to capture and process the spatial domain information defined as the 3D FC. One of the essential elements of this type of fuzzy system is the 3D fuzzy set used for modeling the 3D uncertainty. A 3D fuzzy set is introduced in Fig.6 by developing a third dimension for spatial information from the traditional fuzzy set. The 3D fuzzy set defined on the universe of discourse X and on the one-dimensional space is given by:

$$\overline{V} = \{(x,z), \mu_{\overline{V}}(x,z) \quad \forall \quad x \in X, \quad z \in Z\} \text{ and } 0 \le \{(x,z), \mu_{\overline{V}}(x,z) \le 1 \qquad (8)$$

When X and Z are discrete, $\overline{V}$ is commonly written as $\overline{V} = \sum_{z \in Z} \sum_{x \in X} \mu_{\overline{V}}(x,z) / (x,z)$

Where $\sum \sum$ denotes union over all admissible x and z. Using this 3D fuzzy set, a 3D fuzzy membership function (3D MSF) is developed to describe a relationship between input x and the spatial variable z with the fuzzy grade u.

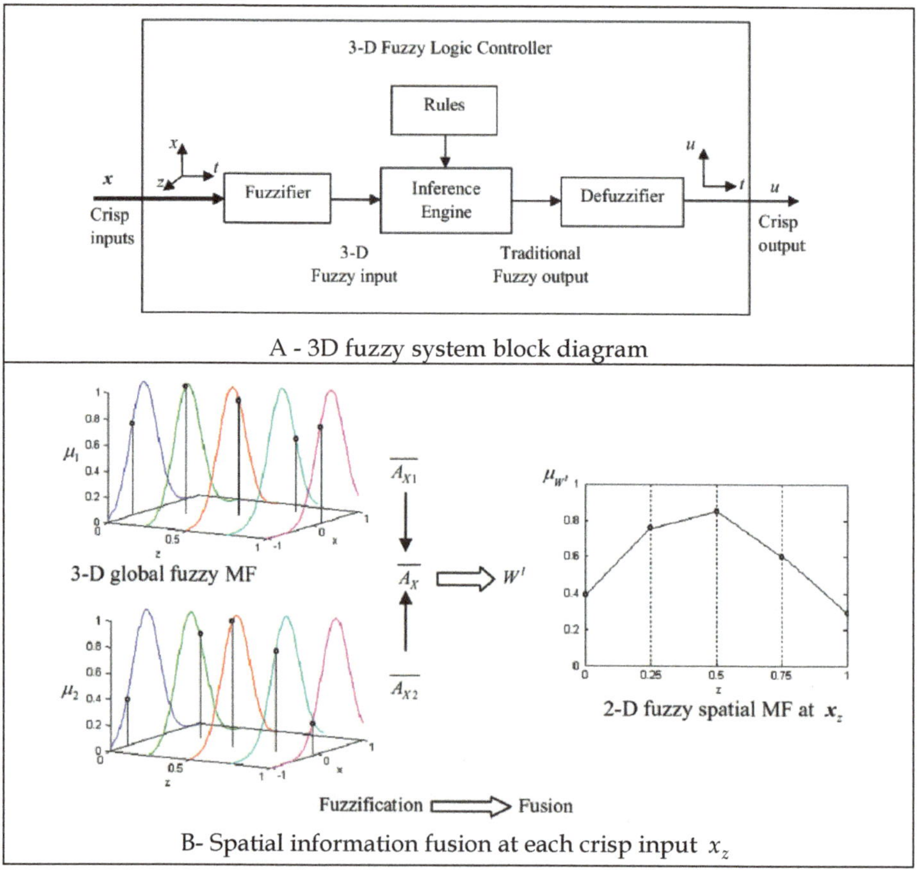

Fig. 7. 3D fuzzy system illustration (Li & Li, 2007)

Theoretically, the 3D fuzzy set or 3D global fuzzy MSF is the assembly of 2D traditional fuzzy sets at every spatial location (Li & Li, 2007). However, the complexity of this global 3D nature may cause difficulty in developing the FC. Practically, this 3D fuzzy MSF is approximately constructed by 2D fuzzy MSF at each sensing location. Thus, a centralized rule based is more appropriate, which avoid the exponential explosion of rules when sensors increase. The new FC has the same basic structure as the traditional one. The 3D FC is composed of fuzzification, rule inference and defuzzification as shown in Fig.7A. Due to its unique 3D nature, some detailed operations of this new FC are different from the traditional one. Crisp inputs from the space domain are first transformed into one 3D fuzzy input via the 3D global fuzzy MSF. This 3D fuzzy input goes through the spatial information fusion and dimension reduction to become a traditional 2D fuzzy input. After that, a traditional fuzzy inference is carried out with a crisp output produced from the traditional defuzzification operation. Similar to the traditional 2D FC, there are two different fuzzifications: singleton fuzzifier and non-singleton.

A singleton fuzzifier is selected as follows: Let $\overline{A}$ be a 3D fuzzy set, x is a crisp input, $x \in X$ and z is a point $z \in Z$ in one-dimensional space Z. The singleton fuzzifier maps x into $\overline{A}$ in X at location z then $\overline{A}$ s a fuzzy singleton with support x' if $\mu_{\overline{A}}(x,z)=1$ for $x=x'$, $z=z'$ and $\mu_{\overline{A}}(x,z)=0$ for all other $x \in X$, $z \in Z$ with $x \neq x'$, $z \neq z'$ if finite sensors are used. This 3D fuzzification is considered as the assembly of the traditional 2D fuzzification at each sensing location. Therefore, for p discrete measurement sensors located at $z_1, z_2, \ldots\ldots, z_p$, $x_z = [x_1(z), x_2(z), \ldots, x_j(z)]$ is defined as J crisp spatial input variables in space domain $Z = \{z_1, z_2, \ldots\ldots, z_p\}$ where $x_j(z_i) \in X_j \subset IR(j=1,2,\ldots,J)$ denotes the crisp input at the measurement location $z = z_i$ for the spatial input variable $x_j(z)$, X_j denotes the domain of $x_j(z_i)$. The variable $x_j(z)$ is marked by " z " to distinguish from the ordinary input variable, indicating that it is a spatial input variable. The fuzzification for each crisp spatial input variable $x_j(z)$ is uniformly expressed as one 3D fuzzy input $\overline{A}_{xj}$ in the discrete form as follows:

$$\overline{A_{X1}} = \sum_{z\in Z} \sum_{x_1(z)\in X_1} \mu_{X1}(x_1(z),z) \,/\, (x_1(z),z)$$

$$\overline{A_{XJ}} = \sum_{z\in Z} \sum_{x_J(z)\in X_J} \mu_{XJ}(x_J(z),z) \,/\, (x_J(z),z)$$

Then, the fuzzification result of J crisp inputs x_z can be represented by:

$$\overline{A_X} = \sum_{z\in Z} \sum_{x_1(z)\in X_1} \sum_{x_2(z)\in X_2} \cdots\cdots \sum_{x_J(z)\in X_J} \{\mu_{X1}(x_1(z),z) * .. * \mu_{XJ}(x_J(z),z)\} \,/\, \{(x_1(z),z) * .. * (x_J(z),z)\} \quad (9)$$

Where * denotes the triangular norm; t-norm (for short) is a binary operation. The t-norm operation is equivalent to logical AND. Also it has been assumed that the membership function $\mu_{\overline{A}_X}$ is separable .

Using the 3D fuzzy set, the γth rule in the rule based is expressed as follows:

$$\overline{R}^{\gamma}: if \quad x_1(z) is \quad \overline{C}_1^{\gamma} and.......and \quad x_J(z) is \quad C_J^{\gamma} then \quad u \quad is \quad G^{\gamma} \tag{10}$$

Where $\overline{R}^{\gamma}$ denotes the γth rule $\gamma = (1,2,....,N)$ $x_j(z),(j=1,2,...,J)$ denotes spatial input variable C_J^{γ} denotes 3D fuzzy set, u denotes the control action $u \in U \subset IR$, G^{γ} denotes a traditional fuzzy set N is the number of fuzzy rules, the inference engine of the 3D FC is expected to transform a 3D fuzzy input into a traditional fuzzy output. Thus, the inference engine has the ability to cope with spatial information. The 3D fuzzy DTM controller is designed to have three operations: spatial information fusion, dimension reduction, and traditional inference operation. The inference process is about the operation of 3D fuzzy set including union, intersection and complement operation. Considering the fuzzy rule expressed as (10), the rule presents a fuzzy relation $\overline{R}^{\gamma}: \overline{C}_1^{\gamma} \times .... \times C_J^{\gamma} \rightarrow G^{\gamma}$ $\gamma = (1,2,....,N)$ thus, a traditional fuzzy set is generated via combining the 3D fuzzy input and the fuzzy relation is represented by rules.
The spatial information fusion is this first operation in the inference to transform the 3D fuzzy input $\overline{A}x$ into a 3D set W^{γ} appearing as a 2D fuzzy spatial distribution at each input x_z. W^{γ} is defined by an extended sup-star composition on the input set and antecedent set. Fig.7B. gives a demonstration of spatial information fusion in the case of two crisp inputs from the space domain Z, $x_z = [x_1(z), x_2(z), ..., x_j(z)]$.

This spatial 3D MSF, is produced by the extended sup-star operation on two input sets from singleton fuzzification and two antecedent sets in a discrete space Z at each input value x_z. An extended sup-star composition employed on the input set and antecedent sets of the rule, is denoted by:

$$W^{\gamma}{}_{\overline{A}x^{o}(\overline{C}_1^{\gamma} \times ... \times \overline{C}_J^{\gamma})} = \overline{A}x^{o}(\overline{C}_1^{\gamma} \times ... \times \overline{C}_J^{\gamma}) \tag{11}$$

The grade of the 3D MSF derived as

$$\mu_{\overline{W^{\gamma}}}(z) = \mu_{\overline{A}X^{o}(\overline{C}_1^{\gamma} \times ... \times \overline{C}_J^{\gamma})}(x_z, z) \tag{12}$$

$\mu_{\overline{W^{\gamma}}}(z) = \sup_{x_1(z) \in X_1, ..., x_J(z) \in X_J} [\mu_{\overline{A}X}(x_z, z) * \mu_{\overline{C}_1^{\gamma}} \times ... \times \mu_{\overline{C}_J^{\gamma}}(x_z, z)]$ where $z \in Z$ and * denotes the t-norm operation.

$$\mu_{\overline{W^{\gamma}}}(z) = \sup_{x_1(z) \in X_1, ..., x_J(z) \in X_J} [\mu_{\overline{A}X1}(x_1(z), z) * ... \quad .. * \mu_{\overline{C}_1^{\gamma}}(x_1(z), z) * .... * \mu_{\overline{C}_J^{\gamma}}(x_J(z), z)]$$
$$......... * \mu_{\overline{A}XJ}(x_J(z), z) * \mu_{\overline{C}_1^{\gamma}}(x_1(z), z) * ...$$

$$\mu_{\overline{W^{\gamma}}}(z) = \{\sup_{x_1(z) \in X_1} [\mu_{\overline{A}X1}(x_1(z), z) \mu_{\overline{C}_1^{\gamma}}(x_1(z), z)]\} * ......$$
$$......... * \{\sup_{x_J(z) \in X_J} [\mu_{\overline{A}XJ}(x_J(z), z) \mu_{\overline{C}_J^{\gamma}}(x_J(z), z)]\}$$

The dimension reduction operation is to compress the spatial distribution information (x_z, μ, z) into 2D information (x_z, μ) as shown in Fig.7B. The set W^{γ} shows an approximate fuzzy spatial distribution for each input x_z in which contains the physical information. The 3D set W^γ is simply regarded as a 2D spatial MSF on the plane (μ,z) for each input x_z . Thus, the option to compress this 3D set W^γ into a 2D set φ^γ is approximately described as the overall impact of the spatial distribution with respect to the input x_z .The traditional inference operation is the last operation in the inference. Where implication and rules' combination are similar to those in the traditional inference engine.

$$\mu_{V}{}^{\gamma}(u) = \varphi^{\gamma} * \mu_{G^\gamma}(u), \quad u \in U \tag{13}$$

Where * stands for a t-norm, $\mu_{G^\gamma}(u)$ is the membership grade of the consequent set of the fired rule $\overline{R}^\gamma$. Finally, the inference engine combines all the fired rules (14) .Where V_γ the output is fuzzy set of the fired rule $\overline{R}^\gamma$, N' denotes the number of fired rules and V denotes the composite output fuzzy set.

$$V = \bigcup_{\gamma=1}^{N'} V_\gamma \tag{14}$$

The traditional defuzzification is used to produce a crisp output. The center of area (COA) is chosen as the defuzzifier due to its simple computation (Yager et al., 1994).

$$u = \frac{\sum_{\gamma=1}^{N'} C^\gamma \mu\varphi^\gamma}{\sum_{\gamma=1}^{N'} \mu\varphi^\gamma} \tag{15}$$

Where $C^\gamma \in U$ is the centroid of the consequent set of the fired rule $\overline{R}^\gamma$ $\gamma = (1,2,....,N')$ which represents the consequent set G^γ in (13), N' is the number of fire rules $N' \leq N$
For Multi-Core CPU system; each core is considered as heat source. The heat conduction Q path is inverse propositional to the distance between the heat sources (16). The nearest hotspot has the highest effect on core temperature increase. Also the far hotspot has the lowest effect on core temperature increase.

$$Q = \frac{\sigma \ A \ \Delta T}{d} \tag{16}$$

Where Q is the heat conducted, σ the thermal conductivity, A the cross-section area of heat path (constant value), ΔT the temperature difference at the hotspots locations, d the length of heat path (the distance between the heat sources). The 3D MSF gain G_{ij} is selected as the inverse the distance between 2 cores hotspots locations

$$MSF_{3D} = \sum MSF_{2D} G_{ij} \tag{17}$$

Where MSF_{2D} the 2D MSF, G_{ij} the correlation gains between core i and core j. G_{ij} is not a constant value as the hotspots locations are changing during the run time. The maximum gain = 1 in case of calculating the correlation gain locally G_{ii} .

The 3D FC is based on 32 variables as follow (Yager et al., 1994):
The inputs 3D fuzzy variable at step n for each core are: 8 frequency deviation variables calculate as per (3). The output: for each core, the output is the core operating frequency at step n+1. The relationships: at step n CPU throughput is proportional to cores operating frequency. The core operating frequency is also proportional to the power consumption. The maximum power consumption leads to the maximum temperature increase.
In order to compare between the 2D FC and the 3D FC responses, the same configuration are reused with the 3D FC. The same the control objectives. The same fuzzy inputs, the same Meta decisions rules, the same rule space , and the same input 2D MSF Normal distribution configurations. Also The output membership functions are tuned per DTM controller. In general we have four outputs MSF: Max - DVFS - TSC MSF - FS. Thus the only design different between the 2D FC and the 3D FC that the 3D FC DTM takes into consideration the surrounding core hotspot temperatures and their operating frequencies. Fig.8. shows the 3D fuzzy DTM controller implementation.
3D-Fuzzy Example:
The number of p sensors = 5; the sensors are located at $z_1, z_2, \ldots\ldots, z_5$ Two crisp input, $x \in X$ and z is a point $z \in Z$ in one-dimensional space . For $p = 5$ discrete measurement sensors located at $z_1, z_2, \ldots\ldots, z_5$, $x_z = [x_1(z), x_2(z)]$ is defined as J is two crisp spatial input variables in space domain $Z = \{z_1, z_2, \ldots\ldots, z_5\}$ where $x_j(z_i) \in X_j \subset IR (j = 1,2)$. The fuzzification for each crisp spatial input variable $x_j(z)$ is uniformly expressed as the 3D fuzzy inputs are $\overline{A}_{x1}$ and $\overline{A}_{x2}$ in the discrete form. As shown in Fig.7B; μ_1 values are the local substitutions of $x_1(z)$ in each 2D MSF at each z location. μ_2 values are the local substitutions of $x_2(z)$ in each 2D MSF at each z location. μ_{W^1} values are the sup-star composition of μ_1 and μ_2 at each z location as shown in Table 1. The sup-star composition in the fuzzy inference engine becomes a sup- minimum composition.

$x_1(z)$	$x_2(z)$	z	μ_1	μ_2	μ_{W^1}
- 0.5	- 0.6	0.0	0.8	0.4	0.4
0.0	0.2	0.5	0.8	0.9	0.8
0.3	0.1	0.25	0.9	1	0.9
0.7	0	0.75	0.6	0.7	0.6
0.2	-0.1	1	0.8	0.3	0.3

Table 1. 3D Fuzzy with Two crisp input example

7. Simulation results

Simulation is used for validating the designed 3D fuzzy DTM controller. The CPU chip selection is based on the on the amount of published information. The IBM POWER processor family is selected based on published information include floor plan, thermal design power (TDP), technology, chip area, and operating frequencies. IBM POWER4 MCM chip is selected chip. The floor plans of the POWER4 processor and the MCM are published

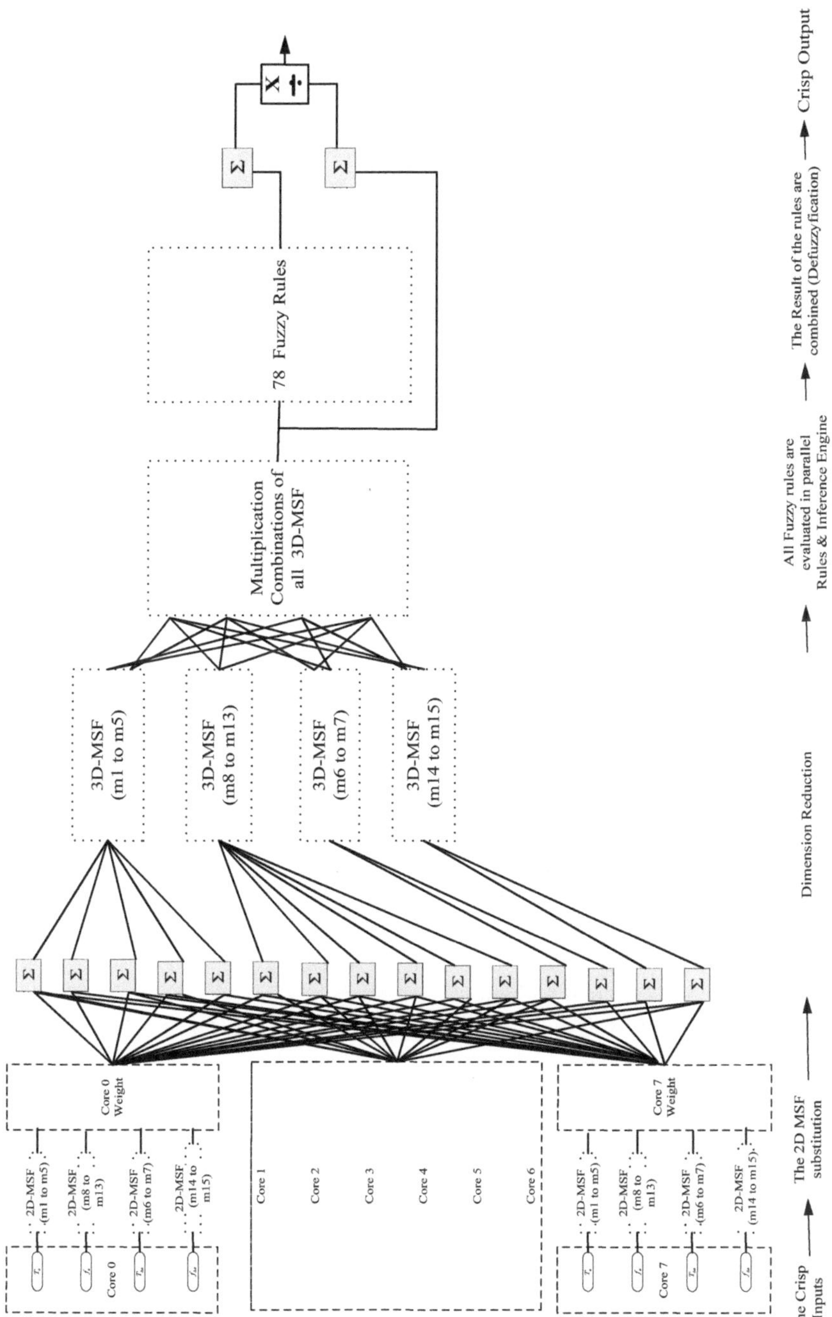

Fig. 8. 3D-Fuzzy controller block diagram

as pictures. The entire processor manufacturers consider the CPU floor plan and its power density map as confidential data. Thus there is major difficulty to build a thermal model based on real CPU chip information. Only old CPU chip thermal data is published. The MCM POWER4 floor plan and power density map are published. The only way to build up a CPU thermal model is the reverse engineering of IBM MCM POWER4 chip Fig.9. The reverse engineering process took a lot of time and efforts. The extracted MCM POWER4 chip is scaled into 45nm technology as POWER4 chip is built on the old 90nm technology (Sinharoy et al., 2005).

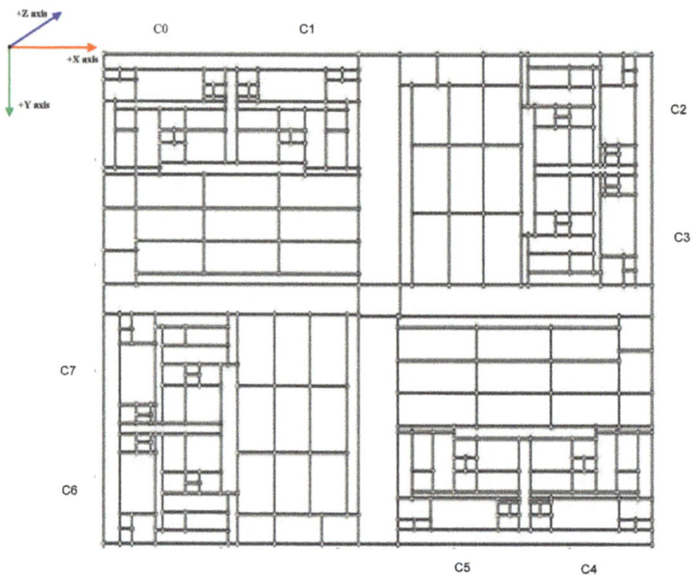

Fig. 9. The extracted IBM POWER4 MCM floor plan

Virginia Hotspot simulator is selected based on simulator features and on line support provided by Hotspot team at Virginia University. The Hotspot 5 simulator uses the duality between RC circuits and thermal systems to model heat transfer in silicon. The Hotspot 5 simulator uses a Runge-Kutta (4th order) numerical approximation to solve the differential equations that govern the thermal RC circuit's operation (LAVA , 2009).

7.1 Simulation analysis

All simulations starts from 814 seconds as the CPU thermal model required 814 seconds to reach $T_{Control}$ 70 °C. Assuming that the CPU output response follows the open loop curve until it reaches 70 °C. At $T_{Control}$, the DTM controller output selects the cores operating frequency. Then each core temperature changes according to its operating frequency. All DTM fuzzy designs tuning are based on their output membership functions (MSF) tuning without changing the fuzzy rules. The DTM evaluation index covers the simulation times between 814 seconds to 1014 seconds. Theses simulation tests 3D-FC1, FC1, 3D-FC2, FC2, 3D-FC3 and FC3 perform both DVFS and TSC together. But these tests FC4, 3D-FC4, 3D-

FC5, and 3D-FC6 perform DVFS only. The DTM controller evaluation index (4) has only two parameters $l = 2$, the frequency and the temperature. Its desired value is $\zeta_t = 2$ or near 2. There are two DTM evaluation index implementations presented in this section. The first DTM implementation assumed that the CPU is required to run 20% of its time at the maximum frequency, 50% of its time at high frequency, 20% of its time at medium frequency and 10% of it is time at low frequency. Also the CPU is required to 30% of its time at high temperature, 40% at medium temperature, and 30% of its time at low temperature. This first DTM requirement evaluation against the DTM controller designs are as follow:
Table 2 shows the percentage of time when the CPU operates at each frequency ranges. Table 3 shows the percentage of time of the CPU operates at each temperature ranges. The best results are highlighted in bold. The DTM evaluation index selected FC3 and 3D-FC6 as the best DTM controller designs as shown in Table 4. The best results are highlighted in bold. Only FC3 and 3D-FC6 controllers have high results in both frequency, and temperature evaluation indexes. As shown in Fig.10A, both DTM controllers' frequency change responses oscillate all times. The 3D-FC6 controller has less number of frequency oscillation and smaller amplitudes. The FC3 controller operates at maximum frequency then it is switched off between 1014 and 1100 seconds. The 3D-FC6 controller is never switched off and operates at high frequency ranges but not on the maximum frequency. From the temperature point of view; both controllers temperatures are oscillating. 3D-FC6 controller has minimum temperature amplitudes at 970 and 1070 seconds as shown in Fig.10B. The 3D-FC6 is always operating on lower temperature than the FC3 controller. Thus the 3D-FC6 controller is better then the FC3 controller. As shown in Table 5, Table 6, Table 7; only FC4, 3D-FC3 and 3D-FC6 controllers have high results in both frequency, and temperature evaluation indexes. As shown in Fig.10 A,C,E, all DTM controllers' frequency change responses oscillate all times. The 3D-FC6 controller has the lowest number of frequency oscillation. The 3D-FC3 controller has smallest frequency changes amplitudes. The 3D-FC3 controller operates at high frequency ranges but not on the maximum frequency. From the temperature point of view; all controller temperature are increasing as shown in Fig.10 B,D,F. The 3D-FC6 temperature is oscillating and has minimum temperature amplitudes at 970 and 1070 seconds. There is no large advantage of any controllers over the others from temperature point of view. Thus the 3D-FC3 is better then the FC4 controller, and the 3D-FC6 controller as the 3D-FC3 controller operates at higher frequency ranges and almost the same temperature ranges.
Some observations are extracted from these two DTM evaluation index implementations as follow: 3D-FC5 vs. 3D-FC6: In the first implementation the DTM evaluation index of both controllers are almost the same from the frequency point of view. The standard deviation of the DVFS membership function (MSF) is the same but the mean is shifted by 0.2. This shift leads to insignificant frequency objective change but also leads to less CPU temperature. In the second implementation the DTM evaluation index values are totally different. So the similarity between any 2 DTM controller responses for a specific DTM design objective is not maintain for other DTM design objective. 2D Fuzzy vs. 3D Fuzzy: These DTM controllers share the same input and output membership functions. The correlation between the CPU cores has significant effect i.e. (FC1 vs. 3D-FC1) and (FC3 vs. 3D-FC3). But for (FC2 vs. 3D-FC2) there is almost no correlation effect in both DTM evaluation index implementations. This means that the selection of non proper membership functions could ignore the correlation effect between the CPU cores. (TSC+DVFS) vs. (DVFS alone): the

DTM temperature design objectives could be fulfilled by TSC+DVFS or by DVFS alone i.e. 3D-FC3 vs. 3D-FC4. The driver for using TSC with DVFS is the CPU thermal throttling limits. So if DVFS can fulfil alone the temperature DTM design objective then there is no need for combining both TSC with DVFS.

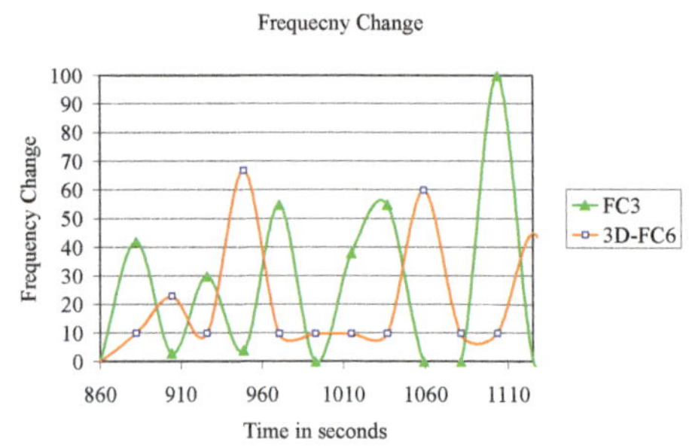

A - frequency comparisons of FC3 and 3D-FC6

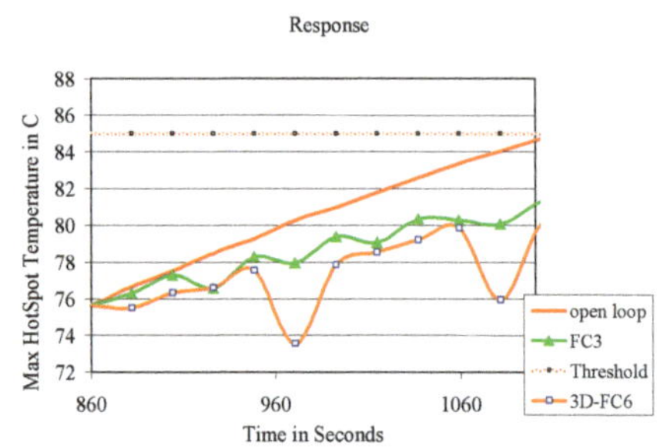

B- temperature comparisons of FC3 and 3D-FC6

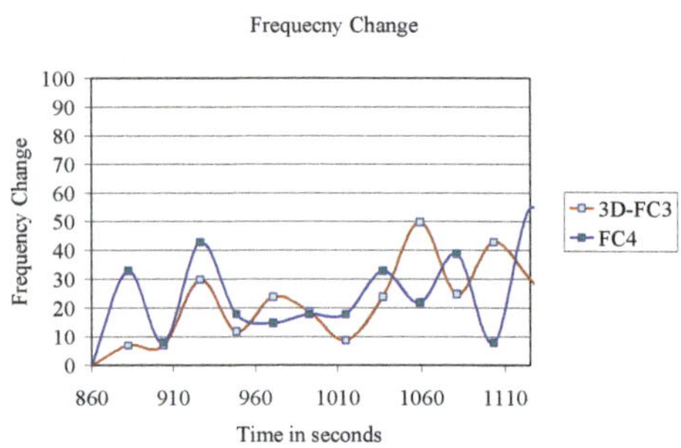

C- frequency comparisons of FC4 and 3D-FC3

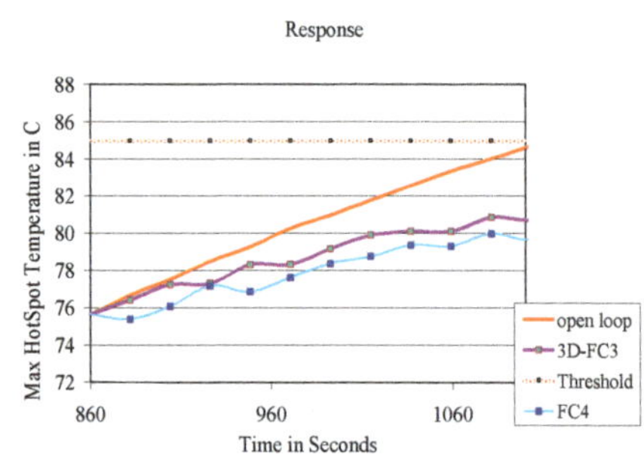

D- temperature comparisons of FC4 and 3D-FC3

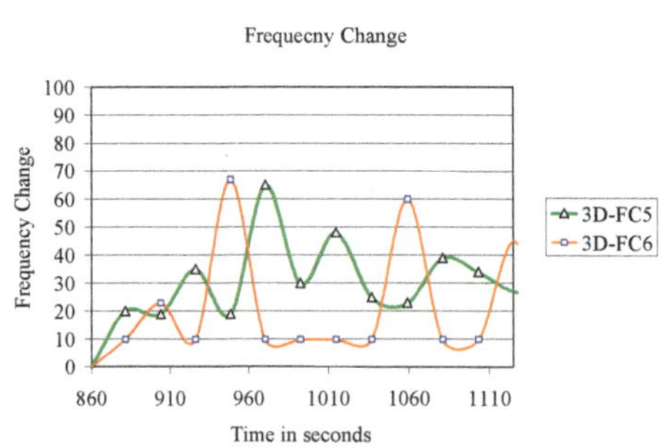

E -frequency comparisons of 3-FC5 and 3D-FC6

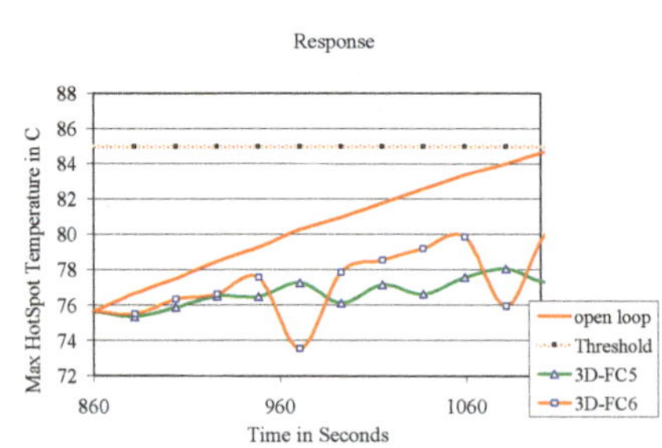

F- temperature comparisons of 3D-FC5 and 3D-FC6

Fig. 10. The Simulation Results

Controller Name	Frequency Ranges % σ_{1j}^{Actual}				Frequency Ranges Values σ_{1j}				λ_1
	(M) j=1	(H) j=2	(m) j=3	(L) j=4	(M) j=1	(H) j=2	(m) j=3	(L) j=4	
$\sigma_{1j}^{Desired}$	20%	50%	20%	10%	1.0	1.0	1.0	1.0	1.00
Switch	0%	100%	0%	0%	0	2	0%	0	0.500
P	10%	0%	22%	22%	2.7	0.0	1	2	1.528
FC1	12%	22%	44%	22%	0.5	0.4	2	2.2	1.315
3D-FC1	0%	10%	33%	11%	0.0	1.1	1.7	1.1	0.972
FC2	0%	100%	0%	0%	0.0	2.0	0.0	0.0	0.500
3D-FC2	0%	89%	11%	0%	0.0	1.8	0.6	0.0	0.123
FC3	22%	22%	10%	0%	1.1	0.4	2.8	0.0	1.083
3D-FC3	0%	78%	22%	0%	0.0	1.6	1.1	0.0	0.667
FC4	0%	66%	33%	0%	0.0	1.3	1.7	0.0	0.750
3D-FC4	22%	10%	22%	0%	1.1	1.1	1.1	0.0	0.833
3D-FC5	0%	10%	33%	11%	0.0	1.1	1.7	1.1	0.972
3D-FC6	0%	78%	0%	22%	0.0	1.6	0.0	2.2	0.944

Table 2. The frequency comparisons of the first implementation

Controller Name	Temperature Ranges % σ_{2j}^{Actual}			Temperature Ranges Values σ_{2j}			λ_2
	(H) j=1	(m) j=2	(L) j=3	(H) j=1	(m) j=2	(L) j=3	
$\sigma_{2j}^{Desired}$	30%	40%	30%	1.0	1.0	1.0	1.00
Switch	0.0%	100%	0.0%	0.0	2.5	0.0	0.83
P	78%	0%	22%	2.6	0.0	0.7	1.11
FC1	11%	89%	0%	0.4	2.2	0.0	0.86
3D-FC1	22%	78%	0%	0.7	1.9	0.0	0.90
FC2	67%	33%	0%	2.2	0.8	0.0	1.02
3D-FC2	10%	44%	0%	1.8	1.1	0.0	0.99
FC3	67%	33%	0%	2.2	0.8	0.0	1.02
3D-FC3	33%	67%	0%	1.1	1.7	0.0	0.93
FC4	44%	10%	0%	1.5	1.4	0.0	0.96
3D-FC4	33%	67%	0%	1.1	1.7	0.0	0.93
3D-FC5	0%	100%	0%	0.0	2.5	0.0	0.83
3D-FC6	33%	10%	11%	1.1	1.4	0.4	0.96

Table 3. The temperature comparisons of the first implementation

Controller Name	Frequency Index λ_1	Temperature Index λ_2	The Evaluation Index ζ_t
Desired	1.00	1.00	2.00
Switch	0.500	0.83	1.33
P	1.528	1.11	2.64
FC1	1.315	0.86	2.23
3D-FC1	0.972	0.90	1.87
FC2	0.500	1.02	1.52
3D-FC2	0.123	0.99	1.11
FC3	1.083	1.02	2.10
3D-FC3	0.667	0.93	1.13
FC4	0.750	0.96	1.71
3D-FC4	0.833	0.93	1.76
3D-FC5	0.972	0.83	1.81
3D-FC6	0.944	0.96	1.90

Table 4. The DTM evaluation index of the first implementation

Controller Name	Frequency Ranges % σ_{1j}^{Actual}				Frequency Ranges Values σ_{1j}				λ_1
	(M) j=1	(H) j=2	(m) j=3	(L) j=4	(M) j=1	(H) j=2	(m) j=3	(L) j=4	
$\sigma_{1j}^{Desired}$	10%	70%	10%	10%	1.0	1.0	1.0	1.0	1.00
Switch	0%	100%	0%	0%	0.0	1.4	0.0	0.0	0.311
P	10%	0%	22%	22%	5.6	0.0	2.2	2.2	2.500
FC1	12	22%	44%	22%	1.1	0.3	4.4	2.2	2.024
3D-FC1	0%	10%	33%	11%	0.0	0.8	3.3	1.1	1.309
FC2	0%	100%	0%	0%	0.0	1.4	0.0	0.0	0.311
3D-FC2	0%	89%	11%	0%	0.0	1.3	1.1	0.0	0.135
FC3	22%	22%	10%	0%	2.2	0.3	5.6	0.0	2.024
3D-FC3	0%	78%	22%	0%	0.0	1.1	2.2	0.0	0.833
FC4	0%	67%	33%	0%	0.0	0.9	3.3	0.0	1.071
3D-FC4	22%	10%	22%	0%	2.2	0.8	2.2	0.0	1.309
3D-FC5	0%	10%	33%	11%	0.0	0.8	3.3	1.1	1.309
3D-FC6	0%	78%	0%	22%	0.0	1.1	0.0	2.2	0.833

Table 5. The frequency comparisons of the second implementation

Controller Name	Temperature Ranges % σ_{2j}^{Actual}			Temperature Ranges Values σ_{2j}			λ_2
	(H) j=1	(m) j=2	(L) j=3	(H) j=1	(m) j=2	(L) j=3	
$\sigma_{2j}^{Desired}$	30%	40%	30%	1.0	1.0	1.0	1.00
Switch	0%	100%	0%	0.0	2.0	0.0	0.67
P	78%	0%	22%	3.9	0.0	0.7	1.54
FC1	111%	89%	0%	0.6	1.8	0.0	0.78
3D-FC1	22%	78%	0%	1.1	1.6	0.0	0.89
FC2	67%	33%	0%	3.3	0.7	0.0	1.33
3D-FC2	10%	44%	0%	2.8	0.9	0.0	1.22
FC3	67%	33%	0%	3.3	0.7	0.0	1.33
3D-FC3	33%	67%	0%	1.7	1.3	0.0	1.00
FC4	44%	10%	0%	2.2	1.1	0.0	1.11
3D-FC4	33%	67%	0%	1.7	1.3	0.0	1.00
3D-FC5	0%	100%	0%	0.0	2.0	0.0	0.67
3D-FC6	33%	10%	11%	1.7	1.1	0.4	1.05

Table 6. The temperature comparisons of the second implementation

Controller Name	Frequency Index λ_1	Temperature Index λ_2	The Evaluation Index ζ_t
Desired	1.00	1.00	2.00
Switch	0.311	0.67	1.02
P	2.500	1.54	4.04
FC1	2.024	0.78	2.80
3D-FC1	1.309	0.89	2.20
FC2	0.311	1.33	1.69
3D-FC2	0.135	1.22	1.82
FC3	2.024	1.33	3.36
3D-FC3	0.833	1.00	1.83
FC4	1.071	1.11	2.18
3D-FC4	1.309	1.00	2.31
3D-FC5	1.309	0.67	1.98
3D-FC6	0.833	1.05	1.88

Table 7. The DTM evaluation index of the second implementation

8. Conclusion

Moore's Law continues with technology scaling, improving transistor performance to increase frequency, increasing transistor integration capacity to realize complex architectures, and reducing energy consumed per logic operation to keep power dissipation within limit. The technology provides integration capacity of billions of transistors; however, with several fundamental barriers. The power consumption, the energy level, energy delay, power density, and floor planning are design challenges. The Multi-Core CPU design increases the CPU performance and maintains the power dissipation level for the same chip area. The CPU cores are not fully utilized if parallelism doesn't exist. Low cost portable cooling techniques exploration has more importance everyday as air cooling reaches its limits "198 Watt". In order to study the Multi-Core CPU thermal problem a thermal model is built. The thermal model floor plan is similar to the IBM MCM POWER4 chip scaled to 45nm technology. This floor plan is integrated to the Hotspot 5 thermal simulator. The CPU open loop thermal profile curve is extracted. The advanced dynamic thermal management (DTM) techniques are mandatory to avoid the CPU thermal throttling. As the CPU is not 100% utilized all time, the thermal spare cores (TSC) technique is proposed. The TSC technique is based on the reservation of cores during low CPU utilization. These cores are not activate simultaneously due to limitations. During thermal crises, these reserved cores are activated to enhance the CPU utilization. The semiconductor technology permits more cores to be added to CPU chip. But the total chip area overhead is up to 27.9 % as per ITRS (ITRS , 2009). That means there is no chip area wasting in case of TSC. From the thermal point of view; the horizontal heat transfer path has up to 30% of CPU chip heat transfer (Stan et al., 2006). The TSC is a big coldspot within the CPU area that handles the horizontal heat transfer path.

The cold TSC also handles the static power as the TSC core is turned off. The TSC is used simultaneous with other DTM technique. From the CPU utilization point of view, the TSC activation is equivalent to the CPU cores DVFS for a low operating frequency range. Fuzzy logic improves the DTM controller response. Fuzzy control handles the CPU thermal process without knowing its transfer function. This simplifies the DTM controller design and reduces design time. The fuzzy control permits the designers to select the appropriate CPU temperature and frequency responses. For the same CPU chip, the DTM response depends on the DTM fuzzy controller design. As the 3D fuzzy permits the preservation of portable device battery but this affects the CPU utilization. Or it permits the high performance computing (HPC). But due to cooling limitation this DTM design is not suitable for the portable devices. The 3D-FC is successfully implemented to the CPU DTM problem. Different DTM techniques are compared using simulation tests. The results demonstrate the effectiveness of the 3D fuzzy DTM controller to the nonlinear Multi-Core CPU thermal problem. The 3D fuzzy DTM takes into consideration the surrounding core hotspot temperatures and operating frequencies. The 3D fuzzy DTM avoids the complexity and maintains the correlations. As the 3D fuzzy DTM controller calculates the correlation between local core hotspot and the surrounding cores hotspots. Then it selects the appropriate local core operating frequency. The Fuzzy DTM controller has better response than the traditional DTM P controller. For the same input rules and the same output membership functions (MSF), the 3D fuzzy logic reduces the CPU temperature better than the 2D fuzzy logic. The fuzzy output MSF is a critical DTM design parameter. The small deviation from the appropriate output membership function affects the DTM controller behavior.

The Fuzzy DTM controller has better response than the traditional DTM P controller. For the same input rules and the same output membership functions (MSF), the 3D Fuzzy logic reduces the CPU temperature better than the 2D Fuzzy logic. The 3D Fuzzy controller takes into consideration multiple temperatures readings distributed over the CPU chip floor plan. The Fuzzy control permits the designers to select the appropriate CPU temperature and frequency responses. For the same CPU chip, the DTM response depends on the Fuzzy controller design. The fuzzy output MSF is a critical DTM design parameter. The small deviation from the appropriate output membership function affects the DTM controller behavior. From the CPU temperature point of view; the TSC looks like a large coldspot. The cold TSC absorb the horizontal heat path as if it is a heatsink pipe. The CPU cooling system behavior depends on the combinations of the operating frequencies and temperatures. The objective of multi-parameters evaluation index is to show the different parameters effect on the CPU response. Thus the designer selects the suitable DTM controller that fulfils his requirements. The multi-parameters evaluation index permits the selection of DTM design that provides the best frequency parameter value without leading to the worst temperature parameter value.

9. References

Chaparro, P. ; Lez, J. G. Cai, Q. & Lez, A. G. (2007). Understanding The Thermal Implications of Multicore Architectures, *IEEE Transactions*, Vol.18, No.8, pp. 109-1065.

Chung, S. W. ; & Skadron, K. (2006). Using on-chip event counters for high-resolution, real-time temperature measurements, Proceedings of *International Conference For Scientific & Engineering Exploration Of Thermal*, Thermomechanical & Emerging Technology, IEEE ITHERM06, pp. 114-120.

Donald, J. ; & Martonosi, M. (2006). Techniques For Multicore Thermal Management Classification & New Exploration, Proceedings of *International Symposium on Computer Architecture*, IEEE ISCA'06, pp. 78-88.

Doumanidis, C. C.; & Fourligkas, N. (2001). Temperature Distribution Control In Scanned Thermal Processing Of Thin Circular Parts, *IEEE Transaction Control System Technolgy*, Vol.9, No.5, (May 2001), pp. 708–717.

Ferreira, A. P.; Moss,D. & Oh, J. C. (2007). Thermal Faults Modeling using an RC model with an Application to Web Farms, Proceedings of *19th Euromicro Conference on Real-Time Systems*,Italy, pp. 113-124.

Huangy, W. ; Stany, M. R. Skadronz,K. Sankaranarayananz, K. Ghoshyz, S. & VelUSAmyz, S (2006). Hotspot: A Compact Thermal Modeling Methodology For Early-Stage Vlsi Design, *IEEE Transactions*, 2006, Vol.5, pp. 501-513.

Gustafson, J. L.(1988). Re-Evaluating Amdahl's Law, ACM Communications, Vol.31, No.5, pp. 82-83.

Kim, D. D.; J. Kim, Cho, C. Plouchart, J.O. & Trzcinski, R. (2008). 65nm SOI CMOS SoC Technology for Low-Power mmWave & RF Platform, *Silicon Monolithic Integrated Circuits in RF Systems*, pp. 46-49.

Kim, S. ; Dick, R. P. & Joseph, R. (2007). Power Deregulation: Eliminating Off-Chip Voltage Regulation Circuitry From Embedded Systems, Proceedings of *the International Conference on Hardware-Software Codesign & System Synthesis*, IEEE/ACM (CODES+ISSS), pp. 105-110.

Li, H. Zhang; X. & Li, S. (2007). A Three-Dimensional Fuzzy Control Methodology For A Class Of Distributed Parameter Systems, IEEE Transactions, *Fuzzy Systems*, Vol.15, No.3, pp. 470-481.

Mccrorie, P. (2008). On-Chip Thermal Analysis Is Becoming M&atory, *Chip Design Magazine*.

Moore, G. E. (1965). Cramming More Components Onto Integrated Circuits, *IEEE Electronics*,Vol.38, No.8, (19 April 1965), pp.114. This Paper Appears Again In *IEEE Solid-State Circuits Newsletter*, 2006, Vol.20, No.3, pp. 33-35.

Ogras, U.Y. et al. (2008). Variation-Adaptive Feedback Control for Networks-on-Chip with Multiple Clock Domains, Proceedings of *International Conference on Design Automation Conference*, IEEE DAC08, pp. 154-159.

Passino, K. M.; & Yurkovich, S. (1998). Fuzzy Control, Addison Wesley Longman.

Patyra, M. J.; Grantner, J.L. & Koster, K. (1996). Digital Fuzzy Logic Controller Design & Implementation, *IEEE Transactions Fuzzy Systems*, Vol.4, No.4, pp. 439-413.

Rao, R. ; & Vrudhula, S. (2007). Performance Optimal Processor Throttling Under Thermal Constraints, Proceedings of *International Conference On Compilers, Architecture, & Synthesis For Embedded Systems, CASES'07*, pp. 211-266.

Sinharoy, B.; Kalla, R. N. Tendler, J. M. & Eickemeyer,R. J. (2005). *POWER5 System Microarchitecture, IBM J. Res. & Dev*. Vol.49 No. 4/5 July/September 2005.

Stan, M. R. ; Skadron, K. Barcella, M. Sankaranarayanan, W. H. K. & Velusamy, S. (2006). Hotspot: A Compact Thermal Modeling Methodology For Early-Stage VLSI Design, *IEEE Transactions*, Vol.14, No.5, pp. 501-513.

Trabelsi, A. ; Lafont, F. Kamoun, M. & Enea, G. (2004). Identification of Nonlinear Multivariable Systems By Adaptive Fuzzy Takagi-Sugeno Model, *International Journal of Computational Cognition*, Vol.2, No.3, pp. 137-18.

Wu, Q. et al. (2004). Formal online methods for voltage/frequency control in multiple clock domain microprocessors, Proceedings of *International Conference on Architectural Support for Programming Languages and Operating Systems*, ASPLOS, Vol.32, No.5, pp. 248-213.

Yager, R. ; & Filev, D. (1994). *Essential Of Fuzzy Modeling & Control*, Wiley, New York 1994, pp. 121.

http://lava.cs.virginia.edu/hotspot

http://www.itrs.net

16

Design of Electronic Equipment Casings for Natural Air Cooling: Effects of Height and Size of Outlet Vent on Flow Resistance

Masaru Ishizuka and Tomoyuki Hatakeyama
Toyama Prefectural University
Japan

1. Introduction

As the power dissipation density of electronic equipment has continued to increase, it has become necessary to consider the cooling design of electronic equipment in order to develop suitable cooling techniques. Almost all electronic equipment is cooled by air convection. Of the various cooling systems available, natural air cooling is often used for applications for which high reliability is essential, such as telecommunications. The main advantage of natural convection is that no fan or blower is required, because air movement is generated by density differences in the presence of gravity. The optimum thermal design of electronic devices cooled by natural convection depends on an accurate choice of geometrical configuration and the best distribution of heat sources to promote the flow rate that minimizes temperature rises inside the casings. Although the literature covers natural convection heat transfer in simple geometries, few experiments relate to enclosures such as those used in electronic equipment, in which heat transfer and fluid flow are generally complicated and three dimensional, making experimental modeling necessary. Guglielmini et al. (1988) reported on the natural air cooling of electronic boards in ventilated enclosures. Misale (1993) reported the influence of vent geometry on the natural air cooling of vertical circuit boards packed within a ventilated enclosure. Lin and Armfield (2001) studied natural convection cooling of rectangular and cylindrical containers. Ishizuka et al. (1986) and Ishizuka (1998) presented a simplified set of equations derived from data on natural air cooling of electronic equipment casings and showed its validity. However, there is insufficient information regading thermal design of practical electronic equipment. For example, the simplified set of equations was based on a ventilation model like a chimney with a heater at the base and an outlet vent on the top, yet in practical electronic equipment, the outlet vent is located at the upper part of the side walls, and the duct is not circular. Therefore, here, we studied the effect of the distance between the outlet vent location and the heat source on the cooling capability of natural-air-cooled electronic equipment casings.

2. Set of equations

Ishizuka et al. (1986) proposed the following set of equations for engineering applications in the thermal design of electronic equipment:

$$Q = 1.78 S_{eq.}\ \Delta T_m^{\ 1.25} + 300 A_o (h / K)^{0.5}\ \Delta T_o^{\ 1.5} \tag{1}$$

$$K = 2.5(1 - \beta) / \beta^2 \tag{2}$$

$$\Delta T_o = 1.3\, \Delta T_m \tag{3}$$

$$S_{eq.} = S_{top} + S_{side} + 1/2\, S_{bottom} \tag{4}$$

where Q denotes the total heat generated by the components, S_{eq} is the equivalent total surface of the casing, ΔT_m is the average temperature rise in the casing, A_o is the outlet vent area of the casing, h is the distance from the heater position to the outlet, K is the flow resistance coefficient arising from air path interruption at the outlet, ΔT_o is the air temperature rise at the outlet vent, and β is the porosity coefficient of the outlet vent. K was approximated as a function of β: Ishizuka et al. (1986, 1987) obtained the following relation for wire nets at low values of the Reynolds number (Re):

$$K = 40(Re(1 - \beta)/\beta^2)^{-0.95} \tag{5}$$

where Re is defined on the basis of the wire diameter used in the wire nets. However, since the effect of Re on K is less prominent than that of β, K can be approximated as a function of β only. Therefore, Eq. (2) is considered to be a reasonable expression for practical applications. The h term in Eq. (1) refers to chimney height, as Eq. (1) assumes a ventilation model like a chimney with a heater at the base and an outlet vent on the top (Fig. 1). However, as practical electronic equipment is nothing like a chimney, we took practical details into account in the following experiments.

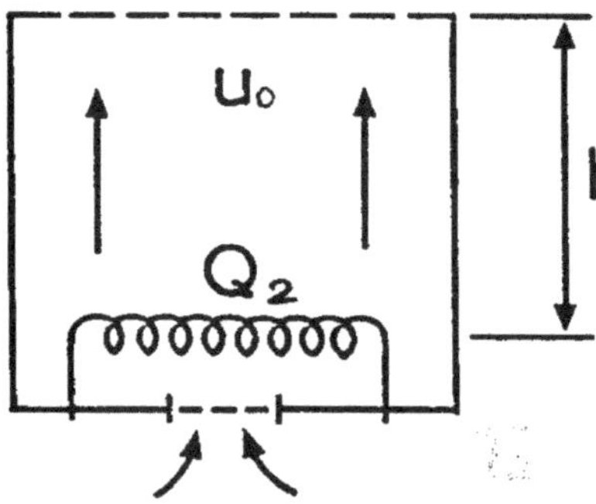

Fig. 1. Ventilation model

3. Experiments

3.1 Experimental apparatus

The experimental casing measured 220 mm long × 230 mm wide × 310 mm high (Fig. 2). The plastic casing material was 10 mm thick and had a thermal conductivity of 0.01 W/(m·K). A wire heater (2.3-mm gauge) was placed inside (Fig. 3). The size, location, and grille patterning of a rectangular vent on one of the side walls were varied (Fig. 4). Experiments

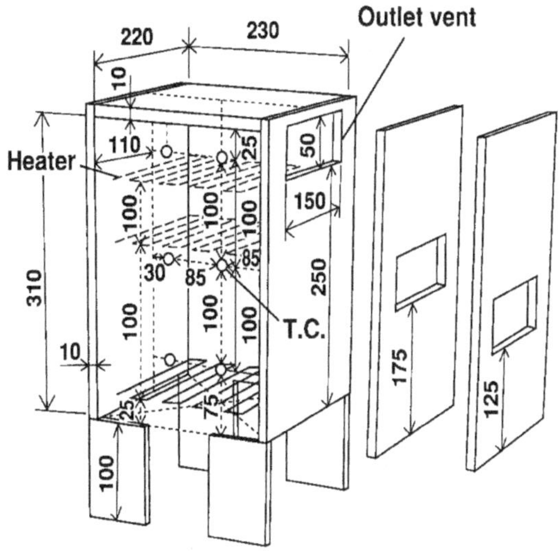

Fig. 2. Experimental casing

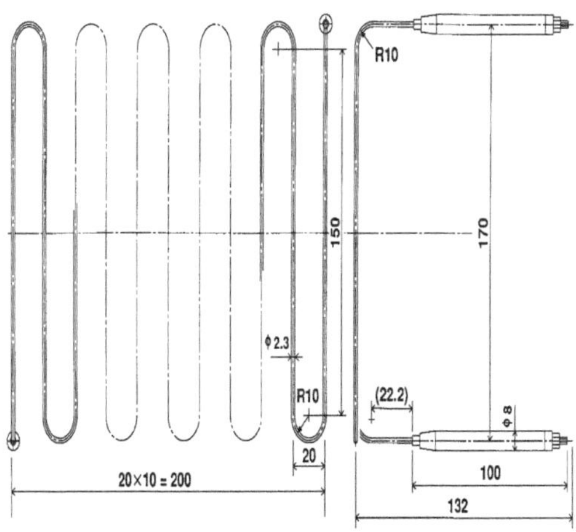

Fig. 3. Heater construction

were performed for three distances between the wall base and the center of the outlet vent (H_v = 275, 200, and 150 mm) and for three heights of the heater above the base (H_h = 25, 125, and 225 mm), with the heater always placed below the outlet vent. The cooling air entered through an opening in the center of the casing base (150 mm × 130 mm) and was exhausted through the upper outlet. The air temperature distribution inside the casing, the room temperature, and the wall temperatures were measured by calibrated K-type thermocouples (±0.1 K, 0.1 mm in diameter). The thermocouples inside the casing were arranged 30, 85, and 115 mm from the inside of the left side wall at each of 75, 175, and 275 mm above the base (Fig. 2). On the walls, thermocouples were placed at the center of the top surface and at three locations down the center line of each side wall, and one thermocouple was placed outside the casing to measure the room temperature. The mean inner air temperature rise, ΔT_m, was calculated from the locally measured values over the whole volume of the casing.

B (mm)	H (mm)	βo
150	50	1.0
100	50	0.67
150	30	0.6
100	30	0.4
50	50	0.33
50	30	0.2

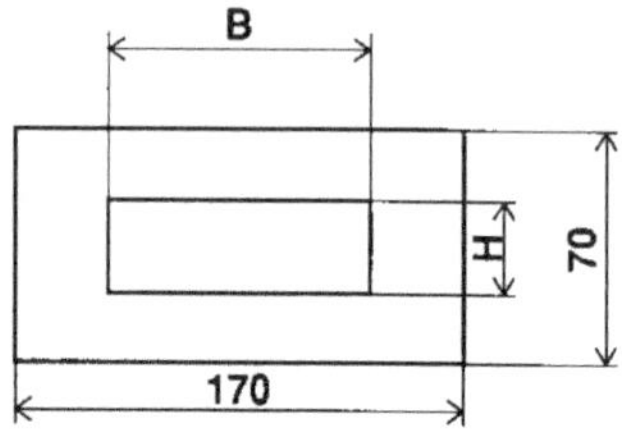

(a) Opening pattern

W (mm)	P (mm)	βo
7	5	0.7
6	4	0.667
4	2	0.49

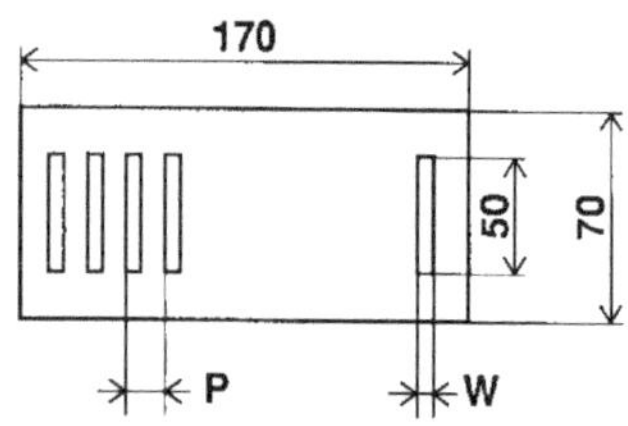

(b) Grille pattern

d (mm)	Mesh No.	βo
0.376	35	0.34
0.254	50	0.25
0.122	100	0.13

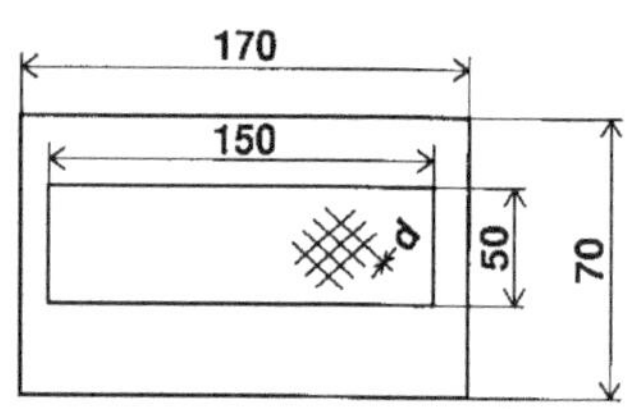

βo = Open area / 150×50

(c) Mesh pattern

Fig. 4. Outlet vent openings

3.2 Estimation of heat removed from casing surfaces

The amount of heat removed from the casing surfaces was estimated by experiment. For this purpose, the casing was considered to be a closed unit with no vents except at the base. Using the natural convective heat transfer equations for individual surfaces presented by Ishizuka et al. (1986), we expressed the amount of heat removed from the casing surfaces, Q_s, as:

$$Q_s = D_1 \Delta T_m^{1.25} \tag{6}$$

Eq. (6) shows the first term on the right-hand side of Eq. (1). Where, D_1 is the constant coefficient. The coefficient D_1 includes a radiative heat transfer factor and determined by the experiment. The amount of heat generation was within the Q_s = 6–40 W range (Fig. 5) when the room temperature T_a was 298 K. The results are shown in Fig.5 and are well expressed by Eq.(7).

$$Q_s = 0.445 \Delta T_m^{\ 1.25} \tag{7}$$

where D_1 was determined to be 0.445. The temperature distribution in the casing was relatively uniform, within ±5%. Hereafter, the amount of heat removed from the outlet vent, Q_v, was calculated as:

$$Q_v = Q - Q_s \tag{8}$$

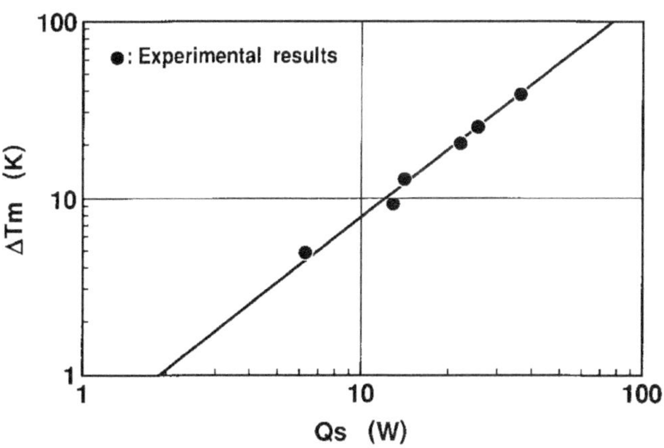

Fig. 5. Relationship between Q_s and ΔT_m

3.3 Influence of outlet vent size on temperature rise in the casing

The experiment was carried out by varying the outlet vent size of the reference casing at outlet vent position H_v = 275 mm and heater position at H_h = 25 mm (Fig. 2). The porosity coefficient β_o (Fig. 4) was defined as the ratio of the open area of each individual vent to the area of the reference vent (150 mm × 50 mm).

At each value of input power, Q, as β_o decreased, ΔT_m increased linearly on the logarithmic plot (Fig. 6). As Q increased, ΔT_m also increased.

3.4 Influence of outlet vent position on mean temperature rise in the casing

The relationship between ΔT_m and outlet vent position H was investigated at two opening sizes with the heater at the bottom. ΔT_m decreased as H increased at both opening sizes (Fig. 7). It decreased faster at lower H.

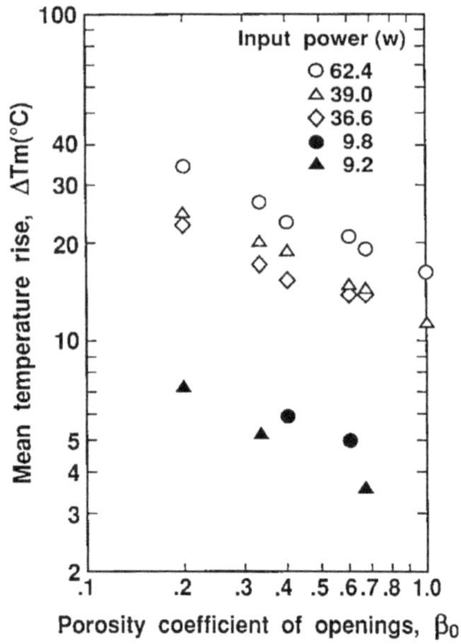

Fig. 6. Influence of vent porosity and input power on mean temperature rise in the casing

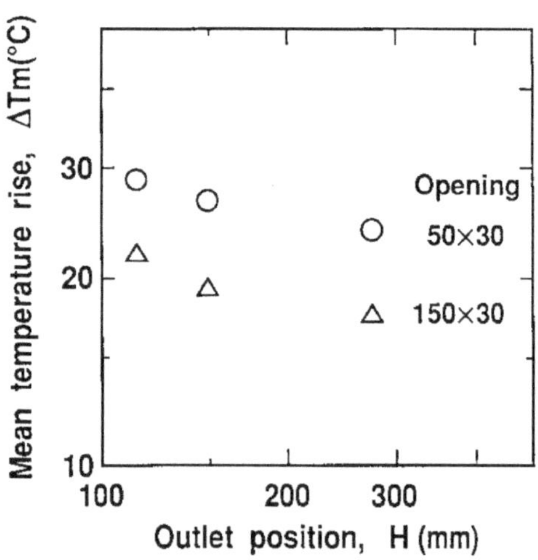

Fig. 7. Influence of outlet vent position on mean temperature rise in the casing

3.5 Influence of distance between outlet vent position and heater position on mean temperature rise in the casing

The relationship between ΔT_m (average of temperatures measured only above the heater position) and the distance between the outlet vent position and the heater position, h, was investigated by varying opening size and input power while the outlet height was fixed at H_v = 275 mm. As h increased, ΔT_m decreased at all values of input power (Fig. 8).

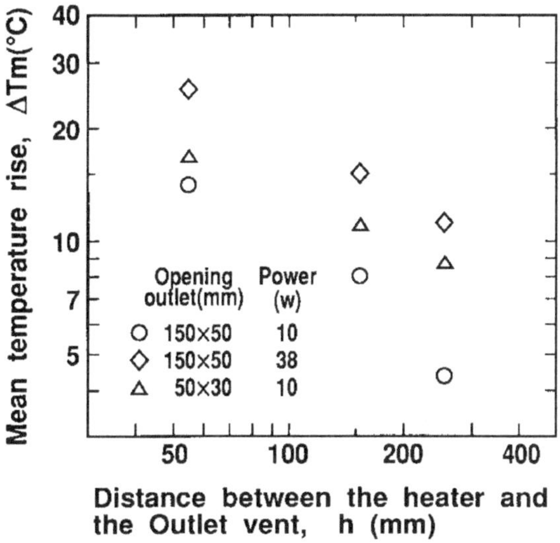

Fig. 8. Influence of distance between outlet vent position and heater position on mean temperature rise in the casing

4. Correlations using non-dimensional parameters

4.1 Flow resistance coefficient *K*

The flow resistance coefficient K was related to Q_v and h. If we assume a uniform temperature distribution and a one-dimensional steady-state flow in a ventilation model as shown in Fig. 1, we can express Eq. (9) for the overall energy balance and Eq. (10) for the balance between flow resistance and buoyancy force:

$$Q_v = \rho c_p A u \Delta T \tag{9}$$

$$(\rho_a - \rho) g h = K \rho u^2 / 2 \tag{10}$$

where Q_v is dissipated power, c_p is specific heat of the air at constant pressure, A is the cross-sectional area of the duct, u is airflow velocity, ΔT is temperature rise, ρ is air density (ρ_a is atmospheric condition), g is acceleration due to gravity, h is the distance between the outlet and the heater, and K is the flow resistance coefficient for the system. Since the pressure change in the system is small, the expression can be rewritten to assume a perfect gas:

$$(\rho_a - \rho) / \rho = (T - T_a) / T_a \tag{11}$$

K is defined in terms of hand Q_v as:

$$K = 2 g h \, \Delta T^3 / (T_a (\rho c_p A / Q_v)^2) \tag{12}$$

In this study, H value was used in spite of h to arrange the present data.

4.2 Porosity coefficient

Generally, β_o is defined as the ratio of the area of the opening to the reference opening (Fig. 4). Here, as the top surface area is an ideal outlet vent area for a casing cooled by natural convection, we defined the porosity coefficient β as:

$$\beta = \text{open area in outlet vent / inner casing top surface area} \tag{13}$$

4.3 Reynolds number Re

The velocity u was obtained using Eq. (9) and the hydrodynamic equivalent diameter of an opening with height A and width B was used as a reference length:

$$L=4\,AB\,/\,2(A+B) \tag{14}$$

Thus, Re is defined as:

$$Re = u\,L/\,\nu \tag{15}$$

4.4 Relationship among K, Re, and b

Re was multiplied by the term $\beta^2/(1-\beta)^2$ to give X for correlation with K, as for wire nets and perforated plates reported by Ishizuka et al. (1986):

$$X = Re(\beta^2/(1-\beta)\,) \tag{16}$$

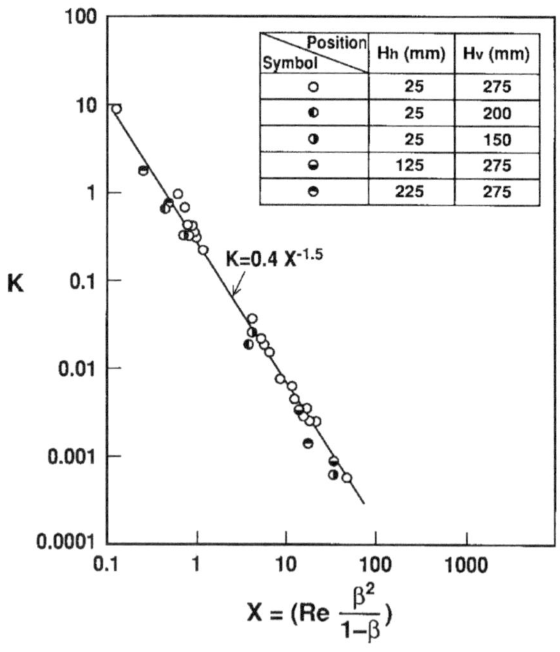

Fig. 9. Relationship among K,Re and β

This multiplier has previously been used for higher values of Re, for example, in the case of forced air convection (Collar 1939). In the empirical correlation of *K* with *X*, all the *K* values obtained under the reference condition (outlet vent in the upper position and heater at the bottom) lie on the line, but others lie slightly below the line (Fig. 9). The reason for this discrepancy is likely to be measurement uncertainty, due to:

1. errors in the estimation of mean temperature rise from the temperature measured at the flow location
2. errors in the estimation of the amount of heat removed from the casing surface
3. the method of estimating the mean temperature rise when the heater position was varied.

The best-fit line was:

$$K = 0.4X^{-1.5} \tag{17}$$

where the coefficient of 0.4 is inherent in this apparatus and is not a general value. This fit indicates that *h* can be considered as chimney height in practical equipment as well, and that the definition of β is reasonable. A more detailed discussion requires 3-dimensional thermo-fluid analysis and more precise measurement. However, we consider that a useful relationship among *K*, *Re*, and β can be determined from a practical point of view.

5. Conclusion

Experimental analysis of the effects of the size of the outlet vent opening and the distance between the outlet vent and the heater location on the flow resistance in a natural-air-cooled electronic equipment case revealed the following relationship among the flow resistance coefficient *K*, Reynolds number *Re*, and the outlet vent porosity coefficient β (defined on the basis of the top surface area):

$$K = BX^{-1.5}$$

where $X = \mathrm{Re}(\beta^2/(1-\beta)$ and B is an inherent coefficient of the casing in question.

6. References

Bergles, A. E., 1990, *Heat Transfer in Electronic and Microelectronic Equipment*, Hemisphere, New York.

Collar, A. R., 1939, The effect of a gauze on the velocity distribution in a uniform duct, *Brit. Aero. Res. Coun., Rep. Mem.* 1867.

Guglielmini, D., Milano, G., and Misale M., 1988, "Natural air cooling of electronic cards in ventilated enclosures," *Second UK National Conference on Heat Transfer*, Vol. I, pp. 199-210.

Ishizuka, M., Miyazaki, Y., and Sasaki, T., 1986, On the cooling of natural air cooled electronic equipment casings, *Bull. JSME*, Vol. 29, No. 247, pp. 119-123.

Ishizuka, M., Miyazaki, Y. and Sasaki, T., 1987, Air resistance coefficients for perforated plates in free convection, *A SME J Heat Transfer*, Vol. 109, pp. 540-543.

Ishizuka, M., 1995, A Thermal Design Approach for Natural Air Cooled Electronic Equipment Casings, ASME-HTD-Vol.303, National Heat Transfer Conference, Portland, USA, pp.65-72.

Wenxian Lin , S.W. Armfield, Natural convection cooling of rectangular and cylindrical containers, *International Journal of Heat and Fluid Flow,* 22 (2001) 72–81

Permissions

The contributors of this book come from diverse backgrounds, making this book a truly international effort. This book will bring forth new frontiers with its revolutionizing research information and detailed analysis of the nascent developments around the world.

We would like to thank Dr. Prof. Vyacheslav S. Vikhrenko, for lending his expertise to make the book truly unique. He has played a crucial role in the development of this book. Without his invaluable contribution this book wouldn't have been possible. He has made vital efforts to compile up to date information on the varied aspects of this subject to make this book a valuable addition to the collection of many professionals and students.

This book was conceptualized with the vision of imparting up-to-date information and advanced data in this field. To ensure the same, a matchless editorial board was set up. Every individual on the board went through rigorous rounds of assessment to prove their worth. After which they invested a large part of their time researching and compiling the most relevant data for our readers. Conferences and sessions were held from time to time between the editorial board and the contributing authors to present the data in the most comprehensible form. The editorial team has worked tirelessly to provide valuable and valid information to help people across the globe.

Every chapter published in this book has been scrutinized by our experts. Their significance has been extensively debated. The topics covered herein carry significant findings which will fuel the growth of the discipline. They may even be implemented as practical applications or may be referred to as a beginning point for another development.

The editorial board has been involved in producing this book since its inception. They have spent rigorous hours researching and exploring the diverse topics which have resulted in the successful publishing of this book. They have passed on their knowledge of decades through this book. To expedite this challenging task, the publisher supported the team at every step. A small team of assistant editors was also appointed to further simplify the editing procedure and attain best results for the readers.

Our editorial team has been hand-picked from every corner of the world. Their multi-ethnicity adds dynamic inputs to the discussions which result in innovative outcomes. These outcomes are then further discussed with the researchers and contributors who give their valuable feedback and opinion regarding the same. The feedback is then collaborated with the researches and they are edited in a comprehensive manner to aid the understanding of the subject.

Apart from the editorial board, the designing team has also invested a significant amount of their time in understanding the subject and creating the most relevant covers. They scrutinized every image to scout for the most suitable representation of the subject and create an appropriate cover for the book.

The publishing team has been involved in this book since its early stages. They were actively engaged in every process, be it collecting the data, connecting with the contributors or procuring relevant information. The team has been an ardent support to the editorial, designing and production team. Their endless efforts to recruit the best for this project, has resulted in the accomplishment of this book. They are a veteran in the field of academics and their pool of knowledge is as vast as their experience in printing. Their expertise and guidance has proved useful at every step. Their uncompromising quality standards have made this book an exceptional effort. Their encouragement from time to time has been an inspiration for everyone.

The publisher and the editorial board hope that this book will prove to be a valuable piece of knowledge for researchers, students, practitioners and scholars across the globe.

List of Contributors

David Sands
University of Hull, UK

Alejandro Crespo-Sosa
Instituto de Física, Universidad Nacional Autonoma de México, México

Michał Szyma´ nski
Institute of Electron Technology, Poland

Etsuji Ohmura
Osaka University, Japan

Nikolay Kazanskiy and Vsevolod Kolpakov
Image Processing Systems Institute, Russian Academy of Sciences, S.P. Korolev Samara State Aerospace University (National Research University), Russia

Shigeru Nozu and Hiroaki Tsuji
Okayama Prefectural University, Japan

Kenji Onishi
Chugoku Rubber Industry Co. Ltd., Japan

Z. Gürdal
Delft University of Technology, The Netherlands

G. Abdelal
Virtual Engineering Centre, University of Liverpool, UK

K.C. Wu
Structural Mechanics and Concepts, NASA Langley Research Center, USA

Z.G. Qu, H.J. Xu, T.S. Wang and W.Q. Tao
Key Laboratory of Thermal Fluid Science and Engineering, MOE, China

T.J. Lu
Key Laboratory of Strength and Vibration, MOE of Xi'an Jiaotong University in Xi'an, China

Murat Karahan
Dumlupinar University, Simav Technical Education Faculty, Turkey

Özcan Kalenderli
Istanbul Technical University, Electrical-Electronics Faculty, Turkey

S. K. Maity
National Metallurgical Laboratory, India

R. Kawalla
TU Bergademie, Germany

G.C. Mavropoulos
Internal Combustion Engines Laboratory, Thermal Engineering Department, School of Mechanical Engineering, National Technical University of Athens (NTUA), Greece

Yu-xing Peng, Zhen-cai Zhu and Guo-an Chen
School of Mechanical and Electrical Engineering, China University of Mining and Technology, Xuzhou, China

Wiesław Zima
Cracow University of Technology, Poland

Jung-Chang Wang
National Taiwan Ocean University, Taiwan, R.O.C.

Sih-Li Chen
National Taiwan University, Taiwan, R.O.C.

M. A. Elsawaf, A. L. Elshafei and H. A. H. Fahmy
Faculty of Engineering, Cairo University, Egypt

Masaru Ishizuka and Tomoyuki Hatakeyama
Toyama Prefectural University, Japan